Springer Proceedings in Mathematics & Statistics

Volume 95

More information about this series at http://www.springer.com/series/10533

Springer Proceedings in Mathematics & Statistics

This book series features volumes composed of select contributions from workshops and conferences in all areas of current research in mathematics and statistics, including OR and optimization. In addition to an overall evaluation of the interest, scientific quality, and timeliness of each proposal at the hands of the publisher, individual contributions are all refereed to the high quality standards of leading journals in the field. Thus, this series provides the research community with well-edited, authoritative reports on developments in the most exciting areas of mathematical and statistical research today.

David Gao • Ning Ruan • Wenxun Xing
Editors

Advances in Global Optimization

Springer

Editors
David Gao
School of Science, Information
Technology and Engineering
Federation University Australia
Ballarat, VIC, Australia

Ning Ruan
School of Science, Information
Technology and Engineering
Federation University Australia
Ballarat, VIC, Australia

Wenxun Xing
Department of Mathematical Sciences
Tsinghua University
Beijing, China

ISSN 2194-1009 ISSN 2194-1017 (electronic)
ISBN 978-3-319-08376-6 ISBN 978-3-319-08377-3 (eBook)
DOI 10.1007/978-3-319-08377-3
Springer Cham Heidelberg New York Dordrecht London

Library of Congress Control Number: 2014953010

Mathematics Subject Classification (2010): 90C26, 90C27, 90C33, 46A20, 91B70

Printed on acid-free paper

Springer is part of Springer Science+Business Media (www.springer.com)

The Third World Congress of Global Optimization

Preface

Global optimization is a multi-disciplinary research field that deals with the analysis, characterization, and computation of global minima and/or maxima of nonlinear, nonconvex, and nonsmooth functions in a continuous or discrete form. Global optimization problems are frequently encountered in modeling real-world systems for a very broad range of applications including aerospace and civil engineering, chemical and process engineering, computational biology and bioinformatics, computational materials science, clustering and pattern recognition, computer vision and robotics, cryptography and data mining, electrical and control engineering, information and technology management, industrial and systems engineering, communication and information network design, intelligent information and security, management science and operations research, mathematical economics and financial engineering, mechanical and structural engineering, neuroscience and cognition, refining and petrochemicals, reliability and quality engineering, signal and image processing, tomography and seismic optimization, production planning and scheduling, transportation and logistics, etc.

With the rapid development and deployment of practical global optimization methodologies in the last 30 years, more and more scientists in diverse disciplines have been using global optimization techniques and algorithms to solve their problems. Global optimization is playing a pivoting role in the development of modern engineering and sciences.

The World Congress on Global Optimization in Engineering & Science (WCGO) is an international conference held biennially. It is supported by the International Society of Global Optimization (iSoGO), which is a professional organization that seeks to promote common understanding of all disciplines in related fields of global optimization, and to advance the theory and methodology for academicians and practitioners. The 1st and 2nd WCGO were successfully held in Changsha, Hunan, China, 2009 and in Chania, Greece 2012, respectively. The 3rd WCGO was held in the Yellow Mountains, Anhui, China during July 8–12, 2013. More than 100 participants from over 20 countries attended this WCGO-III.

This special volume is dedicated to this 3rd WCGO, which contains selected papers from over 100 submissions for the congress. These papers are grouped in eight sections: Mathematical Programming, Combinatorial Optimization, Duality Theory, Topology Optimization, Variational Inequalities and Complementarity Problems, Numerical Optimization, Stochastic Models and Simulation, Complex Simulation, and Supply Chain Analysis. The completion of this book would not have been possible without the assistance of many of our colleagues. We wish to express our sincere appreciation to all those who helped. We are also deeply grateful to selected anonymous referees who provided prompt and insightful reviews for all the submissions. Their constructive comments have greatly contributed to the quality of the volume. Our special thanks go to Eve Mayer and her term at Springer for their great enthusiasm and professional help in expediting the publication of this book.

Ballarat, VIC, Australia David Gao
Ballarat, VIC, Australia Ning Ruan
Beijing, China Wenxun Xing

Contents

Part II Combinatorial Optimization

Part III Duality Theory

Part IV Topology Optimization

Part V Variational Inequalities and Complementarity Problems

Part I
Mathematical Programming

On a Reformulation of Mathematical Programs with Cardinality Constraints

Oleg Burdakov, Christian Kanzow, and Alexandra Schwartz

Abstract Mathematical programs with cardinality constraints are optimization problems with an additional constraint which requires the solution to be sparse in the sense that the number of nonzero elements, i.e. the cardinality, is bounded by a given constant. Such programs can be reformulated as a mixed-integer ones in which the sparsity is modeled with the use of complementarity-type constraints. It is shown that the standard relaxation of the integrality leads to a nonlinear optimization program of the striking property that its solutions (global minimizers) are the same as the solutions of the original program with cardinality constraints. Since the number of local minimizers of the relaxed program is typically larger than the number of local minimizers of the cardinality-constrained problem, the relationship between the local minimizers is also discussed in detail. Furthermore, we show under which assumptions the standard KKT conditions are necessary optimality conditions for the relaxed program. The main result obtained for such conditions is significantly different from the existing optimality conditions that are known for the somewhat related class of mathematical programs with complementarity constraints.

Keywords Cardinality constraints • Complementarity constraints • Global minima • Local minima • Stationary points • M-stationarity

1 Introduction

Consider the *cardinality-constrained optimization problem*

$$\min_x \; f(x) \quad \text{s.t.} \quad x \in X, \; \|x\|_0 \leq \kappa, \tag{1}$$

O. Burdakov (✉)
Department of Mathematics, Linköping University, 58183 Linköping, Sweden
e-mail: oleg.burdakov@liu.se

C. Kanzow • A. Schwartz
Institute of Mathematics, University of Würzburg, Emil-Fischer-Str. 30, 97074 Würzburg, Germany
e-mail: kanzow@mathematik.uni-wuerzburg.de; schwartz@mathematik.uni-wuerzburg.de

D. Gao et al. (eds.), *Advances in Global Optimization*, Springer Proceedings in Mathematics & Statistics 95, DOI 10.1007/978-3-319-08377-3_1

where $f : \mathbb{R}^n \to \mathbb{R}$ denotes a function whose smoothness is specified below in each separate case, $\kappa > 0$ is a given natural number, $\|x\|_0$ denotes the number of nonzero elements in the vector $x \in \mathbb{R}^n$, and $X \subseteq \mathbb{R}^n$ is a subset that contains any further constraints on x. Throughout this manuscript, we assume that $\kappa < n$ since otherwise the cardinality constraint would vanish.

One of the first studies of cardinality-constrained problems is due to Bienstock [1], who assumes f to be quadratic and X to be described by some linear (and box) constraints. He then rewrites (1) as a mixed-integer problem and solves this mixed-integer formulation by a tailored branch-and-bound algorithm, where the cardinality constraint is replaced by a surrogate constraint.

All subsequent works follow a similar pattern in the sense that they view (1) as a mixed-integer problem for which a global minimum has to be found. This mixed-integer formulation is, in general, an NP-hard problem (for an exception, see [2]). The existing solution procedures therefore apply branch-and-bound techniques, evolutionary methods, concave minimization, or apply some heuristic ideas in order to solve the corresponding mixed-integer formulation, see [3–9] and references therein for a list of existing methods. Several of these papers focus on the compressed sensing and the portfolio selection problem which are probably the most prominent instances of cardinality-constrained optimization problem (for some other applications, we refer to [10]). The ideas from these papers admit extension to the more general setting of the form (1).

Here we follow another idea and rewrite the cardinality-constrained problem as a continuous optimization problem. To this end, we also begin with a (somewhat different) reformulation of (1) as a mixed-integer problem and then observe that the standard relaxation of this mixed-integer problem still has the same solutions as in the case of the underlying cardinality constraints. Of course, the price we have to pay is that our continuous optimization problem is nonconvex, and that the one-to-one correspondence of the solutions is true only for the solutions in the sense of global minima. Nevertheless, some useful results can also be given for the local minima.

The cardinality-constrained problem (1) is actually a single-criterion approach to solving a bi-criteria *sparse optimization problem* in which both $f(x)$ and $\|x\|_0$ are to be minimized. An alternative single-criterion approach is to combine the two criteria in one, so that the resulting problem looks as follows:

$$\min_x \; f(x) + \rho\|x\|_0 \quad \text{s.t.} \quad x \in X, \tag{2}$$

where $\rho > 0$ is kind of a penalty parameter. One of the most popular approaches to solving this problem is based on replacing the $\|\cdot\|_0$-term by the $\|\cdot\|_1$-norm. This has the major advantage that the resulting optimization problem is convex (provided that f and X are convex), although nonsmooth. There exists an enormous activity in this area (see, e.g., [9, 10]).

While the sparse optimization problem (2) differs from the cardinality-constrained optimization problem (1), it is interesting to observe that the very recent talk [11] also presents a reformulation of problem (2) as a continuous optimization problem with a complementarity-type constraint. Hence the central idea used in [11], obtained completely independent from our approach, is identical to ours. Apart from this central idea, however, there is no further coincidence simply because we are dealing with two different kind of optimization problems.

The organization of this paper is as follows. Our reformulation of the cardinality-constrained optimization problem (1) as a continuous optimization problem is presented in Sect. 2. There, we also discuss the relation between the local minimizers. Section 3 shows under which assumptions the standard KKT-conditions can be assumed to hold at a solution of our reformulated problem; it also illustrates this result using some preliminary numerical experiments. We then close with some discussion in Sect. 4 where we point out some further research topics in the context of cardinality-constrained optimization problem that are currently under investigation by the authors.

Note that this paper does not contain any proofs, since they will be provided in a separate (full-length) paper. It will also contain some additional results as well as an algorithmic approach for the solution of problem (1) that is motivated by our reformulation.

2 Reformulations Based on Complementarity Constraints

This section presents a reformulation of the cardinality-constrained problem (1) as a smooth optimization problem. We discuss a relation between their global and local minimizers in Sects. 2.1 and 2.2, respectively. We further illustrate the potential of our complementarity-based reformulation in Sect. 2.3, where the portfolio selection problem is considered. In that section, our approach is extended to modeling, in a smooth way, not only sparsity, but also a semi-continuous nature of variables.

In order to obtain a suitable reformulation of the cardinality-constrained problem (1), we first consider the mixed-integer problem

$$\begin{aligned} \min\nolimits_{x,y} \ f(x) \text{ s.t. } & x \in X \\ & e^T y = n - \kappa, \\ & x_i y_i = 0 \quad \forall i = 1, \dots, n, \\ & y \in \{0,1\}^n, \end{aligned} \tag{3}$$

where $e := (1, \dots, 1)^T$.

Next, we consider the following standard relaxation of the mixed-integer problem (3)

$$\begin{aligned} \min\nolimits_{x,y \in \mathbb{R}^n} \ f(x) \text{ s.t. } & x \in X \\ & e^T y = n - \kappa, \\ & x_i y_i = 0 \quad \forall i = 1, \dots, n, \\ & 0 \le y \le e, \end{aligned} \tag{4}$$

where the binary constraints are replaced in the usual way by simple box constraints.

Note that, alternatively, we may replace the equality constraint $e^T y = n - \kappa$ in the two programs (3) and (4) by the inequality constraint $e^T y \geq n - \kappa$ without destroying the subsequent results.

2.1 Relation Between Global Minimizers

According to the following result, the two problems (1) and (3) have the same solutions in x in the sense of global minimizers.

Theorem 1. *A vector $x^* \in \mathbb{R}^n$ is a solution of problem* (1) *if and only if there exists a vector $y^* \in \mathbb{R}^n$ such that the pair (x^*, y^*) is a solution of the mixed-integer problem* (3).

The next result states that the relaxed problem (4) is still equivalent to the original cardinality-constrained problem (1) in the sense of the corresponding global minimizers.

Theorem 2. *A vector $x^* \in \mathbb{R}^n$ is a solution of problem* (1) *if and only if there exists a vector $y^* \in \mathbb{R}^n$ such that the pair (x^*, y^*) is a solution of the relaxed problem* (4).

An immediate consequence of the previous observation is the following existence result.

Theorem 3. *Suppose that the feasible set $\mathscr{F} := \{x \in X \mid \|x\|_0 \leq \kappa\}$ of the cardinality-constrained problem* (1) *is nonempty and X is compact. Let the function $f : \mathbb{R}^n \to \mathbb{R}$ be continuous on $\mathscr{F}$. Then both problem* (1) *and the relaxed problem* (4) *have a nonempty solution set.*

The formulation (4) will be of central importance for our subsequent discussion.

2.2 Relation Between Local Minimizers

In view of Theorem 2, there is a one-to-one correspondence between the solutions of the original problem (1) and the solutions of the relaxed problem (4). Our next aim is to investigate the relation between the local minimizers of these two optimization problems. The next result shows that every local minimizer of the given cardinality-constrained problem yields a local minimizer of the relaxed problem (4).

Theorem 4. *Let $x^* \in \mathbb{R}^n$ be a local minimizer of* (1). *Then there exists a vector $y^* \in \mathbb{R}^n$ such that the pair (x^*, y^*) is also a local minimizer of* (4).

Note that if $\|x^*\|_0 = \kappa$, then the vector y^* in Theorem 4 is unique (see Proposition 1 below). If $\|x^*\|_0 < \kappa$, then y^* is not unique. Unfortunately, the converse of Theorem 4 is not true in general. This is shown by the following counterexample.

Example 1. Consider the three-dimensional problem

$$\min_x \|x - a\|_2^2 \quad \text{s.t.} \quad \|x\|_0 \leq \kappa, \; x \in \mathbb{R}^3 \tag{5}$$

with $a := (1,2,3)^T$ and $\kappa := 2$. It is easy to see that this problem has a unique global minimizer at

$$x^* := (0,2,3)^T$$

as well as two local minimizers at

$$x^1 := (1,0,3)^T \quad \text{and} \quad x^2 := (1,2,0)^T.$$

On the other hand, the relaxed problem (4) has a unique global minimum at

$$x^* := (0,2,3)^T, \; y^* := (1,0,0)^T$$

(this is consistent with Theorem 2), but the number of local minimizers is larger, namely, they are

$$\begin{aligned}
&x^1 := (1,0,3)^T, \; y^1 := (0,1,0)^T, \\
&x^2 := (1,2,0)^T, \; y^2 := (0,0,1)^T, \\
&x^3 := (1,0,0)^T, \; y^3 := (0,t,1-t)^T \quad \forall t \in (0,1), \\
&x^4 := (0,2,0)^T, \; y^4 := (t,0,1-t)^T \quad \forall t \in (0,1), \\
&x^5 := (0,0,3)^T, \; y^5 := (t,1-t,0)^T \quad \forall t \in (0,1), \\
&x^6 := (0,0,0)^T, \; y^6 := (t_1,t_2,t_3)^T \quad \forall t_i > 0 \text{ such that } t_1 + t_2 + t_3 = 1.
\end{aligned}$$

Note that the corresponding y^i is neither unique nor binary for $i = 3,4,5,6$, i.e. for all those x^i which are not local minimizers of (1). ◇

Let (x^*, y^*) be a local minimizer of problem (4). One may think that if y^* is binary, then x^* is a local minimizer of problem (1). Unfortunately, this idea is not true in general. We demonstrate this by a simple modification of the previous counterexample.

Example 2. Consider once again the three-dimensional cardinality-constrained problem from (5), but this time with $a := (1,2,0)^T$ and the cardinality number $\kappa := 1$. Here, it is easy to see that the pair (x^*, y^*) with $x^* := (0,0,0)^T, y^* := (1,1,0)^T$ is a local minimizer of the corresponding relaxed problem (4) with a binary vector y^*, while x^* is not a local minimizer of (1). Note, however, that the vector y^* is not unique in this case. ◇

The previous two examples illustrate that the relation between the local minimizers of the two problems (1) and (4) are not as easy as for the global minimizers. A central observation in this context is that the cardinality constraint was active in the *equivalent* local minimizers which, in view of the subsequent result, is equivalent to the uniqueness of y^* defined by x^*.

Proposition 1. *Let* (x^*, y^*) *be a local minimizer of problem* (4). *Then* $\|x^*\|_0 = \kappa$ *holds if and only if* y^* *is unique. In this case, the components of* y^* *are binary.*

We finally have a special case of the converse of Theorem 4.

Theorem 5. *Let* (x^*, y^*) *be a local minimizer of problem* (4) *satisfying* $\|x^*\|_0 = \kappa$. *Then* x^* *is a local minimizer of the cardinality-constrained problem* (1).

Regarding the additional assumption $\|x^*\|_0 = \kappa$ used in Theorem 5, we note that, in practice, this condition is typically satisfied at the global minimizers of the cardinality-constrained optimization problem (1).

2.3 Continuous Reformulation of Portfolio Selection Problem

Consider the following mean-variance portfolio selection problem with cardinality and minimum buy-in threshold

$$\begin{aligned} &\min\ x^T Q x \\ &\text{s.t.}\ \mu^T x \geq \rho, \quad e^T x = 1, \quad \|x\|_0 \leq \kappa, \\ &\qquad x_i \in \{0\} \cup [a_i, b_i] \quad \forall i = 1, \ldots, n. \end{aligned} \tag{6}$$

For the origin of this problem, its properties and approaches to solving it, see, e.g., [10] and references therein. Note that $x_1, \ldots, x_n$ are semi-continuous variables. The next result presents a continuous reformulation of the problem.

Theorem 6. *A vector* $x^* \in R^n$ *is a solution to the portfolio selection problem* (6) *if and only if there exists a vector* $y^* \in R^n$ *such that the pair* (x^*, y^*) *is a solution to the problem*

$$\begin{aligned} &\min\ x^T Q x \\ &\text{s.t.}\ \mu^T x \geq \rho, \quad e^T x = 1, \quad e^T y \geq n - \kappa, \quad 0 \leq y \leq e, \\ &\qquad x_i y_i = 0 \quad \forall i = 1, \ldots, n, \\ &\qquad a_i(1 - y_i) \leq x_i \leq b_i(1 - y_i) \quad \forall i = 1, \ldots, n. \end{aligned}$$

In the rest of this paper, we focus on the relaxed program (4).

3 Stationarity Conditions

Based on the previous results, it is a very natural idea to solve the cardinality-constrained optimization problem (1) via the relaxed program (4). The latter is a continuous optimization problem and therefore, in principle, allows the application of standard software. Of course, our relaxed program is nonconvex, and there is no guarantee that standard solvers will find a solution. In general, the best one can expect is that these solvers find a KKT point. In the moment, however, it is not even clear whether the usual KKT conditions are necessary optimality conditions at a (local or global) minimizer of the relaxed program since the constraints are still difficult. The aim of this section is therefore to show that this is indeed the case under very reasonable assumptions.

To this end, we begin with some preliminary discussions on constraint qualifications in Sect. 3.1. We then show in Sect. 3.2 that the feasible set of the relaxed program (4) satisfies a suitable constraint qualification, thus guaranteeing that the KKT conditions are indeed necessary optimality conditions. Some very preliminary numerical results are then presented in Sect. 3.3.

3.1 Some Preliminary Discussions

Let us consider a standard nonlinear program (NLP for short)

$$\begin{aligned} \min_x \ f(x) \text{ s.t. } g_i(x) &\leq 0 \quad \forall i = 1,\dots,m \\ h_i(x) &= 0 \quad \forall i = 1,\dots,p \end{aligned} \tag{7}$$

within this subsection, where $f, g_i, h_i : \mathbb{R}^n \to \mathbb{R}$ are continuously differentiable functions. Let

$$X := \{x \in \mathbb{R}^n \mid g_i(x) \leq 0 \ (i = 1,\dots,m),\ h_i(x) = 0 \ (i = 1,\dots,p)\}$$

denote the feasible set of the NLP (7). Then the *(Bouligand) tangent cone* at any feasible point $x \in X$ is defined by

$$\mathcal{T}_X(x) := \left\{d \in \mathbb{R}^n \mid \exists \{x^k\} \subseteq X, \exists \{t_k\} \downarrow 0 :\ x^k \to x \text{ and } \frac{x^k - x}{t_k} \to d\right\},$$

whereas the corresponding *linearization cone* of X at the feasible point x is described by

$$\mathcal{L}_X(x) := \left\{d \in \mathbb{R}^n \mid \nabla g_i(x)^T d \leq 0 \ (i : g_i(x) = 0),\ \nabla h_i(x)^T d = 0 \ (i = 1,\dots,p)\right\}.$$

While it is not difficult to see that the inclusion $\mathcal{T}_X(x) \subseteq \mathcal{L}_X(x)$ always holds, the converse is not true in general and gives rise to the following definition: The *Abadie constraint qualification (Abadie CQ)* is said to hold at a feasible point $x \in X$ if $\mathcal{T}_X(x) = \mathcal{L}_X(x)$.

The Abadie CQ is a relatively weak constraint qualification and often satisfied. For example, it holds automatically provided that all constraint function g_i and h_i are (affine-) linear. Moreover, if one of the more prominent CQs like the linear independence constraint qualification (LICQ) or the Mangasarian–Fromovitz

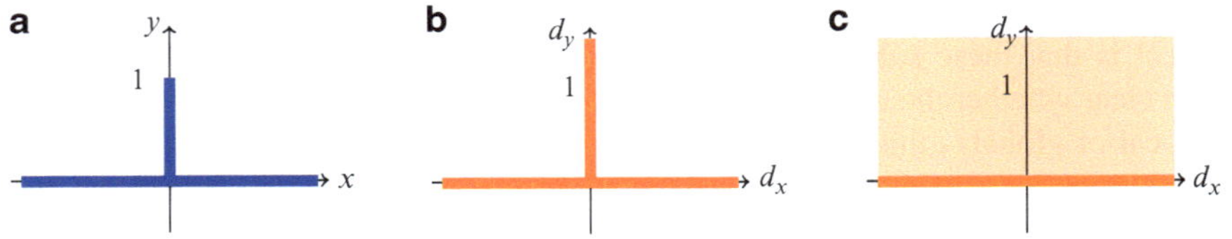

Fig. 1 Feasible set and tangent cones for NLP (8). (**a**) $\mathscr{F}$. (**b**) $\mathscr{T}_X((0,0)^T)$. (**c**) $\mathscr{L}_X((0,0)^T)$

constraint qualification (MFCQ) holds at $x \in X$, then the Abadie CQ is also satisfied. For more details on a whole bunch of CQs, we refer the reader to [12].

Despite these facts, however, it turns out that the Abadie CQ typically does not hold for the class of cardinality-constrained optimization problems. To see this, consider the following example:

$$\min_{x,y} \ f(x,y) \quad \text{s.t.} \quad xy = 0, \ y \geq 0, \ y \leq 1, \tag{8}$$

where $x, y \in \mathbb{R}$ are single variables and the objective function $f : \mathbb{R}^2 \to \mathbb{R}$ plays no role in this case. This nonlinear program is prototypical for the relaxed program (4) that results from a cardinality-constrained optimization problem. The feasible set of (8) and the corresponding tangent and linearization cones at the origin are depicted in Fig. 1. It follows that the tangent cone is a strict subset of the linearization cone, hence the Abadie CQ does not hold at the origin.

We therefore have to look for a weaker constraint qualification. To this end, let us denote by

$$C^\circ := \{d \in \mathbb{R}^n \mid d^T x \leq 0 \quad \forall x \in C\}$$

the *polar cone* of a given cone $C \subseteq \mathbb{R}^n$. Then the *Guignard constraint qualification (Guignard CQ)* is said to hold at a feasible point $x \in X$ of the nonlinear program NLP provided that $\mathscr{L}_X(x)^\circ = \mathscr{T}_X(x)^\circ$ holds. Obviously, the Abadie CQ implies the Guignard CQ, whereas the converse is not true in general. This can be seen immediately by using the example from (8) which obviously satisfies the Guignard CQ.

The Guignard CQ is (in a certain sense) the weakest constraint qualification that exists for nonlinear programs, see, once again, [12] and references therein for a more detailed discussion. In particular, it guarantees that a local minimizer of NLP (7) satisfies the corresponding KKT conditions. For the sake of completeness, we state this formally in the following result.

Theorem 7. *Let x^* be a local minimizer of NLP* (7) *where the Guignard CQ holds. Then there exist Lagrange multipliers $\lambda_i \in \mathbb{R}$ $(i = 1, \ldots, m)$ and $\mu_i \in \mathbb{R}$ $(i = 1, \ldots, p)$ such that the following KKT conditions are satisfied at $x = x^*$:*

$$\nabla f(x) + \sum_{i=1}^{m} \lambda_i \nabla g_i(x) + \sum_{i=1}^{p} \mu_i \nabla h_i(x) = 0,$$
$$\lambda_i \geq 0,\ g_i(x) \leq 0,\ \lambda_i g_i(x) = 0 \quad \forall i = 1, \ldots, m,$$
$$h_i(x) = 0 \quad \forall i = 1, \ldots, p.$$

The previous discussion motivates that the Guignard CQ has a chance to hold for the relaxed program (4), whereas any stronger condition is likely to be violated. Of course, whether or not the Guignard CQ holds depends also on the linear constraint $e^T y = n - \kappa$ (or $e^T y \geq n - \kappa$) as well as on the abstract set X. The following section shows that everything is fine provided that X is a polyhedral convex set, but also indicates that we have to expect severe difficulties for nonlinear sets X.

3.2 Necessary Optimality Conditions

In order to be able to prove the existence of Lagrange multipliers in minimizers of the reformulated problem (4), we consider the special case, where X is polyhedral convex, i.e.

$$X = \{x \in \mathbb{R}^n \mid a_i^T x \leq \alpha_i\ (i = 1, \ldots, m),\ b_i^T x = \beta_i\ (i = 1, \ldots, p)\}.$$

It turns out that, in this case, the Guignard CQ is satisfied in every feasible point and thus every local minimizer of (4) is a KKT point.

Theorem 8. *Let* $(x^*, y^*) \in Z$ *be an arbitrary feasible point of* (4)*. Then the Guignard CQ holds in* (x^*, y^*)*.*

As mentioned before, stronger constraint qualifications such as the Abadie CQ do not hold in general. Even the Guignard CQ does not necessarily hold anymore when X is not polyhedral convex. This is illustrated by the following example.

Example 3. Consider the convex, but nonpolyhedral, set

$$X := \{x \in \mathbb{R}^2 \mid (x_1 - \tfrac{1}{2})^2 + (x_2 - 1)^2 \leq 1\}$$

and $f(x) = x_1 + cx_2$ with $c > 0$ (Fig. 2). When we choose $\kappa = 1$ and c sufficiently large, the unique solution of the corresponding reformulated problem (4) is $x^* = (\frac{1}{2}, 0)^T$ and $y^* = (0, 1)^T$. However, it is easy to verify that (x^*, y^*) is not a KKT point of (4) and thus the Guignard CQ cannot be satisfied in (x^*, y^*). ◇

Example 3 and the previous discussion indicate that a formal proof of Theorem 8 cannot be completely trivial. It is indeed somewhat involved.

In addition, we would like to stress that the result itself is a bit surprising. To this end, note that the relaxed problem (4) contains the constraints

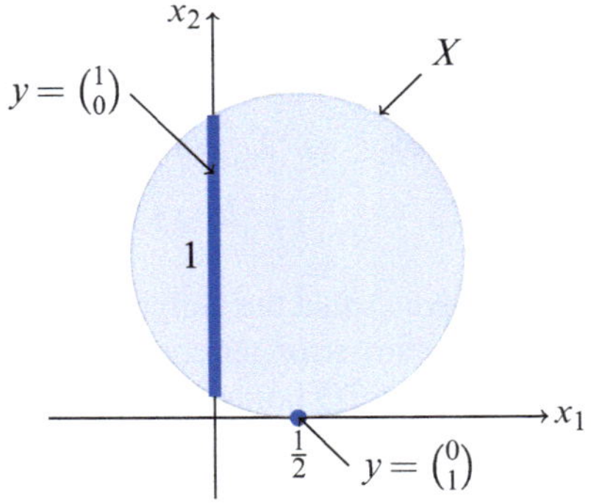

Fig. 2 Illustration of Example 3

$$x_i y_i = 0, \quad y_i \geq 0 \quad \forall i = 1, \ldots, n,$$

which are called "half-complementarity" conditions in [11]. Indeed, if we assume, without loss of generality, that the polyhedral set X is given in standard form by

$$X = \{x \mid Ax = b, \ x \geq 0\}$$

for some matrix $A \in \mathbb{R}^{p \times n}$, then we have the "full complementarity" conditions

$$x_i y_i = 0, \quad x_i \geq 0, \quad y_i \geq 0 \quad \forall i = 1, \ldots, n,$$

and our relaxed program (4) becomes a special case of a *mathematical program with complementarity constraints* (MPCC for short)

$$\begin{aligned}
\min_{x,y} \quad & f(x, y) \\
\text{s.t.} \quad & g_i(x, y) \leq 0 \quad \forall i = 1, \ldots, m, \\
& h_i(x, y) = 0 \quad \forall i = 1, \ldots, p, \\
& G_i(x, y) \geq 0, \ \ H_i(x, y) \geq 0, \ \ G_i(x, y) H_i(x, y) = 0 \quad \forall i = 1, \ldots, l
\end{aligned}$$

for some smooth functions $f, g_i, h_i, G_i, H_i : \mathbb{R}^n \times \mathbb{R}^m \to \mathbb{R}$. A simple example from [13] shows that the Guignard CQ may not hold for MPCCs even if all constraint functions g_i, h_i, G_i, H_i are linear (as in our case). The reason that the Guignard CQ holds for our relaxed problem (4) is due to the fact that the constraints have a very special structure that can be exploited in a better way than for MPCCs with more general linear constraints.

3.3 Some Numerical Illustrations

To test whether the relaxed formulation (4) is of any numerical use, we implemented this reformulation in MATLAB and used the sparse SQP method SNOPT to solve the resulting nonlinear program.

First we tried to solve the problem from Example 1 with this method. To this end, we initialized $y^0 = e$ and chose 10.000 random initial values for x^0 in the cube $[0, 3]^3$. In over 80 % of these test runs, SNOPT found the global minimizer x^*. The local minimizers x^1 and x^2 were each found in approximately 6 % of the runs. For the remaining 8 % of the initial points, SNOPT mostly ended up in x^5.

Then, we applied the same method to the problem from Example 2, now choosing x^0 randomly from $[0, 2]^3$. This time, we ended up in the global minimizer $x = (0, 2, 0)^T$ in about 87 % and in the local minimizer $x = (1, 0, 0)^T$ in approximately 11 % of the runs. The point $x^* = (0, 0, 0)^T$, which was a local minimum of the relaxed problem (4) but not of the original problem (1), was found for less than 2 % of the initial points.

Finally, we also applied the method to the nonpolyhedral problem from Example 3. For this example, the initial values for x^0 were randomly chosen in $[-\frac{1}{2}, \frac{3}{2}] \times [0, 2]$. Remember that in this case the global minimizer $x^* = (\frac{1}{2}, 0)^T$ with $y^* = (0, 1)^T$ is not a KKT point of the relaxed formulation (4), whereas the local minimizer $x = (0, 1 - \frac{\sqrt{3}}{2})^T$ with $y = (1, 0)^T$ is. Nonetheless, we end up in the global minimizer in about 55 % of the runs.

All in all, the numerical results look promising. Of course the previous examples are all very small and hence the question remains whether we still get such positive results for bigger ones. On the other hand, the algorithm is still very elementary and does not yet incorporate any strategies to cope with the "half-complementarity" conditions or to detect local minima of the relaxed formulation (4), which are not local minima of the original problem (1).

4 Concluding Remarks

So far, we have only shown that solutions of the relaxed formulation (4) are KKT points whenever the set X is polyhedral convex. For the future, we plan to exploit the close connection between (4) and MPCC in order to derive appropriate stationarity concepts and related constraint qualifications for the nonpolyhedral case.

Closely related to the previous point is the development of a suitable solution method for the relaxed problem (4). At the moment, we are working on an approach similar to the regularization methods for MPCCs. Here, of course, the question arises of what kind of limit points we can expect from such an algorithm and what properties the regularized subproblems will have.

References

1. Bienstock, D.: Computational study of a family of mixed-integer quadratic programming problems. Math. Program. **74**, 121–140 (1996)
2. Gao, J., Li, D.: A polynomial case of the cardinality-constrained quadratic optimization problem. J. Glob. Optim. **56**, 1441–1455 (2013)
3. Bertsimas, D., Shioda, R.: Algorithm for cardinality-constrained quadratic optimization. Comput. Optim. Appl. **43**, 1–22 (2009)
4. Chang, T.-J., Meade, N., Beasley, J.E., Sharaiha, Y.M.: Heuristics for cardinality constrained portfolio optimisation. Comput. Oper. Res. **27**, 1271–1302 (2000)
5. Di Lorenzo, D., Liuzzi, G., Rinaldi, F., Schoen, F., Sciandrone, M.: A concave optimization-based approach for sparse portfolio selection. Optim. Methods Softw. **27**, 983–1000 (2012)
6. Murray, W., Shek, H.: A local relaxation method for the cardinality constrained portfolio optimization problem. Comput. Optim. Appl. **53**, 681–709 (2012)
7. Shaw, D.X., Liu, S., Kopman, L.: Lagrangian relaxation procedure for cardinality-constrained portfolio optimization. Optim. Methods Softw. **23**, 411–420 (2008)
8. Streichert, F., Ulmer, H., Zell, A.: Evolutionary algorithms and the cardinality constrained portfolio optimization problem. In: Operations Research Proceedings, vol. 2003, pp. 253–260. Springer, Berlin (2004)
9. Tropp, J.A., Wright, S.J.: Computational methods for sparse solution of linear inverse problems. Proc. IEEE **98**, 948–958 (2010)
10. Sun, X., Zheng, X., Li, D.: Recent advances in mathematical programming with semi-continuous variables and cardinality constraints. J. Oper. Res. Soc. China **1**, 55–77 (2013)
11. Mitchell, J., Pang, J.-S., Waechter, A., Bai, L., Feng, M., Shen, X.: Complementarity formulations for L_0 norm optimization problems. In: Presentation at the 11-th Workshop on Advances in Continuous Optimization, Florence, 26–28 June 2013
12. Bazaraa, M.S., Shetty, C.M.: Foundations of Optimization. Lecture Notes in Economics and Mathematical Systems. Springer, Berlin (1976)
13. Scheel, H., Scholtes, S.: Mathematical programs with complementarity constraints: stationarity, optimality, and sensitivity. Math. Oper. Res. **25**, 1–22 (2000)

The Orthogonal Complement of Faces for Cones Associated with the Cone of Positive Semidefinite Matrices

Qinghong Zhang

Abstract It is known that the minimal cone for the constraint system of a conic linear optimization problem is a key component in obtaining strong duality without any constraint qualification. In the particular case of semidefinite optimization, an explicit expression for the dual cone of the minimal cone allows for a dual program of polynomial size that satisfies strong duality. This is achieved due to the fact that we can express the orthogonal complement of a face of the cone of positive semidefinite matrices completely in terms of a system of semidefinite inequalities. In this paper, we extend this result to cones that are either faces of the cone of positive semidefinite matrices or the dual cones of faces of the cone of positive semidefinite matrices. The newly proved result was used in Zhang (4OR 9:403–416, 2011). However, a proof was not given in Zhang (4OR 9:403–416, 2011).

Keywords Semidefinite optimization • Duality • Face • Complementary face

1 Introduction

It is well known that if the standard primal and dual semidefinite programs satisfy the Slater conditions, then both the primal and dual programs have optimal solutions and there is a zero duality gap between them. However, such Slater-type conditions are not always true for semidefinite primal-dual pairs. There has been interest in a unified duality theory without any Slater-type conditions in conic linear optimization and semidefinite optimization, see [1–8]. The idea behind the construction of a primal-dual pair, which guarantees strong duality (that is zero duality gap and dual attainment), is to use the so-called minimal cone to replace the cone which appears in the original problem so that the generalized Slater condition holds for the consistent primal program. The process of eliminating the possibility of a "duality gap" for consistent programs is called a regularization in the literature.

Q. Zhang (✉)
Department of Mathematics and Computer Science,
Northern Michigan University,
Marquette, MI 49855, USA
e-mail: qzhang@nmu.edu

D. Gao et al. (eds.), *Advances in Global Optimization*, Springer Proceedings
in Mathematics & Statistics 95, DOI 10.1007/978-3-319-08377-3_2

In [2], a regularization for an abstract convex program was studied and a theoretical algorithm for computing the minimal cone was also developed. In [4], an exact duality model called the *Extended Lagrange-Slater dual* $(ELSD)$ was derived, and the zero duality gap was proved for a consistent program pair (D)–$(ELSD)$ without any Slater-type conditions, where (D) is the standard dual semidefinite program. Unlike the dual in [2], where the minimal cone is used explicitly, $(ELSD)$ can be written explicitly in terms of equality and inequality constraints. In [3, 5], the relation between these two approaches was discussed. The equivalence between the dual formulated using the minimal cone and $(ELSD)$ was obtained under the assumption that (D) is consistent. Though $(ELSD)$ was originally obtained by working directly with semidefinite programming primal-dual pair in [4], it is the expression of the orthogonal complement of a face of the cone of positive semidefinite matrices completely in terms of a system of semidefinite inequalities that gives the possibility to write the strong dual problem in polynomial times, which plays an critical role in formulating an embedding problem that can be used to solve a semidefinite optimization problem without the Slater-type condition [9].

In [10], a recursive algorithm is discussed to obtain the minimal cone for the constraint system of conic linear optimization. As an example of this process, semidefinite optimization problems are studied and $(ELSD)$ are obtained using this recursive algorithm. In the formulation of the strong dual of a semidefinite optimization problem, an expression of the orthogonal complement of a face of a cone, that is the dual cone of a face of the cone of positive semidefinite matrices, in terms of a system of semidefinite inequalities is used in [10]. However, a proof was not given there. In this paper, we give a complete proof of this result together with a conclusion that gives an expression of the orthogonal complement of a face of a cone that is a face of the cone of positive semidefinite matrices.

In the rest of this section, we introduce some basic concepts in convex analysis. For other concepts and notation used in this paper, the reader is referred to [11]. Let $\mathcal{R}$ denote the set of all real numbers, $\mathcal{R}_+$ the set of all nonnegative numbers, and $\mathcal{R}^{m\times n}$ the set of all $m \times n$ matrices. Let $\mathcal{V}$ be an inner product vector space with an inner product denoted by $\langle x, y\rangle$ for $x, y \in \mathcal{V}$. For any set $\mathcal{D}$ in $\mathcal{V}$, we use $ri\ \mathcal{D}$ to denote the relative interior of $\mathcal{D}$. Let $\mathcal{K}$ be a convex cone in $\mathcal{V}$. $\mathcal{K}$ can be used to define a partial order in $\mathcal{V}$: $x \succeq_{\mathcal{K}} y$ ($x \succeq y$ if $\mathcal{K}$ is apparent from the context) if and only if $x - y \in \mathcal{K}$. A subcone $\mathcal{K}_1$ of $\mathcal{K}$ is called a face of $\mathcal{K}$ if $x \in \mathcal{K}_1$, $x \succeq y \succeq 0$ implies $y \in \mathcal{K}_1$, where 0 represents the zero vector in $\mathcal{V}$. For a given face $\mathcal{K}_1$ of $\mathcal{K}$, the complementary (or conjugate) face of $\mathcal{K}_1$ is defined to be $\mathcal{K}_1^c \equiv \{z \in \mathcal{K}^* \mid \langle z, x\rangle = 0 \text{ for all } x \in \mathcal{K}_1\} = \mathcal{K}^* \cap \mathcal{K}_1^\perp$, where $\mathcal{K}^*$ is the dual cone of $\mathcal{K}$, that is, $\mathcal{K}^* = \{y \in \mathcal{V} \mid \langle x, y\rangle \geq 0 \text{ for all } x \in \mathcal{K}\}$, and $\mathcal{K}^\perp = \mathcal{K}^* \cap (-\mathcal{K}^*)$. The complementary face of a face in $\mathcal{K}^*$ is defined similarly. We write $\mathcal{K}_1^{c\perp}$ for $(\mathcal{K}_1^c)^\perp$. If $\mathcal{C}$ is a convex set of $\mathcal{K}$, then the minimal face of $\mathcal{C}$, denoted by $F(\mathcal{C}, \mathcal{K})$, is the smallest face of $\mathcal{K}$ containing $\mathcal{C}$. Let n be a positive integer. We let $\mathcal{S}^{n\times n}$ denote the real linear vector space of $n \times n$ symmetric matrices. An inner product in this space is defined by $\langle U, W\rangle \equiv U \bullet W = Tr\ (UW)$, where UW denotes ordinary, dimension-compatible matrix multiplication for U and $W \in \mathcal{S}^{n\times n}$, and $Tr\ (U)$ represents the trace of U. $U \succeq W$ means that $U - W$ is positive semidefinite. We use $\mathcal{S}_+^{n\times n}$ to denote the cone of all $n \times n$ positive semidefinite matrices.

2 Description of Orthogonal Complement of a Face

In general, a description of the orthogonal complement of a face of a cone in a finite dimensional space cannot be obtained. However, if the cone of positive semidefinite matrices is considered, a description of the orthogonal complement of a face in terms of matrix inequalities is available due to the following theorem proved by Ramana et al. [5, Lemma 2.1]. This theorem makes a strong dual of a semidefinite problem explicit in terms of inequality and equality constraints.

Lemma 1. *[5, Lemma 2.1] Suppose that* $\mathcal{C}$ *is a convex cone and* $\mathcal{C} \subseteq \mathcal{S}_+^{n\times n}$. *Let* $\mathcal{K} \equiv \{W + W^T \mid W \in \mathcal{R}^{n\times n}$ *and* $U \succeq WW^T$ *for some* $U \in \mathcal{C}\}$. *Then* $F(\mathcal{C}, \mathcal{S}_+^{n\times n})^{c\perp} = \mathcal{K}$.

We would like to see if this description can be extended to cones, which are either a face of $\mathcal{S}_+^{n\times n}$ or the dual cone of a face of $\mathcal{S}_+^{n\times n}$. We start with the description of a face and the dual cone of a face for the cone of positive semidefinite matrices.

Lemma 2. *Suppose that* $\mathcal{P}$ *is a face of* $\mathcal{S}_+^{n\times n}$. *Then there is* $r \in \mathcal{N}$ *(the set of all natural numbers) and* $Q \in \mathcal{R}^{n\times n}$ *with* $Q^TQ = I$, *such that*

$$\mathcal{P} = \left\{ Q \begin{pmatrix} B & 0 \\ 0 & 0 \end{pmatrix} Q^T \mid B \in \mathcal{S}_+^{r\times r} \right\}.$$

Proof. Let $U \in ri(\mathcal{P})$. Then there is a $Q \in \mathcal{R}^{n\times n}$ and $D = \begin{pmatrix} d_1 & 0 & \cdots & 0 \\ 0 & d_2 & \cdots & 0 \\ \cdots & \cdots & \cdots & \cdots \\ 0 & 0 & \cdots & d_r \end{pmatrix}$ with $d_i > 0$ for $i = 1, 2, \ldots, r$, such that $U = Q \begin{pmatrix} D & 0^{r\times(n-r)} \\ 0^{(n-r)\times r} & 0^{(n-r)\times(n-r)} \end{pmatrix} Q^T$, where $0^{m\times n}$ represents 0 matrix with m rows and n columns. We sometimes use 0 to represent the 0 matrix if the dimension of the matrix is clear in the context. Of course,

$$\mathcal{P} \supseteq \left\{ Q \begin{pmatrix} B & 0 \\ 0 & 0 \end{pmatrix} Q^T \mid B \in \mathcal{S}_+^{r\times r} \right\} \tag{1}$$

due to the assumption that $U \in ri(\mathcal{P})$ and that $\mathcal{P}$ is a face of $\mathcal{S}_+^{n\times n}$. Next we prove that

$$\mathcal{P} \subseteq \left\{ Q \begin{pmatrix} B & 0 \\ 0 & 0 \end{pmatrix} Q^T \mid B \in \mathcal{S}_+^{r\times r} \right\}. \tag{2}$$

We choose any $M \in \mathcal{P}$. Then $M = Q(Q^T MQ)Q^T$. We need to prove that $Q^T MQ$ has the form $\begin{pmatrix} B & 0 \\ 0 & 0 \end{pmatrix}$ with $B \in \mathcal{S}_+^{r\times r}$. We achieve this by using the assumption that $U \in ri(\mathcal{P})$. In other words, there exists a $\lambda < 0$, such that $(1-\lambda)U + \lambda M \in \mathcal{P}$. Therefore,

$$(1-\lambda)Q\begin{pmatrix} D & 0 \\ 0 & 0 \end{pmatrix}Q^T + \lambda Q\begin{pmatrix} M_{11} & M_{12} \\ M_{12}^T & M_{22} \end{pmatrix}Q^T \in \mathcal{P} \subseteq \mathcal{S}_+^{n\times n}, \tag{3}$$

where $Q^T MQ = \begin{pmatrix} M_{11} & M_{12} \\ M_{12}^T & M_{22} \end{pmatrix}$ with $M_{11} \in \mathcal{S}^{r\times r}$. Because $M \in \mathcal{P}$, we know that $M_{22} \in \mathcal{S}_+^{(n-r)\times(n-r)}$, and hence, $M_{22} = 0$ by (3), which further implies that $M_{12} = 0$. $M_{11} \in \mathcal{S}_+^{r\times r}$ follows from the assumption that $M \in \mathcal{P}$. Therefore, (2) holds. By combining (1) and (2), we know the conclusion of the theorem is true.

Lemma 3. *Suppose that $\mathcal{P}$ is a face of $\mathcal{S}_+^{n\times n}$. Then there is $r \in \mathcal{N}$ and $Q \in \mathcal{S}^{n\times n}$ with $QQ^T = I$, such that*

$$\mathcal{P}^* = \left\{ Q\begin{pmatrix} B_{11} & B_{12} \\ B_{12}^T & B_{22} \end{pmatrix}Q^T \mid B_{11} \in \mathcal{S}_+^{r\times r}, B_{12} \in \mathcal{R}^{r\times(n-r)}, \right.$$
$$\left. \text{and } B_{22} \in \mathcal{S}^{(n-r)\times(n-r)} \right\}.$$

Proof. By Lemma 2, we know that there is $Q \in \mathcal{R}^{n\times n}$ and $r \in \mathcal{N}$, such that

$$\mathcal{P} = \left\{ Q\begin{pmatrix} B & 0 \\ 0 & 0 \end{pmatrix}Q^T \mid B \in \mathcal{S}_+^{r\times r} \right\}.$$

It is easy to see that

$$\mathcal{P}^* \supseteq \left\{ Q\begin{pmatrix} B_{11} & B_{12} \\ B_{12}^T & B_{22} \end{pmatrix}Q^T \mid B_{11} \in \mathcal{S}_+^{r\times r}, B_{12} \in \mathcal{R}^{r\times(n-r)}, \right.$$
$$\left. \text{and } B_{22} \in \mathcal{S}^{(n-r)\times(n-r)} \right\}.$$

We now prove the converse inclusion. We choose any $N \in \mathcal{P}^*$. Let $N = Q(Q^T NQ)Q^T$ and $Q^T NQ = \begin{pmatrix} N_{11} & N_{12} \\ N_{12}^T & N_{22} \end{pmatrix}$ with $N_{11} \in \mathcal{S}^{r\times r}$, $N_{12} \in \mathcal{R}^{r\times(n-r)}$, and $N_{22} \in \mathcal{S}^{(n-r)\times(n-r)}$. Then $N_{11} \in \mathcal{S}_+^{r\times r}$. Otherwise, there is $B \in \mathcal{S}_+^{r\times r}$ such that $B \bullet N < 0$ contradicting the assumption that $N \in \mathcal{P}^*$. Therefore, N takes the form defined in the lemma.

Of course, $\mathcal{P}^*$ is a cone which contains $\mathcal{S}_+^{n\times n}$. As a cone, $\mathcal{P}^*$ has its own faces. The next lemma gives the form of a face of $\mathcal{P}^*$.

Lemma 4. *Let $\mathcal{P}$, Q, r be as in Lemma 3. Then a subset $\mathcal{Q}$ of $\mathcal{P}^*$ is a face of $\mathcal{P}^*$ if and only if there is a face $\mathcal{F}$ of $\mathcal{S}_+^{r\times r}$ such that*

$$\mathcal{Q} = \left\{ Q \begin{pmatrix} B_{11} & B_{12} \\ B_{12}^T & B_{22} \end{pmatrix} Q^T \,\middle|\, B_{11} \in \mathcal{F}, B_{12} \in \mathcal{R}^{r\times(n-r)}, \right.$$

$$\left. \textit{and } B_{22} \in \mathcal{S}^{(n-r)\times(n-r)} \right\}. \quad (4)$$

Proof. By Lemma 3 and in a straightforward manner, we can prove that $\mathcal{Q}$ defined in (4) is a face of $\mathcal{P}^*$.

Let $\mathcal{D}$ be a face of $\mathcal{P}^*$. Assume that $Q \begin{pmatrix} 2D_{11} & P_{12} \\ P_{12}^T & P_{22} \end{pmatrix} Q^T \in \mathcal{D}$ with $D_{11} \in \mathcal{S}^{r\times r}$, $P_{12} \in \mathcal{R}^{r\times(n-r)}$, and $P_{22} \in \mathcal{S}^{(n-r)\times(n-r)}$. By Lemma 3, we know that $D_{11} \in \mathcal{S}_+^{r\times r}$. Now let $D_{12} \in \mathcal{R}^{r\times(n-r)}$ and $D_{22} \in \mathcal{S}^{(n-r)\times(n-r)}$ be any matrices. If we can prove that $Q \begin{pmatrix} D_{11} & D_{12} \\ D_{12}^T & D_{22} \end{pmatrix} Q^T \in \mathcal{D}$, then we know that

$$\left\{ Q \begin{pmatrix} D_{11} & B_{12} \\ B_{12}^T & B_{22} \end{pmatrix} Q^T \,\middle|\, B_{12} \in \mathcal{R}^{r\times(n-r)} \text{ and } B_{22} \in \mathcal{S}^{(n-r)\times(n-r)} \right\} \subseteq \mathcal{D}. \quad (5)$$

If we can further prove that

$$\mathcal{F} = \left\{ B_{11} \,\middle|\, Q \begin{pmatrix} B_{11} & B_{12} \\ B_{12}^T & B_{22} \end{pmatrix} Q^T \in \mathcal{D} \right.$$

$$\left. \text{for some } B_{12} \in \mathcal{R}^{r\times(n-r)} \text{ and } B_{22} \in \mathcal{S}^{(n-r)\times(n-r)} \right\} \quad (6)$$

is a face of $\mathcal{S}_+^{r\times r}$, then we know that $\mathcal{D}$ is of the form of (4).

Now let's prove (5). By Lemma 3, we know that $Q \begin{pmatrix} D_{11} & D_{12} \\ D_{12}^T & D_{22} \end{pmatrix} Q^T \in \mathcal{P}^*$ and $Q \begin{pmatrix} D_{11} & P_{12} - D_{12} \\ P_{12}^T - D_{12}^T & P_{22} - D_{22} \end{pmatrix} Q^T \in \mathcal{P}^*$. Since

$$Q \begin{pmatrix} D_{11} & D_{12} \\ D_{12}^T & D_{22} \end{pmatrix} Q^T + Q \begin{pmatrix} D_{11} & P_{12} - D_{12} \\ P_{12}^T - D_{12}^T & P_{22} - D_{22} \end{pmatrix} Q^T = Q \begin{pmatrix} 2D_{11} & P_{12} \\ P_{12}^T & P_{22} \end{pmatrix} Q^T \in \mathcal{D}$$

and $\mathcal{D}$ is a face of $\mathcal{P}^*$, we obtain that $Q \begin{pmatrix} D_{11} & D_{12} \\ D_{12}^T & D_{22} \end{pmatrix} Q^T \in \mathcal{D}$. Therefore, (5) holds.

Now we prove that $\mathcal{F}$ is a face of $\mathcal{S}_+^{r\times r}$. Let $E_{11} \in \mathcal{S}_+^{r\times r}$ and $F_{11} \in \mathcal{S}_+^{r\times r}$ such that $E_{11} + F_{11} \in \mathcal{F}$. Then for any matrices $G_{12} \in \mathcal{R}^{r\times(n-r)}$ and $G_{22} \in \mathcal{S}^{(n-r)\times(n-r)}$, we know that $Q \begin{pmatrix} E_{11} + F_{11} & G_{12} \\ G_{12}^T & G_{22} \end{pmatrix} Q^T \in \mathcal{D}$. Since

$$Q\begin{pmatrix} E_{11} & \frac{G_{12}}{2} \\ \frac{G_{12}^T}{2} & \frac{G_{22}}{2} \end{pmatrix} Q^T + Q\begin{pmatrix} F_{11} & \frac{G_{12}}{2} \\ \frac{G_{12}^T}{2} & \frac{G_{22}}{2} \end{pmatrix} Q^T = Q\begin{pmatrix} E_{11}+F_{11} & G_{12} \\ G_{12}^T & G_{22} \end{pmatrix} Q^T \in \mathcal{D},$$

and $Q\begin{pmatrix} E_{11} & \frac{G_{12}}{2} \\ \frac{G_{12}^T}{2} & \frac{G_{22}}{2} \end{pmatrix} Q^T \in \mathcal{P}^*$ and $Q\begin{pmatrix} F_{11} & \frac{G_{12}}{2} \\ \frac{G_{12}^T}{2} & \frac{G_{22}}{2} \end{pmatrix} Q^T \in \mathcal{P}^*$, we obtain that $Q\begin{pmatrix} E_{11} & \frac{G_{12}}{2} \\ \frac{G_{12}^T}{2} & \frac{G_{22}}{2} \end{pmatrix} Q^T \in \mathcal{D}$ and $Q\begin{pmatrix} F_{11} & \frac{G_{12}}{2} \\ \frac{G_{12}^T}{2} & \frac{G_{22}}{2} \end{pmatrix} Q^T \in \mathcal{D}$ due to the assumption that $\mathcal{D}$ is a face of $\mathcal{P}^*$. Therefore, $E_{11} \in \mathcal{F}$ and $F_{11} \in \mathcal{F}$, which implies that $\mathcal{F}$ is a face of $\mathcal{S}_+^{r\times r}$.

Now, we are ready to give an expression of the orthogonal complement of a face of a cone, which is the dual cone of a face of the cone of positive semidefinite matrices.

Theorem 1. *Let $\mathcal{P}$ be a face of $\mathcal{S}_+^{n\times n}$. Then $\mathcal{P}^*$ is a cone that contains $\mathcal{S}_+^{n\times n}$. Let $\mathcal{C}$ be a subcone of $\mathcal{P}^*$. $F(\mathcal{C},\mathcal{P}^*)$ represents the minimal face of $\mathcal{P}^*$ which contains $\mathcal{C}$. $F(\mathcal{C},\mathcal{P}^*)^c = \mathcal{P} \cap F(\mathcal{C},\mathcal{P}^*)^\perp$ is the complementary face of $F(\mathcal{C},\mathcal{P}^*)$. Then $F(\mathcal{C},\mathcal{P}^*)^{c\perp} = \left\{W + W^T \mid W \in \mathcal{R}^{n\times n} \text{ and } U \succeq_{\mathcal{P}^*} WW^T \text{ for some } U \in \mathcal{C}\right\}$.*

Proof. By Lemma 3, there is a Q and $r \in \mathcal{N}$ such that

$$\mathcal{P}^* = \left\{ Q\begin{pmatrix} B_{11} & B_{12} \\ B_{12}^T & B_{22} \end{pmatrix} Q^T \mid B_{11} \in \mathcal{S}_+^{r\times r}, B_{12} \in \mathcal{R}^{r\times(n-r)}, \right.$$
$$\left. \text{and } B_{22} \in \mathcal{S}^{(n-r)\times(n-r)} \right\}.$$

Let

$$\mathcal{C}_{11} = \left\{ C_{11} \,\middle|\, Q\begin{pmatrix} C_{11} & C_{12} \\ C_{12}^T & C_{22} \end{pmatrix} Q^T \in \mathcal{C} \right.$$
$$\left. \text{for some } C_{12} \in \mathcal{R}^{r\times(n-r)} \text{ and } C_{22} \in \mathcal{S}^{(n-r)\times(n-r)} \right\}. \quad (7)$$

Then $\mathcal{C}_{11} \subseteq \mathcal{S}_+^{r\times r}$. By Lemma 4, we know there is a face $\mathcal{E}$ of $\mathcal{S}_+^{r\times r}$ such that

$$F(\mathcal{C},\mathcal{P}^*) = \left\{ Q\begin{pmatrix} B_{11} & B_{12} \\ B_{12}^T & B_{22} \end{pmatrix} Q^T \mid B_{11} \in \mathcal{E}, B_{12} \in \mathcal{R}^{r\times(n-r)}, \right.$$
$$\left. \text{and } B_{22} \in \mathcal{S}^{(n-r)\times(n-r)} \right\}. \quad (8)$$

Because $F(\mathcal{C},\mathcal{P}^*)$ is the minimal face of $\mathcal{P}^*$ containing $\mathcal{C}$, we further obtain that $\mathcal{E} = F(\mathcal{C}_{11},\mathcal{S}_+^{r\times r})$. Therefore, we obtain $F(\mathcal{C},\mathcal{P}^*)^c = \mathcal{P} \cap F(\mathcal{C},\mathcal{P}^*)^\perp = Q\begin{pmatrix} F(\mathcal{C}_{11},\mathcal{S}_+^{r\times r})^c & 0 \\ 0 & 0 \end{pmatrix} Q^T$. Hence

$$F(\mathcal{C},\mathcal{P}^*)^{c\perp} = \left\{ Q \begin{pmatrix} B_{11} & B_{12} \\ B_{12}^T & B_{22} \end{pmatrix} Q^T \mid B_{12} \in \mathcal{R}^{r\times(n-r)}, \right.$$
$$\left. B_{22} \in \mathcal{S}^{(n-r)\times(n-r)}, \text{ and } B_{11} \in F(\mathcal{C}_{11}, \mathcal{S}_+^{r\times r})^{c\perp} \right\}. \quad (9)$$

Since by Lemma 1 $F(\mathcal{C}_{11}, \mathcal{S}_+^{r\times r})^{c\perp} = \{W_1 + W_1^T \mid W_1 \in \mathcal{R}^{r\times r} \text{ and } U_{11} \succeq W_1 W_1^T \text{ for some } U_{11} \in \mathcal{C}_{11}\}$, for any element in $F(\mathcal{C},\mathcal{P}^*)^{c\perp}$, it can be written as $Q \begin{pmatrix} W_1 + W_1^T & B_{12} \\ B_{12}^T & B_{22} \end{pmatrix} Q^T$, where $B_{12} \in \mathcal{R}^{r\times(n-r)}$ and $B_{22} \in \mathcal{S}^{(n-r)\times(n-r)}$. Therefore,

$$Q \begin{pmatrix} W_1 & 0 \\ B_{12}^T & \frac{B_{22}}{2} \end{pmatrix} Q^T + Q \begin{pmatrix} W_1 & 0 \\ B_{12}^T & \frac{B_{22}}{2} \end{pmatrix}^T Q^T = W + W^T,$$

where $W = Q \begin{pmatrix} W_1 & 0 \\ B_{12}^T & \frac{B_{22}}{2} \end{pmatrix} Q^T$.

Since $U_{11} \in \mathcal{C}_{11}$, so there is U_{12} and U_{22} such that $U = Q \begin{pmatrix} U_{11} & U_{12} \\ U_{12}^T & U_{22} \end{pmatrix} Q^T \in \mathcal{C}$. Because $WW^T = Q \begin{pmatrix} W_1 W_1^T & W_1 B_{12} \\ B_{12}^T W_1^T & B_{12}^T B_{12} + \frac{B_{22}^2}{4} \end{pmatrix} Q^T$ and $U_{11} \succeq W_1 W_1^T$, we obtain that $U = Q \begin{pmatrix} U_{11} & U_{12} \\ U_{12}^T & U_{22} \end{pmatrix} Q^T \succeq_{\mathcal{P}^*} WW^T$.

Remark 1. Theorem 1 was used in [10]. But a proof was not given in [10]. Here we provide a complete proof of this result.

Now we briefly discuss an expression of the orthogonal complement of a face of a cone that is a face of the cone of positive semidefinite matrices. As in the discussion above, we let $\mathcal{P}$ be a face of $\mathcal{S}_+^{n\times n}$. $\mathcal{P}$ itself is a cone. Let $\mathcal{C}$ be a subcone of $\mathcal{P}$. Then $F(\mathcal{C},\mathcal{P})$ represents the minimal face of $\mathcal{P}$ containing $\mathcal{C}$ and $F(\mathcal{C},\mathcal{S}_+^{n\times n})$ represents the minimal face of $\mathcal{S}_+^{n\times n}$ containing $\mathcal{C}$. Since $\mathcal{P}$ is a face of $\mathcal{S}_+^{n\times n}$, we can easily prove that $F(\mathcal{C},\mathcal{P}) = F(\mathcal{C},\mathcal{S}_+^{n\times n})$. Therefore, $F(\mathcal{C},\mathcal{P})^c = \mathcal{P}^* \cap F(\mathcal{C},\mathcal{P})^\perp \supseteq \mathcal{S}_+^{n\times n} \cap F(\mathcal{C},\mathcal{S}_+^{n\times n})^\perp = F(\mathcal{C},\mathcal{S}_+^{n\times n})^c$, which further implies that $F(\mathcal{C},\mathcal{P})^{c\perp} \subseteq F(\mathcal{C},\mathcal{S}_+^{n\times n})^{c\perp}$. By Lemma 1, we obtain that $F(\mathcal{C},\mathcal{S}_+^{n\times n})^{c\perp} = \mathcal{K} \equiv \{W + W^T \mid W \in \mathcal{R}^{n\times n} \text{ and } U \succeq WW^T \text{ for some } U \in \mathcal{C}\}$. Therefore, $F(\mathcal{C},\mathcal{P})^{c\perp} \subseteq \{W + W^T \mid W \in \mathcal{R}^{n\times n} \text{ and } U \succeq WW^T \text{ for some } U \in \mathcal{C}\}$. The converse inclusion may not be true. However, in a similar way we can prove the following theorem which gives an expression of the orthogonal complement of a face of $\mathcal{P}$.

Theorem 2. *$F(\mathcal{C},\mathcal{P})^{c\perp} = \{W + W^T \mid W \in \mathcal{R}^{n\times n}$ and $U \succeq_{\mathcal{P}} WW^T$ for some $U \in \mathcal{C}\}$.*

Proof. The proof is similar to that of Theorem 1.

3 Conclusion Remarks

An expression in terms of a system of semidefinite inequalities for the orthogonal complement of a face of the cone of positive semidefinite matrices plays an important role in formulating a dual of a polynomial size with a strong duality property for a semidefinite optimization problem. In this paper, we have extended this expression to the orthogonal complement of a face of cones, which are either a face of the cone of positive semidefinite matrices or the dual cone of a face of the cone of positive semidefinite matrices.

References

1. Borwein, J.M., Wolkowicz, H.: Facial reduction for a cone-convex programming problem. J. Aust. Math. Soc. Ser. A **30**, 369–380 (1981)
2. Borwein, J.M., Wolkowicz, H.: Regularizing the abstract convex program. J. Math. Anal. Appl. **83**, 495–530 (1981)
3. Luo, Z., Sturm, J.F., Zhang, S.: Duality results for conic convex programming. Technical Report, Econometric Institute Report No. 9719/A, Econometric Institute, Erasumus University, Rotterdam (1997)
4. Ramana, M.V.: An exact duality theory for semidefinite programming and its complexity implications. Math. Program. Ser. B **77**, 129–162 (1997)
5. Ramana, M.V., Tunçel, L., Wolkowicz, H.: Strong duality for semidefinite programming. SIAM J. Optim. **7**, 641–662 (1997)
6. Sturm, J.F.: Primal-dual interior point approach to semidefinite programming. Ph.D. thesis, Erasmus University (1997)
7. Wolkowicz, H.: Some applications of optimization in matrix theory. Linear Algebra Appl. **40**, 101–118 (1981)
8. Zhang, Q., Chen, G., Zhang, T.: Duality formulation in semidefinite programming. J. Ind. Manag. Optim. **6**, 881–893 (2010)
9. Zhang, Q.: Embedding methods for semidefinite programming. Optim. Methods Softw. **27**, 461–482 (2012)
10. Zhang, Q.: The minimal cone for conic linear programming. 4OR **9**, 403–416 (2011)
11. Rockafellar, R.T.: Convex Analysis. Princeton University Press, Princeton (1970)

Optimality of Bilevel Programming Problems Through Multiobjective Reformulations

Roxin Zhang

Abstract A bilevel optimization problem consists of minimizing an upper level objective function subject to the constraints that involve the solution mapping of the lower level optimization problem parameterized by the upper level decision variable. The global equivalence between a general bilevel programming problem and a multiobjective optimization problem with nonconvex ordering cone is established and optimality conditions of the bilevel problem are obtained using Mordukhovich extremal principles.

Keywords Bilevel programming • Multiobjective programming • Extremal principle

1 Introduction

A bilevel programming problem is an optimization problem with two distinct and non-symmetric levels of decision making. At the upper level the decision maker, called the leader, chooses a strategy variable x to optimize its objective function that involves with another decision variable y chosen by the lower level decision maker, called the follower. The decisions of the follower is determined through minimizing its own objective function that contains the leader's decision variable.

Assume the decision variables of the leader and the follower are $x \in \mathbb{R}^n$ and $y \in \mathbb{R}^m$, respectively, and the objective functions f of the leader and g of the follower are lower semicontinuous from $\mathbb{R}^n$ to $\overline{\mathbb{R}} := \mathbb{R} \cup \{\infty\}$. For a given closed subset $C \subset \mathbb{R}^{n+m}$, the bilevel programming problem under our considerations takes the form

$$(BLP) \quad \begin{cases} \underset{x,y}{\text{minimize}} & f(x, y) \\ \text{subject to} & y \in R(x) \end{cases}$$

R. Zhang (✉)
Northern Michigan University, Marquette, MI, USA
e-mail: rzhang@nmu.edu

D. Gao et al. (eds.), *Advances in Global Optimization*, Springer Proceedings in Mathematics & Statistics 95, DOI 10.1007/978-3-319-08377-3_3

where $R(x)$ is the rational response mapping of the lower-level problem defined as the solution set to the following parametric optimization problem

$$\begin{cases} \underset{y}{\text{minimize}} & g(x, y) \\ \text{subject to} & (x, y) \in C \end{cases}$$

In leading to an eventual multicriterion optimization formulation, we make the observation that the pair of decision variables $(\bar{x}, \bar{y})$ is a solution to our principal bilevel programming problem (BLP) if and only if the following set of criteria is true:

1. (Feasibility) $g(\bar{x}, \bar{y}) \leq g(\bar{x}, y)$ for all $(\bar{x}, y) \in C$;
2. (Optimality) for all (x, y) satisfying $g(x, y) \leq \min_y \{g(x, y) : (x, y) \in C\}$, one has $f(\bar{x}, \bar{y}) \leq f(x, y)$.

Our main goal in the paper is to establish a set of optimality conditions to the bilevel programming problem (BLP) through a reformulation of (BLP) to a general multiobjective programming problem by using Mordukhovich extremal principles. A rich collection of research has been devoted to the studies of bilevel programming and multiobjective bilevel programming problems both in the development of theory and applications. The methods of optimality range from penalty function, KKT-condition, and various reformulations (see [1–4] and references therein). However to our best knowledge, the work in reformulations through multiobjective optimization problems is rather limited. Our work is motivated largely by the research of Fliege and Vicente [5] yet from a complete different perspective. Their reformulation is in the domain space of the objective mappings and ours is in the range space. Furthermore, the tools used to derive optimality conditions are also different.

2 Preliminaries

Our goal is to formulate the bilevel programming problem as a multiobjective programming problem. In a multiobjective problem, we are interested in optimizing an objective mapping with a multi-dimensional range space. An order in the range space is defined through a set Θ and the objective is to find a point $\bar{z}$ such that no other points in the constraint set of the domain space give better objective values. Our aim is to establish a multiobjective problem so the multi-dimensional order is given in the range space of certain mapping in contrast to the practice of building an order in the (x, y)-product space of the decisions of the leader and the follower.

Here we first give a general definition of a (set, function)-optimality concept due to Mordukhovich.

Definition 1. Let $\Theta \subset \mathbb{R}^r$ be an order set, $\phi : \mathbb{R}^n \to \mathbb{R}^r$ be a mapping and $\Lambda \subset \mathbb{R}^n$. We say $\bar{z} \in \mathbb{R}^n$ is a (ϕ, Θ)-extremal point subject to the constraint $z \in \Lambda$ if there is a sequence $a_k \to 0$ and a neighborhood U of $\bar{z}$ such that

$$\phi(z) - \phi(\bar{z}) \notin \Theta + a_k \quad \forall\, z \in \Lambda \cup U,\ \forall\, k. \tag{1}$$

The concept of optimality in this definition covers a number of traditional optimality/efficiency concepts. For a closed cone K (not necessarily convex), we are interested in the generalized Pareto preference relation $\prec$ defined in $\mathbb{R}^r$ by K as:

$$v \prec w \quad \Longleftrightarrow \quad v - w \in K,\ v \neq w \quad \forall\, v, w \in \mathbb{R}^r. \tag{2}$$

For a given mapping $\phi : \mathbb{R}^n \to \mathbb{R}^r$ and a set $\Lambda \subset \mathbb{R}^n$, many classical multiobjective programming problems are based on the concept of Pareto efficiency in that K is assumed to be closed and convex, and the point $\bar{z} \in \Lambda$ is optimal (efficient) for the mapping ϕ with respect to the preference $\prec$ if there does not exist feasible $z \in \Lambda$ near $\bar{z}$ such that $\phi(z) \prec \phi(\bar{z})$.

Throughout this paper, our multiobjective programming problem is in the format defined below.

Definition 2. Given $\phi : \mathbb{R}^n \to \mathbb{R}^r$, $\Lambda \subset \mathbb{R}^n$ and a closed ordering cone K, we say $\bar{z}$ K-minimizes ϕ (globally) over $z \in \Lambda$ if and only if there exists no $z \in \Lambda$ such that $\phi(z) \prec \phi(\bar{z})$ with $\prec$ defined as in (2), in other words,

$$\phi(z) - \phi(\bar{z}) \notin K \quad \forall\, z \in \Lambda \text{ with } \phi(z) \neq \phi(\bar{z}). \tag{3}$$

The local K-minimal points are defined analogously.

Notice that we do not impose convexity on the cone K in the above definition. In general, neither of the concepts of K-minimal points as in Definition 2 and that of the (ϕ, K)-extremal points as in Definition 1 imply each other due to the nonconvexity of K. However for the special ordering cone constructed later in this paper the former implies the latter. From an application point of view, it is more helpful to observe the negation of the condition (3) that signifies a non-solution $\bar{z}$.

Before proceeding with the reformulations of the problem (BLP) and the development of optimalities, we recall some of the fundamental tools in nonsmooth variational analysis and generalized differentiation concepts [6, 7].

Definition 3. Given a set $\Omega \subset \mathbb{R}^n$ and $z \in \Omega$, the ε-normal cone to Ω at z is the set

$$\hat{N}_\varepsilon(z; \Omega) := \left\{ v \in \mathbb{R}^n : \limsup_{z' \to z} \frac{\langle v, z' - z \rangle}{\|z' - z\|} \leq \varepsilon \right\}.$$

With $\varepsilon = 0$, the cone $\hat{N}(z; \Omega) := \hat{N}_0(z; \Omega)$ is called the Fréche normal cone to Ω at z, and the basic (limiting) normal cone to Ω at $\bar{z} \in \Omega$ is the set

$$N(z; \Omega) := \operatorname*{Lim\,sup}_{\substack{z \to \bar{z} \\ \varepsilon \downarrow 0}} \hat{N}_\varepsilon(z; \Omega)$$

where Limsup is the Painlevé-Kuratowski outer limit for set-valued mappings that contains all the cluster points of $N_\varepsilon(z; \Omega)$ as $z \to \bar{z}$ and $\varepsilon \downarrow 0$.

Two generalized differentiation concepts for functions and set-valued mappings can be defined through the basic normal cones.

Definition 4. Given an extended real-valued lower semicontinuous function $f : \mathbb{R}^n \to \overline{\mathbb{R}}$ and a set-valued mapping $F : \mathbb{R}^n \rightrightarrows \mathbb{R}^m$, the basic subdifferential of f at $z \in \mathbb{R}^n$ is the set

$$\partial f(z) := \{v \in \mathbb{R}^n : (v, -1) \in N((z, f(z); \operatorname{epi} f))\}$$

where $\operatorname{epi} f := \{(z, \mu) : \mu \geq f(z)\}$ is the epigraph of f. The coderivative of F at $(z, v) \in \operatorname{gph} F$ is the set-valued mapping $D^* F(z, v) : \mathbb{R}^m \rightrightarrows \mathbb{R}^n$ such that

$$D^* F(z, v)(v^*) := \{z^* : (z^*, -v^*) \in N((z, v); \operatorname{gph} F)\}$$

where $\operatorname{gph} F := \{(z, v) : v \in F(z)\}$ is the graph of F.

For a single valued mapping $\phi : \mathbb{R}^n \to \mathbb{R}^r$, we write $D^*\phi(\bar{z})(v^*)$ in place of $D^*\phi(\bar{z}, \phi(\bar{z}))(v^*)$. If in addition, ϕ is continuously differentiable at $\bar{z}$, we will have

$$D^*\phi(\bar{z})(v^*) = \nabla\phi(\bar{z})^\top v^* \quad \forall\, v^* \in \mathbb{R}^r$$

where $\nabla\phi(\bar{z})$ is the Jacobian matrix of ϕ at $\bar{z}$. When ϕ is locally Lipschitz continuous around $\bar{z}$, then $D^*\phi(\bar{z})(v^*) = \partial\langle v^*, \phi\rangle(\bar{z})$. In the following proposition, we present a result for the optimality conditions of a local extremal point due to Mordukhovich [6, 8]. Let δ_Λ be the indicator function of the set Λ defined as $\delta_\Lambda(z) = 0$ if $z \in \Lambda$ and $+\infty$ otherwise.

Proposition 1. *Let $\bar{z}$ be a local (ϕ, Θ)-extremal point subject to $x \in \Lambda$, where $\phi : \mathbb{R}^n \to \mathbb{R}^r$ is a mapping continuous around $\bar{z}$ relative to Λ, and where the sets $\Lambda \subset \mathbb{R}^n$ and $\Theta \subset \mathbb{R}^r$ are locally closed around $\bar{z}$ and $0 \in \Theta$, respectively. Then there exists $v^* \in \mathbb{R}^r$, not equal to 0, such that*

$$0 \in D^*(\phi + \delta_\Lambda)(\bar{z})(v^*) \quad \text{and} \quad v^* \in N(0; \Theta). \tag{4}$$

If in addition ϕ is Lipschitz continuous around $\bar{z}$ relative to Λ, then

$$0 \in \partial\langle v^*, \phi + \delta_\lambda\rangle(\bar{z}) \quad \text{and} \quad v^* \in N(0; \Theta). \tag{5}$$

3 Multiobjective Reformulations

In this section we will reformulate the bilevel programming problem (BLP) as an multiobjective optimization problem. First we construct a cone K and a mapping $\Phi : \mathbb{R}^n \times \mathbb{R}^m \to \mathbb{R}^r$ such that the upper-lower level decision pair $\bar{z} := (\bar{x}, \bar{y}) \in \mathbb{R}^n \times \mathbb{R}^m$ is not an optimal solution to the bilevel problem (BLP) if and only if there exists $z = (x, y) \in C$ such that

$$\Phi(z) - \Phi(\bar{z}) \in K \quad \text{and} \quad \Phi(z) \neq \Phi(\bar{z}).$$

Of course a point $z = (x, y)$ is not a solution to (BLP) implies one of the two or both possibilities: (a) z is not feasible; or (b) z is not optimal. Following this idea we construct our order cone K_0 as a union of two cones. Specifically let $\mathbb{R}_- := \{\lambda \in \mathbb{R} : \lambda \leq 0\}$ and define

$$K_0 := (\{0\}^n \times \mathbb{R}_- \times \mathbb{R} \times \mathbb{R}) \cup (\mathbb{R}^n \times \mathbb{R} \times \mathbb{R}_- \times \mathbb{R}_-) \subset \mathbb{R}^{n+3}. \tag{6}$$

Apparently K_0 is a union of two closed convex cone but itself may not be convex. For a decision pair $z = (x, y) \in \mathbb{R}^n \times \mathbb{R}^m$, f and g as defined in (BLP), we also define a vector-valued objective mapping $\Phi : \mathbb{R}^{n+m} \to \mathbb{R}^{n+3}$ as

$$\Phi(x, y) := (x, g(x, y), g(x, y) - m(x), f(x, y))$$

where $m(x) := \inf\{g(x, y') : y' \in C(x)\}$ and $C(x) := \{y \in \mathbb{R}^m : (x, y) \in C\}$.

Observe that if $\bar{y} \in R(\bar{x})$, the rational response set of the follower, for a given $\bar{x} \in \mathbb{R}^n$, then $\Phi(\bar{x}, \bar{y}) = (\bar{x}, g(\bar{x}, \bar{y}), 0, f(\bar{x}, \bar{y}))$ and consequently, $\Phi(x, y) - \Phi(\bar{x}, \bar{y}) \in K_0$ if and only if

$$\bar{x} = x, \; g(x, y) < g(\bar{x}, \bar{y}) \quad \text{or} \quad g(x, y) \leq \inf_{y'} g(x, y'), \; f(x, y) < f(\bar{x}, \bar{y}).$$

Based on Definition 2, we state our multiobjective problem below that we will prove to be equivalent to the bilevel programming problem (BLP).

$$K_0\text{-minimize} \quad \Phi(x, y) \quad \text{subject to} \quad (x, y) \in C, \tag{7}$$

where $\bar{z} := (\bar{x}, \bar{y})$ solves (7) if and only if there is no $z = (x, y) \in C$ satisfying $\Phi(z) - \Phi(\bar{z}) \in K_0$ with $\Phi(z) \neq \Phi(\bar{z})$. First we need the following lemma.

Lemma 1. *If a point $\bar{z}$ solves the problem (7), then it is a (Φ, K_0)-extremal point as in Definition 1.*

Proof. Set $a_k := (0^n, -1/k, -1/k, -1/k)$ in (1).

The equivalence between the problems (BLP) and (7) is stated in the following proposition.

Proposition 2. *The vector $\bar{z} := (\bar{x}, \bar{y}) \in \mathbb{R}^n \times \mathbb{R}^m$ is a solution to the bilevel programming problem (BLP) if and only if it is a solution to the multiobjective programming problem (7).*

Proof. *Assume $(\bar{x}, \bar{y})$ solves (BLP). We have $m(\bar{z}) = g(\bar{x}, \bar{y})$ and*

$$\Phi(x, y) = (x, g(x, y), g(x, y) - m(x), f(x, y)) \quad \forall\, (x, y) \in C$$
$$\Phi(\bar{x}, \bar{y}) = (\bar{x}, g(\bar{x}, \bar{y}), 0, f(\bar{x}, \bar{y})).$$

Recall that the cone $K_0 = K_1 \cup K_1$ *where* $K_1 := (\{0\}^n \times \mathbb{R}_- \times \mathbb{R} \times \mathbb{R})andK_1 := (\mathbb{R}^n \times \mathbb{R} \times \mathbb{R}_- \times \mathbb{R}_-)$. *If* $(\bar{x}, \bar{y})$ *were not a solution to (7), then there exists* $(x_0, y_0) \in C$ *such that* $\Phi(x_0, y_0) - \Phi(\bar{x}, \bar{y}) \in K$ *and* $\Phi(x_0, y_0) \neq \Phi(\bar{x}, \bar{y})$. *Consequently we must have*

$$x_0 = \bar{x} \quad \textit{and} \quad g(x_0, y_0) \leq g(\bar{x}, \bar{y}), \quad \textit{or} \tag{8}$$

$$g(x_0, y_0) - m(x_0) \leq 0 \quad \textit{and} \quad f(x_0, y_0) \leq f(\bar{x}, \bar{y}). \tag{9}$$

The strict inequality $g(\bar{x}, y_0) < g(\bar{x}, \bar{y})$ *in (8) is impossible since* $(\bar{x}, \bar{y})$ *is feasible to the problem (BLP), and obviously the first inequality in (9) cannot be strict. Therefore we must have* $f(x_0, y_0) < f(\bar{x}, \bar{y})$. *This contradicts with the assumption that* $(\bar{x}, \bar{y})$ *solves (BLP).*

Conversely, assume $\bar{z} := (\bar{z}, \bar{y})$ *solves (7). If* $(\bar{x}, \bar{y})$ *were not a solution to (BLP), then either* $\bar{y} \notin R(\bar{x})$ *or* $f(\bar{z})$ *is not optimal among feasible points* (x, y). *In other words, we must have either the existence of* $y_1 \in C(\bar{x})$ *such that* $g(\bar{x}, y_1) < g(\bar{x}, \bar{y})$, *or there exists* $(x_2, y_2) \in C$ *such that*

$$g(x_2, y_2) \leq m(x_2) \quad \textit{and} \quad f(x_2, y_2) < g(\bar{x}, \bar{y}).$$

It follows that one of the following must be true:

(a) $\exists x_1 := \bar{x}$ *such that* $g(z_1) < g(\bar{z})$ *which implies* $\Phi(z_1) - \Phi(\bar{z}) \in K_1$ *and* $\Phi(z_1) \neq \Phi(\bar{z})$;

(b) $g(z_2) - m(x_2) \leq 0$ *and* $f(z_2) < f(\bar{z})$ *which implies* $\Phi(z_2) - \Phi(\bar{z}) \in K_2$ *and* $\Phi(z_2) \neq \Phi(\bar{z})$

where $z_1 = (x_1, y_1)$ *and* $z_1 = (x_2, y_2)$. *Therefore* $\bar{z} = (\bar{x}, \bar{y})$ *cannot be a solution to (7), a contradiction.*

4 Optimality Conditions

Recall that our ordering cone K_0 is defined as $K_0 := K_1 \cup K_2$ where

$$K_1 := \{0\}^n \times \mathbb{R}_- \times \mathbb{R} \times \mathbb{R} \quad \text{and} \quad K_2 := \mathbb{R}^n \times \mathbb{R} \times \mathbb{R}_- \times \mathbb{R}_-. \tag{10}$$

Since both K_1 and K_2 are closed polyhedral convex cones in the $\mathbb{R}^{n+3}$ they coincide with their contingent cones at 0, namely $T(0; K_1) = K_1$ and $T(0; K_2) = K_2$. Moreover, since the basic normal cone to a direct product of sets is equal to the direct product of the basic normal cones [6, 7], we have

$$N(0; K_1) = \mathbb{R}^n \times \mathbb{R}_+ \times \{0\} \times \{0\} \quad \text{and} \quad N(0; K_2) = \{0\}^n \times \{0\} \times \mathbb{R}_+ \times \mathbb{R}_+$$

where $\mathbb{R}_+ := \{\mu \in \mathbb{R} : \mu \geq 0\}$.

First we wish to compute the basic normal cone $N(0; K)$. Let K^* be the negative polar cone of K defined by $K^* := \{v : \langle v, w\rangle \leq 0, w \in K\}$. Motivated by Henrion and Outrata [9], we observe that the normal cone $N(0; K)$ can be expressed as $N(0; K) = A \cup B \cup C \cup D$ where

$$\begin{array}{ll} A := K_1^* \cap K_2^*, & B := \bigcup_{v\in K_1\setminus K_2} \hat{N}(v; K1) \\ C := \bigcup_{v\in K_2\setminus K_1} \hat{N}(v; K_2), & D := \bigcup_{v\in (K_1\cap K_2)\setminus\{0\}} \hat{N}(v; K_1 \cup K_2). \end{array} \tag{11}$$

The computations of each of the above mentioned four sets are as follows.

$$A = N(0; K_1)\cap N(0; K_2) = (\mathbb{R}^n\times\mathbb{R}_+\times\{0\}\times\{0\})\cap(\{0\}^n\times\{0\}\times\mathbb{R}_+\times\mathbb{R}_+) = \{0\}^{n+3}.$$

To compute the set D, notice that $K_1 \cap K_2 = \{0\}^n \times \mathbb{R}_- \times \mathbb{R}_- \times \mathbb{R}_-$, and for any $v_0 \in (K_1 \cap K_2)\setminus\{0\}$, we have $v_0 = (0, -a^2, -b^2, -c^2) \in \mathbb{R}^{n+3}$ for some $a, b, c \in \mathbb{R}$ with $a^2 + b^2 + c^2 \neq 0$. Consequently,

$$\begin{aligned} \hat{N}(v_0; K_1 \cup K_2) &\subset \hat{N}(v_0; K_1) \cap \hat{N}(v_0; K_2) \\ &= \hat{N}(v_0; \{0\}^n \times \mathbb{R}_- \times \mathbb{R} \times \mathbb{R}) \cap \hat{N}(v_0; \mathbb{R}^n \times \mathbb{R} \times \mathbb{R}_- \times \mathbb{R}_-) \\ &\subset (\mathbb{R}^n \times \mathbb{R} \times \{0\} \times \{0\}) \cap (\{0\} \times \{0\} \times \mathbb{R} \times \mathbb{R}) = \{0\}. \end{aligned}$$

Therefore $D = \{0\} \in \mathbb{R}^{n+3}$. Now we turn to the other twos sets B and C. Let $v_1 \in K_1\setminus K_2$, without loss of generality, we have $v_1 = (0, -a^2, 1 + b^2, 1 + c^2)$ for some $a, b, c \in \mathbb{R}$ and

$$\begin{aligned} \hat{N}(v_1; K_1) &= \hat{N}((0, -a^2, 1 + b^2, 1 + c^2); \{0\}^n \times \mathbb{R}_- \times \mathbb{R} \times \mathbb{R}) \\ &= \begin{cases} \mathbb{R}^n \times \mathbb{R}_+ \times \{0\} \times \{0\}, & \text{if } a = 0, \\ \mathbb{R}^n \times \{0\} \times \{0\} \times \{0\}, & \text{if } a \neq 0. \end{cases} \end{aligned}$$

It follows that $\bigcup_{v\in K_1\setminus K_2} \hat{N}(v; K_1) = \mathbb{R}^n\times\mathbb{R}_+\times\{0\}\times\{0\}$. Similarly, for $v_2 \in K_2\setminus K_1$, we have $v_2 = (x, 1 + a^2, -b^2, -c^2)$ for some $x \in \mathbb{R}^n$ and $a, b, c \in \mathbb{R}$ and

$$\begin{aligned} \hat{N}(v_2; K_2) &= \hat{N}((x, 1 + a^2, -b^2, -c^2); \mathbb{R}^n \times \mathbb{R} \times \mathbb{R}_- \times \mathbb{R}_-) \\ &= \begin{cases} \{0\}^n \times \{0\} \times \mathbb{R}_+ \times \mathbb{R}_+, & \text{if } b = 0, c = 0 \\ \{0\}^n \times \{0\} \times \mathbb{R}_+ \times \{0\}, & \text{if } b = 0, c \neq 0, \\ \{0\}^n \times \{0\} \times \{0\} \times \mathbb{R}_+, & \text{if } b \neq 0, c = 0, \\ \{0\}^n \times \{0\} \times \{0\} \times \{0\}, & \text{if } b \neq 0, c \neq 0. \end{cases} \end{aligned}$$

Therefore $\bigcup_{v\in K_2\setminus K_1} \hat{N}(v; K_2) = \{0\}^n \times \{0\} \times \mathbb{R}_+ \times \mathbb{R}_+$. Combine all of the above discussions, we have the following lemma.

Lemma 2. *For the cone K_0 as defined in (6), we have*

$$N(0; K_0) = (\mathbb{R}^n \times \mathbb{R}_+ \times \{0\} \times \{0\}) \cup (\{0\}^n \times \{0\} \times \mathbb{R}_+ \times \mathbb{R}_+).$$

Define the vector-valued indicator function of the set C by $\Delta_C := 0^n$ if $x \in C$ and $\Delta_C(x) = +\infty^n$ otherwise with the superscripts specifying the dimension of the space. Now we are ready to state our local optimality condition for (BLP).

Proposition 3. *Let f, g, and C be given as in (BLP) and $\bar{z} = (\bar{x}, \bar{y})$ be a local optimal solution. Assume in addition that f and g are continuous around $\bar{z}$ and C is locally closed around $\bar{z}$. Then there exists $(x^*, \mu^*, \eta^*, \lambda^*) \in \mathbb{R}^{n+3}$ satisfying either one of the following conditions (12) or (13)*

$$x^* \in \mathbb{R}^n, \quad \mu^* \geq 0, \quad \eta^* = 0, \quad \lambda^* = 0 \tag{12}$$

$$x^* = 0, \quad \mu^* = 0, \quad \eta^* \geq 0, \quad \lambda^* \geq 0, \tag{13}$$

such that

$$0 \in D^*(\Phi + \Delta_C)(\bar{z})(x^*, \mu^*, \eta^*, \lambda^*).$$

If, in addition, f and g are Lipschitz continuous around $\bar{z}$, then

$$0 \in \partial \left[\langle x^*, x \rangle + \mu^* g + \eta^*(g - m) + \lambda^* f \right] (\bar{z}) + N(\bar{z}, C).$$

Proof. Direct application of Theorem 4.2 of [8].

Corollary 1. *Assume $\bar{z}$ solves (BLP) and in addition to the assumptions in Proposition 3, assume f and g are continuously differentiable around $\bar{z}$. Then there exists $(x^*, \mu^*, \eta^*, \lambda^*) \in \mathbb{R}^{n+3}$ satisfying either one of the conditions (12) and (13)*

$$0 \in \nabla \left[\langle x^*, x \rangle + \mu^* g + \eta^*(g - m) + \lambda^* f \right] (\bar{z}) + N(\bar{z}, C).$$

5 Conclusions

In this paper, we successfully reformulated a general bilevel programming problem into a multiobjective optimization problem with a nonconvex ordering cone. The optimality conditions were obtained as applications of Mordukhovich extremal principles.

References

1. Bard, J.: Practical Bilevel Optimization, Algorithms and Applications. Kluwer Academic, Dordrecht (1998)
2. Colson, B., Marcotte, P., Savard, G.: Bilevel programming: a survey. 4OR **3**(2), 87–107 (2005)
3. Dempe, S.: Foundations of Bilevel Programming. Kluwer Academic, Dordrech (2002)
4. Dempe, S.: Annotated bibliography on bilevel programming and mathematical programs with equilibrium constraints. Optimization **52**, 333–359 (2003)
5. Fliege, J., Vicente, L.N.: Multicriteria approach to bilevel optimization. J. Opt. Theory Appl. **131**(2), 209–225 (2006)
6. Mordukhovich, B.S.: Variational Analysis and Generalized Differentiation I and II. Springer, Berlin (2006)
7. Rockafellar, R.T., Wets, R.: Variational Analysis. Grundlehren der Mathematischen Wissenschafte, vol. 317. Springer, Berlin (1998)
8. Mordukhovich, B.S.: Methods of variational analysis in multiobjective optimization. Optimization **58**(4), 413–430 (2009)
9. Henrion, R., Outrata, J.: On calculating the normal cone to a finite union of convex polyhedra. Optimization **57**, 57–78 (2008)

Global Sufficient Conditions for Nonconvex Cubic Minimization Problem with Box Constraints

Yanjun Wang, Zhian Liang, and Linsong Shen

Abstract In this paper, we focus on deriving some sufficient conditions for global solutions to cubic minimization problems with box constraints. Our main tool is an extension of the global subdifferential, L-normal cone approach, developed by Jeyakumar et al. (J. Glob. Optim., 2007; Math. Program. Ser. A 110, 2007), and underestimator functions. By applying these tools to characteristic global solutions, we provide some sufficient conditions for cubic programming problem with box constraints. An example is given to demonstrate that the sufficient conditions can be used effectively for identifying global minimizers of certain cubic minimization problems with box constraints.

Keywords Cubic minimization problem • Global sufficient conditions • Box constraints

1 Introduction

In this paper, we will focus on deriving the global sufficient conditions of the following cubic minimization problem with box constraints:

$$(CP): \begin{cases} \min f(x) = \sum_{i=1}^{n} b_i x_i^3 + \frac{1}{2} x^T A x + a^T x \\ \text{s.t. } x \in D = \prod_{i=1}^{n} [u_i \ \ v_i] \end{cases}$$

where $x = (x_1, \cdots, x_n)^T$ is the vector of decision variables, $b_i \in R$ and $a \in R^n$ are given. $A = (a_{ij}) \in S^n$ where S^n is the set of all symmetric $n \times n$ matrices.

The cubic optimization problem has spawned a variety of applications, especially in cubic polynomial approximation optimization [1], convex optimization [2], engineering design, and structural optimization [3]. Moreover, research results about

Y. Wang (✉) • Z. Liang • L. Shen
Department of Applied Mathematics, Shanghai University of Finance & Economics, Shanghai 200433, People's Republic of China
e-mail: wangyj@mail.shufe.edu.cn

D. Gao et al. (eds.), *Advances in Global Optimization*, Springer Proceedings in Mathematics & Statistics 95, DOI 10.1007/978-3-319-08377-3_4

cubic optimization problem can be applied to quadratic programming problems, which have been widely studied because of its broad applications, to enrich quadratic programming theory.

Several general approaches can be used to establish optimality conditions for solutions to optimization problems [4–8]. These approaches can be broadly classified into three groups: convex duality theory, local subdifferentials by linear functions, global L-subdifferential and L-normal cone by quadratic functions. The third approach, which we extend in this paper, is often adopted to develop optimality conditions for special optimization forms: quadratic minimization with box or binary constraints, quadratic minimization with quadratic constraints, bivalent quadratic minimization with inequality constraints, etc.

In this paper we focus our attention on cubic minimization problems (CP). We will consider the box constraints. Our main tool is an extension of global L-subdifferential and L-normal cone approach from quadratic function to cubic function forms. By exploring some fundamental properties of the problems, we establish some sufficient conditions under which a feasible point will be a global solution to (CP).

The layout of this paper is as follows. In Sect. 2, we introduce some basic definitions and propositions. In Sect. 3, we extend global L-subdifferential approach and present the sufficient optimality conditions for global solutions to (CP) with box constraints. And an example is given, respectively, to show the effectiveness of the proposed global sufficient conditions. Conclusion is in Sect. 4.

2 L-Subdifferentials and Cubic Minimization Problem

We begin with some basic definitions and notations that will be used.

Definition 2.1 (L-Subdifferentials, [4]). Let $f : R^n \to R$ and $x_0 \in R^n$. An element $l \in L$ is called an L-subgradient of f at a point x_0 if

$$f(x) \geq f(x_0) + l(x) - l(x_0), \forall x \in R^n$$

The set $\partial_L f(x_0)$ of all L-subgradients of f at x_0 is referred to as L-subdifferential of f at x_0.

Note that if L is the set of all linear functions defined on R^n, then for any real–valued convex function f defined on R^n, $\partial_L f(x) = \partial f(x)$, where $\partial f(x)$ is the subdifferential in the sense of convex analysis [5, 6].

Definition 2.2 (L-Normal Cones, [4]). For a set $D \subset R^n$ and $x_0 \in D$, the L-normal cone of D at x_0 with respect to L is given by

$$N_{L,D}(x_0) := \{l \in L : l(y) - l(x_0) \leq 0 \text{ for all } y \in D\}.$$

Observe that if L is the set of all linear functions defined on R^n, then $N_{L,D}(x_0) = N_D(x_0)$, the normal cone in the sense of convex analysis [5,6].

Throughout the rest of the paper, we use the following specific choice of L defined by

$$L := \left\{ \sum_{i=1}^{n} b_i x_i^3 + \frac{1}{2} x^T Q x + d^T x | Q = diag(c_1, \cdots, c_n), c_i \in R, d \in R^n \right\} \tag{2.1}$$

where $b_i \in R$ $(i = 1, \cdots, n)$ as defined before.

Proposition 2.3. *Let* $f(x) = \sum_{i=1}^{n} b_i x_i^3 + \frac{1}{2} x^T A x + a^T x$, $\bar{x} = (\bar{x}_1, \cdots, \bar{x}_n)^T \in R^n$. *Then,*

$$\partial_L f(\bar{x}) = \left\{ \sum_{i=1}^{n} b_i x_i^3 + \frac{1}{2} x^T Q x + d^T x \,\middle|\, \begin{array}{c} A \succeq Q, Q = diag(c_1, \cdots, c_n), c_i \in R, \\ d = (A - Q)\bar{x} + a. \end{array} \right\} \tag{2.2}$$

The proof can be referred to [7].

Next we generalize the definition of quadratic underestimators [8] to cubic underestimators, and then explain the subdifferential from another point of view.

Definition 2.4. The function $h : R^n \to R$ is an overestimator of the function $f : R^n \to R$ at $\bar{x}$ over D if for each $x \in D$, $f(x) \leq h(x)$, and $f(\bar{x}) = h(\bar{x})$. The function $h : R^n \to R$ is an underestimator of the function $f : R^n \to R$ at $\bar{x}$ over D if for each $x \in D$, $f(x) \geq h(x)$, and $f(\bar{x}) = h(\bar{x})$.

Proposition 2.5. *Let* $f(x) = \sum_{i=1}^{n} b_i x_i^3 + \frac{1}{2} x^T A x + a^T x$, *and given* $\bar{x} = (\bar{x}_1, \cdots, \bar{x}_n)^T \in R^n$. *Suppose that there exists a diagonal matrix* Q *such that* $A - Q \succeq 0$. *Let* $l(x) = \sum_{i=1}^{n} b_i x_i^3 + \frac{1}{2} x^T Q x + ((A - Q)\bar{x} + a)^T x$. *Then the function* $h(x) = l(x) - l(\bar{x}) + f(\bar{x})$ *is a cubic underestimator over* R^n.

Proof. Note that in this proposition, $l(x)$ is also an L-subgradient of $f(x)$ at $\bar{x}$, so the proof easily follows from Proposition 2.3 and Definition 2.4. □

If $D \subset R^n$, and $\bar{x} \in D$, then we can easily conclude that the function $h(x)$ as shown in Proposition 2.5 is an underestimator of $f(x)$ at $\bar{x}$ over D.

3 Sufficient Conditions for the Solution to Cubic Programming with Box Constraints

We define four index sets $I_1, \cdots, I_4$ according to the given point $\bar{x}$ at first and make sure that $\cup_{i=1}^{4} I_i = \{1, 2, \cdots, n\}$.

$$I_1 = \{i | \forall x_i \in [u_i, v_i], \text{it holds that } b_i(x_i - \bar{x}_i) \geq 0, \text{and } b_i \neq 0\}$$

$$I_2 = \{i \mid \forall x_i \in [u_i, v_i], \text{it holds that } b_i(x_i - \bar{x}_i) \le 0, \text{and } b_i \ne 0\}$$

$$I_3 = \{i \mid \bar{x}_i \in (u_i, v_i), \text{and } b_i \ne 0\}, I_4 = \{i \mid b_i = 0\}$$

Clearly, $I_1 \cup I_2 \cup I_3 \cup I_4 = \{1, \cdots, n\}$.

Now we define n functions $\omega_i^S(x_i)$ according to the index i as follows:

$$\omega_i^S(x_i) := \begin{cases} min_{x_i \in [u_i, v_i]}(x_i^2 + x_i\bar{x}_i + \bar{x}_i^2), & if \quad i \in I_1 \\ max_{x_i \in [u_i, v_i]}(x_i^2 + x_i\bar{x}_i + \bar{x}_i^2), & if \quad i \in I_2 \\ c, \qquad if \ i \in I_4 \end{cases} \tag{3.1}$$

When $i \in I_3$, we define:

$$\omega_i^S(x_i) := \begin{cases} min_{b_i(x_i - \bar{x}_i) \ge 0, x_i \in [u_i, v_i]}(x_i^2 + x_i\bar{x}_i + \bar{x}_i^2) = \omega_{S_i}^1, \text{ if } b_i(x_i - \bar{x}_i) \ge 0; \\ max_{b_i(x_i - \bar{x}_i) \le 0, x_i \in [u_i, v_i]}(x_i^2 + x_i\bar{x}_i + \bar{x}_i^2) = \omega_{S_i}^2, \text{ if } b_i(x_i - \bar{x}_i) < 0; \end{cases} \tag{3.2}$$

So for any i, we always have the inequality

$$b_i(x_i - \bar{x}_i)\omega_i^S(x_i) \le b_i(x_i - \bar{x}_i)(x_i^2 + x_i\bar{x}_i + \bar{x}_i^2), \quad \forall x_i \in [u_i \ \ v_i]$$

We also define

$$\tilde{\chi}_i := \begin{cases} -1, & \bar{x}_i = u_i, \\ 1, & \bar{x}_i = v_i, \\ (a + A\bar{x})_i & \bar{x}_i \in (u_i, v_i). \end{cases} \tag{3.3}$$

If there exists a matrix $Q = diag(\alpha_1, \cdots \alpha_n)$, $\alpha_i \in R$ such that $A - Q \succeq 0$, then we can define:

$$\hat{\alpha}_i = max\{0, -\alpha_i\}, \quad i = 1, 2, \cdots, n, \tag{3.4}$$

We are ready to present the sufficient conditions of a global solution to (CP).

Theorem 3.1. *For (CP), let $\bar{x} \in D$. If there exists a matrix $Q = diag(\alpha_1, \cdots, \alpha_n)$ such that $A - Q \succeq 0$, and $\forall i \in I_1 \cup I_2 \cup I_4$, it holds that*

$$\frac{1}{2}\hat{\alpha}_i(v_i - u_i) + \tilde{\chi}_i[(a + A\bar{x})_i + b_i\omega_i^S(x_i)] \le 0, \tag{SC1}$$

also $\forall i \in I_3$, if $b_i > 0$, it holds that

$$\begin{cases} \frac{1}{2}\hat{\alpha}_i(v_i - \bar{x}_i) - [(a + A\bar{x})_i + b_i\omega_{S_i}^1] \le 0 \\ \frac{1}{2}\hat{\alpha}_i(\bar{x}_i - u_i) + [(a + A\bar{x})_i + b_i\omega_{S_i}^2 \le 0, \end{cases} \tag{SC2}$$

otherwise, it holds that

$$\begin{cases} \frac{1}{2}\hat{\alpha}_i(v_i - \bar{x}_i) - [(a + A\bar{x})_i + b_i\omega_{S_i}^2] \leq 0 \\ \frac{1}{2}\hat{\alpha}_i(\bar{x}_i - u_i) + [(a + A\bar{x})_i + b_i\omega_{S_i}^1] \leq 0. \end{cases} \tag{SC3}$$

Then $\bar{x}$ is a global solution to problem (CP).

Proof. Assume that there exists a matrix $Q = diag(\alpha_1, \cdots \alpha_n)$, such that $A - Q \succeq 0$ and the conditions (SC1) (SC2) (SC3) are true. Let $\beta = a + (A - Q)\bar{x}$, then by Proposition 2.3,

$$l(x) = \sum_{i=1}^{n} b_i x_i^3 + \frac{1}{2}x^T Qx + \beta^T x \in \partial_L f(\bar{x}),$$

i.e.,

$$f(x) - f(\bar{x}) \geq l(x) - l(\bar{x}), \forall x \in R^n \tag{3.5}$$

A first observation regarding (3.5) is that if $\bar{x}$ is a global solution to $l(x)$ on D, then $\bar{x}$ must be a global minimizer to $f(x)$ on D. So it is sufficient to provide the condition which can guarantee a feasible point $\bar{x}$ is a global solution of $l(x)$ on D.

Clearly, $\bar{x}$ is a global solution of $l(x)$ on D if and only if for every $x \in D, l(x) - l(\bar{x}) \geq 0$, i.e.

$$\sum_{i=1}^{n}(b_i(x_i - \bar{x}_i)(x_i^2 + x_i\bar{x}_i + \bar{x}_i^2) + \frac{1}{2}\alpha_i(x_i - \bar{x}_i)^2 + (a + A\bar{x})_i(x_i - \bar{x}_i)) \geq 0. \tag{3.6}$$

Note that $b_i(x_i - \bar{x}_i)\omega_i^S(x_i)$ is an underestimator of the function $b_i(x_i - \bar{x}_i)(x_i^2 + x_i\bar{x}_i + \bar{x}_i^2)$ at the point $\bar{x}$ over the interval $[u_i \ \ v_i]$. Then the sufficient conditions which can make (3.6) hold are: for every i and for every $x \in D$,

$$b_i(x_i - \bar{x}_i)\omega_i^S(x_i) + \frac{1}{2}\alpha_i(x_i - \bar{x}_i)^2 + (a + A\bar{x})_i(x_i - \bar{x}_i) \geq 0 \tag{3.7}$$

So our next goal is to show (3.7) is equivalent to (SC1) or (SC2) or (SC3) according to the index i.

First, by (3.4) it is obvious that $\forall i = 1, 2, \cdots, n$,

$$\hat{\alpha}_i \geq -\alpha_i, \ \ \hat{\alpha}_i \geq 0.$$

Also we can observe that (SC1) is equivalent to $\forall x_i \in [u_i, v_i]$,

$$\frac{1}{2}\hat{\alpha}_i(x_i - u_i) + \tilde{\chi}_i[(a + A\bar{x})_i + b_i\omega_i^S(x_i)] \leq 0 \tag{3.8}$$

and also equivalent to $\forall x_i \in [u_i, v_i]$,

$$-\frac{1}{2}\hat{\alpha}_i(x_i - v_i) + \tilde{\chi}_i[(a + A\bar{x})_i + b_i\omega_i^S(x_i)] \leq 0 \tag{3.9}$$

Next we will show the equivalence between (3.7) and (SCI) when $i \in I_1 \cup I_2 \cup I_3$ and the equivalence between (3.7)and (SC2–SC3) when $i \in I_3$.

(1) When $i \in I_1 \cup I_2 \cup I_4$, we will show that under the following three cases.

Case 1. $\bar{x}_i = u_i$,.
In this case we have $\tilde{\chi}_i = -1$. Then if (SC1) holds, together with (3.8) we must have

$$-[\frac{1}{2}\hat{\alpha}_i(x_i - u_i) + \tilde{\chi}_i((a + A\bar{x})_i + b_i\omega_i^S(x_i))](x_i - u_i) \geq 0$$

Then for (3.7), we have

$$\frac{1}{2}\alpha_i(x_i - \bar{x}_i)^2 + [(a + A\bar{x})_i + b_i\omega_i^S(x_i)](x_i - \bar{x}_i)$$

$$\geq -\frac{1}{2}\hat{\alpha}_i(x_i - \bar{x}_i)^2 + [(a + A\bar{x})_i + b_i\omega_i^S(x_i)](x_i - \bar{x}_i) \geq 0$$

i.e. (3.7) holds. [In fact, if (3.7) holds, (SC1) is also true. We omit the proof. So (SC1) is equivalent to (3.7).]

Case 2. $\bar{x}_i = v_i$.
In this case $\tilde{\chi}_i = 1$. If (SC1) holds, by (3.9) we must have

$$[-\frac{1}{2}\hat{\alpha}_i(x_i - v_i) + \tilde{\chi}_i((a + A\bar{x})_i + b_i\omega_i^S(x_i))](x_i - v_i) \geq 0$$

Then for (3.7), we have

$$\frac{1}{2}\alpha_i(x_i - \bar{x}_i)^2 + [(a + A\bar{x})_i + b_i\omega_i^S(x_i)](x_i - \bar{x}_i)$$

$$\geq -\frac{1}{2}\hat{\alpha}_i(x_i - \bar{x}_i)^2 + [(a + A\bar{x})_i + b_i\omega_i^S(x_i)](x_i - \bar{x}_i) \geq 0$$

[In fact, in this case if (3.7) holds, (SC1) is also true. The proof is also omitted. So (SC1) is equivalent to (3.7).]

Case 3. $\bar{x}_i \in (u_i, v_i)$.
In this case $b_i = 0, \tilde{\chi}_i = (a + A\bar{x})_i$ obviously. If (SC1) holds, then $\alpha_i \geq 0, (a + A\bar{x})_i = 0$, so (3.7) holds. Conversely, if (3.7) holds, similarly, we also have $\alpha_i \geq 0, (a + A\bar{x})_i = 0$. So (SC1) holds.

(2) When $i \in I_3$, we will show that under the following two cases.

If $b_i > 0$, since $i \in I_3$, then $\omega_i^S(x_i)$ is a piecewise function. So (3.7) holds if and only if

$$\begin{cases} \frac{1}{2}\alpha_i(x_i - \bar{x}_i) + (a + A\bar{x})_i + b_i S_i^{min} \geq 0, & x_i \in [\bar{x}_i, v_i] \\ \frac{1}{2}\alpha_i(x_i - \bar{x}_i) + (a + A\bar{x})_i + b_i S_i^{max} \leq 0, & x_i \in [u_i, \bar{x}_i] \end{cases}$$

Also note that the above conditions are equivalent to $(SC1)$. Therefore, (3.7) is equivalent to (SC2).

If $b_i < 0$, noting that $\omega_i^S(x_i)$ is still a piecewise function, then (3.7) holds if and only if

$$\begin{cases} \frac{1}{2}\alpha_i(x_i - \bar{x}_i) + (a + A\bar{x})_i + b_i S_i^{max} \geq 0, & x_i \in [\bar{x}_i, v_i] \\ \frac{1}{2}\alpha_i(x_i - \bar{x}_i) + (a + A\bar{x})_i + b_i S_i^{min} \leq 0, & x_i \in [u_i, \bar{x}_i] \end{cases}$$

Clearly, the above conditions are equivalent to $(SC3)$. Hence, (3.7) is equivalent to $(SC3)$.

Combining the proof of (1) and (2), we can see that if conditions (SC1–SC3) hold, then $\bar{x}$ must be the global minimizer of $l(x)$ on D. Observing (3.5), we can conclude that $\bar{x}$ is a global minimizer of (CP). □

Next we will present an example to illustrate the effectiveness of the sufficient conditions provided in Theorem 3.1.

Example. Consider the following problem

$$\begin{cases} \min\ f(x) = \sum_{i=1}^{4} b_i x_i^3 + \frac{1}{2}x^T A x + a^T x \\ s.t.\ \ x \in D := \prod_{i=1}^{4}[-1, 1] \end{cases}$$

where $b = (\frac{2}{3}, 1, 0, -2)^T$,$a = (4, \frac{9}{2}, -1, -1)^T$,

$$A = \begin{pmatrix} -1 & 2 & 0 & 1 \\ 2 & -1 & 1 & 0 \\ 0 & 1 & 6 & -1 \\ 1 & 0 & -1 & -2 \end{pmatrix}$$

Identify whether the feasible point $\bar{x} = (-1, -1, \frac{1}{2}, 1)^T$ is a global minimizer.

For this problem, we can take a matrix $Q = diag(-4, -4, 0, -4)$. Note that

$$A - Q = \begin{pmatrix} 3 & 2 & 0 & 1 \\ 2 & 3 & 1 & 0 \\ 0 & 1 & 6 & -1 \\ 1 & 0 & -1 & 2 \end{pmatrix}$$

is a diagonal dominant matrix, so $A - Q \succeq 0$. Also by computing, we have $\hat{\alpha} = (4, 4, 0, 4)^T$,$\omega_1^S(x_1) = \frac{3}{4}$,$\omega_2^S(x_2) = \frac{3}{4}$,$\omega_3^S(x_3) = c$, $\omega_4^S(x_4) = \frac{3}{4}$. Therefore $(a+A\bar{x})_1+b_1\omega_1^S(x_1) = \frac{9}{2}$, $(a+A\bar{x})_2+b_2\omega_2^S(x_2) = \frac{17}{4}$, $(a+A\bar{x})_3+b_3\omega_3^S(x_3) = 0$, and $(a + A\bar{x})_4 + b_4\omega_4^S(x_4) = -6$. From (3.3), we have $\tilde{\chi} = (-1, -1, 0, 1)^T$. Substituting the above values to the sufficient conditions in Theorem 3.1, we have for $i = 1$, $(SC1)= -\frac{1}{2} < 0$, then (SC1) holds; for $i = 2$,$(SC1)= -\frac{1}{4} < 0$, then (SC1) holds; for $i = 3$, $(SC3)= 0 = 0$, then (SC3) holds; for $i = 4$, $(SC1)= -2 < 0$, then (SC1) holds. We can see that $\bar{x} = (-1, -1, \frac{1}{2}, 1)^T$ satisfies the sufficient conditions in Theorem 3.1. Hence we can conclude that this point is a global minimizer.

4 Conclusion

In this paper we derive some sufficient conditions to cubic programming with box constraints. Our main tools are the L-subdifferential, L-normal cone, underestimator, and overestimator functions. The results presented in this paper are easily extended to another special cubic problem where the constraint is binary, i.e. $x_i \in \{u_i, v_i\}$. Another important observation is that in the cubic objective function there is no three-time cross term. So the results of this paper just can be applied to some special cubic programming. In future research, we will focus on developing some optimality conditions for more general cubic programming problems.

Acknowledgements This research was supported by NSFC (11271243), Innovation Program of Shanghai Municipal Education Commission (12ZZ071), and Shanghai Pujiang Program (11PJC059).

References

1. Canfield, R.A.: Multipoint cubic surrogate function for sequential approximate optimization. Struct. Multidiscip. Optim. **27**, 326–336 (2004)
2. Nesterov, Y.: Accelerating the cubic regularization of Newton's method on convex problems. Math. Program. **112**(1), 159–181 (2008)
3. Lin, C.-S., Chang, P.-R., Luh, J.Y.S.: Formulation and optimization of cubic polynomial joint trajectories for industrial robots. IEEE Trans. Autom. Control **28**(12), 1066–1074 (1983)
4. Jeyakumar, V., Rubinov, A.M., Wu, Z.Y.: Nonconvex quadratic minimization with quadratic constraints: global optimality conditions. Math. Program. Ser. A **110**(3), 521–541 (2007)
5. Bertsekas, D.P., Nedic, A., Ozdaglar, A.E.: Convex Analysis and Optimization. Athena Scientific and Tsinghua University Press, Belmont (2006)
6. Hiriart-Urruty, J.B., Lemarechal, C.: Convex Analysis and Minimization Algorithms. Springer, Berlin (1993)
7. Wang, Y., Liang, Z.: Global optimality conditions for cubic minimization problem with box or binary constraints. J. Glob. Optim. **47**(4), 583–595 (2010)
8. Jeyakumar, V., Huy, N.Q.: Global minimization of difference of quadratic and convex functions over box or binary constraints. Optim. Lett. **2**, 223–238 (2008)

An Outcome Space Branch-and-Bound Algorithm for a Class of Linear Multiplicative Programming Problems

Yuelin Gao, Nihong Zhang, and Xiaohua Ma

Abstract This article presents an outcome space branch-and-bound algorithm for globally solving a class of linear multiplicative programming problem. In this algorithm, the lower bound is found by solving a separable relaxation programming problem. A convex quadratic programming problem is constructed so as to improve the ability to set the upper bound. The convergence of the algorithm is proved. Numerical experiments are reported to show the feasibility and effectiveness of the proposed algorithm.

Keywords Global optimization • Linear multiplicative programming • Branch-and-bound • Outcome space

1 Introduction

In this paper, we consider the linear multiplicative programming problem as the following form:

$$(LMP)\begin{cases} \min & w(x) = f(x) + \sum_{j=1}^{p} f_j(x)g_j(x), \\ \text{s.t.} & x \in D = \{x \in R^n \mid Ax \leq b\}. \end{cases} \tag{1}$$

where $f(x), f_j(x), g_j(x)$ for $j = 1, 2, \cdots, p$ are linear functions defined on R^n. $D \subset R^n$ is a nonempty bounded convex polyhedron.

Y. Gao (✉)
Institute of Information & System Science, Beifang University of Nationalities, Yinchuan 750021, China

School of Mathematics and Computer Science, Ningxia University, Yinchuan 750021, China
e-mail: gaoyuelin@163.net

N. Zhang • X. Ma
Institute of Information & System Science, Beifang University of Nationalities, Yinchuan 750021, China

D. Gao et al. (eds.), *Advances in Global Optimization*, Springer Proceedings in Mathematics & Statistics 95, DOI 10.1007/978-3-319-08377-3_5

The problem (LMP) has spawned a wide variety of important applications, especially in financial optimization [1] and optimal packing and layout [2]. It is difficult to solve the global optimal solution of the problem (LMP), because it usually possesses many local optimal solutions that are not the global optimal solution. In the last decade, many global optimization algorithms have been proposed for solving the problem (LMP), such as dual simplex method [3], outer-approximation [4,5], cutting-plane method [6,7], heuristic method [8], Linearization method [9], branch-and-bound method [5, 10, 11], etc.

In this paper, an outcome space branch-and-bound algorithm is proposed for globally solving a class of linear multiplicative problems (LMP), the convergence of the algorithm is proved and numerical experiments are given to show the feasibility and effectiveness of the proposed algorithm.

This paper is organized as follows. In Sect. 2 we give a separable relaxation programming problem to obtain the lower bound of the problem (LMP) in the outcome space and construct a convex quadratic programming problem to determine the upper bound of the global optimal value for the problem (LMP). In Sect. 3 we specifically describe the outcome space branch-and-bound algorithm and show the convergence of the algorithm. Finally, the numerical experiments are given to illustrate the feasibility and effectiveness of the proposed algorithm.

2 Bounding

A relaxation programming problem of the problem (LMP) is given to find its lower in the outcome space as follows.

Solve the following linear programming problem:

$$\begin{cases} \min & f(x), \\ \text{s.t.} & x \in D. \end{cases} \text{and} \begin{cases} \max & f(x), \\ \text{s.t.} & x \in D. \end{cases} \tag{2}$$

The optimal values are noted as y_1, y_2, respectively, the optimal solutions noted as x^1, x^2, respectively. Obviously, x^1, x^2 are feasible solutions of the problem (LMP).

Let $Q = \{x^1, x^2\}$. Q represents feasible solutions set of the problem (LMP) known at present.

In a similar way, solve (3) and (4) as follows:

$$\begin{cases} \min & f_j(x), \\ \text{s.t.} & x \in D. \end{cases} \text{and} \begin{cases} \max & f_j(x), \\ \text{s.t.} & x \in D. \end{cases} \tag{3}$$

$$\begin{cases} \min & g_j(x), \\ \text{s.t.} & x \in D. \end{cases} \text{and} \begin{cases} \max & g_j(x), \\ \text{s.t.} & x \in D. \end{cases} \tag{4}$$

The optimal values of (3) are noted as y_{i1}^1, y_{i1}^2, optimal solutions are noted as x_{i1}^1, x_{i1}^2 for $j = 1, 2, \cdots, p$, respectively. Obviously, x_{i1}^1, x_{i1}^2 are feasible solutions of the problem (LMP). Let $Q = Q \bigcup \{x_{i1}^1, x_{i1}^2 : j = 1, 2, \cdots, p\}$. The optimal values of (4) are noted as y_{i2}^1, y_{i2}^2, optimal solutions are noted as x_{i2}^1, x_{i2}^2. Let $Q = Q \bigcup \{x_{i2}^1, x_{i2}^2 : j = 1, 2, \cdots, p\}$. Let $y_0 = f(x), y_{i1} = f_i(x), y_{i2} = g_i(x)$ for $j = 1, 2, \cdots, p$,

$$\Omega = \{y \in R^{2p+1} \mid y_1 \le y_0 \le y_2, y_{j1}^1 \le y_{j1} \le y_{j1}^2, y_{j2}^1 \le y_{j2} \le y_{j2}^2, j = 1, 2, \cdots, p\}.$$

The separable relaxation programming problem of the original problem is given in the outcome space R^{2p+1} as follows:

$$SRP(\Omega) \begin{cases} \min & \bar{W}(Y) = Y_0 + \sum_{j=1}^{p} y_{j1} y_{j2}, \\ \text{s.t.} & y \in \Omega. \end{cases} \tag{5}$$

Its optimal value is a lower bound of the global optimal value of the problem (LMP), and its optimal solution denotes $\bar{y} = (\bar{y}_0, \bar{y}_{11}, \bar{y}_{21}, \cdots, \bar{y}_{p1}, \bar{y}_{p2})^T \in R^{2p+1}$.

Obviously the problem (5) can be separated into $p + 1$ simple optimization problems for $j = 1, 2, \cdots, p$ as following:

$$\begin{cases} \min\ y_0, \\ \text{s.t.}\ \ y_0 \in \Omega_0 = [y_0^1, y_0^2] \end{cases} \text{and} \begin{cases} \min\ y_{j1} y_{j2}, \\ \text{s.t.}\ \ y_j \in \Omega_j = [y_{j1}^1, y_{j1}^2] \times [y_{j2}^1, y_{j2}^2] \end{cases} \tag{6}$$

The optimal solutions of the problems in (6) can be achieved at a point y_1 and $(\bar{y}_{i1}, \bar{y}_{i2})$ where $\bar{y}_{i1}\bar{y}_{i2} = argmin\{y_{i1}^1 y_{i2}^1, y_{i1}^1 y_{i2}^2, y_{i1}^2 y_{i2}^1, y_{i1}^2 y_{i2}^2\}, j = 1, 2, \cdots, p$, respectively. Hence, the global optimal solution of the problem $SRP(\Omega)$ is $\bar{y} = (\bar{y}_0, \bar{y}_{11}, \bar{y}_{21}, \cdots, \bar{y}_{p1}, \bar{y}_{p2})^T \in R^2 p + 1$.

Without loss of generality, suppose that Ω^k be a subhyperrectangle of Ω

$$\Omega^k = \prod_{j=0}^{p} \Omega_j^k = [y_0^{k1}, y_0^{k2}] \times \prod_{j=1}^{p} [y_{j1}^{k1}, y_{j1}^{k2}] \times [y_{j2}^{k1}, y_{j2}^{k2}].$$

Thus the global optimal solution of the problem $SRP(\Omega^k)$ is $\bar{y}^k = (\bar{y}_0^k, \bar{y}_{11}^k, \bar{y}_{12}^k, \cdots, \bar{y}_{p1}^k, \bar{y}_{p2}^k)$, whose global optimal value is a lower bound of the global optimal value of the problem (LMP) over the Ω^k in the outcome space R^{2p+1}.

Construct the following convex quadratic programming so as to effectively search the optimal solution for the problem (LMP):

$$\begin{cases} \min\ F(x) = (f(x) - \bar{y}_0^k)^2 + \sum_{j=1}^{p}(f_j(x) - \bar{y}_{j1}^k)^2 + \sum_{j=1}^{p}(g_j(x) - \bar{y}_{j2}^k)^2, \\ \text{s.t. } y_0^{k1} \le f(x) \le y_0^{k2}, \\ \quad\ y_{j1}^{k1} \le f_j(x) \le y_{j1}^{k2}, j = 1, \cdots, p, \\ \quad\ y_{j2}^{k1} \le g_j(x) \le y_{j2}^{k2}, j = 1, \cdots, p, \\ \quad\ x \in D = \{x \in R^n : Ax \le b\}. \end{cases} \tag{7}$$

Let $\Psi_k = \{x \in R^n : y_0^{k1} \le f(x) \le y_0^{k2}, y_{j1}^{k1} \le f_j(x) \le y_{j1}^{k2}, y_{j2}^{k1} \le g_j(x) \le y_{j2}^{k2}, j = 1, \cdots, p, \}$.

If the problem (7) doesn't have any solution, which implies $D \cap \Psi_K = \Phi$, then Ω^k is deleted; otherwise, we assume that x^k be an optimal solution of the problem (7), obviously x^k is also a feasible solution of the problem (LMP). Let $Q = Q \cup \{x^k\}$, simultaneously, if the optimal value of the problem (7) equals 0, then x^k is the global optimal solution of the problem (LMP) on Ω^k in the outcome space, and Ω^k is deleted. Let $\alpha_k = \min\{w(x) : x \in Q\}$ and it is an upper bound of the global optimal value for the problem (LMP), $x^* \in \min\{w(x) : x \in Q\}$ is the current best feasible solution of the problem (LMP). If so, we can gradually improve the lower bound and the upper bound of the global optimal value of the problem (LMP), and update feasible solutions in the process of branch-and-bound.

3 Branch-and-Bound Algorithm and Its Convergence

For the proposed algorithm, at the k_{th} step, Q represents the feasible solution set found at present. T represents the hyperrectangle set rest at present, α_k and β_k are noted the best lower bound and the best upper bound of the global optimal value of the problem (LMP) at present, respectively.

Algorithm OSA Statement:

Step1 (initialization)
Give an initial hyperrectangle $\Omega^0 = \Omega$, and find the optimal solution y_0 and the optimal value β_0 of the problem (SRP). Set the initial lower bound $\beta_k = \beta_0$ and the initial upper bound $\alpha_k = \min\{w(x) : x \in Q\}$ and the feasible solution set Q find at present, and find a current optimal solution $x^* \in argmin\{w(x) : x \in Q\}$. Let $T = \{\Omega^0\}$ and $k = 0$.

Step2 (termination rule)
If $\alpha_k = \beta_k$, then stop, and outcome the global optimal solution x^* and the global optimal value $w(x^*)$ of problem (LMP); otherwise, go to Step3.

Step3 (selection rule)
Select the hyperrectangle Ω^k satisfying $\beta(\Omega^k) = \beta_k$ in T, let $T = T \backslash \{\Omega^k\}$.

Step4 (partition rule)
Subdivide Ω^k into the two hyperrectangles by the midpoint of the longest edge of Ω^k, which are noted as Ω^{k1}, Ω^{k2}.

Step5 Solve the problem $SRP(\Omega^{ki})$, its optimal value is noted as $\beta(\Omega^{ki})$ and its optimal solution is noted as $\bar{y}^{ki} = (\bar{y}_0^{ki}, \bar{y}_{11}^{ki}, \bar{y}_{12}^{ki}, \cdots, \bar{y}_{p1}^{ki}, \bar{y}_{p2}^{ki})$. Solve the problem (7). If the problem (7) doesn't have any solution, then Ω^{ki} is deleted. If the problem (7) has solution, then its optimal value is noted as F and its optimal solution is noted as $x(\Omega^{ki})$. Let $Q = Q \cup \{x(\Omega^{ki})\}$ and $T = T \cup \{\Omega^{ki}\}$.

Step6 (bounding rule)
Lower bounding: $\beta_k = \min\{\beta : \beta = \beta(H), H \in T\}$;
Upper bounding: $\alpha_k = \min\{w(x), x \in Q\}$.
The current best feasible solution: $x^* \in \arg\min\{w(x) : x \in Q\}$.

Step7 (deleting rule)

$$T = T \backslash \{H \in T : \beta(H) \geq \alpha_k\}.$$

Step8 Let $k = k + 1$, return to Step2.

Convergence of the Proposed Algorithm:

Theorem. *If the algorithm terminates after finite steps, then x^* as the algorithm terminates is the global optimal solution of problem (LMP); if the algorithm does not terminate after finite steps, then every accumulation point $\{x^k\}$ is a global optimal solution of the problem (LMP).*

Proof. If the algorithm terminates after finite steps, then the conclusion is obvious. □

Assume that the algorithm does not terminate after finite steps, then it generates a sequence $\{x^k\}$. According to the iterative process and assumptions of the algorithm, we have

$$\beta_k \leq w^* \leq \alpha_k = w(x_k), k = 0, 1, 2, \cdots \tag{8}$$

where w^* denotes the global optimal objective value of the problem (LMP), β_k and α_k denote the lower bound and the upper bound of the problem (LMP), respectively.

Since $\{\alpha_k\}$ is a decreasing sequence and $\{\beta_k\}$ is an increasing sequence, both $\{\alpha_k\}$ and $\{\beta_k\}$ are convergent. Take limit for both sides of (8), i.e.

$$\lim \beta_k \leq w^* \leq \lim \alpha_k = \lim w(x_k). \tag{9}$$

Set $\lim \beta_k = \beta_*, \lim \alpha_k = \alpha_*$, then (9) turns into

$$\beta_* \leq w^* \leq \lim w(x_k) = \alpha_*. \tag{10}$$

Without loss of generality, assume that $\Omega^{k+1} \subset \Omega^k, \beta_k = \beta(\Omega^k), \alpha_k = \alpha(\Omega^k)$. Since the given bisection process is an exhaustive subdivision process, $\lim \Omega_k = \lim x^k = x^*$, thus

$$\beta_* = \lim \beta(\Omega^k) = w^* = \lim w(x_k) = \alpha_*. \tag{11}$$

Therefore, every accumulation of $\{x^k\}$ is the global optimal solution of the problem (LMP).

4 Numerical Experiments

We show the feasibility and effectiveness of the proposed algorithm by the simple example from [7].

Example 1. By use of the proposed algorithm in this article, solve the example from [7], i.e.

$$\begin{cases} \min \quad w(x) = 3x_1 - 4x_2 + (x_1 + 2x_2 - 1.5)(2x_1 - x_2 + 4) \\ \qquad\qquad +(x_1 - 2x_2 + 8.5)(2x_1 + x_2 - 1), \\ \text{s.t.} \quad 5x_1 - 8x_2 \geq -24, \\ \qquad 5x_1 + 8x_2 \leq 44, \\ \qquad 6x_1 - 3x_2 \leq 15, \\ \qquad 4x_1 + 5x_2 \geq 10, \\ \qquad x_1 \geq 0. \end{cases}$$

where its feasible set is noted as D, $f(x) = 3x_1 - 4x_2$, $f_1(x) = x_1 + 2x_2 - 1.5$, $g_1(x) = 2x_1 - x_2 + 4$, $f_2(x) = x_1 - 2x_2 + 8.5$, $g_2(x) = 2x_1 + x_2 - 1$.

We construct the hyperrectangle $\Omega_0 = [y_1, y_2] \times \prod_{j=1}^{2} ([y_{j1}^1, y_{j1}^2] \times [y_{j2}^1, y_{j2}^2]), i.e.$

$$\Omega_0 = \begin{bmatrix} -12.0000, & -7.5000 \\ 1.0000, & 9.0000 \\ 2.0000, & 11.0000 \\ 1.0000, & 9.0000 \\ 1.0000, & 10.0000 \end{bmatrix}$$

Solve the following problem $SRP(\Omega_k)$for $k = 0$ and $p = 2$ so as to determine an initial lower bound

$$SRP(\Omega_k) \begin{cases} \lim \quad \bar{w}(y) = y_0 + \sum_{j=1}^{p} y_{j1} y_{j2}, \\ \text{s.t.} \quad y \in \Omega_k. \end{cases} \tag{12}$$

We can get the optimal solution

$$y^{0*} = (y_0^{0*}, y_{11}^{0*}, y_{12}^{0*}, \cdots y_{p1}^{0*}, y_{p2}^{0*}) = (-12.000; 1.000; 2.000; 1.000; 1.000),$$

and the global optimal value $\beta_0 = -9.000$ on (13) which is a lower bound of the global optimal value for *Example 1* on Ω_0.

Solve the convex quadratic programming problem as following for $k = 0$ and $p = 2$:

$$\begin{cases} \min \quad F(x) = (f(x) - \bar{y}_0^{k*})^2 + \sum_{j=1}^{p} (f_j(x) - \bar{y}_{j1}^{k*})^2 + \sum_{j=1}^{p} (g_j(x) - \bar{y}_{j2}^{k*})^2, \\ \text{s.t.} \quad y_1^{k1} \le f(x) \le y_2^{k2}, \\ \qquad y_{j1}^{k1} \le f_j(x) \le y_{j1}^{k2}, j = 1, \cdots, p, \\ \qquad y_{j2}^{k1} \le g_j(x) \le y_{j2}^{k2}, j = 1, \cdots, p, \\ \qquad x \in D = \{x \in R^n : Ax \le b\}. \end{cases} \tag{13}$$

We can get the optimal value $F_0 = -13.500$ and the optimal solution $x^* = [0.000, 3.000]$. x^* is the current best feasible solution of the *Example 1*, thereby the current best upper bound $\alpha_0 = -2.500$ of the *Example 1* and set $Q = \{x^*\}$.

We choose Ω_0 and partition it into Ω_{11}, Ω_{12}, by the partitioning rule in the proposed algorithm, i.e.

$$\Omega_{11} = \begin{bmatrix} -12.0000, & -7.5000 \\ 1.0000, & 9.0000 \\ 2.0000, & 6.5000 \\ 1.0000, & 9.0000 \\ 1.0000, & 10.0000 \end{bmatrix}, \Omega_{12} = \begin{bmatrix} -12.0000, & -7.5000 \\ 1.0000, & 9.0000 \\ 6.5000, & 11.0000 \\ 1.0000, & 9.0000 \\ 1.0000, & 10.0000 \end{bmatrix}$$

For solving (12) and (13) over Ω_{12}, if (12) has no solution, then Ω_{12} is deleted; similarly, else, we solve (12) and (13) over Ω_{11}, the optimal value $\beta_1 = -9.000$ of (12) is obtained and the optimal value $F_1 = 13.500$ of (13) and the global optimal value $\alpha_1 = -2.500$ are obtained from (13). In the first iteration, the optimal value of (13) equals to 13.500, and its global optimal value equals -2.500, the optimal solution $x^* = [0.000, 3.000]$, its lower bound equals to -9.000; then we choose Ω_{11} as the next partitioned rectangle up to the tenth iteration, $\Omega_{10,1}$, which is shown in Flow chart, is subdivided into $\Omega_{11,1}$ and $\Omega_{11,2}$, which is shown in box 5 and 6 of the Flow chart.

$$\Omega_{10,1} = \begin{bmatrix} -12.0000, & -9.7500 \\ 5.0000, & 9.0000 \\ 4.5000, & 6.5000 \\ 1.0000, & 5.0000 \\ 1.0000, & 5.5000 \end{bmatrix} \Rightarrow$$

$$\Omega_{11,1} = \begin{bmatrix} -12.0000, & -9.7500 \\ 5.0000, & 9.0000 \\ 4.5000, & 6.5000 \\ 1.0000, & 5.0000 \\ 1.0000, & 3.2500 \end{bmatrix}, \Omega_{11,2} = \begin{bmatrix} -12.0000, & -9.7500 \\ 5.0000, & 9.0000 \\ 4.5000, & 6.5000 \\ 1.0000, & 5.0000 \\ 3.2500, & 5.5000 \end{bmatrix}$$

By solving (12) and (13) over $\Omega_{11,1}$, the optimal value of (13), the global optimal value and the lower bound [i.e., an optimal value of (12)] are achieved such that $F_{11,1} = 4.313, \alpha_{11,1} = -2.500, \beta_{11,1} = -2.750$, respectively; similarly, we solve (12) and (13) over $\Omega_{11,2}$, the optimal value of (13), the global optimal value and the lower bound [i.e., optimal value of (12)] are attained and correspondingly be denoted as $F_{11,2} = 3.801, \alpha_{11,2} = 0.954, \beta_{11,2} = 6.813$, respectively. It is evident to see that $F_{11,1}$ is bigger than $F_{11,2}$, satisfying the pruning rule, so we delete $\Omega_{11,1}$; while $\beta_{11,2}$ is bigger than $\alpha_{11,2}$ over $\Omega_{11,2}$, satisfying deleting rule in the Algorithm OSA, so we delete $\Omega_{11,2}$. At present, the hyperrectangle set T becomes empty set, thus the termination rule is satisfied, the global optimal value of the *Example 1* is -2.500, the lower bound of optimal value of the *Example 1* is -2.750, and its optimal solution is $x^* = (0.000, 3.000)$.

Example 1 [7] Solving a kind of multiplicative programs by the proposed algorithm as following:

$$(TP) \quad \begin{cases} \min & \sum_{i=1}^{p} (c_i^t x \times d_i^t x), \\ \text{s.t.} & Ax \geq b, x \geq 0. \end{cases}$$

where $c_i, d_i \in R^n (i = 1, 2, \cdots, p), A \in R^{n \times m}$, and $b \in R^m$. All elements of c_i, d_i, A, and b are randomly generated from the range [1, 100].

The algorithm was coded in MATLAB (7.8) language and run on a Pentium IV, CPU3.20GHZ, 1.00 GB RAM microcomputer. Let $\varepsilon = 10^{-5}$ and the proposed algorithm terminates as $\beta_k - \alpha_k < 10^{-5}$.

Table 1 shows the comparison of two algorithms for (TP). Here OSA represents the presented algorithm in Sect. 3; OAM is the outer-approximation algorithm in [7]. The main objective of this section is to investigate the computer performance of the proposed optimization algorithm. The following indices the performance of the Algorithm OSA: T is the average time of CPU (in second); SD (in parenthesis) is the standard deviation of T; R is the maximal number of the rest of hyperrectangle via pruning from hyperrectangle set each time, while C represents the average number of cuts in [7].

For each size of (p, n, m), the table contains T, SD, and R that we solved the needed for solving different random instances. The results of OAM are taken from [7], in which its examples were carried out on a SUN SPARC-2 computer (27.5MIPS). Table 1 shows the results of OSA for (TP), when (p, n, m) ranges from $(2, 20, 30)$ to $(5, 50, 70)$. T, SD, and R of ten examples are presented. We see some results from Table 1 as follows:

Table 1 The comparison with results reported in [7]

p	2	2	2	3	3	3
n	20	50	50	20	50	50
m	30	30	70	30	30	70
OSA T	9.9	15.8	30.6	29.1	63.9	71.1
(SD)	(4.1)	(5.8)	(6.2)	(5.8)	(13.9)	(20.2)
OAM T	5.4	25.9	55.6	49.3	202.7	1,087.7
(SD)	(1.8)	(5.1)	(14.8)	(33.1)	(74.2)	(900.4)
OSA R	3	2	2	2	4	2
OAM C	21.2	21.6	19.6	29.5	30.2	32.3
p	4	4	5	5	5	
n	20	50	20	50	50	
m	30	30	30	30	70	
OSA T	48.9	98.2	101.0	202.5	281.8	
(SD)	(17.1)	(38.6)	(56.8)	(40.6)	(60.8)	
OAM T	416.5	3,897.6				
(SD)	(233.2)	(2,158.6)				
OSA R	2	2	2	3	3	
OAM C	38.8	42.7	–	–	–	

Table 1 shows that n and m have some substantial influence on R (and consequently on T), because R is directly related to the speed of pruning the hyper-rectangle by the Algorithm OSA in the outcome space.

The comparison between the two algorithms indicates that although initially the computing time requirements of Algorithm OSA grow faster due to higher computational costs for solving problem and problem (13), its growth rate (except $(p,n,m) = (2,20,30)$) tends to be far less than that exhibited of [7], as the number of variables and constraints increases. On the other hand, we see from Table 1 that the algorithm OAM was very sensitive to the size of p; while the algorithm OSA only has some relationship with p. Notice that we test the numerical results of the case when $p = 5$, which didn't have in [7].

5 Concluding Remarks

The paper gives an outcome space branch-and-bound algorithm for globally solving a class of linear multiplicative programming problem. In this algorithm, the lower bound is given by solving a separable relaxation programming problem. Furthermore, in order to efficiently search for the global optimal solution to the problem, the paper constructs a convex quadratic programming problem to resaise the ability of upper-bounding. The convergence of the proposed algorithm is proved. Numerical experiments show that the proposed algorithm can solve the middle-scale large problems. Our proposed algorithm seems better than the text comparison algorithm, but due to the computing environments differing, we cannot say the text comparison algorithm is not good, the results of their calculations are also very good, but we can say that the proposed algorithm is feasible and effective in computation.

Acknowledgements The work is supported by the Foundation of National Natural Science China (11161001) and by the Scientific Project Foundation of Beifang University of Nationalities (2013XY Z025).

References

1. Maranas, C.D., Androulakis, I.P., Floudas, C.A., Berger, A.J., Mulvey, J.M.: Solving long-term financial Planning problems via global optimization. J. Econ. Dyn. Control **21**, 1405–1425 (1997)
2. Kuno, T.: Globally determining a minimum-aera hyperrectangle enclosing the projection of a bigger dimensional set. Oper. Res. Lett. **13**, 295–303 (1993)
3. Schaible, S., Sodini, C.: Finite algorithm for generalized linear multiplicative programming. J. Optim. Theory Appl. **87**(2), 441–455 (1995)
4. Kuno, T., Yajima, Y., Konno, H.: An outer approximation method for minimizing the product of several convex functions on a convex set. J. Glob. Optim. **3**(3), 325–335 (1993)
5. Yuelin, G., Guorong, W., Weimin, M.: A new global optimization approach for convex multiplicative programming. Appl. Math. Comput. **216**, 1206–1218 (2010)
6. Benson, H.P., Boger, G.M.: Outcome-space cutting-plane algorithm for linear multiplicative programming. J. Optim. Theory Appl. **104**, 301–322 (2000)
7. Konno, H., Kuno, T., Yajima. Y.: Global minimization of a generalized convex multiplicative function. J. Glob. Optim. **4**, 47–62 (1994)
8. Liu, X.J., Umegaki, T., Yamamoto, Y.: Heuristic methods for linear multiplicative programming. J. Glob. Optim. **4**, 433–447 (1999)
9. Chun-Feng, W., San-Yang, L.: A new linearization method for generalized linear multiplicative programming. Comput. Oper. Res. **38**, 1008–1013 (2011)
10. Ryoo, H.S., Sahinidis, N.V.: Global optimization of multiplicative programs. J. Glob. Optim. **26**, 387–418 (2003)
11. Kuno, T.: A finite branch-and-bound algorithm for linear multiplicative programming. Comput. Optim. Appl. **20**, 119–35 (2001)

A Modified Cut-Peak Function Method for Global Optimization

Sun Li and Wang Yuncheng

Abstract We present a cut-peak function method for finding a global minimizer of the bound constrained optimization problems. A simple cut-peak function and choice function are defined at a local minimizer. By minimizing the choice function, a global descent of the original objective function is assured. Since the pattern search method does not require the gradient of the choice function, smoothing technique is not employed. The new algorithm is simple to implement and numerical results indicate its efficiency.

1 Introduction

The global optimization algorithms play an important role in real-world applications, but the definition of an efficient algorithm for these problems is an open question. In literature, many different approaches have been proposed to solve this class of problems. One of these is the function modification approach, such as the filled function methods [1–4], the tunneling methods [5], and the cut-peak function methods [6,7]. These methods have two main phases to achieve the solution procedure,

Phase 1: (Local Search) Start from a given point and use certain local optimization method to locate a feasible local minimizer x_k^*.

Phase 2: (Global Search) With a constructed function, a different minimizer with lower cost value is found in another valley. When obtaining such a point, let it be the next iteration and return to Phase 1. If such a point does not found, stop the algorithm and return the current iteration x_k^* as the global solution to the original problem.

S. Li (✉) • W. Yuncheng
College of Information Science and Engineering, Shandong Agricultural University, Taian 271018, China

Postdoctoral Station of Agricultural Resources and Environment, Shandong Agricultural University, Taian 271018, China
e-mail: sunlishi@hotmail.com; ycwang@sdau.edu.cn

D. Gao et al. (eds.), *Advances in Global Optimization*, Springer Proceedings in Mathematics & Statistics 95, DOI 10.1007/978-3-319-08377-3_6

The cut-peak function method has been initially introduced by Wang [1]. In this method, a cut-peak function and a choice function are defined at a local minimizer, and minimizing the choice function assures a global descent of the original objective function. Since the choice functions are defined on the exponential and logarithmic terms or they are not smooth, these result in the additional employment of the smoothing technique. Later, Huang et al. [2] explain that it's possible to cut down the global minimizer, when all the other minimizers of the concerned problem are far away from the current local minimizer. They give a revised cut-peak function at the current local minimizer x_k^*, that is, $w(x_k^*, x) := f(x_k^*)$. In this method, the smoothing technique is also applied to overcome the difficulty from the non-differentiable of the choice function. In the global search phase, a different minimizer with lower cost value is found in another valley with the cut-peak function. But the cut-peak function is a horizontal line connecting the incumbent minimizer x_k^*, and the gradient of the choice function is zero in a certain neighborhood of x_k^*. This leads to a serious of problems, such as the definition of descent direction and the selection of the step size.

In this paper, we define a simple cut-peak function and we employ the pattern search method to minimize the choice function. Since the pattern search method does not require the gradient of the objective function, we need not smooth the choice function with certain technique, this decreases the nonlinearity of the problem. The newly proposed method is simple to implement and numerical results show that this method works well.

This paper is organized as follows. Some basic concepts and notations are stated in the next section and we present our new algorithm. The numerical results are shown in Sect. 3.

2 The Cut-Peak Function Algorithm

2.1 The Cut-Peak Function and the Choice Function

Consider the global minimization problem

$$\begin{aligned} &\min f(x) \\ &s.t.\ l \leq x \leq u \end{aligned} \tag{1}$$

where $f(x)$ is continuously differentiable and l, u are given vectors in R^n.

Definition 1. *$w(r, x_k^*, x)$ is called a cut-peak function of $f(x)$ at point x_k^* with a positive parameter r, if the following two conditions are satisfied:*

(i) x_k^* is the unique maximum point of $w(r, x_k^*, x)$ and $w(r, x_k^*, x_k^*) = f(x_k^*)$;
(ii) for any direction $d \in R^n$, $w(r, x_k^*, x_k^* + \lambda d)$ is strictly decreasing with respect to the step length λ, and

$$\lim_{\lambda \to +\infty} w(r, x_k^*, x_k^* + \lambda d) = f(x_k^*) - c(r) > -\infty$$

where $c(r)$ is a positive scalar with respect to the given vector r and is called the maximum cut of $w(r, x_k^*, x)$ at x_k^*.

The cut-peak function defined in [1] is as follows:

$$w(r, x_k^*, x) = f(x_k^*) - \frac{rt^2}{1+t^2},$$

and

$$w(r, x_k^*, x) = f(x_k^*) - r(1 - e^{-t^2})$$

where $t = \|x - x_k^*\|$.

Definition 2. $F(r, x_k^*, x) = \min\{f(x), w(r, x_k^*, x)\}$ *is called a choice function of $f(x)$ crossing through the point x_k^*.*

It should be noted that the cut-peak function cuts the hyperplane of $f(x)$ that locates above the cut-peak function $w(r, x_k^*, x)$ and replaces it by $w(r, x_k^*, x)$ itself. This is the reason we name it cut-peak function.

2.2 *The Modified Cut-Peak Function Method*

In this paper, we give a modified cut-peak function, that is,

$$w(\varepsilon, x_k^*) = f(x_k^*) - t\varepsilon, \tag{2}$$

where $t = \|x - x_k^*\|$. Then, the choice function is

$$F(\varepsilon, x_k^*) = \min\{f(x_k^*), w(\varepsilon, x_k^*)\}. \tag{3}$$

It's easy to verify that the cut-peak function defined by (2) satisfies the conditions of Definition 1. Hence, it holds all the properties of the cut-peak function in [5], which indicates the global convergence of the newly present algorithm.

Now, we're ready to present our algorithm to solve problem (1).

2.2.1 The Modified Cut-Peak Function Method

Step 0 **(Initialization)**
Given an initial point $x_0, l \leq x \leq u$. Set $k := 0$, and choose the positive constant ε.

Step 1 **(Local Search)**
Obtain a local minimizer of the original problem (1) by using a local search procedure, starting from the initial point $x_k^{(0)} = x_0$. Let x_k^* be the local minimizer obtained.

Step 2 **(Global Search)**
Start from the current local minimizer x_k^*, find a local minimizer $\widetilde{x_k}$ of the choice function (3) with the pattern search method.
If $f(\widetilde{x_k}) < f(x_k^*)$, then set $x_{k+1}^{(0)}$, $k := k + 1$, go to Step 1.
Else, stop the algorithm and output x_k^* as the final solution.

3 Numerical Results

In this section, some numerical results are reported. The code is written in Matlab 7.0 with double precision. For each problem, the local minimizations have been performed by using the projected gradient algorithm [8–10] and the termination condition is the Euclidean norm of the gradient at the current point x_k^* below 10^{-5}, namely

$$\|\nabla f(x_k^*)\| \leq 10^{-5}.$$

In the global optimization toolbox of Matlab, pattern search solver can solve optimization problems which is linear, nonlinear, or with bound constraints. We compare our new algorithm (MCPF) with pattern search method (PSM) in the following 4 examples.

In all the following examples, we take $\varepsilon = 10^{-5}$. The numerical results of all examples are presented in their respective tables with the initial point x_0. The symbols used in the tables are given as follows:

IT: The iteration number of both the global search phase and local search phase;
IF: The number of function evaluation needed to satisfy the stopping criterion;
IC: The number of the global search;
GM: The obtained global minimizer;
FF: The obtained optimal function value.

Example 1. The six-hump camelback problem

$$f(x) = 4x_1^2 - 2.1x_1^4 + \frac{1}{3}x_1^6 + x_1x_2 - 4x_2^2 + 4x_2^4, \quad l = (-3, -1.5)^T, u = (3, 1.5)^T.$$

The global minimizers are $(-0.089842, 0.712656)^T$ and $(0.089842, -0.712656)^T$, and the cost value is $f(x^*) = -1.031628$ (Table 1).

Table 1 Computational results for Example 1

Algorithm	x_0	IT	IF	IC	GM	FF
MCPF	$(1,1)^T$	54	199	1	$(0.898, -0.7127)^T$	-1.0316
MCPF	$(2,2)^T$	64	215	1	$(0.898, -0.7127)^T$	-1.0316
MCPF	$(3,3)^T$	58	203	1	$(0.898, -0.7127)^T$	-1.0316
PSM	$(1,1)^T$	54	186	0	$(0.898, -0.7127)^T$	-1.0316
PSM	$(2,2)^T$	58	190	0	$(0.898, -0.7127)^T$	-1.0316
PSM	$(3,3)^T$	54	176	0	$(0.898, -0.7127)^T$	-1.0316

Example 2. The two-dimensional Shubert problem I

$$f(x) = \{\sum_{i=1}^{5} i cos((i+1)x_1 + i)\}\{\sum_{i=1}^{5} i cos((i+1)x_2 + i)\},$$
$$l = (-10, -10)^T, u = (10, 10)^T.$$

The global minimizer of Example 2 is $(-1.42513, -0.80032)^T$ and the cost value is $f(x^*) = -186.730909$ (Table 2).

Table 2 Computational results for Example 2

Algorithm	x_0	IT	IF	IC	GM	FF
MCPF	$(2,2)^T$	66	219	1	$(4.8581, -0.8003)^T$	-186.7309
PSM	$(2,2)^T$	50	129	0	$(-7.0835, 10.0000)^T$	-48.5068

Example 3. The two-dimensional Shubert problem II

$$f(x) = \{\textstyle\sum_{i=1}^{5} i cos((i+1)x_1 + i)\}\{\textstyle\sum_{i=1}^{5} i cos((i+1)x_2 + i)\}$$
$$+\tfrac{1}{2}[(x_1 + 1.42513)^2 + (x_2 + 0.80032)^2], \;\; l = (-10, -10)^T, u = (10, 10)^T.$$

The global minimizer of Example 3 is $(-1.42513, -0.80032)^T$ and the cost value is $f(x^*) = -186.730909$ (Table 3).

Table 3 Computational results for Example 3

Algorithm	x_0	IT	IF	IC	GM	FF
MCPF	$(2,2)^T$	31	159	0	$(4.8581, -0.8003)^T$	-186.7309
PSM	$(2,2)^T$	68	237	0	$(-0.8005, 4.8568)^T$	-170.5306

Example 4. The two-dimensional Shubert problem III

$$f(x) = \{\sum_{i=1}^{5} i cos((i+1)x_1 + i)\}\{\sum_{i=1}^{5} i cos((i+1)x_2 + i)\} \\ +(x_1 + 1.42513)^2 + (x_2 + 0.80032)^2, \quad l = (-10, -10)^T, u = (10, 10)^T.$$

The global minimizer of Example 4 is $(-1.42513, -0.80032)^T$ and the cost value is $f(x^*) = -186.730909$ (Table 4).

Table 4 Computational results for Example 4

Algorithm	x_0	IT	IF	IC	GM	FF
MCPF	$(4, 4)^T$	69	241	1	$(-7.0809, 4.8556)^T$	-122.7247
PSM	$(4, 4)^T$	58	188	0	$(-0.1960, -0.8003)^T$	-122.0653

Acknowledgements The work was supported in part by the National Science Foundation of China (10901094,11301307), the Excellent Young Scientist Foundation of Shangdong Province (BF2011SF024, BF2012SF025), and Young Teacher Foundation of Shandong Agricultural University (23744).

References

1. Ge, R.: A filled function method for finding a global minimizer of a function of several variables. Math. Prog. **46**, 191–204 (1990)
2. Liang, Y., Zhang, L., Li, M., Han, B.: A filled function method for global optimization. J. Comput. Appl. Math. **205**, 16–31 (2007)
3. Wu, Z., Zhang, L., Bai, F.: NewModified function method for global optimization. J. Optim. Theory Appli. **125**, 181–203 (2005)
4. Wu, Z., Zhang, L.: Global descent methods for unconstrained global optimization. J. Glob. Optim. **50**, 379–396 (2011)
5. Zhang, L., Fang, L.,Wang, Y.: A new tunnel function method for global optimization. Pac. J. Optim. **46**, 125–138 (2008)
6. Wang, Y., Fang, W., Wu, T.: A cut-peak function method for global optimization. J. Comput. Appl. Math. **230**, 135–142 (2009)
7. Huang, Z., Miao, X., Wang, P.: A revised cut-peak function method for box constrained continuous global optimization. Appl. Math. Comput. **194**, 224–233 (2007)
8. Sun, L., He, G., Wang, Y., Fang, L.: An active set quasi-Newton method with projected search for bound constrained minimization. Comput. Math. Appl. **58**, 161–170 (2009)
9. Sun, L., He, G.,Wang, Y., Zhou, C.: An accurate active set Newton method for large scale bound constrained optimization. Appl. Math. **56**, 297–314 (2011)
10. Sun, L., He, G., Zhou, C.: An active set strategy based on the multiplier function or the gradient. Appl. Math. **55**, 291–304 (2010)

Modified Filled Function Method for Global Discrete Optimization

You-Lin Shang, Zhen-Yang Sun, and Xiang-Yi Jiang

Abstract We present a modified definition of the filled function for discrete nonlinear programming problem and give a filled function satisfying our definition. The properties of the proposed filled function and the method using this filled function to solve discrete nonlinear programming problem are discussed in this paper. The results of preliminary numerical experiments are also reported.

Keywords Integer programming • Global discrete optimization • Local minimizer • Global minimizer • Filled function

1 Introduction

For continuous global optimization problem, many deterministic methods have been proposed to search for a globally optimal solution of a given function of several variables, including filled function method [1], tunneling method [2], etc. The filled function method was first put forward by Ge's paper [1], and many other filled functions have been put forward afterwards [3–6]. The idea of this method is to construct a filled function $P(x)$ and by minimizing $P(x)$ to escape from a given local minimizer x_1^* of the original objective function $f(x)$.

With regard to discrete nonlinear programming problem, the approaches of continuity are presented by Ge's paper [7] and Zhang's paper [8]. This paper modified the conditions of the filled function in paper [3], and making it satisfy the discrete nonlinear programming problem.

The paper is organized as follows: Sect. 2 lists problem and an algorithm related to discrete nonlinear programming problem. In Sect. 3, we present a filled function which is modified from that of the unconstrained global optimization problem. Next,

The National Natural Science Foundation of China (No. 10971053).

Y.-L. Shang (✉) • Z.-Y. Sun • X.-Y. Jiang
Mathematics Department, Henan University of Science and Technology, Luoyang 471023, China
e-mail: mathshang@sina.com; 651245654@qq.com; 361314899@qq.com

D. Gao et al. (eds.), *Advances in Global Optimization*, Springer Proceedings in Mathematics & Statistics 95, DOI 10.1007/978-3-319-08377-3_7

in Sect. 4, a filled function algorithm is presented and the results of preliminary numerical experiments are reported. In Sect. 5, an example of supply chain is given. Finally, conclusions are included in Sect. 6.

2 Problem and an Algorithm

We consider the following discrete nonlinear programming problem

$$(DP) \quad \begin{cases} \min f(x), \\ s.t. x \in \Omega, \end{cases} \tag{2.1}$$

where $\Omega \subseteq I^n$ is a bounded and closed set. I^n is an integer set in R^n. $f(x) = +\infty$ as $x \bar{\in} \Omega$.

The local minimizer of $f(x)$ for problem (DP) is obtained by the following algorithm.

Algorithm 1 [9]

Step 1. Choose any integer $x_0 \in \Omega$.
Step 2. If x_0 is a local minimizer of $f(x)$ over Ω, then stop; otherwise we search the neighborhood of x_0. Then we obtain a $x \in N(x_0) \cap \Omega$ and have $f(x) < f(x_0)$.
Step 3. Let $x_0 := x$, go to step 2.

3 A Filled Function and Its Properties

This section gives a new definition of the filled function of $f(x)$ at its local minimizer x_1^* for problem (DP) on the basis of paper [9] as follows:

Definition 3.1. $P(x, x_1^*)$ is called a filled function of $f(x)$ at a local minimizer x_1^* for problem (DP) if $P(x, x_1^*)$ has the following properties:

(i) $P(x, x_1^*)$ has no local minimizer in the set $S_1 \setminus \{x_0\}$. The prefixed point x_0 is in the set S_1 and is not necessarily local minimizer of $P_{x_1^*}(x)$;
(ii) If x_1^* is not a global minimizer of $f(x)$, then there exists a local minimizer x_1 of $P(x, x_1^*)$, such that $f(x_1) < f(x_1^*)$, that is $x_1 \in S_2$.

Note that the Definition 3.1 is different from the definition in paper [9]. In Definition 3.1, x_0 is not necessarily the local minimizer of $P(x, x_1^*)$. And in this paper, we define sets $S_1 = \{x \in \Omega : f(x) \geq f(x_1^*)\}$ and $S_2 = \{x \in \Omega : f(x) < f(x_1^*)\}$, x_1^* is the current local minimizer of problem (DP).

Therefore, we can present a filled function of $f(x)$ at its local minimizer x_1^* for problem (DP) as follows:

$$P(x, x_1^*, A, \rho) = \eta(\|x - x_0\|) \cdot \varphi[A(f(x) - f(x_1^*) + \rho)], \tag{3.1}$$

where $A > 0$ and $\rho > 0$ are two parameters, prefixed point $x_0 \in \Omega$ satisfies the condition $f(x_0) \geq f(x_1^*)$ and functions $\eta(t)$ and $\varphi(t)$ need to satisfy the following conditions different from that of functions $\bar{\eta}(t)$ and $\bar{\varphi}(t)$ in paper [3]:

(i) $\eta(t)$ and $\varphi(t)$ are strictly monotone increasing function for any $t \in [0, +\infty)$ and any $t \in (-\infty, +\infty)$ respectively;
(ii) $\eta(0) = 0$ and $\eta(t_1 + t_2) \geq \eta(t_1) + \eta(t_2)$ for any $t_1, t_2 \in [0, +\infty)$.
(iii) $\varphi(0) = 0$ and $\lim_{t \to +\infty} \varphi(t) = B > 0$.

Without loss of generality, let $f(x)$ is not a constant function in the Ω, we choose parameter ρ satisfies:

$$0 < \rho \leq f(x_1^*) - \underline{f}\,, \tag{3.2}$$

where $\underline{f} \leq \min_{x \in \Omega} f(x)$.

We construct the following auxiliary discrete nonlinear programming problem (AP) relate to the problem (DP) as follows:

$$(AP) \qquad \begin{cases} \min P(x, x_1^*, A, \rho), \\ \quad s.t. x \in \Omega. \end{cases} \tag{3.3}$$

In the following we will prove that the function $P(x, x_1^*, A, \rho)$ is really a filled function of $f(x)$ at local minimizer x_1^* satisfying Definition 3.1. First, we list a lemma in the paper [9] as follows:

Lemma 3.1 ([9]). *For any integer point x, if $x \neq x_0$, then there exists an integer point y_0, such that*

$$\|y_0 - x_0\| \leq \|x - x_0\| - 1. \tag{3.4}$$

Theorem 3.1. *$P(x, x_1^*, A, \rho)$ has no local minimizer in the integer set $S_1 \setminus \{x_0\}$ if $A > 0$ satisfies the following condition*

$$\varphi(A\rho) > \frac{\eta(L)}{\eta(L) + \eta(1)} B, \tag{3.5}$$

where $L = \max_{x \in \Omega} \|x - x_0\|$.

Proof. For any $x \in S_1$ and $x \neq x_0$, from Lemma 3.1 [9], we know that there exists an integer point $y_0 \in N(x) \cap \Omega$, such that $\|y_0 - x_0\| \leq \|x - x_0\| - 1$.

Consider the following two cases:

(1) If $f(y_0) < f(x_1^*)$, then $f(y_0) < f(x)$, we have $\varphi[A(f(y_0) - f(x_1^*) + \rho)] < \varphi[A(f(x) - f(x_1^*) + \rho)]$, and $\eta(\|y_0 - x_0\|) \leq \eta(\|x - x_0\| - 1) < \eta(\|x - x_0\|)$. Therefore,

$$P(y_0, x_1^*, A, \rho) = \eta(\|y_0 - x_0\|) \cdot \varphi[A(f(y_0) - f(x_1^*) + \rho)]$$
$$< \eta(\|x - x_0\|) \cdot \varphi[A(f(x) - f(x_1^*) + \rho)] = P(x, x_1^*, A, \rho),$$

that is, x is not a local minimizer of function $P(x, x_1^*, A, \rho)$.

(2) If $f(y_0) \geq f(x_1^*)$, then

$$\begin{aligned}
&P(y_0, x_1^*, A, \rho) - P(x, x_1^*, A, \rho) \\
&= \eta(\|y_0 - x_0\|) \cdot \varphi[A(f(y_0) - f(x_1^*) + \rho)] - \eta(\|x - x_0\|) \cdot \varphi[A(f(x) - f(x_1^*) + \rho)] \\
&\leq \eta(\|x - x_0\| - 1) \cdot \varphi[A(f(y_0) - f(x_1^*) + \rho)] - \eta(\|x - x_0\|) \cdot \varphi[A(f(x) - f(x_1^*) + \rho)] \\
&\leq [\eta(\|x - x_0\|) - \eta(1)] \cdot \varphi[A(f(y_0) - f(x_1^*) + \rho)] - \eta(\|x - x_0\|) \cdot \varphi[A(f(x) - f(x_1^*) + \rho)] \\
&= \eta(\|x - x_0\|)\{\varphi[A(f(y_0) - f(x_1^*) + \rho)] - \varphi[A(f(x) - f(x_1^*) + \rho)]\} \\
&\quad - \eta(1) \cdot \varphi[A(f(y_0) - f(x_1^*) + \rho)] \\
&\leq \eta(L) \cdot [B - \varphi(A\rho)] - \eta(1) \cdot \varphi(A\rho) = \eta(L) \cdot B - [\eta(L) + \eta(1)] \cdot \varphi(A\rho).
\end{aligned}$$

It follows from (3.5) that $P(y_0, x_1^*, A, \rho) - P(x, x_1^*, A, \rho) < 0$.

It is shown that x is also not a local minimizer of function $P(x, x_1^*, A, \rho)$. □

Theorem 3.2. *If the set S_2 is nonempty, then function $P(x, x_1^*, A, \rho)$ does have local minimizer in the set S_2.*

Proof. Let x^* is a global minimizer of function $f(x)$. Since $S_2 \neq \phi$, $f(x^*) < f(x_1^*)$. It follows from (3.2) that $f(x^*) - f(x_1^*) + \rho \leq 0$. Therefore $P(x^*, x_1^*, A, \rho) = \eta(\|x^* - x_0\|) \cdot \varphi[A(f(x^*) - f(x_1^*) + \rho)] \leq 0$.

On the other hand, for any $y \in S_1 \setminus \{x_0\}$, we have $P(y, x_1^*, A, \rho) = \eta(\|y - x_0\|) \cdot \varphi[A(f(y) - f(x_1^*) + \rho)] > 0$.

Therefore, the global minimizer of $P(x, x_1^*, A, \rho)$ is in the set S_2, that is, $P(x, x_1^*, A, \rho)$ does have local minimizer in the set S_2. □

From Theorems 3.1 and 3.2, the function $P(x, x_1^*, A, \rho)$ satisfies the conditions of Definition 3.1, i.e., function $P(x, x_1^*, A, \rho)$ is a filled function of $f(x)$ at its local minimizer x_1^* for discrete nonlinear programming problem (DP).

Prefixed point $x_0 \in S_1$ has the following two properties:

Theorem 3.3. *The prefixed point $x_0 \in S_1$ is a local minimizer of $P(x, x_1^*, A, \rho)$ if $x_0 \in S_1$ is a local minimizer of $f(x)$.*

Proof. Since $x_0 \in S_1$ is a local minimizer of $f(x)$, there exists a neighborhood $N(x_0)$, for any $x \in N(x_0) \cap \Omega$, we have $f(x) \geq f(x_0)$ and

$$\begin{aligned} P(x, x_1^*, A, \rho) &= \eta(\|x - x_0\|) \cdot \varphi[A(f(x) - f(x_1^*) + \rho)] \\ &\geq \eta(\|x_0 - x_0\|) \cdot \varphi[A(f(x_0) - f(x_1^*) + \rho)] = P(x_0, x_1^*, A, \rho), \end{aligned}$$

that is, $x_0 \in S_1$ is a local minimizer of $P(x, x_1^*, A, \rho)$. □

Theorem 3.4. *Suppose $P(x, x_1^*, A, \rho)$ is a filled function of $f(x)$ at local minimizer x_1^*, and x_1^* is already a global minimizer of $f(x)$, then x_0 is the unique local minimizer of $P(x, x_1^*)$.*

Proof. Since x_1^* is already a global minimizer of $f(x)$, then for any $x \in N^0(x_0) \cap \Omega$, we have $f(x) - f(x_1^*) + \rho > 0$ and

$$\varphi[A(f(x) - f(x_1^*) + \rho)] > 0. \tag{3.6}$$

It follows from (3.6) that

$$\begin{aligned} P(x, x_1^*, A, \rho) &= \eta(\|x - x_0\|) \cdot \varphi[A(f(x) - f(x_1^*) + \rho)] \\ &> 0 = \eta(\|x_0 - x_0\|) \cdot \varphi[A(f(x_0) - f(x_1^*) + \rho)] = P(x_0, x_1^*, A, \rho), \end{aligned}$$

that is, x_0 is a local minimizer of function $P(x, x_0, A, \rho)$.

On the other hand, x_1^* is already a global minimizer of $f(x)$, then $S_2 = \phi$ and $\Omega = S_1 \cup S_2 = S_1$, by condition (i) of Definition 3.1, we know that $P(x, x_1^*, A, \rho)$ has no local minimizer in the set S_1 except prefixed point $x_0 \in S_1$, i.e., x_0 is the unique local minimizer of $P(x, x_1^*, A, \rho)$. □

4 Filled Function Algorithm and Numerical Results

In this section, we put our filled function in the following algorithm to solve the discrete nonlinear programming problem (DP).

- **Algorithm 2** (The filled function method)

  Step 1. Choose:

  (a) choose functions $\eta(t)$ and $\varphi(t)$ satisfy the conditions in Sect. 3 of this paper;
  (b) choose a constant $N_L > 0$ as the tolerance parameter for terminating the minimization process of problem (DP);

  Step 2. Input:

  (a) input an integer point $x_0 \in \Omega$;

(b) input a sufficiently small constant ρ satisfying the condition (3.2) and $A > 0$ satisfying the condition (3.5).

Step 3. Starting from the point x_0, obtain a local minimizer x_1^* of $f(x)$ over Ω.

(a) if x_0 is a local minimizer of $f(x)$ over Ω, let $x_1^* = x_0$ and go to Step 4;
(b) if x_0 is not a local minimizer of $f(x)$ over Ω, search the neighborhood $N(x_0)$ and obtain a point $x \in N(x_0) \cap \Omega$ such that $f(x) < f(x_0)$;
(c) let $x_0 = x$ and go to (a) of Step 3.

Step 4. Construct the filled function $P_{A,x_1^*,x_0}(x)$ as follows:

$$P_{A,x_1^*,x_0}(x) = \eta(\|x - x_0\|) \cdot \varphi(A[f(x) - f(x_1^*) + \rho]).$$

Step 5. Let $N = 0$.
Step 6. If $N \geq N_L$, then go to Step 11.
Step 7. Set $N = N + 1$. Choose an initial point on the set Ω. Starting from this point, minimize $P_{A,x_1^*,x_0}(x)$ on the set Ω using any local minimization method. Suppose that x' is an obtained local minimizer.
Step 8. If $x' = x_0$, go to Step 6; otherwise, go to Step 9.
Step 9. Minimize $f(x)$ on the set Ω from the initial point x', and obtain a local minimizer x_2^* of $f(x)$.
Step 10. Let $x_1^* = x_2^*$ and go to Step 4.
Step 11. Output x_1^* and $f(x_1^*)$ as an approximate global minimal solution and global minimal value of problem (DP) respectively.

- **Explanations of Algorithm 2**

(1) Since we generally do not know the value of the $\underline{f}$, it is impossible that we choose ρ satisfying the (3.2), but we can choose ρ is sufficiently small such that it satisfies the following condition

$$0 < \rho \leq \begin{cases} \min |f(x_1) - f(x_2)| \,, \\ s.t. \; f(x_1) \neq f(x_2) \,, \\ x_1, x_2 \in \Omega \,, \end{cases} \tag{4.1}$$

i.e., ρ satisfies the (3.4). In particular, when $f(x)$ is a polynomial function with integer coefficients, we can choose $\rho = 1$.

(2) When the problem (DP) was known, we can obtain L, and when the functions $\eta(x)$ and $\varphi(x)$ are known, we will obtain $\eta(L)$, $\eta(1)$ and B. Therefore, we can choose $A > 0$ satisfying the condition

$$A\rho > \varphi^{-1}(\frac{\eta(L)}{\eta(L) + \eta(1)}B).$$

Example 1. An unconstrained discrete nonlinear programming problem

$$\min\ f(x) = (x_1 - 1)^2 + (x_n - 1)^2 + n\sum_{i=1}^{n-1}(n-i)(x_i^2 - x_{i+1})^2,$$

$$s.t.\ |x_i| \le 5, x_i \text{ integer}, i = 1, 2, \cdots, n.$$

This problem has 11^n feasible points and many local minimizers (4, 6, 7, 10 and 12 local minimizers for n =2, 3, 4, 5 and 6, respectively), but only one global minimum solution: $x^*_{global} = (1, 1, \cdots, 1)$ with $f(x^*_{global}) = 0$, for all n. We considered three sizes of the problem: n =2, 3 and 5. There were about 1.21×10^2, 1.331×10^3, 1.611×10^5 feasible points, for n =2, 3, 5 respectively.

Example 2.

$$\min\ f(x) = \sum_{i=1}^{n-1}[100(x_{i+1} - x_i^2)^2 + (1 - x_i)^2],$$

$$s.t.\ |x_i| \le 5, x_i \text{ integer}, i = 1, 2, \cdots, n.$$

This problem is a box constrained/unconstrained discrete nonlinear programming problem. It has 11^n feasible points and many local minimizers (5, 6, 7, 9, and 11 local minimizers for n = 2, 3, 4, 5, and 6, respectively), but only one global minimum solution: $x^*_{global} = (1, 1, \cdots, 1)$ with $f(x^*_{global}) = 0$, for all n. We considered three cases of the problem: n = 4, 5, and 6. There were about 1.464×10^4, 1.611×10^5, 1.772×10^6 feasible points, for n = 4, 5, 6, respectively.

In the following, the proposed solution algorithm is programmed in FORTRAN Release for working on the WINDOWS XP system with 900 MHz CPU. The FORTRAN 95 subroutine is used as the local neighborhood search scheme to obtain local minimizers of $f(x)$ in step 3 and the local minimizers of $P(x, x_1^*, A, \rho)$ in step 7. We choose $\varphi(t) = t/(1+t)$, $\eta(t) = t$, and $A > 0$ satisfies (3.7), the tolerance parameter $N_L = 10^n + 1$, n is the variable number of $f(x)$.

The iterative results of the computational for Example 1 are summarized in Tables 1 and 2 for n =2, 5, respectively and the iterative results of the computational for Example 2 are summarized in Tables 3 and 4 for n =5, 6 respectively. The symbols used are shown as follows:

n: The number of variables;

T_S: The number of initial points to be chosen;

k: The number of times that the local minimization process of the problem (DP);

x^k_{ini}: The initial point for the kth local minimization process of problem (DP);

x^k_{f-lo}: The minimizer for the kth local minimization process of problem (DP);

Table 1 Results of numerical Example 1

$n = 2, \rho = 0.05, A = 283, N_L = 10^2 + 1$							
T_S	k	x^k_{ini}	x^k_{f-lo}	$f(x^k_{f-lo})$	x^k_{p-lo}	$f(x^k_{p-lo})$	QIN
1	1	(−5,−3)	(0,0)	2	(1,1)	0	5
	2	(1,1)	(1,1)	0			$\geq 10^2$+1
2	1	(5,5)	(2,3)	7	(1,1)	0	1
	2	(1,1)	(1,1)	0			$\geq 10^2$+1
3	1	(−4,3)	(−2,3)	15	(1,1)	0	8
	2	(1,1)	(1,1)	0			$\geq 10^2$+1
4	1	(−1,−4)	(0,0)	2	(1,1)	0	11
	2	(1,1)	(1,1)	0			$\geq 10^2$+1

Table 2 Results of numerical Example 1

$n = 5, \rho = 0.05, A = 448, N_L = 10^5 + 1$							
T_S	k	x^k_{ini}	x^k_{f-lo}	$f(x^k_{f-lo})$	x^k_{p-lo}	$f(x^k_{p-lo})$	QIN
1	1	(−1,3,-4,3,2)	(0,0,0,0,0)	2	(1,1,1,1,1)	0	12
	2	(1,1,1,1,1)	(1,1,1,1,1)	0			$\geq 10^5$+1
2	1	(2,−2,1,0,0)	(0,0,0,0,0)	2	(1,1,1,1,1)	0	21
	2	(1,1,1,1,1)	(1,1,1,1,1)	0			$\geq 10^5$+1
3	1	(−2,2,0,1,1)	(−1,1,1,1,1)	4	(0,0,0,0,0)	2	4
	2	(0,0,0,0,0)	(0,0,0,0,0)	2	(1,1,1,1,1)	0	8
	3	(1,1,1,1,1)	(1,1,1,1,1)	0			$\geq 10^5$+1

Table 3 Results of numerical Example 2

$n = 5, \rho = 0.05, A = 448, N_L = 10^5 + 1$							
T_S	k	x^k_{ini}	x^k_{f-lo}	$f(x^k_{f-lo})$	x^k_{p-lo}	$f(x^k_{p-lo})$	FIN
1	1	(−2,−3,−1,−4,5)	(0,0,0,−2,4)	412	(0,0,0,0,0)	4	5
	2	(0,0,0,0,0)	(1,1,1,1,1)	0	(1,1,1,1,1)	0	41
	3	(1,1,1,1,1)	(1,1,1,1,1)	0			$\geq 10^5$+1
2	1	(−4,−2,−3,−1,5)	(0,0,0,−2,4)	412	(1,1,1,1,1)	0	0
	2	(1,1,1,1,1)	(1,1,1,1,1)	0			$\geq 10^5$+1
3	1	(0,0,−2,0,0)	(0,0,0,0,0)	4	(1,1,1,1,1)	0	46
	2	(1,1,1,1,1)	(1,1,1,1,1)	0			$\geq 10^5$+1
4	1	(−4,−2,−3,−1,5)	(0,0,0,−2,4)	412	(1,1,1,2,4)	101	0
	2	(1,1,1,2,4)	(1,1,1,1,1)	0	(1,1,1,1,1)	0	12
	3	(1,1,1,1,1)	(1,1,1,1,1)	0			$\geq 10^5$+1

Table 4 Results of numerical Example 2

$n = 6, \rho = 0.05, A = 490, N_L = 10^6 + 1$							
T_S	k	x_{ini}^k	x_{f-lo}^k	$f(x_{f-lo}^k)$	x_{p-lo}^k	$f(x_{p-lo}^k)$	FIN
1	1	(−3,−5,−4,−1,−2,5)	(0,0,0,0,−2,4)	413	(1,1,1,1,2,4)	101	1
	2	(1,1,1,1,2,4)	(0,0,0,0,0,0)	5	(1,1,1,1,1,1)	0	60
	3	(1,1,1,1,1,1)	(1,1,1,1,1,1)	0	(1,1,1,1,1,1)	0	62
	4	(1,1,1,1,1,1)	(1,1,1,1,1,1)	0			$\geq 10^6$+1
2	1	(−3,−1,−2,−5,−4,5)	(0,0,0,0,−2,4)	413	(1,1,1,1,1,1)	0	60
	2	(1,1,1,1,1,1)	(1,1,1,1,1,1)	0			$\geq 10^6$+1
3	1	(0,0,0,0,0,0)	(0,0,0,0,0,0)	5	(1,1,1,1,1,1)	0	0
	2	(1,1,1,1,1,1)	(1,1,1,1,1,1)	0			$\geq 10^6$+1
4	1	(0,−5,0,0,0,−5)	(0,0,0,0,0,0)	5	(1,1,1,1,1,1)	0	64
	2	(1,1,1,1,1,1)	(1,1,1,1,1,1)	0			$\geq 10^6$+1

$f(x_{f-lo}^k)$: The minimum of the x_{f-lo}^k;
x_{p-lo}^k: The minimizer for the kth local minimization process of problem (AP);
$f(x_{p-lo}^k)$: The minimum of the x_{p-lo}^k;
QIN: The iteration number for the kth local minimization process of problem (AP).

5 Example of Supply Chain

In this section, we give a case of supply chain and calculate it with the algorithm.

The various assumptions involved in this paper are below described. Products only can be transferred from each supplier to all plants, from each plant to all distributors, and from each distributor to all customer zones. Each supplier has a restriction on the available raw materials. Customer demand is deterministic and is necessary to satisfy all customer demands. Different customer zones price is predictable, and the function concerning response time, sales income, total cost can be fitted according to data analysis. The unit cost of different transport mode affecting response time can be got.

To help understand the model, all the symbols in the model are defined as follow:
Sets:

L set of suppliers
N set of available plants
DC set of available distribution centers
CM set of customers market
TM set of transport mode

Parameters:

S_l supply capacity of Supplier l
K_n production capacity of Plant n
F_n build and operation cost of Plant n
W_e process capacity of Distribution Center e
H_e build and operation cost of Distribution Center e
D_j demand quantity in Customer Market j
P_j product price in Customer Market j
$[T_{j1}, T_{j2}]$ response time interval in market zone j
$C^i_{l,n}$ unit transport cost from Supplier l to Plant n via transport mode i
$C^i_{n,e}$ unit transport cost from Plant n to Distribution Center e via transport mode i
$C^i_{e,j}$ unit transport cost from Distribution Center e to Customer Market j via transport mode i
C_{Total} expected total cost of supply chain network
$F(T_j)$ function between response time T_j in customer market j and sales income
$G(T_1, T_2, \cdots, T_j)$ function between all response time T_j and total cost of supply network

Variables:

$X^i_{l,n}$ transport quantity from Supplier l to Plant n via transport mode i
$Y^i_{n,e}$ unit transport cost from Plant n to Distribution Center e via transport mode i
$Z^i_{e,j}$ unit transport cost from Distribution Center e to Customer Market j via transport mode i
U_n decided whether building Plant n or not
V_e decided whether building Distribution Center e or not
T_j response time in customer market j

Formulation: The mathematical formulation is as follows:

$$Minf(v) = -\{\sum_{j=1} F(T_j)P_j \sum_i \sum_e Z^i_{e,j} - [\sum_n F_n U_n + \sum_e H_e V_e + \sum_l \sum_n \sum_i C^i_{l,n} X^i_{l,n} + \sum_n \sum_e \sum_i C^i_{n,e} Y^i_{l,n} + \sum_e \sum_j \sum_i C^i_{e,j} Z^i_{e,j} + C_{total} G(T_1, T_2, \ldots, T_j)\} \quad (5.1)$$

s.t.

$$\sum_n \sum_i X^i_{l,n} \leq S_1, \forall l \in L \quad (5.2)$$

$$\sum_e \sum_i Y^i_{n,e} \leq K_n U_n,, \forall n \in N \quad (5.3)$$

$$\sum_j \sum_i Z^i_{e,j} \leq W_e V_e, \forall e \in DC \quad (5.4)$$

$$\sum_e \sum_i Z^i_{e,j} \geq D_j, \forall j \in CM \quad (5.5)$$

$$\sum_l \sum_i X_{l,n}^i = \sum_e \sum_i Y_{n,e}^i, \forall n \in N \tag{5.6}$$

$$\sum_n \sum_i Y_{n,e}^i = \sum_j \sum_i Z_{e,j}^i, \forall e \in DC \tag{5.7}$$

$$T_{j1} \leq T_j, \forall j \in CM \tag{5.8}$$

$$U_n \in \{0,1\}, n \in N; V_e \in \{0,1\}, e \in DC \tag{5.9}$$

$$X_{l,n}^i \geq 0, Y_{n,e}^i \geq 0, Z_{e,j}^i \geq 0 \tag{5.10}$$

function (5.1) is to maximize total profit computed by minimizing the opposite subtracting total cost from total revenue.

A case in paper 11 (Cao Cuizheng, 2009) is cited for the model. The transportation unit cost is provided as follows: S, F, DC, CM represent supplier, Factory, distribution center, customer market respectively. Tm1, Tm2 are different transport modes (Table 5).

The optimal result which is calculated by this method is demonstrated in Table 6.

As Table 6 depicted, the profit is 30,590,000 and goods transferred from plant 3 to distribution center 1 and from distribution center 2 to market 2 is via transport mode 2. This reveals that our method can get a good result by considering transport mode factors.

Table 5 Transportation unit cost between facilities

Facility	Tm1	Tm2	Facility	Tm1	Tm2	Facility	Tm1	Tm2
$S_1 \to F_1$	200	300	$S_3 \to F_3$	100	150	$F_3 \to DC_2$	400	450
$S_1 \to F_2$	400	500	$F_1 \to DC_1$	200	250	$F_3 \to DC_3$	200	250
$S_1 \to F_3$	300	600	$F_1 \to DC_2$	500	400	$DC_1 \to CM_1$	900	800
$S_2 \to F_1$	200	100	$F_1 \to DC_3$	700	600	$DC_1 \to CM_2$	800	850
$S_2 \to F_2$	100	150	$F_2 \to DC_1$	400	300	$DC_2 \to CM_1$	600	700
$S_2 \to F_3$	400	300	$F_2 \to DC_2$	100	200	$DC_2 \to CM_2$	1000	900
$S_3 \to F_1$	800	750	$F_2 \to DC_3$	300	300	$DC_3 \to CM_1$	1000	1200
$S_3 \to F_2$	600	500	$F_3 \to DC_1$	600	400	$DC_3 \to CM_4$	900	800

Table 6 Transportation unit cost between facilities

Item	Value	Item	Value
Profit	30,590,000	$Factory2 \to DC2(Tm1)$	500
$Supplier1 \to Factory2(Tm1)$	350	$Factory3 \to DC1(Tm2)$	0
$Supplier2 \to Factory3(Tm1)$	100	$DC1 \to Market2(Tm1)$	400
$Supplier2 \to Factory2(Tm1)$	150	$DC2 \to Market1(Tm1)$	450
$Supplier3 \to Factory3(Tm1)$	300	$DC2 \to Market2(Tm2)$	50

6 Conclusions

This paper gives a modified definition of the filled function of $f(x)$ at its local minimizer for global discrete nonlinear programming problem (DP), and present a filled function which has two parameters and modify the conditions of functions $\bar{\eta}(t)$ and $\bar{\varphi}(t)$ in paper [3]. The results of preliminary numerical experiments are also reported.

References

1. Ge, R.P.: A filled function method for finding a global minimizer of a function of several variables. Math. Program. **46**, 191–204 (1990)
2. Levy, A.V., Montalvo, A.: The tunneling algorithm for the global minimization of function. SIAM J. Sci. Stat. Comput. **6**(1), 15–29 (1985)
3. Ge, R.P., Qin, Y.F.: The global convexized filled functions for globally optimization. Appl. Math. Comput. **35**, 131–158 (1990)
4. Lucid, S., Piccialli, V.: New classes of globally convexized filled functions for global optimization. J. Glob. Optim. **24**, 219–236 (2002)
5. Ge, R.P., Qin, Y.F.: A class of filled functions for finding a global minimizer of a function of several variables. J. Optim. Theory Appl. **54**(2), 241–252 (1987)
6. Zhang, L.S., Ng, C., Li, D., Tian, W.W.: A new filled function method for global optimization. J. Glob. Optim. **28**, 17–43 (2004)
7. Ge, R.P., Huang, H.: A continuous approach to nonlinear integer programming. Appl. Math. Comput. **34**, 39–60 (1989)
8. Zhang, L.S., Gao, F., Yao, Y.R.: Continuity methods for nonlinear integer programming. OR Trans. **2**(2), 59–66 (1998)
9. Zhu, W.X.: A filled function method for nonlinear integer programming. Chin. Acta Math. Appl. Sin. **23**(4), 481–487 (2000)
10. Hui, H.: An improved response time-constrained MINLP model of supply chain network design. In: LISS 2011-International Conference on Logistics, Informatics and Services Science, pp. 154–157 (2011)

Constrained Global Optimization Using a New Exact Penalty Function

Fangying Zheng and Liansheng Zhang

Abstract The aim of this paper is to propose a global algorithm model for continuous constrained nonlinear programming based on a new simple and exact penalty function. Under weak assumptions, we show that the optimizer obtained by the algorithm is converged to the global minimizer of the original problem.

Keywords Nonlinear programming • Continuous constrained optimization • Global optimization • Penalty function

1 Introduction

In this paper, we consider the following general continuous constrained minimization problem:

$$(P) \qquad s.t. \quad \begin{aligned} & G - \min f(x) \\ & F_j(x) = 0, j \in E \\ & g_l(x) \leq 0, l \in I \end{aligned} \tag{1}$$

where $f : R^n \to R, F_j : R^n \to R, j \in E, g_l : R^n \to R, l \in I$, E, I are the index of equality constraint functions and inequality constraint functions, respectively. $E = \{1, \ldots, m\}, I = \{1, \ldots, k\}$, "$G - \min$" denotes the global minimization. We assume that $f, F_j, j \in E, g_l, l \in I$ are continuously differentiable functions.

Global optimization has many applications in engineering, economics, and applied sciences. This motivated a growing attention in the search for global rather than local solutions of nonlinear optimization problems. In the last decades, most

F. Zheng (✉)
Department of Mathematical Sciences, Zhejiang Sci-Tech University, Hangzhou 310018, China
e-mail: zfy@zstu.edu.cn

L. Zhang
Department of Mathematics, Shanghai University, Shanghai 200444, China
e-mail: zhangls@staff.edu.cn

D. Gao et al. (eds.), *Advances in Global Optimization*, Springer Proceedings in Mathematics & Statistics 95, DOI 10.1007/978-3-319-08377-3_8

research papers have been devoted to solving the global minimization problems of unconstrained problems or problems with simple constraints. There are many algorithmic methods, either deterministic or probabilistic have been studied, for example [1]. Recently Birgin et al. proposed a global algorithm by using Augmented Lagrangian penalty function to deal with the general constraints (see [2]). At the same time, resorting to a non-differentiable exact penalty approach, Di Pillo et al. presented a global approach to solve the constrained programming with general nonlinear constraints and simple convex constraints (see [3]).

Although the traditional penalty function method is a popular method, there are some disadvantages. Generally, if the penalty function is exact and smooth, then it is not simple, and if the penalty function is simple and smooth, then it is not exact, where "simple" means that the penalty function only includes the functions of the primal problem, and can't include their gradients. For example, the Augmented Lagrangian penalty function in [2] is smooth and simple, but is not exact, and the penalty function in [3] is exact and simple, but is not smooth.

On the other hand, a new exact penalty function is given in [4] for the equality constrained minimization problem $(\bar{P})$, where a new approach is presented by adding one variable to the problem $(\bar{P})$.

$$(\bar{P}) \quad L - \min_{x\in S} f(x), \quad S = \{x \in [u,v] : F_j(x) = 0, j \in E\}, \tag{2}$$

where $[u,v]$ is a box on R^n with nonempty interior, $[u,v] = \{x \in R^n : u \le x \le v\}$, and $(\{-\infty\} \cup R)^n \le u < v \le (\{+\infty\} \cup R)^n$, $f : D \to R$ and $F_j : D \to R$, $j \in E$ are continuously differentiable in an open set D containing $[u,v]$. Then fix $w_j \in R$, $j \in E$ and consider the following equivalent problem:

$$L - \min_{(x,\varepsilon)\in S_{\varepsilon 0}} f(x), \quad S_{\varepsilon 0} = \{(x,\varepsilon) \in [u,v] \times [0,\bar{\varepsilon}] : F_j(x) = \varepsilon w_j, j \in E, \varepsilon = 0\}, \tag{3}$$

Let

$$\bar{f_\sigma}(x,\varepsilon) = \begin{cases} f(x), & if\ \varepsilon = 0, x \in S, \\ f(x) + \frac{1}{2\varepsilon}\frac{\Delta(x,\varepsilon)}{1-q\Delta(x,\varepsilon)} + \sigma\beta(\varepsilon), & if\ 0 < \varepsilon \le \bar{\varepsilon}, \Delta(x,\varepsilon) < q^{-1}, \\ +\infty, & otherwise\ \ (\varepsilon = 0, x \bar{\in} S\ or\ \varepsilon > 0, \Delta(x,\varepsilon) \ge q^{-1}). \end{cases} \tag{4}$$

with the constrained violation measure is

$$\Delta(x,\varepsilon) = \|F(x) - \varepsilon w\|^2 = \sum_{j\in E}(F_j(x) - \varepsilon w_j)^2,$$

where $\bar{\varepsilon} > 0, q > 0$ is fixed, $\sigma > 0$ is a penalty parameter and $\beta : [0,\bar{\varepsilon}) \to [0,+\infty)$ is continuous and continuously differentiable on $(0,\bar{\varepsilon}]$ with $\beta(0) = 0$. Obviously, the penalty function $\bar{f_\sigma}(x,\varepsilon)$ is continuously differentiable on $[u,v] \times (0,\bar{\varepsilon}]$, but not continuously differentiable on $[u,v] \times [0,\bar{\varepsilon}]$.

The corresponding penalty problem $(\bar{P}_\sigma)$ is

$$(\bar{P}_\sigma) \quad L - \min_{(x,\varepsilon)\in[u,v]\times[0,\bar{\varepsilon}]} \bar{f}_\sigma(x,\varepsilon), \tag{5}$$

The most important new idea is that the penalty function is considered as a function of x and ε simultaneously, with the property that under appropriate assumptions, for sufficiently large $\sigma > 0$, every local minimizer $(x_\sigma, \varepsilon_\sigma)$ of $(\bar{P}_\sigma)$ with finite $\bar{f}_\sigma(x_\sigma, \varepsilon_\sigma)$ has the form $(x_\sigma, 0)$, and x_σ is a local minimizer of the primal problem $(\bar{P})$.

Motivated by [4], another exact penalty function is proposed in [5], which is defined as follows.

$$f_\sigma(x,\varepsilon) = \begin{cases} f(x), & if \ \varepsilon = 0, x \in S, \\ f(x) + \varepsilon^{-\alpha}\Delta(x,\varepsilon) + \sigma\varepsilon^{\beta}, & if \ 0 < \varepsilon \le \bar{\varepsilon}, \\ +\infty, & if \ \varepsilon = 0, x \notin S, \end{cases} \tag{6}$$

where $\Delta(x,\varepsilon) = \sum_{j\in E}(F_j(x) - \varepsilon^{\gamma} w_j)^2 + \sum_{l\in I}(\max(0, g_l(x) - \varepsilon^{\gamma} w_l))^2, \gamma, \alpha, \beta > 1, w_j \in R_+, \ for \ j \in I$ are fixed, $\sigma > 0$ is a penalty parameter.

The corresponding penalty problem is as follows:

$$(P_\sigma) \qquad \min_{x\in R^n\times[0,\bar{\varepsilon}]} f_\sigma(x,\varepsilon) \tag{7}$$

Compared with the penalty function in [4], the penalty function in [5] has the following properties: Firstly, it is more simple in form; secondly, it is generalized to the general constrained programming which contained equality and inequality constraints; finally, under weak assumptions, there exists a threshold value $\sigma_0 > 0$ of the penalty parameter σ such that, for any $\sigma \ge \sigma_0$, any global solution of penalty problem (P_σ) is a global solution of primal problem (P). Furthermore the penalty function $f_\sigma(x,\varepsilon)$ is continuously differential on $R^n \times (0,\bar{\varepsilon}]$. Therefore we can use continuously unconstrained global optimization approaches to solve penalty problem (P_σ).

In this paper, a new approach based on the exact penalty function method introduced in [5] is used to solve the global optimizers of constrained minimization problems. It is shown in [5] that, under some mild assumptions, any local minimizer of the penalty problem has the form $(x^*, 0)$, where x^* is the local minimizer of the original problem when the penalty parameter is sufficiently large.

Before formal discussions, we give the following assumptions and results.

Assumption 1. *The MFCQ holds at every global solution x^* of problem (P).*

Assumption 2. *There exists a global minimizer of problem (P_σ) on $R^n \times [0,\bar{\varepsilon}]$.*

Theorem 1. *There exists a threshold value $\sigma_0 > 0$, such that for any $\sigma \ge \sigma_0$, if $(x^*, 0)$ is a global solution of problem (P_σ), then x^* is a global solution of primal problem (P).*

Proof. From Theorem 4.2 in paper [5], we know that there exists a threshold value $\sigma_0 >$, when $\sigma \geq \sigma_0$, problem (P_σ) has local minimizers which have the form $(x^*, 0)$. So assuming that $(x^*, 0)$ is a global solution of problem (P_σ), then for $\forall x \in S$, we have

$$f(x^*) = f_\sigma(x^*, 0) \leq f_\sigma(x, 0) = f(x)$$

That means x^* is a global solution of primal problem (P).

In the paper, given a vector $x \in R^n$, we denote by $\|x\|_2$ its 2-norm, and by $x^+ = \max\{0, x\}$ the n-vector $(\max\{0, x_1\}, \ldots, \max\{0, x_n\})$.

2 Global Optimization Algorithm

In this section, we describe the GO (Global Optimization) algorithm model for finding a global solution of problem (P) using the new exact penalty function $f_\sigma(x, \varepsilon)$, and analyze its convergence properties.

Algorithm 2.1 GO Algorithm Model

Step 1. $k := 0, \sigma_0 > 0, \gamma > \alpha = \beta > 1, \delta^{(0)} > 0, \theta \in (0, 1), \bar{\varepsilon} > 0, (x^{(0)}, \varepsilon_0) \in R^n \times (0, \bar{\varepsilon}), \rho > 1$.

Step 2. Compute $(x^{(k)}, \varepsilon_k) \in R^n \times [0, \bar{\varepsilon}]$, such that

$$f_{\sigma_k}(x^{(k)}, \varepsilon_k) \leq f_{\sigma_k}(x, \varepsilon) + \delta^{(k)}, \forall (x, \varepsilon) \in R^n \times [0, \bar{\varepsilon}] \tag{8}$$

Step 3. If $\varepsilon_k = 0$ and $\Delta(x^{(k)}, \varepsilon_k) = 0$, set $\sigma_{k+1} := \sigma_k$ and go to step 5.

Step 4. If $\frac{1}{\sigma_k}(\|\nabla f(x^{(k)})\|_2 + \Delta(x^{(k)}, \varepsilon_k)) > \|\frac{\partial \Delta(x^{(k)}, \varepsilon_k)}{\partial x}\|_2$, set $\sigma_{k+1} := \rho\sigma_k, \delta^{(k+1)} := \delta^{(k)}, k := k + 1$ and go to step 2.
Else set $\sigma_{k+1} := \sigma_k$ and go to step 5.

Step 5. Set $\delta^{(k+1)} = \theta\delta^{(k)}$ and go to step 2.

In the algorithm, at step 2, $(x^{(k)}, \varepsilon_k)$ is a $\delta^{(k)}$- global minimizer of problem (P_σ) which can be obtained by using any global unconstrained optimization method, such as filled function method. At step 3, we check feasibility of $x^{(k)}$ and if $x^{(k)}$ is feasible, we reduce the value of $\delta^{(k)}$ in order to find a better approximation of the global solution of (P_σ). Step 4 is intended to determine whether the updating of the penalty parameter is timely. When the value $\delta^{(k)}$ is sufficiently small, then the algorithm will be stopped and $(x^{(k)}, \varepsilon_k)$ can be considered an approximation of a global minimizer of problem (P_σ) and $x^{(k)}$ can be considered an approximation of a global minimizer of problem (P).

Theorem 2. *Assume that the sequence $\{(x^{(k)}, \varepsilon_k)\}$ is produced by GO algorithm and well defined and admits a limit point (x^*, ε_*). Then $x^* \in S, \varepsilon_* = 0$.*

Proof. We consider two different cases:

Case(1) There exists an index k^*, such that for any $k \geq k^*, \sigma_k = \sigma_*$. By contradiction, we assume that there exists an accumulation point $x^* \notin S$.
In this case, according to step 4, for sufficiently large k, we have

$$\frac{1}{\sigma_k}(\|\nabla f(x^{(k)})\|_2 + \Delta(x^{(k)}, \varepsilon_k)) \leq \|\frac{\partial \Delta(x^{(k)}, \varepsilon_k)}{\partial x}\|_2 \tag{9}$$

Let $k \to +\infty$, then $(x^{(k)}, \varepsilon_k) \to (x^*, \varepsilon_*)$ and by step 5, when $k \to +\infty, \delta^{(k)} \to 0$, according to step 2, we have that (x^*, ε_*) is a global minimizer of $f_\sigma(x, \varepsilon)$, then from Theorem 4.1, 4.2 in paper [5], we have $\lim_{k\to+\infty} \varepsilon_k = \varepsilon_* = 0$. For $x^* \notin S$, according to the definition of $f_\sigma(x, \varepsilon)$, we know that $f_\sigma(x^*, \varepsilon_*) = +\infty$. This contradicts the fact that (x^*, ε_*) is a global minimizer of $f_\sigma(x, \varepsilon)$. Therefore $x^* \in S$.

Case(2) $\lim_{k\to+\infty} \sigma_k = +\infty$
By contradiction, we assume that $\lim_{k\to+\infty} \varepsilon_k = \bar{\varepsilon} \neq 0$
According to the definition of $\{(x^{(k)}, \varepsilon_k)\}$, for $\forall x \in S, \varepsilon = 0$, we have

$$f_{\sigma_k}(x^{(k)}, \varepsilon_k) \leq f_{\sigma_k}(x, 0) + \delta^{(k)}.$$

That is

$$f(x^{(k)}) + \varepsilon_k^{-\alpha} \Delta(x^{(k)}, \varepsilon_k) + \sigma_k \varepsilon_k^\beta \leq f(x) + \delta^{(k)} \tag{10}$$

Arranging (10), we have

$$\sigma_k \varepsilon_k^\beta \leq f(x) - f(x^{(k)}) - \varepsilon_k^{-\alpha} \Delta(x^{(k)}, \varepsilon_k) + \delta^{(k)}.$$

Let $k \to +\infty$, from the above inequality, we have

$$\lim_{k\to+\infty} \sigma_k \varepsilon_k^\beta \leq \lim_{k\to+\infty} (f(x) - f(x^{(k)}) - \varepsilon_k^{-\alpha} \Delta(x^{(k)}, \varepsilon_k) + \delta^{(k)}) \tag{11}$$

In (11), for $\lim_{k\to+\infty} \sigma_k = +\infty$ and $\lim_{k\to+\infty} \varepsilon_k = \bar{\varepsilon} \neq 0$, therefore we have $\lim_{k\to+\infty} \sigma_k \varepsilon_k^\beta = +\infty$ and the value of the right side of the inequality (11) is finite, this is impossible. So the assumption of $\lim_{k\to+\infty} \varepsilon_k = \varepsilon_* \neq 0$ is not correct and $\varepsilon_* = 0$. On the other hand, according to the definition of $f_\sigma(x, \varepsilon)$ and $\{(x^{(k)}, \varepsilon_k)\}$, we know $x^* \in S$.

Theorem 3. *Every accumulation point x^* of a sequence $\{(x^{(k)}\}$ produced by GO algorithm is a global minimizer of problem (P).*

Proof. According to the definition of $\{(x^{(k)}, \varepsilon_k)\}$, for any z global minimizer of problem (P) we have

$$f(x^{(k)}) \leq f_{\sigma_k}(x^{(k)}, \varepsilon_k) \leq f_{\sigma_k}(z, 0) + \delta^{(k)} = f(z) + \delta^{(k)}$$

Let $k \to +\infty$, we have

$$f(x^*) \leq f(z).$$

From Theorem 2, we know $x^* \in S$, then x^* is a global minimizer of problem (P).

In the next theorem, if Assumption 1 holds, we conclude that the updating times of the penalty parameter σ is finite.

Theorem 4. *Assuming that Assumption 1 holds. Let* $\{(x^{(k)}, \varepsilon_k)\}$ *and* σ_k *be the sequences produced by GO algorithm. Then there exists an index* k_0 *and a value* $\sigma_{k_0} < +\infty$, *such that for any* $k \geq k_0$, *we have* $\sigma_k = \sigma_{k_0}$.

Proof. By contradiction, assuming $\lim_{k\to} \sigma_k = +\infty$. Then from step 3 and step 4 of the GO algorithm, there exists a subsequence $\{(x^{(k)}, \varepsilon_k)\}_K$, such that for all $k \in K, x^{(k)} \notin S$ and the following inequality

$$\frac{1}{\sigma_k}(\|\nabla f(x^{(k)})\|_2 + \Delta(x^{(k)}, \varepsilon_k)) > \|\frac{\partial \Delta(x^{(k)}, \varepsilon_k)}{\partial x}\|_2$$

is satisfied. Let $k \in K, k \to +\infty$, we have

$$\lim_{k \in K, k \to +\infty} \|\frac{\partial \Delta(x^{(k)}, \varepsilon_k)}{\partial x}\|_2 = 0. \tag{12}$$

It implies

$$\lim_{k \in K, k \to +\infty} \frac{\partial \Delta(x^{(k)}, \varepsilon_k)}{\partial x} = 0. \tag{13}$$

By construction, from (13), we have

$$\lim_{k \in K, k \to +\infty} \{\sum_{l=1}^{k} u_l^{(k)} \nabla g_l(x^{(k)}) + \sum_{j=1}^{m} v_j^{(k)} \nabla F_j(x^{(k)})\} = 0 \tag{14}$$

where

$$u_l^{(k)} = \frac{\max(0, g_l(x^{(k)}) - \varepsilon_k^\gamma \omega_l)}{\sqrt{\Delta(x^{(k)}, \varepsilon_k)}} \quad v_j^{(k)} = \frac{F_j(x^{(k)}) - \varepsilon_k^\gamma \omega_j}{\sqrt{\Delta(x^{(k)}, \varepsilon_k)}} \tag{15}$$

From (15), we can obtain

$$\|(u^{(k)}, v^{(k)})^T\|_2 = 1, \quad \lim_{k \in K, k \to +\infty} x^{(k)} = x^*, \quad \lim_{k \in K, k \to +\infty} u^{(k)} = u^* \geq 0 \tag{16}$$

By Theorem 3, we have $x^* \in S$. By continuity of $g(x)$, for k sufficiently large, we have

$$\{l : g_l(x^{(k)}) - \varepsilon_k^\gamma \omega_l < 0\} \supseteq \{l : g_l(x^*) < 0\}$$

From which, we have

$$u_i^* = 0, \forall l \in \{l : g_l(x^*) < 0\} \tag{17}$$

Then from (14), we have

$$\lim_{k \in K, k \to +\infty} \{\sum_{l=1}^{k} u_l^{(k)} \nabla g_l(x^{(k)}) + \sum_{j=1}^{m} v_j^{(k)} \nabla F_j(x^{(k)})\}$$

$$= \sum_{l \in I_0(x^*)} u_l^* \nabla g_l(x^*) + \sum_{j \in E} v_j^* \nabla F_j(x^*) = 0 \tag{18}$$

By (16) and (17), we obtain

$$(u_l^*, l \in I_0(x^*)) \neq 0,$$

and

$$u_l^* \geq 0, \quad l \in I_0(x^*)$$

On the other hand, by Theorem 3, we have that x^* is a global minimizer of problem (P). Then by (18) we get a contradiction with Assumption 1.

3 Conclusions

From GO algorithm model, we can obtain the global minimizer of the constrained optimization problems based on a new simple exact penalty function which is differentiable on the sets$\{(x, \varepsilon) \in R^n \times [0, \bar{\varepsilon}] : \varepsilon = 0, x \in S\}$ and $\{(x, \varepsilon) \in R^n \times [0, \bar{\varepsilon}] : \varepsilon \in (0, \bar{\varepsilon}]\}$. So we can solve the global optimizer of penalty problem (P_σ) by any differentiable algorithm for unconstrained optimization.

Acknowledgements This research was partially supported by the National Natural Science Foundation of China (10571116 and 51075421), and supported by Science Foundation of Zhejiang Sci-Tech University (ZSTU) under Grant No. 1206830-Y.

References

1. Ge, R.P.: The theory of filled function method for finding global minimizers of nonlinearly constrained minimization problems. J. Comput. Math. **5**, 1–9 (1987)
2. Birgin, E.G., Floudas, C.A., Martínez, J.M.: Global minimization using an Augmented Lagrangian method with variable lower-level constraints. Math. Program. Ser. A **125**, 139–162 (2010)
3. Di Pillo, G., Lucidi, S., Rinaldi, F.: An approach to constrained global optimization based on exact penalty functions. J. Glob. Optim. **54**(2), 251–260 (2012)
4. Huyer, W., Neumaier, A.: A new exact penalty function. SIAM J. Optim. **3**(4), 1141–1158 (2003)
5. Zheng, F.Y., Zhang, L.S.: New simple exact penalty function for constrained minimization. Appl. Math. Mech. **33**(7), 951–962 (2012)

Part II
Combinatorial Optimization

A Multiobjective State Transition Algorithm for Single Machine Scheduling

Xiaojun Zhou, Samer Hanoun, David Yang Gao, and Saeid Nahavandi

Abstract In this paper, a discrete state transition algorithm is introduced to solve a multiobjective single machine job shop scheduling problem. In the proposed approach, a non-dominated sort technique is used to select the best from a candidate state set, and a Pareto archived strategy is adopted to keep all the non-dominated solutions. Compared with the enumeration and other heuristics, experimental results have demonstrated the effectiveness of the multiobjective state transition algorithm.

Keywords Discrete state transition algorithm • Multiobjective optimization • Single machine scheduling

1 Introduction

The multiobjective optimization is encountered in many real-world applications [1]. For a specific policy, the decision maker may find it advantageous for one goal but disadvantageous for others. A traditional way to deal with this issue is to impose a priori preference reflecting the relative importance of different objectives; however, the final solution just indicates a decision maker's satisfaction, and it might be dissatisfactory for other decision makers.

To ameliorate the problem, the concept of *Pareto optimality* and other relevant concepts are introduced. These are defined as follows:

(1) *Pareto dominance*: A feasible solution $\boldsymbol{x} = (x_1, \cdots, x_n)$ is said to Pareto dominate another feasible solution $\boldsymbol{y} = (y_1, \cdots, y_n)$, denoted as $\boldsymbol{x} \prec \boldsymbol{y}$, if

$$f_i(\boldsymbol{x}) \leq f_i(\boldsymbol{y}), \forall i \in \{1, \cdots, k\}, \text{ and } \exists j \in \{1, \cdots, k\}, f_j(\boldsymbol{x}) < f_j(\boldsymbol{y}), \quad (1)$$

where $f_i(\boldsymbol{x})$ is the ith objective function, k is the number of objectives.

X. Zhou (✉) • D.Y. Gao
School of Science, Information Technology and Engineering,
University of Ballarat, Ballarat, VIC 3353, Australia
e-mail: tiezhongyu2010@gmail.com; d.gao@ballarat.edu.au

S. Hanoun • S. Nahavandi
Centre for Intelligent Systems Research, Deakin University, Geelong, VIC, Australia
e-mail: samer.hanoun@deakin.edu.au; saeid.nahavandi@deakin.edu.au

D. Gao et al. (eds.), *Advances in Global Optimization*, Springer Proceedings in Mathematics & Statistics 95, DOI 10.1007/978-3-319-08377-3_9

(2) *Pareto optimality*: A feasible solution $\boldsymbol{x}^*$ is said to be Pareto optimal if and only if

$$\neg\exists \boldsymbol{x} \in S, \boldsymbol{x} \prec \boldsymbol{x}^*, \tag{2}$$

where S is the feasible space.

(3) *Pareto optimal set*: The Pareto optimal set, denoted as P^*, is defined by

$$P^* = \{\boldsymbol{x}^* \in S | \neg\exists \boldsymbol{x} \in S, \boldsymbol{x} \prec \boldsymbol{x}^*\}. \tag{3}$$

(4) *Pareto front*: The Pareto front, denoted as Pf^*, is defined by

$$Pf^* = \{(f_1(\boldsymbol{x}^*), \cdots, f_k(\boldsymbol{x}^*)) | \boldsymbol{x}^* \in P^*\}. \tag{4}$$

The introduction of *Pareto optimality* allows us to find a set of Pareto optimal solutions simultaneously, independent of the decision maker's priori preference.

In the past few decades, evolutionary-based and nature-inspired multiobjective optimization techniques have drawn considerable attention for scheduling problems [2–7]. In this paper, we introduce a recently new heuristics called state transition algorithm [8–11] as the basic search engine for the multiobjective optimization. A non-dominated sort approach is used to select the best from a candidate state set, and the best state is stored using a Pareto archive strategy. Experimental results have testified the effectiveness of the proposed algorithm.

2 Problem Description

In the field of joinery manufacturing, jobs with similar materials can be scheduled together to minimize the amount of materials used; therefore, reducing the cost.

For example, based on the cost savings matrix shown in Table 1, pairing Job1 and Job2 will provide saving in the cost equivalent to 4 units.

Additionally, based on the jobs' processing times and due dates as shown in Table 2, and for any given sequence and pair of jobs, not only the total cost saving C is affected but also the total tardiness time T, which is calculated as:

Table 1 The cost savings matrix for five jobs having the same material

	Job1	Job2	Job3	Job4	Job5
Job1	0	4	2.64	4.08	3.9
Job2	4	0	3.64	4.72	4.23
Job3	2.64	3.64	0	2.65	2.87
Job4	4.08	4.72	2.65	0	3.84
Job5	3.9	4.23	2.87	3.84	0

Table 2 Due dates and processing times for a set of five jobs

Job	Due date (days)[a]	Processing Time (h)
Job1	8	17:40
Job2	2	24:00
Job3	11	19:20
Job4	3	25:00
Job5	3	14:40

[a]Number of operational hours = 8 h per day

$$T = \sum_{j=1}^{n} \max\{0, c_j - d_j\} \tag{5}$$

where c_j and d_j are the completion time and the due time of job j, respectively.

The goal of this paper is to determine the optimal sequence with pairing, in order to maximize the total cost savings and minimize the total tardiness time.

It is obvious that finding the permutation of the sequence $\{1, 2, \cdots, n\}$ with pairing becomes a solution to the multiobjective single machine scheduling problem; however, not without the necessity to discuss the number of pairs for any fixed sequence of jobs.

Given a sequence $s = (1, 2, \cdots, n)$, for $n = 3$, we have two possible pairing options (1-2)-3 and 1-(2-3); for $n = 4$, we have two possible pairing options (1-2)-(3-4) and 1-(2-3)-4, as pairing options (1-2)-3-4 and 1-2-(3-4) are discarded; for $n = 5$, we have three possible options (1-2)-(3-4)-5, (1-2)-3-(4-5) and 1-(2-3)-(4-5), as options (1-2)-3-4-5, 1-2-(3-4)-5, 1-2-3-(4-5) and 1-(2-3)-4-5 are discarded.

If $P_1(n)$ denotes the number of pairs with the first two jobs pairing, and $P_2(n)$ denotes the complement of $P_1(n)$, then we have the following theorem:

Theorem 1.

$$P_1(n+1) = P(n-1),\ P_2(n+1) = P_1(n), n \geq 3 \tag{6}$$

where $P(n) = P_1(n) + P_2(n)$ is the total number of pairs. For example, $P_1(2) = 1$, $P_2(2) = 0$, $P_1(3) = 1$, $P_2(3) = 1$, $P_1(4) = 1$, $P_2(4) = 1$, $P_1(5) = 2$, $P_2(5) = 1$, we have $P_1(4) = P(2)$, $P_2(4) = P_1(3)$, $P_1(5) = P(3)$, $P_2(5) = P_1(4)$.

Figure 1 shows the growth trend of the number of pairs with the sequence size; however, only small size job scheduling problems are considered in this study. Considering that $P(10) = 12 \ll 10! = 3{,}628{,}800$, a complete enumeration approach is used for pairing and only the permutation of a sequence is focused.

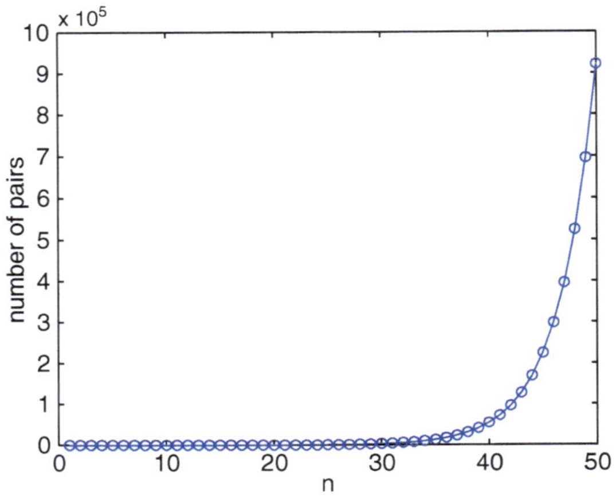

Fig. 1 Growth trend relative to the sequence size

3 Discrete State Transition Algorithm

In the case a solution to a specific optimization problem is described as a state, then the transformation to update the solution becomes a state transition. Without loss of generality, the unified form of discrete state transition algorithm can be described as:

$$\begin{cases} \boldsymbol{x}_{k+1} = A_k(\boldsymbol{x}_k) \bigoplus B_k(\boldsymbol{u}_k) \\ y_{k+1} = f(\boldsymbol{x}_{k+1}) \end{cases}, \tag{7}$$

where $\boldsymbol{x}_k \in \mathscr{Z}^n$ stands for a current state, corresponding to a solution of a specific optimization problem; $\boldsymbol{u}_k$ is a function of $\boldsymbol{x}_k$ and historical states; $A_k(\cdot)$, $B_k(\cdot)$ are transformation operators, which are usually state transition matrixes; $\bigoplus$ is an operation, which is admissible to operate on two states; and f is the cost function or evaluation function.

The following three transformation operators are defined to permute current solution [10]:

(1) Swap Transformation

$$\boldsymbol{x}_{k+1} = A_k^{swap}(m_a)\boldsymbol{x}_k, \tag{8}$$

where $A_k^{swap} \in \mathbb{R}^{n \times n}$ is the swap transformation matrix, m_a is the swap factor, a constant integer used to control the maximum number of positions to be exchanged, while the positions are random. Figure 2 shows an example of the swap transformation with $m_a = 2$.

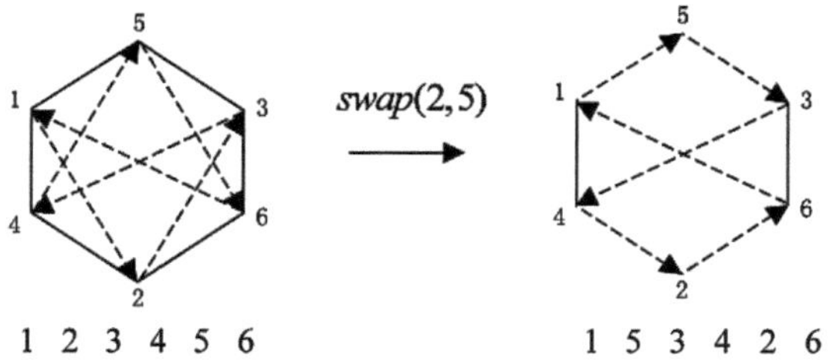

Fig. 2 Illustration of the swap transformation

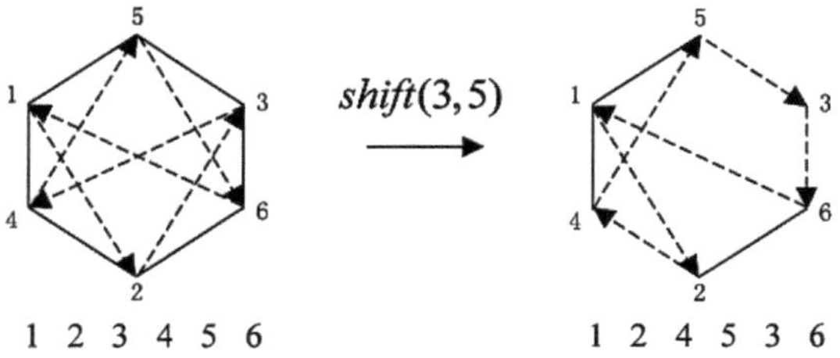

Fig. 3 Illustration of the shift transformation

(2) Shift Transformation

$$\boldsymbol{x}_{k+1} = A_k^{shift}(m_b)\boldsymbol{x}_k, \tag{9}$$

where $A_k^{shift} \in \mathbb{R}^{n \times n}$ is the shift transformation matrix, m_b is the shift factor, a constant integer used to control the maximum length of consecutive positions to be shifted. Note that both the selected position to be shifted after and positions to be shifted are chosen randomly. Figure 3 shows an example of the shift transformation with $m_b = 1$.

(3) Symmetry Transformation

$$\boldsymbol{x}_{k+1} = A_k^{sym}(m_c)\boldsymbol{x}_k, \tag{10}$$

where $A_k^{sym} \in \mathbb{R}^{n \times n}$ is the symmetry transformation matrix, m_c is the symmetry factor, a constant integer used to control the maximum length of subsequent positions as center. Note that both the component before the subsequent positions and consecutive positions to be symmetrized are created randomly. Figure 4 shows an example of the symmetry transformation with $m_c = 0$.

4 Pareto Archived Strategy Based on DSTA

In state transition algorithm, the times of transformation are called search enforcement (*SE*); as a result, after each transformation operator, a candidate state set S is generated.

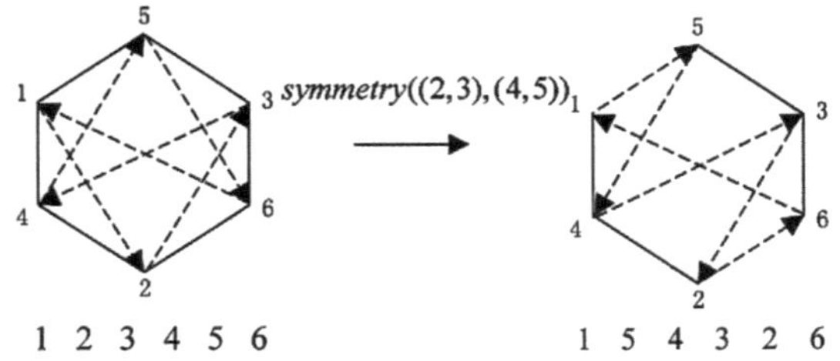

Fig. 4 Illustration of symmetry transformation

4.1 Non-dominated Sort

We use a sorting approach similar to the fast-non-dominated-sort proposed in [12], described as follows:

for each $s \in S$ **do**
  $n_s \leftarrow 0$
  for each $t \in S$ **do**
    if $t \prec s$ **then**
      $n_s \leftarrow n_s + 1$
    end if
  end for
end for

where n_s is the domination count, representing the number of solutions dominating solution s. After the non-dominated sort, the state with the least count will be stored as incumbent *best* for the next transformation operator.

4.2 Pareto Archived Strategy

We adopt a simple Pareto archived strategy to select current *best* as follows:

for each $A_i \in \mathscr{A}$ **do**
  if $best \prec A_i$ **then**
    $\mathscr{A} \leftarrow \mathscr{A} - A_i$
  else if $A_i \prec best$ **then**
    $\mathscr{A} \leftarrow \mathscr{A}$
  else
    $\mathscr{A} \leftarrow \mathscr{A} \bigcup best$
  end if
end for

where $\mathscr{A}$ is the archive keeping all non-dominated solutions.

4.3 Pseudocodes of the Proposed Algorithm

The core procedure of the proposed algorithm can be outlined in pseudocodes:

repeat
  $State \leftarrow operator(best, SE, n)$
  $best \leftarrow update\_best(best, SE, n, data)$
  $Paretoset \leftarrow update\_archive(Paretoset, best)$
until the maximum number of iterations is met

where $State$ is the state set; $operator$ stands for the three transformation operators, which are carried out sequentially; $update\_best$ is corresponding to the non-dominated sort, and $update\_archive$ corresponds to the Pareto archived strategy. The $data$ is the known information (cost saving matrix, due dates, and processing times) about a specific scheduling problem.

5 Experimental Results

In order to test the performance of the proposed multiobjective state transition algorithm, two typical examples are used for comparison. In the following experiments, $SE = 20, m_a = 2, m_b = 1, m_c = 0$ are adopted for parameter settings. The maximum number of iterations for are 100 and 1,000, respectively, for the two examples. The known data for Example 1, 2 are given in Tables 1 and 2, Tables 3 and 4, respectively, and the corresponding results can be found in Tables 5 and 6. It is worth noting that the pairing methodology used with complete enumeration and Cuckoo Search (CS) is based on a greedy approach by first selecting the pair that produces the highest cost savings, and then repeating the same procedure for the remaining set of pairs in the sequence [7]. We can find that for Example 1,

Table 3 The cost savings matrix for ten jobs having the same material

	Job1	Job2	Job3	Job4	Job5	Job6	Job7	Job8	Job9	Job10
Job1	0	2.73	2.1	2.16	2.66	3.6	2.46	2.7	2.46	2.8
Job2	2.73	0	2	1.6	4.3	3.69	2.3	3.5	2.76	3.6
Job3	2.1	2	0	1.4	3.51	3.33	2.52	3.68	2.52	2.46
Job4	2.16	1.6	1.4	0	2.17	2.32	2.72	3.04	2.04	2.97
Job5	2.66	4.3	3.51	2.17	0	3.6	4.05	4.41	2.7	2.64
Job6	3.6	3.69	3.33	2.32	3.6	0	2.58	4.7	3.44	2.94
Job7	2.46	2.3	2.52	2.72	4.05	2.58	0	2.6	2.88	2.82
Job8	2.7	3.5	3.68	3.04	4.41	4.7	2.6	0	3.64	3.57
Job9	2.46	2.76	2.52	2.04	2.7	3.44	2.88	3.64	0	3.76
Job10	2.8	3.6	2.46	2.97	2.64	2.94	2.82	3.57	3.76	0

Table 4 Due dates and processing times for a set of ten jobs

Job	Due date (days)	Processing time (h)
Job1	11	14:00
Job2	2	18:00
Job3	13	15:00
Job4	14	8:20
Job5	11	17:20
Job6	9	16:00
Job7	4	19:40
Job8	6	23:20
Job9	10	20:00
Job10	10	19:20

Table 5 Comparison results for the set of jobs presented in Tables 1 and 2

Approach	Optimal solutions	T	C
Complete Enumeration	(2-5)-(1-4)-3	13	8.31
	(5-2)-(1-4)-3	13	8.31
	(2-5)-(4-1)-3	13	8.31
	(2-4)-(5-1)-3	15	8.62
CS [7]	(2-5)-(1-4)-3	13	8.31
	(5-2)-(1-4)-3	13	8.31
	(2-5)-(4-1)-3	13	8.31
	(2-4)-(5-1)-3	15	8.62
STA	(5-2)-(1-4)-3	13	8.31
	(2-5)-(1-4)-3	13	8.31
	(2-5)-(4-1)-3	13	8.31
	(5-2)-(4-1)-3	13	8.31
	(2-4)-(5-1)-3	**15**	**8.62**

STA obtained a solution which can dominate the optimal solutions by enumeration and CS. From both examples, it is easy to find that some additional optimal solutions are achieved by STA, as indicated by the bold values.

6 Conclusion

A multiobjective state transition algorithm is presented for a single machine job shop scheduling problem. In this paper, a complete enumeration approach is used for pairing the jobs in a fixed sequence. Compared with a greedy-based approach

Table 6 Comparison results for the set of jobs presented in Tables 3 and 4

Approach	Optimal solutions	T	C
Complete Enumeration	(5-7)-(2-6)-(1-3)-(4-10)-(8-9)	39	16.45
	(5-7)-(2-6)-(1-3)-(4-8)-(10-9)	40	16.64
	(5-7)-(2-6)-(1-3)-(4-8)-(9-10)	40	16.64
	(5-7)-(2-6)-(1-4)-(3-8)-(10-9)	41	17.34
	(5-7)-(2-6)-(1-4)-(3-8)-(9-10)	41	17.34
	(5-2)-(7-4)-(6-1)-(3-8)-(10-9)	43	18.06
	(5-2)-(7-4)-(6-1)-(3-8)-(9-10)	43	18.06
	(2-5)-(7-4)-(6-1)-(3-8)-(10-9)	43	18.06
	(2-5)-(7-4)-(6-1)-(3-8)-(9-10)	43	18.06
CS [7]	2-(7-5)-(6-1)-3-(4-10)-(8-9)	39	14.26
	(5-7)-(2-6)-(1-3)-(4-8)-(9-10)	40	16.64
	(5-7)-(2-6)-(1-4)-(3-8)-(10-9)	41	17.34
	(2-5)-(7-4)-(6-1)-(3-8)-(10-9)	43	18.06
	(2-5)-(7-4)-(6-1)-(3-8)-(9-10)	43	18.06
	(5-2)-(7-4)-(6-1)-(3-8)-(10-9)	43	18.06
STA	(5-7)-(2-6)-(1-3)-(4-10)-(8-9)	39	16.45
	(5-7)-(2-6)-(1-3)-(4-10)-(9-8)	**39**	**16.45**
	(5-7)-(2-6)-(1-3)-(4-8)-(10-9)	40	16.64
	(5-7)-(2-6)-(1-3)-(4-8)-(9-10)	40	16.64
	(5-7)-(2-6)-(1-4)-(3-8)-(10-9)	41	17.34
	(5-7)-(2-6)-(1-4)-(3-8)-(9-10)	41	17.34
	(5-2)-(7-4)-(6-1)-(3-8)-(10-9)	43	18.06
	(5-2)-(7-4)-(6-1)-(3-8)-(9-10)	43	18.06
	(2-5)-(7-4)-(6-1)-(3-8)-(10-9)	43	18.06
	(2-5)-(7-4)-(6-1)-(3-8)-(9-10)	43	18.06

used with both the complete enumeration method and the CS, experimental results show the effectiveness of the proposed algorithm in obtaining the true set of all Pareto optimal solutions.

Acknowledgements This research work is conducted between Deakin University and Ballarat University under the Collaboration Research Network (CRN) initiative. The problem studied in this paper is related to the Australian Research Council (ARC) linkage project number LP0991175.

References

1. Marler, R.T., Arora, J.S.: Survey of multi-objective optimization methods for engineering. Struct. Multidiscip. Optim. **26**(6), 369–395 (2004)
2. Coello, C.A.C.: A comprehensive survey of evolutionary-based multiobjective optimization techniques. Knowl. Inf. Syst. **1**(3), 129–156 (1999)

3. Lei, D.M.: A Pareto archive particle swarm optimization for multi-objective job shop scheduling. Comput. Ind. Eng. **54**(4), 960–971 (2008)
4. Al-Anzi, F.S., Allahverdi, A.: A self-adaptive differential evolution heuristic for two-stage assembly scheduling problem to minimize maximum lateness with setup times. Eur. J. Oper. Res. **182**(1), 80–94 (2007)
5. Xia, W.J., Wu, Z.M.: An effective hybrid optimization approach for multi-objective flexible job-shop scheduling problems. Comput. Ind. Eng. **48**(2), 409–425 (2005)
6. Hanoun, S., Nahavandi, S., Kull, H.: Pareto archived simulated annealing for single machine job shop scheduling with multiple objectives. In: The Sixth International Multi-Conference on Computing in the Global Information Technology (ICCGI), pp. 99–104 (2011)
7. Hanoun, S., Creighton, D., Nahavandi, S. Kull, H.: Solving a multiobjective job shop scheduling problem using Pareto Archived Cuckoo Search. In: Proceedings of the 2012 17th IEEE International Conference on Emerging Technologies & Factory Automation, pp. 1–8 (2012)
8. Zhou, X.J., Yang, C.H., Gui, W.H.: Initial version of state transition algorithm. In: International Conference on Digital Manufacturing and Automation (ICDMA), pp. 644–647 (2011)
9. Zhou, X.J., Yang, C.H., Gui, W.H.: A new transformation into state transition algorithm for finding the global minimum. In: International Conference on Intelligent Control and Information Processing (ICICIP), pp. 674–678 (2011)
10. Yang, C.H., Tang, X.L., Zhou, X.J., Gui, W.H.: State transition algorithm for traveling salesman problem. In: the Proceedings of the 31st Chinese Control Conference (CCC), pp. 2481–2485 (2012)
11. Zhou, X.J., Yang, C.H., Gui, W.H.: State transition algorithm. J. Ind. Manag. Optim. **8**(4), 1039–1056 (2012)
12. Deb, K., Pratap, A., Agarwal, S., Meyarivan, T.: A fast and elitist multiobjective genetic algorithm: NSGA-II. IEEE Trans. Evol. Comput. **6**(2), 182–197 (2002)

Model Modification in Scheduling of Batch Chemical Processes

Xiaojun Zhou, David Yang Gao, and Chunhua Yang

Abstract This paper addresses the model modification in scheduling of batch chemical processes, which is widely used in current literatures. In the modified model, the capacity, storage constraints are modified and the allocation, sequence constraints are simplified. It is shown that the modified model can lead to fewer decision variables, fewer constraints, resulting in low computational complexity. Experimental results with two classical examples are given to demonstrate the effectiveness of the proposed formulation and approach.

Keywords Batch chemical processes • Process scheduling • Model modification

1 Introduction

The objective of process scheduling is to determine the optimal production plan utilizing the available resources over a given time horizon while satisfying the production requirements at the end of the horizon.

In recent decades, the scheduling of batch chemical processes has received considerable attention [1, 2]. To formulate a mathematical model for the scheduling problem, the first major issue is how to deal with the time. Existing scheduling

X. Zhou (✉)
School of Science, Information Technology and Engineering, University of Ballarat, Ballarat, VIC 3353, Australia

School of Information Science and Engineering, Central South University, Changsha 410083, China
e-mail: tiezhongyu2010@gmail.com

D.Y. Gao
School of Science, Information Technology and Engineering, University of Ballarat, Ballarat, VIC 3353, Australia
e-mail: d.gao@ballarat.edu.au

C. Yang
School of Information Science and Engineering, Central South University, Changsha 410083, China
e-mail: ychh@csu.edu.cn

D. Gao et al. (eds.), *Advances in Global Optimization*, Springer Proceedings in Mathematics & Statistics 95, DOI 10.1007/978-3-319-08377-3_10

formulations for time representation can be classified into two categories: discrete-time representation and continuous-time representation, and it was found that different representations would lead to different number of decision variables and constraints, resulting in different complexities [3].

The goal of this paper is to propose a modified model for the scheduling of batch chemical processes based on the continuous-time representation. The capacity, storage constraints are modified and the allocation, sequence constraints are simplified. It is shown that the simplified model has fewer decision variables, fewer constraints, resulting in low computational complexity. Two examples are given to demonstrate the effectiveness of the proposed formulation and approach.

2 Problem Description

Considering the following chemical process, as shown in Fig. 1, the process involves the production of a single product through three processing tasks. Raw materials are stored as the feeds, through mixing, reaction, and purification, we will get the final product.

We use the State-Task Network (STN) representation, as illustrated in Fig. 2, in which, the states, denoted by circles, represent the feeds, intermediates, and final products; here, s1 stands for the feeds, s2 and s3 are intermediates, and s4 are the final products; task, denoted by rectangles, represent the processing operations, which transform materials from input states to output states.

The scheduling problem for batch chemical processes can be stated as follows: For given (1) production recipe; (2) available units and their capacity limits; (3) available storage capacity; (4) production requirement; and (5) the time horizon. Our goal is to find (1) the optimal sequence of tasks taking place in each unit; (2) the amount of material being processed at each time in each unit; and (3) the processing time of each task in each unit.

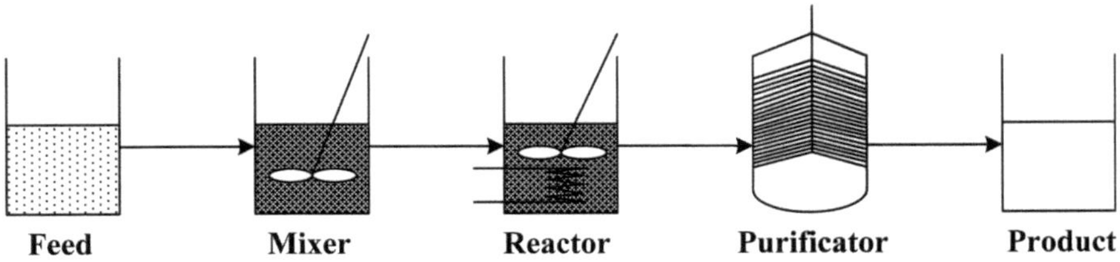

Fig. 1 Plant flow sheet for the chemical process

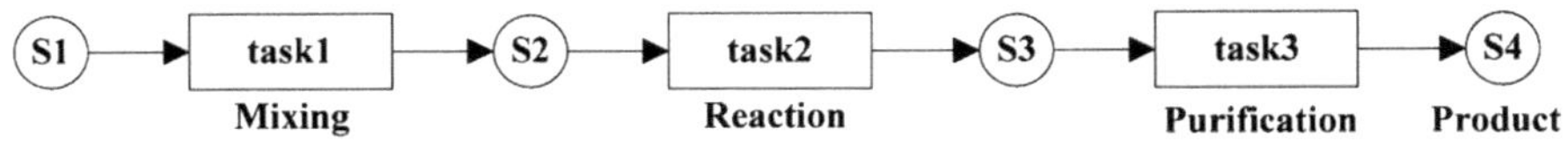

Fig. 2 STN representation for the chemical process

3 Model Modification

In this study, "unit-specific event" based continuous-time representation is used.

3.1 Original Model

The original model requires the following indices, sets, parameters, and variables:
Indices

i: tasks;
j: units;
n: event points, representing the beginning of a task;
s: states.

Sets

I: tasks;
I_j: tasks which can be performed in unit j;
I_s: tasks which process state s, either produce or consume;
J: units;
J_i: units which are suitable for performing task i;
N: event points within the time horizon;
S: sets of all involved states.

Parameters

V_{ij}^{min}: minimum amount of material processed by task i required to start operating unit j;
V_{ij}^{max}: maximum capacity of the specific unit j when processing task i;
$ST(s)^{max}$: available maximum storage capacity for state s;
$r(s)$: market requirement for state s at the end of time horizon;
ρ_{si}^{p}, ρ_{si}^{c}: proportion of state s produced, consumed from task i, respectively;
α_{ij}: constant term of processing time of task i and unit j;
β_{ij}: variable term of processing time of task i and unit j, expressing the time required by the unit to process one unit of material performing task i;
H: time horizon;
$price(s)$: price of state s.

Variables

$wv(i,n)$: binary variables that assign the beginning of task i at event point n;
$yv(j,n)$: binary variables that assign the utilization of unit j at event point n;
$B(i,j,n)$: amount of material undertaking task i in unit j at event point n;
$d(s,n)$: amount of state s being delivered to the market at event n;
$ST(s,n)$: amount of state s at event point n;

$T^s(i, j, n)$: time that task i starts in unit j at event point n;
$T^f(i, j, n)$: time that task i finishes in unit j while it starts at event point n.

Based on the notation, constraints can be described as
Allocation constraints

$$\sum_{i \in I_j} wv(i, n) = yv(j, n), \forall j \in J, n \in N \tag{1}$$

Capacity constraints

$$V_{ij}^{min} wv(i, n) \leq B(i, j, n) \leq V_{ij}^{max} wv(i, n), \forall\ i \in I, j \in J_i, n \in N \tag{2}$$

Storage constraints

$$ST(s, n) \leq ST(s)^{max}, \forall\ s \in S, n \in N \tag{3}$$

Material balances

$$ST(s, n) = ST(s, n-1) - d(s, n) + \sum_{i \in I_s} \rho_{si}^p \sum_{j \in J_i} B(i, j, n-1) + \sum_{i \in I_s} \rho_{si}^c \sum_{j \in J_i} B(i, j, n), \quad \forall s \in S, n \in N \tag{4}$$

Demand constraints

$$\sum_{n \in N} d(s, n) \geq r(s), \forall\ s \in S \tag{5}$$

Duration constraints

$$T^f(i, j, n) = T^s(i, j, n) + \alpha_{ij} wv(i, n) + \beta_{ij} B(i, j, n), \forall\ i \in I, j \in J_i, n \in N \tag{6}$$

Sequence constraints: same task in the same unit

$$T^s(i, j, n+1) \geq T^f(i, j, n) - H(2 - wv(i, n) - yv(j, n)), \forall\ i \in I, j \in J_i, n \in N, n \neq N \tag{7}$$

$$T^s(i, j, n+1) \geq T^s(i, j, n), \forall\ i \in I, j \in J_i, n \in N, n \neq N \tag{8}$$

$$T^f(i, j, n+1) \geq T^f(i, j, n), \forall\ i \in I, j \in J_i, n \in N, n \neq N \tag{9}$$

Sequence constraints: different tasks in the same unit

$$T^s(i, j, n+1) \geq T^f(i', j, n) - H(2 - wv(i', n) - yv(j, n)), \quad \forall\ i, i' \in I_j, j \in J, i \neq i', n \in N, n \neq N \tag{10}$$

Sequence constraints: different tasks in different units

$$T^s(i,j,n+1) \geq T^f(i',j',n) - H(2 - wv(i',n) - yv(j',n)),$$
$$\forall\, i \in I_j, i' \in I_{j'}, j, j' \in J, i \neq i', n \in N, n \neq N \qquad (11)$$

Sequence constraints: completion of previous tasks

$$T^s(i,j,n+1) \geq \sum_{n' \in N, n' \leq n} \sum_{i' \in I_j} (T^f(i',j,n') - T^s(i',j,n')),$$
$$\forall\, i \in I, j \in J_i, n \in N, n \neq N \qquad (12)$$

Time horizon constraints

$$T^s(i,j,n) \leq H, \forall\, i \in I, j \in J_i, n \in N \qquad (13)$$

$$T^f(i,j,n) \leq H, \forall\, i \in I, j \in J_i, n \in N \qquad (14)$$

Objective: maximization of profit

$$\sum_s \sum_n price(s) d(s,n), \forall\, s \in S, n \in N \qquad (15)$$

3.2 Simplified Model

We modify the capacity, storage constraints and simplify the allocation, sequence constraints. For this purpose, we redefine $wv(i, j, v)$ to denote whether or not task i in unit j at event point n, and define $T(j,n)$ to denote the starting time at event n in unit j, or $T^s(j,n)$ to denote the starting time at event n in unit j and $T^f(j,n)$ to denote the finishing time at event n in unit j.

The constraints for the simplified model is given as follows: Allocation constraints

$$\sum_{i \in I_j} wv(i,j,n) \leq 1, \forall j \in J, n \in N \qquad (16)$$

we remove $y(j,n)$ in the allocation constraints.

Capacity constraints

$$V_{ij}^{min} wv(i,j,n) \leq B(i,j,n) \leq V_{ij}^{max} wv(i,j,n), \forall\, i \in I, j \in J_i, n \in N, V_{ij}^{min} \neq 0 \qquad (17a)$$

$$wv(i,j,n) \leq B(i,j,n) \leq V_{ij}^{max} wv(i,j,n), \forall\, i \in I, j \in J_i, n \in N, V_{ij}^{min} = 0 \qquad (17b)$$

if $wv(i,j,n) = 1$, $B(i,j,n)$ should not be 0.

Storage constraints

$$0 \le ST(s,n) \le ST(s)^{max}, \forall\ s \in S, n \in N \tag{18}$$

the storage should not be negative.

Duration constraints

$$T(j,n+1) = T(j,n) + \alpha_{ij} wv(i,j,n) + \beta_{ij} B(i,j,n),\\ \forall\ i \in I, j \in J_i, n \in N \tag{19}$$

or

$$T^f(j,n) = T^s(j,n) + \alpha_{ij} wv(i,j,n) + \beta_{ij} B(i,j,n),\\ \forall\ i \in I, j \in J_i, n \in N \tag{20}$$

Time horizon constraints

$$T(j,1) \ge 0, T(j,N+1) \le H, \forall\ j \in J, n \in N \tag{21}$$

or

$$T^s(j,1) \ge 0, T^f(j,N) \le H, \forall\ j \in J, n \in N \tag{22}$$

Consecutive constraints

$$T(j,n+1) \ge T(j',n+1) - H(1 - wv(i',n)), \forall\ i' \in I_{j'}, j, j', \in J, n \in N \tag{23}$$

or

$$T^s(j,n+1) \ge T^f(j',n) - H(1 - wv(i',n)), \forall\ i' \in I_{j'}, j, j', \in J, n \in N \tag{24}$$

4 Case Studies

In the following, two examples are given to illustrate the effectiveness of the proposed formulation and approach. The modeling language used in this study is YALMIP [4], which is implemented as a free toolbox for MATLAB. The STN representation for Example 1 was given in Fig. 2 and known data can be found in Table 1. The optimal event time for this example is shown in Table 2, where $T(j,\cdot)$ represents the starting time in unit j at a given event. Figure 3 gives the Gantt chart for the simplified model, and comparisons for different models in Example 1 are shown in Table 3, in which, NC is the number of constraints, NV and NIV represent

Table 1 Known data for Example 1

Unit	Capacity	Suitability	Mean processing time
Mixer	100	Task1	4.5
Reactor	75	Task2	3.0
Purificator	50	Task3	1.5
State	Storage capacity	Initial amount	Price
s1	Unlimited	Unlimited	0.0
s2	100	0.0	0.0
s3	100	0.0	0.0
s4	Unlimited	0.0	1.0

Table 2 Optimal event time for the simplified model in Example 1

Time	Event 1	Event 2	Event 3	Event 4	Event 5	End
$T(1,\cdot)$	0	4.6673	8.1455	8.1455	8.1455	8.1455
$T(2,\cdot)$	4.6673	4.6673	8.1455	8.1455	10.5696	10.5696
$T(3,\cdot)$	8.5696	8.5696	8.5696	8.5696	10.5696	12.0000

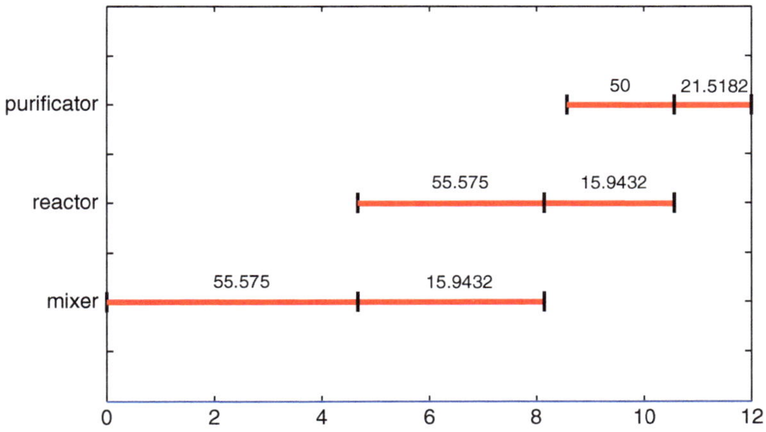

Fig. 3 Gantt chart for the simplified model in Example 1

Table 3 Comparisons for different models in Example 1

Index	Proposed	Ierapetritou [1]	Zhang [5]	Schilling [6]
NC	56	108	263	220
NV	43	105	187	157
NIV	15	15	48	46
OBJ	71.5182	71.518	71.45	71.47

the number of continuous and binary variables, respectively. The data for Example 2 are given in Fig. 4 and Table 4, and corresponding results can be found in Fig. 5 and Table 5, respectively.

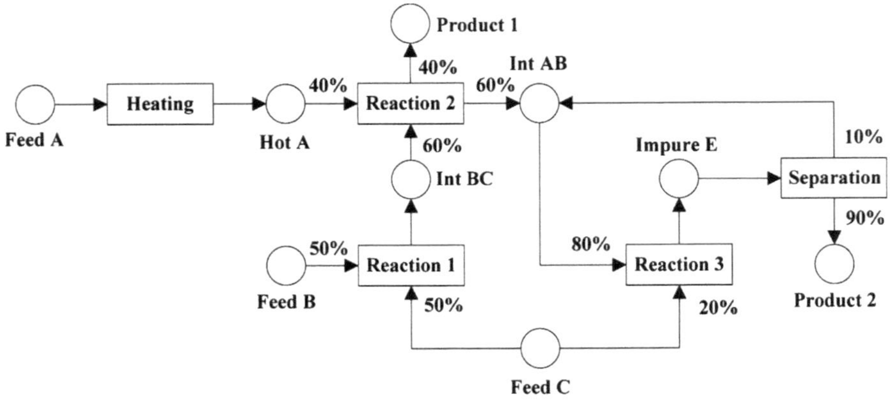

Fig. 4 STN representation for Example 2

Table 4 Known data for Example 2

Unit	Capacity	Suitability	Mean processing time
Heater	100	Heating	1
Reactor1	50	Reaction 1,2,3	2.0, 2.0, 1.0
Reactor2	80	Reaction 1,2,3	2.0, 2.0, 1.0
Separator	200	Separation	2.0
State	Storage capacity	Initial amount	Price
Feed A	Unlimited	Unlimited	0.0
Feed B	Unlimited	Unlimited	0.0
Feed C	Unlimited	Unlimited	0.0
HotA	100	0.0	0.0
IntAB	200	0.0	0.0
IntBC	150	0.0	0.0
ImpureE	200	0.0	0.0
Product1	Unlimited	0.0	10.0
Product2	Unlimited	0.0	10.0

5 Conclusion

This paper focuses on the model modification in scheduling of batch chemical processes. The modified model is not only more accurate but also possesses fewer decision variables and fewer constraints. Experimental results have demonstrated the effectiveness of the proposed formulation and approach.

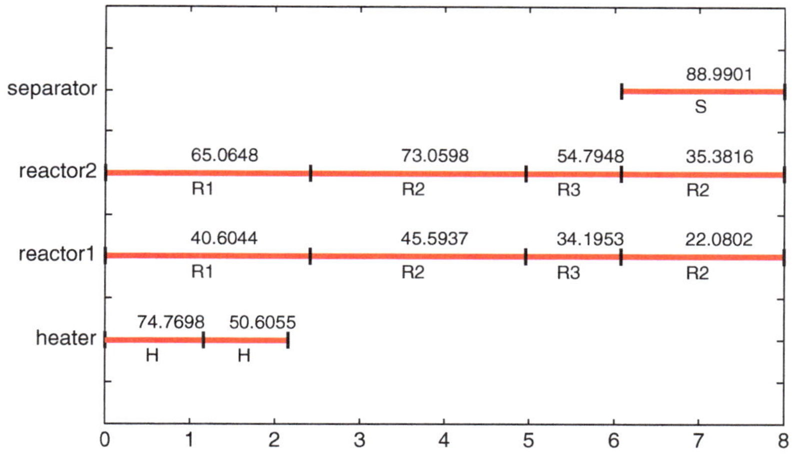

Fig. 5 Gantt chart for the simplified model in Example 2

Table 5 Comparisons for different models in Example 2

Index	Proposed	Ierapetritou [1]	Zhang [5]	Schilling [6]
NC	138	374	741	587
NV	84	260	497	386
NIV	40	40	147	130
OBJ	1,505.372	1,503.15	1,497.69	1,488.05

References

1. Ierapetritou, M.G., Floudas, C.A.: Effective continuous-time formulation for short-term scheduling. 1. Multipurpose batch processes. Ind. Eng. Chem. Res. **37**(11), 4341–4359 (1998)
2. Floudas, C.A., Lin, X.: Mixed integer linear programming in process scheduling: modeling, algorithms, and applications. Ann. Oper. Res. **139**(1), 131–162 (2005)
3. Floudas, C.A., Lin, X.: Continuous-time versus discrete-time approaches for scheduling of chemical processes: a review. Comput. Chem. Eng. **28**(11), 2109–2129 (2004)
4. Lofberg, J.: YALMIP: A toolbox for modeling and optimization in MATLAB. In: IEEE International Symposium on Computer Aided Control Systems Design, pp. 284–289 (2004)
5. Zhang, X.: Algorithms for optimal scheduling using nonlinear models. Ph.D. Thesis, University of London (1995)
6. Schilling, G., Pantelides, C.C.: A simple continuous-time process scheduling formulation and a novel solution algorithm. Comput. Chem. Eng. **20**, S1221–S1226 (1996)

An Approximation Algorithm for the Two-Stage Distributionally Robust Facility Location Problem

Chenchen Wu, Donglei Du, and Dachuan Xu

Abstract In this paper, we introduce a model of distributionally robust facility location problem (DRFLP) under moment constraints up to the second order. We show, via duality theory of moment problems, that the linear relaxation of the DRFLP is equivalent to that of the standard uncapacitated facility location problem (UFLP). Consequently, any LP-based approximation algorithm for the UFLP implies an approximation algorithm for the DRFLP with the same approximation ratio.

Keywords Facility location problem • Approximation algorithm • Distributionally robust optimization

1 Introduction

Facility location problem (FLP) is one of the classical NP-hard combinatorial optimization problems. In the uncapacitated facility location problem (UFLP), we are given a facility set with some opening cost and a client set with some demand. We want to open some facilities, and then connect all clients to some open facilities such that the facility opening cost and connection cost is minimized. The first

C. Wu
College of Science, Tianjin University of Technology, Tianjin 300384, People's Republic of China
e-mail: chenchen86711@gmail.com

D. Xu (✉)
Department of Applied Mathematics, Beijing University of Technology, 100 Pingleyuan, Chaoyang District, Beijing 100124, People's Republic of China
e-mail: xudc@bjut.edu.cn

D. Du
Department of Applied Mathematics, Beijing University of Technology, 100 Pingleyuan, Chaoyang District, Beijing 100124, People's Republic of China

Faculty of Business Administration, University of New Brunswick, Fredericton, NB, E3B 9Y2 Canada
e-mail: ddu@unb.ca

D. Gao et al. (eds.), *Advances in Global Optimization*, Springer Proceedings in Mathematics & Statistics 95, DOI 10.1007/978-3-319-08377-3_11

constant approximation algorithm for the UFLP is due to Shmoys et al. [1], while Li [2] gives an LP based 1.488-approximation algorithm which is the currently best ratio for the UFLP. For other variants of FLP, we refer to [3–7] and references therein.

Due to the uncertainty of the client demands, it is natural to investigate the stochastic version of the UFLP [8–10], where the client demands are assumed to follow a known distribution. However, the exact distribution is practically hard to secure. A less restricted practice is to obtain instead the moment information up to certain order. This last observation motivates us to consider a model of two-stage distributionally robust facility location problem (DRFLP), which only assumes knowing the moments up to the second order, namely, the mean and the variance.

Formally, in the DRFLP, we are given the facility set $\mathscr{F}$ and the client set $\mathscr{C}$. Each facility can be opened at two different stages with different cost. If facility i is opened at the first stage, the opening cost is f_i^I. The opening cost of facility i at the second stage is f_i^{II}. The demand of each client $j \in \mathscr{C}$ is $d_j \geq 0$ which is a random variable. We denote the demand vector $d = (d_1, \ldots, d_{|\mathscr{C}|})$. Instead of knowing the exact joint distribution $F(d)$ of d, we are given some marginal moment information of each d_j $(j \in \mathscr{C})$ up to the second order. The connection cost between $i \in \mathscr{F}$ and $j \in \mathscr{C}$ is c_{ij} which is assumed to be a metric. We can formulate the DRFLP as follows.

$$
\begin{aligned}
\min \sum_{i \in \mathscr{F}} f_i^I y_i^I &+ \sup_{F \sim \Omega} E_F[Q(d, y^I)] \\
\text{s. t. } y_i^I \in \{0, 1\}, &\qquad \forall i \in \mathscr{F},
\end{aligned} \tag{1}
$$

where Ω is the distribution set satisfying the marginal moment information, and

$$
\begin{aligned}
Q(d, y^I) := \min \sum_{i \in \mathscr{F}} f_i^{II} y_i^{II} &+ \sum_{i \in \mathscr{F}} \sum_{j \in \mathscr{C}} d_j c_{ij} x_{ij} \\
\text{s. t. } \sum_{i \in \mathscr{F}} x_{ij} \geq 1, &\qquad \forall j \in \mathscr{C}, \\
x_{ij} \leq y_i^I + y_i^{II}, &\qquad \forall i \in \mathscr{F}, j \in \mathscr{C}, \\
x_{ij}, y_i^{II} \in \{0, 1\}, &\qquad \forall i \in \mathscr{F}, j \in \mathscr{C}.
\end{aligned} \tag{2}
$$

In the above programs, the variable y_i^I denotes whether the facility i is opened at the first stage, that is, $y_i^I = 1$ represents the facility i is open at the first stage, otherwise $y_i^I = 0$. Moreover, when the demand vector d is realized, the variable y_i^{II} denotes whether facility i is opened at the second stage, that is, $y_i^{II} = 1$ represents the facility i is open at the first stage, otherwise $y_i^{II} = 0$. The variable x_{ij} denotes whether the client j is connected to the facility i, that is, $x_{ij} = 1$ represents the client j is connected to the facility i, otherwise $x_{ij} = 0$. Let $\tilde{Q}(d, y^I)$ denote the LP relaxation for (2). Denote $\Delta(y_i^I)$ as the feasible set of $\tilde{Q}(d, y^I)$, i.e.,

$$\Delta(y_i^I) := \left\{ (x_{ij}, y_i^{II}) : \begin{array}{ll} \sum_{i\in\mathscr{F}} x_{ij} \geq 1, & \forall j \in \mathscr{C} \\ x_{ij} \leq y_i^I + y_i^{II}, & \forall i \in \mathscr{F}, j \in \mathscr{C} \\ x_{ij}, y_i^{II} \geq 0, & \forall i \in \mathscr{F}, j \in \mathscr{C} \end{array} \right\}.$$

In this work, we consider two types of Ω: one is given the first marginal moments, and the other is given the first and second marginal moments. We show that the LP relaxation of the DRFLP is equivalent to that of the standard uncapacitated facility location problem (UFLP) for both versions. Consequently, any LP-based approximation algorithm for the UFLP implies an approximation algorithm for the DRFLP with the same approximation ratio. Moreover, the relaxations for these two types of Ω are equivalent. The main approach we used is the duality theory for the moment problems [11].

The organization of this paper is as follows. We present the LP relaxations for the two types of Ω in Sects. 2 and 3, respectively, followed by some discussions in Sect. 4.

2 The Formulation with the First Marginal Moments

In this section, we consider the case where we know the first moments of the random demand vector d; that is, $\Omega = \{F \text{ is a distribution of } d : E[d_j] = \mu_j, \forall j \in \mathscr{C}\}$. For a given binary variable y^I, we consider the moment problem

$$\sup_{F\sim\Omega} E_F[\tilde{Q}(d, y^I)],$$

which can be rewritten as follows.

$$\begin{aligned} \sup_{F\sim\Omega} & \int_{d\geq 0} \tilde{Q}(d, y^I)\mathrm{d}F(d) \\ \text{s.t.} & \int_{d\geq 0} \mathrm{d}F(d) = 1, \\ & \int_{d\geq 0} d_j \mathrm{d}F(d) = \mu_j, \qquad \forall j \in \mathscr{C}, \\ & \mathrm{d}F(d) \geq 0. \end{aligned} \tag{3}$$

The corresponding dual problem is

$$\begin{aligned} \inf_{\theta,\alpha} \ & \theta + \sum_{j\in\mathscr{C}} \alpha_j \mu_j \\ \text{s. t. } & \theta + \sum_{j\in\mathscr{C}} d_j \alpha_j \geq \tilde{Q}(d, y^I), \qquad \forall d \geq 0. \end{aligned} \tag{4}$$

The constraint in (4) is equivalent to

$$\min_{d\geq 0}\left\{\theta+\sum_{j\in\mathscr{C}}d_j\alpha_j-\min_{x,y^{II}\in\Delta}\left(\sum_{i\in\mathscr{F}}f_i^{II}y_i^{II}+\sum_{i\in\mathscr{F}}\sum_{j\in\mathscr{C}}d_jc_{ij}x_{ij}\right)\right\}\geq 0$$

$$\Leftrightarrow \min_{d\geq 0}\max_{x,y^{II}\in\Delta(y^I)}\left\{\theta+\sum_{j\in\mathscr{C}}d_j\alpha_j-\left(\sum_{i\in\mathscr{F}}f_i^{II}y_i^{II}+\sum_{i\in\mathscr{F}}\sum_{j\in\mathscr{C}}d_jc_{ij}x_{ij}\right)\right\}\geq 0. \quad (5)$$

The function $g(d;x,y^I) = \theta+\sum_{j\in\mathscr{C}}d_j\alpha_j-\left(\sum_{i\in\mathscr{F}}f_i^{II}y_i^{II}+\sum_{i\in\mathscr{F}}\sum_{j\in\mathscr{C}}d_jc_{ij}x_{ij}\right)$ is convex with respect to d and concave with respect to x, y^{II}. Therefore, we can interchange the operators min and max in (5) above to obtain

$$\max_{x,y^{II}\in\Delta(y^I)}\min_{d\geq 0}\left\{\theta+\sum_{j\in\mathscr{C}}d_j\alpha_j-\left(\sum_{i\in\mathscr{F}}f_i^{II}y_i^{II}+\sum_{i\in\mathscr{F}}\sum_{j\in\mathscr{C}}d_jc_{ij}x_{ij}\right)\right\}\geq 0$$

$$\Leftrightarrow \max_{x,y^{II}\in\Delta(y^I)}\left(\theta-\sum_{i\in\mathscr{F}}f_i^{II}y_i^{II}+\sum_{j\in\mathscr{C}}\min_{d_j\geq 0}d_j\left(\alpha_j-\sum_{i\in\mathscr{F}}c_{ij}x_{ij}\right)\right)\geq 0. \quad (6)$$

The above program implies that $\alpha_j \geq \sum_{i\in\mathscr{F}}c_{ij}x_{ij}$ which further leads to

$$\max_{x,y^{II}\in\Delta(y^I)}\left(\theta-\sum_{i\in\mathscr{F}}f_i^{II}y_i^{II}\right)\geq 0.$$

Hence, the dual program (4) is equivalent to

$$\begin{aligned}
&\inf\ \theta+\sum_{j\in\mathscr{C}}\alpha_j\mu_j\\
&\text{s. t. } \theta-\sum_{i\in\mathscr{F}}f_i^{II}y_i^{II}\geq 0,\\
&\qquad \alpha_j\geq\sum_{i\in\mathscr{F}}c_{ij}x_{ij}, \qquad \forall j\in\mathscr{C},\\
&\qquad x,y^{II}\in\Delta(y^I).
\end{aligned} \quad (7)$$

Substituting the above program into (1), we get the following relaxation,

$$\min_{\theta,\alpha,x,y^I,y^{II}}\ \sum_{i\in\mathscr{F}}f_i^Iy_i^I+\theta+\sum_{j\in\mathscr{C}}\alpha_j\mu_j$$

$$\begin{aligned}
\text{s. t.}\quad & \theta - \sum_{i\in\mathscr{F}} f_i^{II} y_i^{II} \geq 0, \\
& \alpha_j \geq \sum_{i\in\mathscr{F}} c_{ij} x_{ij}, \qquad \forall j \in \mathscr{C}, \\
& \sum_{i\in\mathscr{F}} x_{ij} \geq 1, \qquad \forall j \in \mathscr{C}, \\
& x_{ij} \leq y_i^I + y_i^{II}, \qquad \forall i \in \mathscr{F}, j \in \mathscr{C}, \\
& y_i^I \in \{0,1\}, x_{ij}, y_i^{II} \geq 0, \qquad \forall i \in \mathscr{F}, j \in \mathscr{C}.
\end{aligned} \tag{8}$$

It is easy to see that we can simplify (8) as follows by setting $\theta := \sum_{i\in\mathscr{F}} f_i^{II} y_i^{II}$ and $\alpha_j := \sum_{i\in\mathscr{F}} c_{ij} x_{ij}$ for each $j \in \mathscr{C}$:

$$\begin{aligned}
\min_{\alpha,x,y^I,y^{II}}\quad & \sum_{i\in\mathscr{F}} f_i^I y_i^I + \sum_{i\in\mathscr{F}} f_i^{II} y_i^{II} + \sum_{j\in\mathscr{C}} \sum_{i\in\mathscr{F}} \mu_j c_{ij} x_{ij} \\
\text{s. t.}\quad & \sum_{i\in\mathscr{F}} x_{ij} \geq 1, \qquad \forall j \in \mathscr{C}, \\
& x_{ij} \leq y_i^I + y_i^{II}, \qquad \forall i \in \mathscr{F}, j \in \mathscr{C}, \\
& y_i^I \in \{0,1\}, x_{ij}, y_i^{II} \geq 0, \qquad \forall i \in \mathscr{F}, j \in \mathscr{C}.
\end{aligned} \tag{9}$$

3 The Formulation with the First and Second Marginal Moments

In this section, we consider the distribution set where the first two marginal moments are known; that is, $\Omega = \{F \text{ is a distribution of } d : E[d_j] = \mu_j, E[d_j^2] = \mu_j^2 + \sigma_j^2, \forall j \in \mathscr{C}\}$. Let us consider the following moment problem.

$$\begin{aligned}
\sup_{F\sim\Omega}\quad & \int_{d\geq 0} \tilde{Q}(d, y^I)\mathrm{d}F(d) \\
\text{s.t.}\quad & \int_{d\geq 0} \mathrm{d}F(d) = 1, \\
& \int_{d\geq 0} d_j \mathrm{d}F(d) = \mu_j, \qquad \forall j \in \mathscr{C}, \\
& \int_{d\geq 0} d_j^2 \mathrm{d}F(d) = \mu_j^2 + \sigma_j^2, \qquad \forall j \in \mathscr{C}, \\
& \mathrm{d}F(d) \geq 0.
\end{aligned} \tag{10}$$

The corresponding dual is

$$\inf_{\theta,\alpha,\beta} \theta + \sum_{j\in\mathscr{C}} \mu_j\alpha_j + \sum_{j\in\mathscr{C}} (\mu_j^2+\sigma_j^2)\beta_j$$

$$\text{s.t. } \theta + \sum_{j\in\mathscr{C}} d_j\alpha_j + \sum_{j\in\mathscr{C}} d_j^2\beta_j \geq \tilde{Q}(d, y^I), \qquad \forall d \geq 0.$$

The constraint of the dual is equivalent to

$$\min_{d\geq 0}\left\{\theta + \sum_{j\in\mathscr{C}} d_j\alpha_j + \sum_{j\in\mathscr{C}} d_j^2\beta_j - \min_{x,y^{II}\in\Delta}\left(\sum_{i\in\mathscr{F}} f_i^{II}y_i^{II} + \sum_{i\in\mathscr{F}}\sum_{j\in\mathscr{C}} d_j c_{ij}x_{ij}\right)\right\} \geq 0,$$

which implies that $\beta_j \geq 0$ for each $j \in \mathscr{C}$. Then, we have

$$\min_{d\geq 0}\max_{x,y^{II}\in\Delta(y^I)}\left\{\sum_{j\in\mathscr{C}} d_j^2\beta_j + \sum_{j\in\mathscr{C}} d_j\left(\alpha_j - \sum_{i\in\mathscr{F}} c_{ij}x_{ij}\right) + \theta - \sum_{i\in\mathscr{F}} f_i^{II}y_i^{II}\right\} \geq 0.$$

The function $g(d; x, y^{II}) := \sum_j d_j^2\beta_j + \sum_j d_j\left(\alpha_j - \sum_{i\in\mathscr{F}} c_{ij}x_{ij}\right) + \theta - \sum_{i\in\mathscr{F}} f_i^{II}y_i^{II}$ is convex with respect to d and concave with respect to x, y^{II}. Similarly as before, we can interchange the operators min and max to obtain

$$\max_{x,y^{II}\in\Delta(y^I)}\left\{\theta - \sum_{i\in\mathscr{F}} f_i^{II}y_i^{II} + \sum_{j\in\mathscr{C}}\min_{d_j\geq 0}\left(d_j^2\beta_j + d_j\left(\alpha_j - \sum_{i\in\mathscr{F}} c_{ij}x_{ij}\right)\right)\right\} \geq 0.$$

Choosing $d_j = -\left(\alpha_j - \sum_i c_{ij}x_{ij}\right)/2\beta_j$ if $\alpha_j \leq \sum_i c_{ij}x_{ij}$, and $d_j = 0$ otherwise, we can further simplify the dual constraint

$$\max_{x,y^{II}\in\Delta(y^I)}\left\{\theta - \sum_{i\in\mathscr{F}} f_i^{II}y_i^{II} - \sum_{j\in\mathscr{C}} \mathbb{I}_{\{\alpha_j\leq\sum_i c_{ij}x_{ij}\}}\frac{(\alpha_j - \sum_{i\in\mathscr{F}} c_{ij}x_{ij})^2}{4\beta_j}\right\} \geq 0,$$

where $\mathbb{I}_{\{\alpha_j\leq\sum_i c_{ij}x_{ij}\}} = 1$ if $\alpha_j \leq \sum_i c_{ij}x_{ij}$, and $\mathbb{I}_{\{\alpha_j\leq\sum_i c_{ij}x_{ij}\}} = 0$, otherwise. Then, we get the relaxation for (1),

$$\min_{x,y^I,y^{II},\theta,\alpha,\beta\geq 0} \sum_{i\in\mathscr{F}} f_i^I y_i^I + \theta + \sum_{j\in\mathscr{C}} \mu_j\alpha_j + \sum_{j\in\mathscr{C}} (\mu_j^2+\sigma_j^2)\beta_j$$

$$\text{s.t.} \qquad \theta - \sum_{j\in\mathscr{C}} \mathbb{I}_{\{\alpha_j\leq\sum_i c_{ij}x_{ij}\}}\frac{\left(\alpha_j - \sum_{i\in\mathscr{F}} c_{ij}x_{ij}\right)^2}{4\beta_j} - \sum_{i\in\mathscr{F}} f_i^{II}y_i^{II} \geq 0,$$

$$\sum_{i\in\mathscr{F}} x_{ij} \geq 1, \qquad \forall j \in \mathscr{C},$$
$$x_{ij} \leq y_i^I + y_i^{II}, \qquad \forall i \in \mathscr{F}, j \in \mathscr{C},$$
$$y_i^I \in \{0,1\}, x_{ij}, y_i^{II} \geq 0, \qquad \forall i \in \mathscr{F}, j \in \mathscr{C}.$$

Obviously, we can set $\theta := \sum_{j\in\mathscr{C}} \mathbb{I}_{\{\alpha_j \leq \sum_i c_{ij}x_{ij}\}} \frac{\left(\alpha_j - \sum_{i\in\mathscr{F}} c_{ij}x_{ij}\right)^2}{4\beta_j} + \sum_{i\in\mathscr{F}} f_i^{II} y_i^{II}$ to simplify the above program as follows:

$$\begin{aligned}
\min_{x,y^I,y^{II},\theta,\alpha,\beta\geq 0} \quad & \sum_{i\in\mathscr{F}} f_i^I y_i^I + \sum_{i\in\mathscr{F}} f_i^{II} y_i^{II} + \sum_{j\in\mathscr{C}} \mu_j \alpha_j \\
& + \sum_{j\in\mathscr{C}} \mathbb{I}_{\{\alpha_j \leq \sum_{i\in\mathscr{F}} c_{ij}x_{ij}\}} \frac{\left(\alpha_j - \sum_{i\in\mathscr{F}} c_{ij}x_{ij}\right)^2}{4\beta_j} + \sum_{j\in\mathscr{C}} (\mu_j^2 + \sigma_j^2)\beta_j \\
\text{s.t.} \quad & \sum_{i\in\mathscr{F}} x_{ij} \geq 1, \qquad \forall j \in \mathscr{C}, \\
& x_{ij} \leq y_i^I + y_i^{II}, \qquad \forall i \in \mathscr{F}, j \in \mathscr{C}, \\
& y_i^I \in \{0,1\}, x_{ij}, y_i^{II} \geq 0, \qquad \forall i \in \mathscr{F}, j \in \mathscr{C}.
\end{aligned}$$

Set $\beta_j := \left(\sum_{i\in\mathscr{F}} c_{ij}x_{ij} - \alpha_j\right)\sqrt{\mu_j^2 + \sigma_j^2}$ if $\alpha_j \leq \sum_{i\in\mathscr{F}} c_{ij}x_{ij}$, and $\beta_j := 0$, otherwise. We obtain the following equivalent formulation.

$$\begin{aligned}
\min_{x,y^I,y^{II},\theta,\alpha} \quad & \sum_{i\in\mathscr{F}} f_i^I y_i^I + \sum_{i\in\mathscr{F}} f_i^{II} y_i^{II} + \sum_{j\in\mathscr{C}} \mu_j \alpha_j \\
& + \sum_{j\in\mathscr{C}} \mathbb{I}_{\{\alpha_j \leq \sum_i c_{ij}x_{ij}\}} \left(\sum_{i\in\mathscr{F}} c_{ij}x_{ij} - \alpha_j\right) \sqrt{\mu_j^2 + \sigma_j^2} \\
\text{s.t.} \quad & \sum_{i\in\mathscr{F}} x_{ij} \geq 1, \qquad \forall j \in \mathscr{C}, \\
& x_{ij} \leq y_i^I + y_i^{II}, \qquad \forall i \in \mathscr{F}, j \in \mathscr{C}, \\
& y_i^I \in \{0,1\}, x_{ij}, y_i^{II} \geq 0, \qquad \forall i \in \mathscr{F}, j \in \mathscr{C}.
\end{aligned}$$

Choosing $\alpha_j := \sum_i c_{ij} x_{ij}$ for each $j \in \mathscr{C}$, we obtain the equivalent minimization problem

$$\begin{aligned}
\min_{\alpha,x,y^I,y^{II}} \quad & \sum_{i\in\mathscr{F}} f_i^I y_i^I + \sum_{i\in\mathscr{F}} f_i^{II} y_i^{II} + \sum_{i\in\mathscr{F}} \sum_{j\in\mathscr{C}} \mu_j c_{ij} x_{ij} \\
\text{s. t.} \quad & \sum_{i\in\mathscr{F}} x_{ij} \geq 1, \qquad \forall j \in \mathscr{C}, \\
& x_{ij} \leq y_i^I + y_i^{II}, \qquad \forall i \in \mathscr{F}, j \in \mathscr{C}, \\
& y_i^I \in \{0,1\}, x_{ij}, y_i^{II} \geq 0, \qquad \forall i \in \mathscr{F}, j \in \mathscr{C}.
\end{aligned} \tag{11}$$

By the above analysis, we can obtain the two types of linear relaxations are the same as the linear relaxation for the UFLP. Moreover, we need give the relationship of integer solution for the DRFLP and UFLP. Indeed, any optimal solution for $Q(\mu, y^I)$ with the demand μ is a feasible solution for $E_F[Q(d, y^I)]$ with the same value. So

$$E_F[Q(d, y^I)] \leq Q(\mu, y^I).$$

Hence, we have that any approximation algorithm based on linear program can imply in the DRFLP with the same approximation ratio. So we have the 1.488-approximation algorithm [2] for the 2-stage DRFLP.

4 Discussion

Note that (11) is the same as (9), implying that the LP relaxation based on the first marginal moments is equivalent to that based on the first two marginal moments. Based on (11), if we apply any LP-based approximation algorithm for the UFLP on the instance with facility cost $\min\{f_i^I, f_i^{II}\}$, we can obtain an approximation algorithm for the DRFLP with the same ratio. Recall that there is an LP-based 1.488-approximation algorithm for the UFLP, implying that there exists a 1.488-approximation algorithm for the DRFLP. It is interesting to study the DRFLP with other distribution set as future research work.

Acknowledgements The research of the first author is supported by NSF of China (No. 11371001). The second author's research is supported by the Natural Sciences and Engineering Research Council of Canada (NSERC) grant 283106. The third author's research is supported by Scientific Research Common Program of Beijing Municipal Commission of Education (No. KM201210005033) and China Scholarship Council.

References

1. Shmoys, D.B., Tardös, E., Aardal, K.I.: Approximation algorithms for facility location problems. In: Proceedings of STOC, pp. 265–274 (1997)
2. Li, S.: A 1.488-approximation algorithm for the uncapacitated facility location problem. Inform. Comput. **222**, 45–58 (2013)
3. Chen, X., Chen, B.: Approximation algorithms for soft-capacitated facility location in capacitated network design. Algorithmica **53**, 263–297 (2007)
4. Shu, J., Teo, C.P., Shen, Z.J.M.: Stochastic transportation-inventory network design problem. Oper. Res. **53**, 48–60 (2005)
5. Zhang, J.: Approximating the two-level facility location problem via a quasi-greedy approach. Math. Program. **108**, 159–176 (2006)
6. Zhang, J., Chen, B., Ye, Y.: A multiexchange local search algorithm for the capacitated facility location problem. Math. Oper. Res. **30**, 389–403 (2005)
7. Zhang, P.: A new approximation algorithm for the k-facility location problem. Theor. Comput. Sci. **384**, 126–135 (2007)
8. Ravi R., Sinha, A.: Hedging uncertainty: approximation algorithms for stochastic optimization problems. Math. Program. Ser. A **108**, 97–114 (2006)
9. Ye, Y., Zhang, J.: An approximation algorithm for the dynamic facility location problem. In: Combinatorial Optimization in Communication Networks, pp. 623–637. Kluwer Academic, Dordrecht (2005)
10. Shmoys, D.B., Swamy, C.: An approximation scheme for stochastic linear programming and its application to stochastic integer programs. J. ACM **53**, 978–1012 (2006)
11. Bertsimas, D., Popescu I,: Optimal inequalities in probability theory: a convex optimization approach. SIAM J. Optim. **15**, 780–804 (2005)

Rainbow Connection Numbers for Undirected Double-Loop Networks

Yuefang Sun

Abstract An edge-colored graph G is rainbow connected if any two vertices are connected by a path whose edges have distinct colors. The rainbow connection number of a connected graph G, denoted by $rc(G)$, is the smallest number of colors that are needed in order to make G rainbow connected. In this paper, we investigate rainbow connection numbers of three subfamilies of undirected double-loop networks, denoted by $rc(G(n; \pm s_1, \pm s_2))$, where $1 \leq s_1 < s_2 < n/2$. We almost determine the precise value for the case that $s_1 = 1, s_2 = 2$. For the case that $n = ks, s_1 = 1, s_2 = s$, where $s \geq 3$ and $k \geq 1$ are integers, we derive that $rc(G(ks; \pm 1, \pm s)) \leq \min\{\lceil \frac{k}{2} \rceil + s, \lceil \frac{s+1}{2} \rceil + k - 1\}$. For the case that $n = 2ks, s_1 = 2, s_2 = s$, where $k \geq 1$ are integers and $s \geq 3$ are odd integers, we have $rc(G(n; \pm s_1, \pm s_2)) \leq \lceil \frac{ks}{2} \rceil + k$.

Keywords Rainbow connection number • Rainbow coloring • Undirected double-loop network

AMS Subject Classification 2010: 05C15, 05C40

1 Introduction

Connectivity is one of the most important concepts in graph theory and its applications, both in a combinatorial sense and in an algorithmic sense. In theoretical computer science, connectivity is a basic measure of reliability of networks. There are many ways to strengthen the connectivity concept, such as requiring hamiltonicity, k-connectivity, requiring the existence of edge-disjoint spanning trees, and so on. An interesting way to strengthen the connectivity requirement, the rainbow connection, was first introduced by Chartrand et al. [1]. Let G be a

Y. Sun (✉)
Department of Mathematics, Shaoxing University,
Zhejiang 312000, People's Republic of China
e-mail: yfsun2013@gmail.com

D. Gao et al. (eds.), *Advances in Global Optimization*, Springer Proceedings in Mathematics & Statistics 95, DOI 10.1007/978-3-319-08377-3_12

nontrivial connected graph on which an edge-coloring $c : E(G) \rightarrow \{1, 2, \cdots, n\}$, $n \in \mathbb{N}$, is defined, where adjacent edges may be colored the same. A path is *rainbow* if no two edges of it are colored the same. An edge-colored graph G is *rainbow connected* if every two distinct vertices are connected by a rainbow path. An edge-coloring under which G is rainbow connected is called a *rainbow coloring*. We define the *rainbow connection number* of a connected graph G, denoted by $rc(G)$, as the smallest number of colors that are needed in order to make G rainbow connected [1].

The concept of rainbow connection number has an interesting application for the secure transmission of information between nodes in the network. Suppose G represents a network. While the information needs to be protected since it relates to security, there must also be procedures that permit access between appropriate nodes. This issue can be addressed by assigning information transfer paths between nodes which may have other nodes as intermediaries while requiring a large enough number of passwords (or firewalls) that is prohibitive to intruders, yet small enough to manage (that is, enough so that one or more paths between every pair of nodes have no password repeated). Thus, an immediate question arises: What is the minimum number of passwords needed that allows one or more secure paths between every two nodes so that the passwords along each path are distinct? Clearly, this number is precisely $rc(G)$. The topic of rainbow connection is fairly interesting and recently quite a lot of papers have been published about it. The reader is referred to a monograph [2] and a recent survey [3] on this topic.

Given an integer $n \geq 3$ and distinct integers $s_1, \ldots, s_k$ between 1 and $n/2$, the *circulant graph* $G(n; \pm s_1, \pm s_2, \ldots, \pm s_k)$ is defined to be the undirected graph with vertex set the additive group $\mathbb{Z}_n$ of integers modulo n, such that each vertex $i \in \mathbb{Z}_n$ is adjacent to $i \pm s_1, i \pm s_2, \ldots, i \pm s_k$, with integers involved modulo n. Note that this graph is $2k$-regular if none of $s_1, \ldots, s_k$ is equal to $n/2$, and $(2k - 1)$-regular otherwise. In the computer science literature, $G(n; \pm s_1, \pm s_2, \ldots, \pm s_k)$ is also called [4, 5] a *distributed loop network*. In particular, $G(n; \pm s_1, \pm s_2)$ with $1 \leq s_1 < s_2 < n/2$ is called a (undirected) *double-loop network*.

Circulant graphs have been studied intensively in computer science and discrete mathematics, as shown in a recent survey [6] and earlier surveys [4, 5]. They have wide applications in many different domains, including small-world networks [7], chemical reactions [8], multi-processor cluster systems [9], discrete cellular neural networks [10], optical networks [11], coding theory for the construction of perfect error-correcting codes [12], etc. In particular, various circulant graphs have been proposed as models for interconnection networks (e.g., ILLIAC-IV [13], Intel Paragon, MICROS). For example, a family of 6-regular circulant graphs have physically been used [14–16] as multiprocessor interconnection networks at the Real-Time Computing Laboratory, The University of Michigan. They are called HARTS (*Hexagonal Architecture for Real-Time Systems*) [15, 16], *C-wrapped hexagonal meshes* [15], or *hexagonal mesh interconnection networks* [17]. A much larger family of 6-regular circulants containing HARTS as a subfamily that are efficient for information dissemination were recently studied in [18], and a similar family of 4-regular circulants were investigated in [19].

Due to the importance of circulant graphs in network design, the significances of reliability and information transmission in networks, it is of interest to understand the rainbow connection numbers of circulant graphs. This motivates our study in this paper, in which we will focus on the rainbow connection numbers of undirected double-loop networks $G(n; \pm s_1, \pm s_2)$. It was proved that computing $rc(G)$ is an NP-Hard problem, and that even deciding whether a graph has $rc(G) = 2$ is NP-Complete [20]. It is also known that deciding whether a given edge-colored graph is rainbow connected is NP-Complete [20]. Then people try to give upper bounds (or precise values) for rainbow connection numbers of graphs, especially special graph classes. In this paper, we try to give upper bounds (or precise values) of three subfamilies of $G(n; \pm s_1, \pm s_2)$. We will almost determine the precise value for the case that $s_1 = 1, s_2 = 2$ (Theorem 3.5). For the case that $n = ks, s_1 = 1, s_2 = s$, where $s, k \geq 3$ are integers, we derive that $rc(G(ks; \pm 1, \pm s)) \leq \min\{\lceil \frac{k}{2} \rceil + s, \lceil \frac{s+1}{2} \rceil + k - 1\}$ (Theorem 3.6). For the case that $n = 2ks, s_1 = 2, s_2 = s$, where $k \geq 1$ is an integer and $s \geq 3$ is an odd integer, we derive that $rc(G(n; \pm s_1, \pm s_2)) \leq \lceil \frac{ks}{2} \rceil + k$ (Theorem 3.7).

2 Preliminaries

We follow the notations and terminology of [21] for those not defined in this paper. Let $V(G)$ and $E(G)$ be the sets of vertices and edges of G, respectively. We define a *clique* in a graph G to be a complete subgraph of G, and a *maximal clique* is a clique that is not contained in a larger clique. The *clique graph* $K(G)$ of G is the intersection graph of the maximal cliques of G; that is, the vertices of $K(G)$ correspond to the maximal cliques of G, and two of these vertices are joined by an edge if and only if the corresponding maximal cliques intersect. For a set S, $|S|$ denotes the cardinality of S. For an edge-coloring c of a graph G, we use $c(H)$ to denote the set of colors of the edges in the subgraph H of G. We call a connected graph G a *clique-cycle-structure*, if it satisfies the following three conditions [22, 23]:

(i) G has at least three maximal cliques;
(ii) each edge belongs to exactly one maximal clique;
(iii) the clique graph $K(G)$ is a cycle.

By condition (ii), any two maximal cliques of G have at most one common vertex. Furthermore, G is formed by its maximal cliques. The *size* of the clique-cycle-structure is the number of its maximal cliques. We call a clique-cycle-structure *odd* if its size is odd, otherwise, it is an *even* clique-cycle-structure. For example, a clique-cycle-structure of size 5 is shown in Fig. 1. Note that a triangle is not a clique-cycle-structure, but a cycle with length $l \geq 4$ is a clique-cycle-structure of

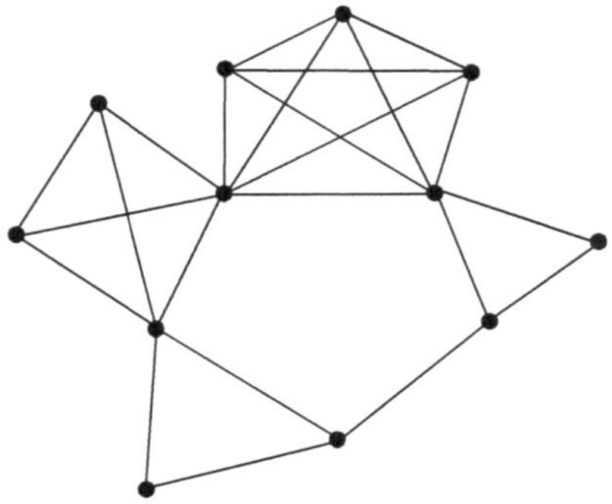

Fig. 1 A clique-cycle-structure of size 5

size l. For a clique-cycle-structure G, if each maximal clique of it is a triangle, then G is a *triangle-cycle-structure*.

We know that if H is a connected spanning subgraph of G, then $rc(H) \leq rc(G)$. Clearly, $rc(G) \geq diam(G)$ where $diam(G)$ denotes the diameter of a graph G. The following simple result on a cycle of length n was first obtained in [1]:

Proposition 2.1. *For each integer* $n \geq 4$, $rc(C_n) = \lceil \frac{n}{2} \rceil$.

In [22], the rainbow connection number of a clique-cycle-structure was given:

Lemma 2.2. *Let G be a clique-cycle-structure of size l, then*

$$rc(G) = \begin{cases} \frac{l}{2} \text{ or } \frac{l}{2} + 1 & l \text{ is even} \\ \frac{l+1}{2} & l \text{ is odd} \end{cases}$$

The following is a special case of a more general result [24] about the connectedness of a circulant graph.

Theorem 2.3 ([24]). $G(n; \pm s_1, \pm s_2)$ *is connected if and only if* $\gcd(n, s_1, s_2) = 1$.

3 Main Results

We first consider the first subfamily of undirected double-loop networks, that is, the case that $s_1 = 1$ and $s_2 = 2$, $G(n; \pm s_1, \pm s_2)$ is connected by Theorem 2.3. We will get our first main result from the following four lemmas.

Lemma 3.1.

$$rc(G(4k; \pm 1, \pm 2)) = k.$$

Proof. Let $V(G(4k; \pm 1, \pm 2)) = \{u_1, u_2, \cdots, u_{4k}\}$. In $G(4k; \pm 1, \pm 2)$, we get $2k$ edge-disjoint triangles: $T_i = \{u_{2i+1}, u_{2i+2}, u_{2i+3}\}$ for $0 \leq i \leq 2k - 1$, where

$u_{4k+1} = u_1$. There is also a cycle $C^1 : u_2, u_4, \cdots, u_{4k}, u_2$ which is edge-disjoint with the above triangles.

We give $G(4k; \pm 1, \pm 2)$ an edge-coloring with k colors as follows: for edges of the above triangles, let $c(T_i) = i + 1$ for $0 \le i \le k - 1$ (this means that all edges in T_i have the same color $i + 1$) and $c(T_i) = i + 1 - k$ for $k \le i \le 2k - 1$; for the remaining edges (that is, edges of C^1), let $c(u_{2j}u_{2j+2}) = j$ for $1 \le j \le k$ and $c(u_{2j}u_{2j+2}) = j - k$ for $k + 1 \le j \le 2k$.

It is not hard to show that $G(4k; \pm 1, \pm 2)$ is rainbow connected under the above coloring, so we have $rc(G(4k; \pm 1, \pm 2)) \le k$. Moreover, it is easy to show that $diam(G(4k; \pm 1, \pm 2)) \ge k$. So the lemma holds. □

Lemma 3.2.

$$rc(G(4k + 2; \pm 1, \pm 2)) = k + 1.$$

Proof. Let $V(G(4k + 2; \pm 1, \pm 2)) = \{u_1, u_2, \cdots, u_{4k+2}\}$. In $G(4k + 2; \pm 1, \pm 2)$, we get $2k + 1$ edge-disjoint triangles: $T_i = \{u_{2i+1}, u_{2i+2}, u_{2i+3}\}$ for $0 \le i \le 2k$, where $u_{4k+3} = u_1$. Let G' be the spanning subgraph of $G(4k + 2; \pm 1, \pm 2)$ which is formed by these triangles. Clearly, it is a triangle-cycle-structure of size $2k + 1$. By Lemma 2.2, we have $rc(G(4k + 2; \pm 1, \pm 2) \le rc(G') = k + 1$. Moreover, we know $diam(G(4k + 2; \pm 1, \pm 2)) = k + 1$. Thus, the conclusion holds. □

Lemma 3.3.

$$rc(G(4k + 1; \pm 1, \pm 2)) = k \text{ or } k + 1.$$

Proof. Let $V(G(4k + 1; \pm 1, \pm 2)) = \{u_1, u_2, \cdots, u_{4k+1}\}$. In $G(4k + 1; \pm 1, \pm 2)$, we get $2k$ edge-disjoint triangles: $T_i = \{u_{2i+1}, u_{2i+2}, u_{2i+3}\}$ for $0 \le i \le 2k - 1$. Let G' be the subgraph of G which is formed by these triangles and the edge $u_1 u_{4k+1}$. Clearly, it is a spanning subgraph of G and is a clique-cycle-structure of size $2k + 1$. By Lemma 2.2, we have $rc(G(4k + 1; \pm 1, \pm 2) \le rc(G') = k + 1$. Moreover, we know $diam(G(4k + 1; \pm 1, \pm 2)) = k$. Thus, the result holds. □

Lemma 3.4.

$$rc(G(4k + 3; \pm 1, \pm 2)) = k + 1 \text{ or } k + 2.$$

Proof. Let $V(G(4k+3; \pm 1, \pm 2)) = \{u_1, u_2, \cdots, u_{4k+3}\}$. In $G(4k+3; \pm 1, \pm 2)$, we get $2k + 1$ edge-disjoint triangles: $T_i = \{u_{2i+1}, u_{2i+2}, u_{2i+3}\}$ for $0 \le i \le 2k$. Let G' be the spanning subgraph of G which is formed by these triangles and the edge $u_1 u_{4k+3}$. Clearly, it is a clique-cycle-structure of size $2k + 2$. By Lemma 2.2, we have $rc(G(4k+3; \pm 1, \pm 2) \le rc(G') = k+2$. Moreover, we know $diam(G(4k + 3; \pm 1, \pm 2)) = k + 1$. □

From Lemmas 3.1–3.4, the following theorem is clear.

Table 1 Vertex sets of cycles C^i and D^j in Theorem 3.6

	D^1	D^2	$\cdots$	D^s
C^1	u_1	u_2	$\cdots$	u_{s+1}
C^2	u_{s+1}	u_{s+2}	$\cdots$	u_{2s+1}
$\cdots$	$\cdots$	$\cdots$	$\cdots$	$\cdots$
C^k	$u_{(k-1)s+1}$	$u_{(k-1)s+2}$	$\cdots$	u_1

Theorem 3.5. *For $n \geq 3$, we have*

$$rc(G(n; \pm 1, \pm 2)) = \begin{cases} k & n = 4k; \\ k \text{ or } k+1 & n = 4k+1; \\ k+1 & n = 4k+2; \\ k+1 \text{ or } k+2 & n = 4k+3. \end{cases}$$

We next consider the second subfamily in which $s_1 = 1, s_2 = s, n = ks$, where $k, s \geq 3$ are integers. Clearly, $G(n; \pm s_1, \pm s_2)$ is connected in this case by Theorem 2.3 and the following result can be derived.

Theorem 3.6. *For $s, k \geq 3$, we have*

$$rc(G(ks; \pm 1, \pm s)) \leq \min\left\{ \lceil \frac{k}{2} \rceil + s, \lceil \frac{s+1}{2} \rceil + k - 1 \right\}.$$

Proof. Let $V(G(ks; \pm 1, \pm s)) = \{u_1, u_2, \cdots, u_{ks}\}$. By the definition of $G(ks; \pm 1, \pm s)$, we define $k + s$ cycles, named C^i and D^j ($1 \leq i \leq k, 1 \leq j \leq s$), as illustrated in Table 1. For example, the vertex set of cycle D^1 is $\{u_1, u_{s+1}, \cdots, u_{(k-1)s+1}\}$. Clearly, $|V(C^i)| = s + 1$ and $|V(D^j)| = k$ for $1 \leq i \leq k, 1 \leq j \leq s$. And C^{i_1} and C^{i_2} are edge-disjoint, D^{j_1} and D^{j_2} are edge-disjoint for $1 \leq i_1, i_2 \leq k, 1 \leq j_1, j_2 \leq s$.

We give $G(ks; \pm 1, \pm s)$ two types of edge-coloring as follows:

Type 1. We first give the cycle $D^j (1 \leq j \leq s)$ a rainbow edge-coloring with $\lceil \frac{k}{2} \rceil$ colors by Proposition 2.1, then give the cycle $C^i (1 \leq i \leq k)$ an edge-coloring with s fresh colors since the edge $u_{(j-1)s+1}u_{js+1}$ has been colored in the cycle D^1.

Type 2. We first give the cycle $C^i (1 \leq i \leq k)$ a rainbow edge-coloring with $\lceil \frac{s+1}{2} \rceil$ colors by Proposition 2.1; then give the cycle $D^j (1 \leq j \leq s)$ an edge-coloring with $k - 1$ fresh colors.

It is not hard to show that $G(ks; \pm 1, \pm s)$ is rainbow connected under each of the above two colorings. As **Type 1** and **Type 2** use $\lceil \frac{k}{2} \rceil + s$ and $\lceil \frac{s+1}{2} \rceil + k - 1$ colors, respectively, we have $rc(G(ks; \pm 1, \pm s)) \leq \min\{\lceil \frac{k}{2} \rceil + s, \lceil \frac{s+1}{2} \rceil + k - 1\}$, and the result holds. □

We finally investigate the third subfamily in which $s_1 = 2, s_2 = s, n = 2ks$ where $k \geq 1, s \geq 3$ are integers. In this case, $G(n; \pm s_1, \pm s_2)$ is connected if and only if s is odd by Theorem 2.3. We can get the following result.

Theorem 3.7. *For an integer $k \geq 1$ and an odd integer $s \geq 3$, we have*

$$rc(G(2ks; \pm 2, \pm s)) \leq \lceil \frac{ks}{2} \rceil + k.$$

Proof. Let $V(G(2ks; \pm 2, \pm s)) = \{u_1, u_2, \cdots, u_{2ks}\}$. We define $s + 2$ cycles:

$$C^i : u_i, u_{i+2}, \cdots, u_{2(ks-1)+i}, u_i (1 \leq i \leq 2)$$

$$D^j : u_j, u_{s+j}, \cdots, u_{(2k-1)s+j}, u_j (1 \leq j \leq s)$$

Clearly, $|V(C^i)| = ks$ and $|V(D^j)| = 2k$ for $1 \leq i \leq 2, 1 \leq j \leq s$. We know that C^1 and C^2 are vertex-disjoint, D^{j_1} and D^{j_2} are vertex-disjoint for $1 \leq j_1, j_2 \leq s$. Furthermore, we have $\bigcup_{1 \leq i \leq 2} V(C^i) = \bigcup_{1 \leq j \leq s} V(D^j) = V(G(2ks; \pm 2, \pm s))$, and it is not hard to show that $V(C^i) \cap V(D^j) \neq \emptyset$ for $1 \leq i \leq 2, 1 \leq j \leq s$. By Proposition 2.1, we first give $C^i (1 \leq i \leq 2)$ a rainbow edge-coloring with $\lceil \frac{ks}{2} \rceil$ colors, then give $D^j (1 \leq j \leq s)$ a rainbow edge-coloring with k fresh colors.

For any two vertices u and v, we only consider the case that $u \in C^{i_1}$ and $v \in D^{j_1}$ since the discussions for the remaining cases are similar. Let $w \in V(C^{i_1}) \cap V(D^{j_1})$. There is a rainbow $u - w$ path P_1 and a rainbow $w - v$ path P_2 in C^{i_1} and D^{j_1}, respectively. By combining P_1 and P_2, we can obtain a rainbow $u - v$ path since P_1 and P_2 are edge-disjoint and $c(P_1) \cap c(P_2) = \emptyset$. Thus, $G(2ks; \pm 2, \pm s)$ is rainbow connected and the result holds. □

Note that in [25], the authors derived that for every connected graph G of order n and minimum degree δ, $rc(G) \leq \frac{3n}{\delta+1} + 3$. From this result, we have $rc(G) \leq \frac{3n}{\delta+1} + 3 = \frac{3n}{5} + 3$ for a 4-regular graph G. By Theorem 3.6, $rc(G(ks; \pm 1, \pm s)) \leq \min\{\lceil \frac{k}{2} \rceil + s, \lceil \frac{s+1}{2} \rceil + k - 1\} \leq \max\{\frac{3k}{2}, \frac{3s}{2}\} \leq \alpha n$, where $\alpha = \max\{\frac{3}{2s}, \frac{3}{2k}\}$ and $n = ks$. Our result improves the upper bound since $\alpha \leq \frac{1}{2}$ for the case that $G = G(ks; \pm 1, \pm s)$. In fact, α can be very small for a sufficiently large k or s. By Theorem 3.7, we have $rc(G(2ks; \pm 2, \pm s)) \leq \lceil \frac{ks}{2} \rceil + k \leq \frac{ks+1}{2} + k \leq \frac{5n}{12} + \frac{1}{2}$, where $n = 2ks$. Thus, our result improves the upper bound for the case that $G = G(2ks; \pm 2, \pm s)$.

References

1. Chartrand, G., Johns, G.L., McKeon, K.A., Zhang, P.: Rainbow connection in graphs. Math. Bohem. **133**, 85–98 (2008)
2. Li, X., Sun, Y.: Rainbow Connections of Graphs. SpringerBriefs in Mathematics. Springer, New York (2012)

3. Li, X., Shi, Y., Sun, Y.: Rainbow connections of graphs–a survey. Graphs Comb. **29**(1), 1–38 (2013)
4. Bermond, J.-C., Comellas, F., Hsu, D.F.: Distributed loop computer networks: a survey. J. Parallel Distrib. Comput. **24**, 2–10 (1995)
5. Hwang, F.K.: A survey on multi-loop networks. Theor. Comput. Sci. **299**, 107–121 (2003)
6. Monakhova, E.A.: A survey on undirected circulant graphs. Discrete Math. Algorithms Appl. **4**(1), 1250002 (30 pp.) (2012)
7. Comellas, F., Mitjana, M., Peters, J.G.: Broadcasting in small-world communication networks. In: 9th Int. Coll. Structural Information and Communication Complexity (SIROCCO 9). Proc. Inform. **13**, 73–85 (2002)
8. Balaban, A.T.: Reaction graphs. In: Bonchev, D., Mekenyan, O. (eds.) Graph Theoretical Approaches to Chemical Reactivity, pp. 137–180. Kluwer Academic, Dordrecht (1994)
9. Muga, F.P., Yu, W.E.S.: A proposed topology for a 192-processor symmetric cluster with a single-switch delay. In: First Philippine Computing Science Congress, Manila, 10 pp. (2000)
10. Nesterenko, B.B., Novotarskiy, M.A.: Cellular neural networks with circulant graphs. Artif. Intell. **3**, 132–138 (2009) (in Russian)
11. Narayanan, L., Opatrny, J., Sotteau, D.: All-to-all optical routing in chordal rings of degree four. Algorithmica **31**(2), 155–178 (2001)
12. Martinez, C., Beivide, R., Stafford, E., Moreto, M., Gabidulin, E.M.: Modeling toroidal networks with the Gaussian integers. IEEE Trans. Comput. **57**(8), 1046–1056 (2008)
13. Barnes, G.H., et al.: The Illiac IV computer. IEEE Trans. Comput. **17**, 746–757 (1968)
14. Chen, M.-S., Shin, K.G., Kandlur, D.D.: Addressing, routing, and broadcasting in hexagonal mesh multiprocessors. IEEE Trans. Comput. **39**(1), 10–18 (1990)
15. Dolter, J.W., Ramanathan, P., Shin, K.G.: Performance analysis of virtual cut-through switching in HARTS: a hexagonal mesh multicomputer. IEEE Trans. Comput. **40**(6), 669–680 (1991)
16. Shin, K.G.: HARTS: a distributed real-time architecture. Computer **24**(5), 25–35 (1991)
17. Albader, B., Bose, B., Flahive, M.: Efficient communication algorithms in hexagonal mesh interconnection networks. IEEE Trans. Parallel Distrib. Syst. **23**(1), 69–77 (2012)
18. Thomson, A., Zhou, S.: Frobenius circulant graphs of valency six, Eisenstein-Jacobi networks, and hexagonal meshes. European J. Combin. **38**, 61–78 (2014)
19. Thomson, A., Zhou, S.: Frobenius circulant graphs of valency four. J. Aust. Math. Soc. **85**, 269–282 (2008)
20. Chakraborty, S., Fischer, E., Matsliah, A., Yuster, R.: Hardness and algorithms for rainbow connectivity. In: 26th International Symposium on Theoretical Aspects of Computer Science STACS 2009, pp. 243–254. Also, see J. Combin. Optim. **21**, 330–347 (2011)
21. Bondy, J.A., Murty, U.S.R.: Graph Theory. Graduate Texts in Mathematics, vol. 244. Springer, New York (2008)
22. Li, X., Sun, Y.: Rainbow connection numbers of line graphs. Ars Combin. **100**, 449–463 (2011)
23. Li, X., Sun, Y.: Upper bounds for the rainbow connection numbers of line graphs. Graphs Comb. **28**, 251–263 (2012)
24. Boesch, F.T., Tindell, R.: Circulants and their connectivity. J. Graph Theory **8**, 487–499 (1984)
25. Chandran, L.S., Das, A., Rajendraprasad, D., Varma, N.M.: Rainbow connection number and connected dominating sets. J. Graph Theory **71**, 206–218 (2012)

A Survey on Approximation Mechanism Design Without Money for Facility Games

Yukun Cheng and Sanming Zhou

Abstract In a facility game one or more facilities are placed in a metric space to serve a set of selfish agents whose addresses are their private information. In a classical facility game, each agent wants to be as close to a facility as possible, and the cost of an agent can be defined as the distance between her location and the closest facility. In an obnoxious facility game, each agent wants to be far away from all facilities, and her utility is the distance from her location to the facility set. The objective of each agent is to minimize her cost or maximize her utility. An agent may lie if, by doing so, more benefit can be obtained. We are interested in social choice mechanisms that do not utilize payments. The game designer aims at a mechanism that is strategy-proof, in the sense that any agent cannot benefit by misreporting her address, or, even better, group strategy-proof, in the sense that any coalition of agents cannot all benefit by lying. Meanwhile, it is desirable to have the mechanism to be approximately optimal with respect to a chosen objective function. Several models for such approximation mechanism design without money for facility games have been proposed. In this paper we briefly review these models and related results for both deterministic and randomized mechanisms, and meanwhile we present a general framework for approximation mechanism design without money for facility games.

Keywords Algorithmic mechanism design • Approximation mechanism design • Facility game • Obnoxious facility • Social choice

---

Y. Cheng (✉)
School of Mathematics and Statistics, Zhejiang University of Finance and Economics, Hangzhou 310018, China
e-mail: ykcheng@amss.ac.cn

S. Zhou
Department of Mathematics and Statistics, The University of Melbourne, Parkville, VIC 3010, Australia
e-mail: smzhou@ms.unimelb.edu.au

D. Gao et al. (eds.), *Advances in Global Optimization*, Springer Proceedings in Mathematics & Statistics 95, DOI 10.1007/978-3-319-08377-3_13

1 Introduction

Algorithmic mechanism design [1] deals with game-theoretic versions of optimization problems such as task scheduling, resource allocation, facility location, etc. which involve one or more selfish agents who are asked to report their private information as part of the input. A mechanism is a function that receives the information reported by the agents, and returns an outcome possibly together with a payment scheme. An agent might lie about her information if doing so increases her own benefit obtained from the outcome of the game. The goal is to design a mechanism that encourages truthfulness or strategy-proofness, on the one hand, and optimizes a related objective function, on the other hand.

In mechanism design with money, the authority can use money as compensation to the agents in order to ensure strategy-proofness. For example, the well-known Vickrey–Clarke–Groves (VCG) mechanism [2] is not only strategy-proof, but also outputs an optimal solution to the problem of maximizing the sum of all agents' utility. However, a disadvantage [3] of this mechanism is that it must return an optimal solution for a given objective function. Since it is often difficult to compute an optimal solution in polynomial time for many combinatorial optimization problems, sometimes the VCG mechanism is not efficient. Meanwhile, as pointed out by Schummer and Vohra [4], "there are many important environments where money cannot be used as a medium of compensation, due to ethical considerations (for instance, in political decision making) or legal considerations (for instance, in the context of organ donations)". Therefore, many researchers are interested in mechanisms without monetary payment.

Procaccia and Tennecholtz [5] first proposed that approximation can be used to obtain strategy-proofness without relying on payment and initiated a case study in approximation mechanism design without money based on facility games. Since then many results have been obtained by various authors on approximation mechanism design without money for facility games. The purpose of this paper is to give a brief review of known results in this area and meanwhile present a general framework based on existing models. This framework will be discussed in the next section. In Sects. 3 and 4, we will give an account of results on different models for classical facility games and obnoxious facility games, respectively. In Sect. 5 we discuss possible research problems in the area.

2 Framework

In any facility game there associates a set $N = \{1, 2, \ldots, n\}$ of agents, where i denotes the ith agent. There is also an underlying *metric space* (Ω, d) whose points are called *locations*, where as usual the metric (distance function) $d : \Omega \times \Omega \to \mathbf{R}$ is non-negative, symmetric and satisfies the triangle inequality. Each agent $i \in N$

has a *true location* $t_i \in \Omega$ that is her private information, and she reports a location $x_i \in \Omega$ that is not necessarily the same as t_i. We call $\mathbf{x} = (x_1, x_2, \ldots, x_m) \in \Omega^n$ a *location profile*.

Assume that k locations in Ω are required to be selected to put k facilities. A deterministic mechanism f outputs k facility locations in Ω according to a given location profile $\mathbf{x}$ without resorting to payments. In other words, f is a function $f : \Omega^n \to \Omega(k)$, where $\Omega(k)$ is the family of non-empty subsets of Ω of cardinality at most k, and $f(\mathbf{x})$ is the set of locations chosen for $\mathbf{x}$ by f where facilities will be put.

Given a mechanism f, each agent i has a utility $u(f(\mathbf{x}), t_i)$ whose value relies on her true location t_i and the output $f(\mathbf{x})$. Since each agent is selfish, she will try her best to maximize her utility. It may be possible for an agent to manipulate the outcome of a mechanism to obtain more benefit by misreporting her location. Therefore, from a game-theoretic perspective, an important goal is to design mechanisms that are *strategy-proof* (SP), in the sense that no agent can ever benefit from reporting a false location regardless of the strategies of other agents. Sometimes we may wish to design mechanisms that are even *group strategy-proof* (GSP), in the sense that whenever a coalition of agents lies, at least one of the members of the coalition does not gain extra benefit from the deviation.

In approximation mechanism design, we are interested in (group) strategy-proof mechanisms that are approximately optimal with respective to a given objective function, where approximation is understood in the usual sense by looking at the worst-case ratio between the optimal objective value and the value of the mechanisms solution to the underlying maximization problem. Based on different conditions, such as the structure of metric spaces, the type of facilities, the number of the facilities, etc., several models of facility games have been proposed. In the following we summarize the major components in facility games.

Metric Space. So far only the following two types of metric spaces have been considered in the literature.

Network Models: In this model a graph G with each edge having a non-negative weight is involved. We may think of G as being realized as a geometric graph (in $\mathbf{R}^3$, for example) such that the weight of each edge represents its length. The metric space Ω is the set of points of G, including both vertices of G and points on its edges, and the distance $d(x, y)$ between $x \in \Omega$ and $y \in \Omega$ is the length of a shortest path connecting x and y in G. We usually write G in place of Ω in this case.

Euclidean Metric Space: In this case $\Omega = \mathbf{R}^m$ for some integer $m \geq 1$ and the distance d is the usual Euclidean distance in $\mathbf{R}^m$.

Number of Facilities. Two cases have been distinguished in the literature:

$k = 1$: In this case the unique facility provides service to all agents.

$k > 1$: In this case a set Y of k locations is required, and an agent is served by the closest facility, namely, a location achieving the distance $d(Y, t_i) := \min_{y \in Y} d(y, t_i)$ between agent i and Y.

Type of Facilities. So far only the following two types of facilities have been considered in the literature:

Desirable Facility: In this case all facilities (e.g., library, school, etc.) are desirable, and each agent wants to be as close to one of the facilities as possible. As such it is reasonable to assume that the utility $u(f(\mathbf{x}), t_i)$ is a monotonically decreasing function of $d(f(\mathbf{x}), t_i)$ with only one peak. Since all facilities are desirable, we may set $cost(f(\mathbf{x}), t_i) := -u(f(\mathbf{x}), t_i)$ and call it the *cost function* of agent i. So far only the simplest case where $u(f(\mathbf{x}), t_i) = -d(f(\mathbf{x}), t_i)$ for each i has been studied in the literature.

Obnoxious Facility: In this case all facilities (e.g. garbage dump, etc.) are obnoxious, and each agent wants to be far away from all facilities. Thus the utility $u(f(\mathbf{x}), t_i)$ may be assumed as a monotonically increasing function of $d(f(\mathbf{x}), t_i)$ with only one dip. The simplest case where $u(f(\mathbf{x}), t_i) = d(f(\mathbf{x}), t_i)$ for each i has received most attention up to now.

According to whether the facilities are desirable or obnoxious, we call a facility game *classical* or *obnoxious*; each agent aims to minimize her cost or maximize her utility, respectively.

Strategy-Proofness. A mechanism f is *strategy-proof* if for any $\mathbf{x} \in \Omega^n$ and every i, we have $u(f(\mathbf{x}), t_i) \leq u(f(\mathbf{x}_{-i}, t_i), t_i)$, where $(\mathbf{x}_{-i}, t_i)$ is obtained from $\mathbf{x}$ by replacing x_i by t_i but keeping all other coordinates. As a stronger requirement, f is called *GSP* if for any $\mathbf{x} \in \Omega^n$ and $I \subseteq N$, we have $u(f(\mathbf{x}), t_i) \leq u(f(\mathbf{x}_{-I}, t_I), t_i)$ for at least one $i \in I$, where $(\mathbf{x}_{-I}, t_I)$ is obtained from $\mathbf{x}$ by replacing x_i by t_i for every $i \in I$ but retaining all other coordinates.

Type of Mechanisms. Two different types of mechanisms have been studied:

Deterministic Mechanism: This was discussed in the beginning of this section.

Randomized Mechanism: A randomized mechanism is a function $f : \Omega^n \to \Delta(\Omega(k))$ where $\Delta(\Omega(k))$ is the set of probability distributions over $\Omega(k)$. In the simplest case, the expected value $E_{Y \sim f}[d(Y, t_i)]$ may be defined as the cost or the utility of agent i in classical facility games or obnoxious facility games, respectively.

Objective Function. The decision maker (or the mechanism designer) is interested in (group) strategy-proof mechanisms that also do well with respect to optimizing a given objective function.

Similar to the k-median and k-center problems [6–8], for classical facility games researchers have so far considered minimizing the *social cost* $SC(f, \mathbf{x}) := \sum_{i=1}^{n} cost(f(\mathbf{x}), t_i)$ or the *maximum cost* $MC(f; \mathbf{x}) := \max_{i=1,\ldots,n} cost(f(\mathbf{x}), t_i)$. Similar to the k-maxisum and k-maximin problems [9–11], for obnoxious facility games researchers have considered maximizing the *obnoxious social welfare* $SW(f; \mathbf{x}) := \sum_{i=1}^{n} u(f(\mathbf{x}), t_i)$ or the *minimum utility* $MU(f; \mathbf{x}) := \min_{i=1,\ldots,n} u(f(\mathbf{x}), t_i)$.

In summary, a facility game consists of: a set N of n agents; a metric space (Ω, d) which may be continuous or discrete; a subset $\{t_1, \ldots, t_n\}$ of Ω, t_i being the true location of agent i; a set of k facilities to be installed at k (not necessarily distinct) locations in Ω; a utility function $u : \Omega(k) \times \Omega \to \mathbf{R}$ taking non-negative values which usually relies on $d(f(\mathbf{x}), t)$, where $x = (x_1, \ldots, x_n) \in \Omega^n$, $t \in \Omega$, and $f : \Omega^n \to \Omega(k)$ is a deterministic mechanism; and a non-negative objective function $F : \Omega(k) \times \Omega^n \to \mathbf{R}$ to be maximized, which is usually defined in terms of $u(f(\mathbf{x}), t_i)$, $1 \leq i \leq n$. We are interested in designing a mechanism f that is strategy-proof or even GSP on the one hand, and on the other hand outputs a good solution for any location profile in the sense that the approximation ratio

$$\sup_{x \in \Omega^n} \frac{\max_{Y \in \Omega(k)} F(Y, \mathbf{x})}{F(f(\mathbf{x}), \mathbf{x})}$$

is as small as possible. Different specification of the components above gives rise to different models for approximation (deterministic) mechanism design without money for facility games.

In approximation randomized mechanism design, the distance function, the utility function, and the objective function are all random, and we can give a similar framework by considering the expected values of the corresponding random variables.

3 Classical Facility Games

3.1 *Single Facility Games*

In the case $k = 1$, the preferences are *single peaked* in the sense that the outcome is less preferred by each agent when it is further from her ideal locations. Beginning with [12], single peaked preferences and their extensions have been extensively studied in the social choice literature. In this subsection, we summarize known results on finding a facility location in different metric spaces that minimizes the social cost or the maximum cost.

If the objective is to minimize the social cost, Procaccia and Tennecholtz [5] proposed a GSP optimal mechanism which returns the location of the median agent as the facility location when all agents are located on a path. This mechanism is GSP since an agent can manipulate the output only by misreporting her location to be on the opposite side of the median. Moreover, the median also minimizes the social cost, because for any location with distance $\epsilon > 0$ to the median, at most $\lfloor n/2 \rfloor$ agents are within distance ϵ to the facility and all other agents are away from the facility by at least ϵ. Similarly, if the graph is a tree, Alon et al. [13] gave a mechanism that outputs the median of the tree as the facility's location. Such a mechanism is also an optimal GSP mechanism.

When all agents are located on a graph G containing a cycle C, Schummer and Vohra [14] showed that if a deterministic mechanism $f : G^n \to G$ is an SP mechanism that is onto G, then there is a cycle dictator, that is, there exists $i \in N$ such that for all $\mathbf{x} \in C^n$, $f(\mathbf{x}) = x_i$. Based on such a characterization, Alon et al. [13] obtained a tight SP lower bound of $n - 1$ on the approximation ratio for any graph G that contains a cycle. For the randomized version, they designed a mechanism which returns a facility location x_i, $i \in N$, with probability $1/n$. This mechanism is SP with approximation ratio $2 - (2/n)$ for any general graph. They showed further that such a mechanism is GSP if and only if the maximum degree of the graph is two.

If the objective is to minimize the maximum cost, the problem of designing an SP mechanism is simpler compared with deterministic mechanisms. Since Schummer and Vohra [14] showed that strategy-proofness can only be obtained by dictatorship, Alon et al. [13] considered the mechanism given by $f(\mathbf{x}) = x_1$ for all $\mathbf{x} \in G^n$, that is, agent 1 is a dictator. It can be proved that such a mechanism is a GSP 2-approximation mechanism. On the other hand, Procaccia and Tennecholtz [5] showed that a deterministic SP mechanism cannot achieve an approximation ratio better than 2 even if the underlying graph G is a path. Thus the dictatorship mechanism of Alon et al. [13] has the best approximation ratio with respect to the maximum cost. Procaccia and Tennenholtz [5] proved that a randomized SP mechanism has approximation ratio at least $3/2$ on a path. They also gave a matching GSP upper bound of $3/2$ by using the *Left-Right-Middle* (LRM) Mechanism, which, for a given $\mathbf{x} \in G^n$, chooses $\min_i x_i$ and $\max_i x_i$ with probability $1/4$, respectively, and chooses the midpoint of the interval $[\min_i x_i, \max_i x_i]$ with probability $1/2$. When the agents are on a circle, Alon et al. [13] proposed a randomized SP mechanism with approximation ratio $3/2$ that combines two mechanisms: the LRM mechanism if the agents are located on one semicircle, and the Random Center Mechanism otherwise. When G is a tree, they showed that there is a randomized SP $(2 - \frac{2}{n+2})$-approximation mechanism that, for a given $\mathbf{x} \in G^n$, outputs x_i for each $i \in N$ with probability $1/(n+2)$ and the center of the tree with probability $2/(n+2)$. They further proved that $2 - O\left(\frac{1}{2^{\sqrt{\log n}}}\right)$ is a lower bound on the approximation ratio for any SP randomized mechanism.

Procaccia and Tennecholtz [5] considered a natural extension of the classical single facility games, in which one facility should be located but each agent controls multiple locations. As before, the objective is to minimize the social cost or the maximum cost. However, the cost of an agent now depends on the objective function. If the objective is to minimize the social cost, the cost of agent i is defined as $cost(y, \mathbf{x}_i) = \sum_{j=1}^{w_i} d(y, x_{ij})$, where y is the location of the facility and $\mathbf{x}_i = (x_{i1}, \ldots, x_{iw_i})$ is the location set controlled by agent i. If the objective is to minimize the maximum cost, the cost of agent i is $cost(y, \mathbf{x}_i) = \max_{j=1,\ldots,w_i} d(y, x_{ij})$. For the social cost, they directly applied the deterministic mechanism by Dekel et al. [15] that returns the median $med(\mathbf{x}')$ of $\mathbf{x}' = (med(\mathbf{x}_1), \ldots, med(\mathbf{x}_n))$. Dekel et al. [15] also showed that this mechanism is a GSP 3-approximation mechanism and provided a matching lower bound. Furthermore, Procaccia and Tennecholtz [5] designed a simple randomized mechanism to return $med(\mathbf{x}_i)$ with probability $\frac{w_i}{\sum_{j \in N} w_j}$. This mechanism is SP, and when $n = 2$ its approximation ratio is $2 + \frac{|w_1 - w_2|}{w_1 + w_2}$. Subsequently, Lu et al. [16] extended the result about the approximation ratio to $3 - \frac{2 \min_{j \in N} w_j}{\sum_{j \in N} w_j}$ for any n and obtained the lower bound 1.33 by solving a related linear programming problem. For the maximum cost, they proposed a GSP 2-approximation deterministic mechanism and a $(3/2)$-approximation randomized mechanism. Since the multiple location setting is the same as the simple setting stated before when $w_i = 1$, $i \in N$, any lower bound for the simple setting holds here as well.

3.2 Two-Facility Games

When the objective function is the social cost and the network is a path, Procaccia and Tennecholtz [5] showed that the mechanism that outputs an optimal solution for a given $\mathbf{x} \in G^n$ is not strategy-proof. They gave the following GSP $(n-1)$-approximation mechanism: choose the leftmost and the rightmost points, and constructed an instance to show that $3/2$ is a lower bound on the approximation ratio for any SP deterministic mechanism. Later, Lu et al. [16] improved such lower bound to 2, designed a randomized $n/2$-approximation mechanism, and explored a lower bound of 1.045 for randomized mechanisms. Moreover, Lu et al. [17] proved that the $(n-1)$-approximation deterministic mechanism given in [5] is asymptotically optimal. They constructed an instance on a path and explored the lower bound $(n-1)/2$ on the approximation ratio by employing two key concepts: partial group strategy-proofness and image set. In the case when all agents are on a circle, they designed a GSP deterministic mechanism with an $(n-1)$-approximation ratio which asymptotically matches the lower bound $(n-1)/2$. Lu et al. [17] also obtained an SP 4-approximation randomized mechanism called the *Proportional Mechanism*: the first facility is allocated uniformly over all reported locations; the second facility is assigned to another reported location with probability proportional to its distance to the first facility.

If the objective is to minimize the maximum cost, only Procaccia and Tennecholtz [5] contributed some positive results in the case when all agents are on a path. For the deterministic version, they applied the same deterministic mechanism as the one for the social cost model. By exploring the characterization of the structure of the optimal solution, they proved that the approximation ratio of such a mechanism is 2, and provided a matching SP lower bound. Furthermore, they designed a randomized SP $5/3$-approximation mechanism. Compared with the deterministic case, the randomized mechanism for this model is much more complicated and the authors applied some new ideas: randomizing over two equal intervals, unbalanced weights at the edges, and correlation between the two facilities. These strategies play a crucial role in satisfying the delicate strategy-proof constraints and break the deterministic lower bound of 2. The lower bound of any randomized SP mechanism is proved to be $3/2$.

3.3 k-Facility Games with $k \geq 3$

For k-facility games with $k \geq 3$, most known results focus on the objective of minimizing the social cost. McSherry and Talwar [19] first used differentially private algorithms as almost strategy-proof approximate mechanisms. The main advantage of such an algorithm is that it can control any agent's influence on the outcome so that any agent has limited motivation to lie. McSherry and Talwar presented a general differentially private mechanism that approximates the optimal social cost within an additive logarithmic term. Unfortunately, the running time of this general mechanism is randomized exponential-time. Subsequently, Gupta et al. [18] presented a computationally efficient differentially private algorithm for several combinatorial optimization problems. Based on [19], Nissim et al. [20] considered *imposing mechanisms* which can penalize liars by restricting the set of allowable post-actions for the agents. They combined the differentially private mechanisms of [19] with an imposing mechanism and obtained a randomized imposing SP mechanism with a running time in k for k-facility location. The mechanism approximates the optimal average social cost, namely the optimal social cost divided by n, within an additive term of roughly $1/n^{\frac{1}{3}}$.

In contrast to [20], Fotakis and Tzamos [21] tried to design an SP mechanism with standard multiplicative notion of approximation. They considered the *winner-imposing* mechanism which chooses k reported locations of agents to build facilities. If an agent's reported location is chosen to put a facility, then she is served by this facility and her service cost is the distance between this facility and her true location. If an agent's reported location is not chosen, then she is served by a facility closest to her true location. Thus the winner-imposing mechanism can penalize an agent without money only if she succeeds in gaining more benefit in the mechanism. Fotakis and Tzamos proved that the winner-imposing version of the Proportional Mechanism in [17] is an SP $4k$-approximation randomized mechanism. Moreover, they addressed the *facility location game* in which there is a uniform facility opening cost, instead of a fixed number of facilities. The authority should place some facilities so as to minimize the social cost and the total facility opening cost. For this game, they showed that the winner-imposing version of Meyerson's randomized algorithm in [22] is an SP 8-approximation mechanism. Meanwhile, they presented a deterministic nonimposing GSP $O(\log n)$-approximation mechanism when all agents are on a path. In addition, Escoffier et al. [23] considered a facility game to locate $n-1$ facilities to n agents. They studied such a game in the general metric space and trees for the social cost and the minimum cost, and provided lower and upper bounds on the approximation ratio of deterministic and randomized SP mechanisms.

4 Obnoxious Facility Games

For obnoxious facility games on a path, the preferences are known as *single-dipped*, meaning that the worse allocation for each agent is the one that places the facility right by their home, and that locations become better as they are further away. In the past a few years, a lot of work [24–28] was focused on characterizations of the strategy-proofness for the single-dipped preference. Cheng et al. [29] initially studied approximation design without money for obnoxious facility games with the objective of maximizing the obnoxious social welfare. In this section, we survey some known results in this domain.

Cheng et al. first proposed GSP mechanisms to locate one facility with respect to different network topologies. In particular, if all agents are on a path, they viewed such a path as an interval with left endpoint a and right endpoint b. Since this model is related to the literature on approximation algorithms for the 1- maxian problem [10, 11, 30, 31] from an algorithmic perspective, it is well known that one of the two endpoints must be an optimal facility location for $\mathbf{x} \in G^n$. Thus they regarded the two endpoints as the candidates for the facility locations and designed a GSP 3-approximation deterministic mechanism, which outputs a if the number n_2 of agents on the right-hand side of the interval is larger than the number n_1 of agents on the left-hand side, and b otherwise. By a similar idea, they presented two GSP deterministic mechanisms, respectively, when all agents are on a tree or a circle, and proved that the approximation ratio of each mechanism is 3. Later, Han et al. [25] provided the matching strategy-proof lower bounds for each model on different networks. Furthermore, when all agents are on an interval, Cheng et al. [29] also gave a randomized mechanism which returns a and b with probability α and $1-\alpha$, respectively, where $\alpha = \frac{2n_1n_2n_2^2}{n_1^2+n_2^2+4n_1n_2}$. They proved that such a randomized mechanism is GSP and has achievable approximation ratio $3/2$. When all agents are on a general network, a GPS 4-approximation deterministic mechanism and a trivial GSP 2-approximation randomized mechanism were derived. In addition, the deterministic mechanism was shown to be asymptotically optimal by using the characterization of strategy-proofness for general networks [26].

Recently, Cheng et al. [32] considered a new model of obnoxious facility games that has a bounded service range. In this model each facility can only serve the agents within its service range due to the limited service ability. Each agent wants to be far away from the facilities. On the other hand, she must stay within at least one facility's range, otherwise she cannot receive any service. Cheng et al. first studied the case when all agents are on an interval, which is normalized as $[0, 1]$, and the service radius is some r with $1/2 \leq r \leq 1$. Compared with the previous model without service range, this new model is more complicated since more than one facilities may be needed and it is no longer true that one of the endpoints must be an optimal solution. According to the value of r, Cheng et al. selected different candidates for the facility locations. To be specific, if $3/4 \leq r \leq 1$, points r or $1-r$ are designated as the facility locations; otherwise, they locate one facility at $1/2$ or two facilities at 0 and 1, respectively. Thus they designed a GSP deterministic mechanism and a GSP randomized mechanism. When $1/2 \leq r < 3/4$ or $3/4 \leq r \leq 1$, the approximation ratio of their deterministic mechanism is $8r-1$ or $\frac{2r+1}{2r-1}$, respectively, and the approximation ratio of their randomized mechanism is $4r$ or $\frac{2r}{2r-1}$, respectively. Meanwhile, they also proved a lower bound for any strategy-proof deterministic mechanism by constructing different instances, which is equal to $4r-1$ if $1/2 \leq r < 3/4$, $1/(2r-1)$ if $3/4 \leq r < 5/6$ and $3r-1$ if $5/6 \leq r < 1$.

5 Conclusion

We reviewed some known results on approximation mechanism design without money for facility games. By comparing our general framework in Sect. 2 and what we surveyed in Sects. 3 and 4, it should be clear that a lot of interesting problems remain open and different models may be considered by specifying the components in the framework. For example, one may investigate various cases where the space of locations is more involved, such as a multi-dimensional Euclidean space or a specific network other than paths, trees and cycles.

For obnoxious facility games, except the results in [25] there are no other results in the case when the objective is to maximize the minimum utility. Han and Du proved that there is no any SP deterministic mechanism with finite approximation ratio for this objective. We believe that results can be obtained by using the differentially private algorithm mentioned in Sect. 3.3 to design almost SP mechanisms.

For facility games with a limited service ability, the only known result is about the obnoxious facility game on interval $[0, 1]$ with a service range $1/2 \leq r \leq 1$. It would be interesting to investigate classical and obnoxious facility games with different types of restrictions to service ability in different metric spaces. In particular, one may consider the obnoxious facility game on $[0, 1]$ when $0 < r < 1/2$. It seems challenging to find a general SP mechanism corresponding to the value of r.

Finally, closing the gap between the lower and upper bounds on the approximation ratios of deterministic or randomized mechanisms for some models is also a significant research problem.

Acknowledgements Research was partially supported by the Nature Science Foundation of China (No. 11301475) and the Nature Science Foundation of Zhejiang Province, China (No. LQ12A01011).

References

1. Nisan, N., Ronen, A.: Algorithmic mechanism design. Game Econ. Behav. **35**(1–2), 166–196 (2001)
2. Nisan, N.: Introduction to mechanism design (for computer scientists). In: Nisan, N., Roughgarden, T., Tardos, E.,Vazirani, V. (eds.) Algorithmic Game Theory, Chap. 9. Cambridge University Press, Cambridge (2007)
3. Rothkopf, M.: Thirteen reasons the Vickrey-Clarke-Groves process is not practical. Oper. Res. **55**(2), 191–197 (2007)
4. Schummer, J., Vohra, R.V.: Mechanism design without money. In: Nisan, N., Roughgarden, T., Tardos, E.,Vazirani, V. (eds.) Algorithmic Game Theory, Chap. 10. Cambridge University Press, Cambridge (2007)
5. Procaccia, A.D., Tennenholtz, M.: Approximate mechanism design without money. In: 10th ACM Conference on Electronic Commerce, pp. 177–186. ACM, New York (2009)
6. Drezner, Z., Hamacher, H.: Facility Location: Applications and Theory. Springer, Berlin (2002)

7. Kariv, O., Hakimi, S.L.: An algorithmic approach to network location poblems. I. The p-centers. SIAM J. Appl. Math. **37**, 441–461 (1979)
8. Kariv, O., Hakimi, S.L.: An algorithmic approach to network location poblems. II. The p-medians. SIAM J. Appl. Math. **37**, 539–560 (1979)
9. Cappanera, P.: A survey on obnoxious facility location problems. Technical Report: TR-99-11 (1999). Available via DIALOG. http://citeseerx.ist.psu.edu/viewdoc/summary?doi=10.1.1.36.2783
10. Tamir, A.: Obnoxious facility location on graphs. SIAM J. Discrete Math. **4**, 550–567 (1991)
11. Zelinka, B.: Medians and peripherian of trees. Arch. Math. **4**, 87–95 (1968)
12. Moulin, H.: On strategy-proofness and single-peakedness. Public Choice **35**, 437–455 (1980)
13. Alon, N., Feldman, M., Proccia, A.D., Tennenholtz, M.: Strategyproof approximation mechanisms for location on networks. Computing Research Repository-CORR, abs/0907.2049 (2009)
14. Schummer, J., Vohra, R.V.: Strategy-proof location on a network. J. Econ. Theory **104**(2), 405–428 (2004)
15. Dekel, O., Fischer, F., Procaccia, A.D.: Incentive compatible regression learning. J. Comput. Syst. Sci. **76**(8), 759–777 (2010)
16. Lu, P., Wang, Y., Zhou, Y.: Tighter boundes for facility games. In: Leonardi, S. (ed.) WINE 2009. Lecture Notes in Computer Science, vol. 5929, pp. 137–148. Springer, Heidelberg (2009)
17. Lu, P., Sun, X., Wang, Y., Zhu, Z.: Asymptotically optimal strategy-proof mechanisms for two-facility games. In: 11th ACM Conference on Electronic Commerce, pp. 315–324. ACM, New York (2010)
18. Gupta, A., Ligett, K., McSherry, F., Roth, A., Talwar, K.: Differentially private combinatorial optimization. In: SODA 2010: Proceedings of the Twenty-First ACM-SIAM Symposium on Discrete Algorithms, pp. 1106–1125 (2010)
19. McSherry, F., Talwar, K.: Mechanism design via differential privacy. In: FOCS 2007: Proceedings of the Forty-Eighth IEEE Symposium on Foundations of Computer Science, pp. 94–103 (2007)
20. Nissim, K., Smorodinsky, R., Tennenholtz, M.: Approximately optimal mehcanism design via differential privacy. Computing Research Repository-CORR, abs/1004.2888 (2010)
21. Fotakis, D., Tzamos, C.: Winner-imposing strategy-proof mechanisms for multiple facility location games. In: Saberi, A. (ed.) WINE 2010. Lecture Notes in Computer Science, vol. 6484, pp. 234–245. Springer, Heidelberg (2010)
22. Meyerson, A.: Online facility location. In: FOCS 2001: Proceedings of the Forty-Second IEEE Symposium on Foundations of Computer Science, pp. 426–431 (2001)
23. Escoffier, B., Gourves, L., Thang, N., Pascual, F., Spanjaard, O.: Strategy-proog mechanisms for facility location games with many facilities. In: Brafman, R.I., Roberts, F.S., Tsoukiàs, A. (eds.) ADT 2011. Lecture Notes in Computer Science, vol. 6992, pp. 67–81. Springer, Heidelberg (2011)
24. Barberà, S., Berga, D., Moreno, B.: Single-dipped preferences. Working paper (2009)
25. Han, Q., Du, D.: Moneyless strategy-proof mechanism on single-dipped policy domain: characerization and applications. Working paper (2012)
26. Ibara, K., Nagamochi, H.: Charactering mechanisms in obnoxious facility game. In: Lin, G. (ed.) COCOA 2012. Lecture Notes in Computer Science, vol. 7402, pp. 301–311. Springer, Heidelberg (2012)
27. Manjunath, V.: Efficient and Strategy-Proof Social Choice When Preferences are Singledipped. Mimeo (2009)
28. Peremans, W., Storcken, T.: Strategy-proofness on single-dipped preferences domains. In: Proceedings of the International Conference, Logic, Game Theory, and Social Choice, pp. 296–313 (1999)
29. Cheng, Y., Yu, W., Zhang, G.: Strategy-proof approximation mechanisms for an obnoxious facility game on networks. Theor. Comput. Sci. **497**, 154–163 (2013)

30. Church, R., Garfinkel, R.: Locating an obnoxious facility on a network. Transp. Sci. **12**, 107–118 (1978)
31. Ting, S.: A linear-time algorithm for maxisum facility location on tree networks. Trans. Sci. **18**, 76–84 (1984)
32. Cheng, Y., Han, Q., Yu,W., Zhang, G.: Obnoxious facility game with a bounded service range. In: Chan, T.-H.H., Lau, L., Trevisan, L. (eds.) TAMC 2013. Lecture Notes in Computer Science, vol. 7876, pp. 272–281. Springer, Heidelberg (2013)

Approximation Algorithms for the Robust Facility Location Problem with Penalties

Fengmin Wang, Dachuan Xu, and Chenchen Wu

Abstract In this paper, we consider the robust facility location problem with penalties, aiming to serve only a specified fraction of the clients. We formulate this problem as an integer linear programming to identify which clients must be served. Based on the corresponding LP relaxation and dual program, we propose a primal-dual 3-approximation algorithm. Combining the greedy augmentation procedure, we further improve the above approximation ratio to 2.

Keywords Facility location problem • Approximation algorithm • Primal-dual • Greedy augmentation

1 Introduction

The classical uncapacitated facility location problem (UFLP), first formulated in the early 1960s, has received widespread attention in the operations research and computer science community [1, 2]. It aims to open some facilities from the given location set to serve all the given clients, so that the sum of facility opening cost and serving cost is minimized. Since the UFLP is one of the classical NP-hard problems, recent works have mainly concentrated on designing approximation algorithms for it [3–11]. Among the existing approximation algorithms for the UFLP, the first constant factor is 3.16 proposed by Shmoys et al. [10]; the currently best factor is 1.488 achieved by Li [8]. It is well known that the lower bound for the UFLP is 1.463 [5].

Due to practical application, various variants of the UFLP are considered in the literatures [12–22]. To model the case when there are a few very distant clients

F. Wang • D. Xu (✉)
Department of Applied Mathematics, Beijing University of Technology, 100 Pingleyuan, Chaoyang District, Beijing 100124, People's Republic of China
e-mail: wfm@emails.bjut.edu.cn; xudc@bjut.edu.cn

C. Wu
College of Science, Tianjin University of Technology, Tianjin 300384, People's Republic of China
e-mail: chenchen86711@gmail.com

D. Gao et al. (eds.), *Advances in Global Optimization*, Springer Proceedings in Mathematics & Statistics 95, DOI 10.1007/978-3-319-08377-3_14

(named outliers) for the majority of the commercial applications of the UFLP, Charikar et al. [14] proposed two variants of the UFLP, i.e., robust facility location problem (RFLP) and facility location problem with penalties (FLPWP). In the RFLP, given n clients and an integer parameter $q < n$, we need to make sure that at least $n - q$ clients are served while leaving out the rest which are called outliers. The objective is to minimize the sum of the opening cost and the connection cost. Charikar et al. [14] presented a primal-dual 3-approximation algorithm for the RFLP. In the FLPWP, each client has a penalty cost and we will provide service to part of the clients while penalizing the rest. The objective is to minimize the sum of the opening cost, the connection cost, and the penalty cost. After the primal-dual 3-approximation algorithm given by Charikar et al. [14] for the FLPWP, Xu and Xu [17, 18] presented an LP-rounding $(2 + 2/e)$-approximation algorithm, and then, combining the power of the primal-dual method and greedy augmentation techniques, they provided an 1.8526-approximation algorithm. Li et al. [23] presented an LP rounding 1.514-approximation algorithm which has the currently best ratio for the FLPWP.

In this paper, we consider the robust facility location problem with penalties (RFLPWP) in which not all clients are required to be served. Given a parameter q, the RFLPWP aims to serve only a specified fraction of the clients, penalize some clients, and ignore at most q outliers. The objective is to minimize the sum of the opening cost, the connection cost, and the penalty cost. We extend the primal-dual method in [7] for the UFLP to a modified instance of the RFLPWP, similar to the one in [14], and obtain a 3-approximation algorithm for the RFLPWP. Combining the greedy augmentation procedure [3, 5], we further improve the above approximation ratio to 2.

The rest of this paper is organized as follows. In Sect. 2, we present some preliminaries including the integer program, the linear programming relaxation, and the dual program for the RFLPWP. In Sect. 3, we provide and analyze the primal-dual algorithm. Finally, some discussions are given in Sect. 4.

2 Preliminaries

In the RFLPWP, given a facility set $\mathscr{F}$ and a client set $\mathscr{C}$, each client j has a penalty cost p_j. The opening cost of facility $i \in \mathscr{F}$ is f_i. The metric connection cost between facility $i \in \mathscr{F}$ and client $j \in \mathscr{C}$ is c_{ij}. We are also given q, the number of the outliers. Our objective is to determine an opening facility set $\hat{\mathscr{F}} \subseteq \mathscr{F}$, while selecting a penalized client set $\hat{\mathscr{P}} \subseteq \mathscr{C}$, an outlier set $\hat{\mathscr{O}} \subseteq \mathscr{C}(\hat{\mathscr{O}} = q)$, and then connect the clients in $\mathscr{C} \backslash (\hat{\mathscr{P}} \bigcup \hat{\mathscr{O}})$ to the opening facilities in $\hat{\mathscr{F}}$, such that the sum of the opening cost, the connection cost, and the penalty cost is minimized.

We introduce four types of $\{0, 1\}$ variables: y_i indicating whether facility i is opened; x_{ij} indicating whether client j is connected to facility i; z_j indicating whether client j is penalized; and r_j indicating whether client j is an extra outlier. The RFLPWP is formulated as

$$
\text{(IP)} \qquad
\begin{aligned}
\min \ & \sum_{i \in \mathscr{F}} f_i y_i + \sum_{i \in \mathscr{F}} \sum_{j \in \mathscr{C}} c_{ij} x_{ij} + \sum_{j \in \mathscr{C}} p_j z_j \\
\text{s. t.} \ & \sum_{i \in \mathscr{F}} x_{ij} + z_j + r_j \geq 1, \quad \forall j \in \mathscr{C}, \\
& x_{ij} \leq y_i, \quad \forall i \in \mathscr{F}, j \in \mathscr{C}, \\
& \sum_{j \in \mathscr{C}} r_j \leq q, \\
& x_{ij}, y_i, z_j, r_j \in \{0, 1\}, \quad \forall i \in \mathscr{F}, j \in \mathscr{C}.
\end{aligned}
\tag{1}
$$

In the above program, the first constraints denote that each client $j \in \mathscr{C}$ is connected to a facility or penalized or ignored as an outlier; the second constraints ensure that if client j is connected to facility i, then this facility must be opened; the third constraints indicate that there are at most q outliers. Relaxing the last constraints, we obtain the LP relaxation.

$$
\text{(LP)} \qquad
\begin{aligned}
\min \ & \sum_{i \in \mathscr{F}} f_i y_i + \sum_{i \in \mathscr{F}} \sum_{j \in \mathscr{C}} c_{ij} x_{ij} + \sum_{j \in \mathscr{C}} p_j z_j \\
\text{s. t.} \ & \sum_{i \in \mathscr{F}} x_{ij} + z_j + r_j \geq 1, \quad \forall j \in \mathscr{C}, \\
& x_{ij} \leq y_i, \quad \forall i \in \mathscr{F}, j \in \mathscr{C}, \\
& \sum_{j \in \mathscr{C}} r_j \leq q, \\
& x_{ij}, y_i, z_j, r_j \geq 0, \quad \forall i \in \mathscr{F}, j \in \mathscr{C}.
\end{aligned}
\tag{2}
$$

Introducing the dual variables α_j, β_{ij} and θ, we obtain the dual of the program (LP)

$$
\text{(DP)} \qquad
\begin{aligned}
\max \ & \sum_{j \in \mathscr{C}} \alpha_j - q\theta \\
\text{s. t.} \ & \alpha_j \leq \beta_{ij} + c_{ij}, \quad \forall i \in \mathscr{F}, j \in \mathscr{C}, \\
& \sum_{j \in \mathscr{C}} \beta_{ij} \leq f_i, \quad \forall i \in \mathscr{F}, \\
& \alpha_j \leq p_j, \quad \forall j \in \mathscr{C}, \\
& \alpha_j \leq \theta, \quad \forall j \in \mathscr{C}, \\
& \alpha_j, \beta_{ij}, \theta \geq 0, \quad \forall i \in \mathscr{F}, j \in \mathscr{C},
\end{aligned}
\tag{3}
$$

where α_j can be viewed as the budget of client j, and β_{ij} as the contribution of client j to facility i.

3 Main Results

In this section, we will first propose a primal-dual algorithm for the RFLPWP, then analyze the algorithm to obtain the approximation ratio of 3.

3.1 The Primal-Dual Algorithm

Algorithm 1 (The primal-dual algorithm).

Step 0 *Constructing a new instance.*
Since there is an unbounded integrality gap for (LP), we guess the most expensive facility cost in the optimal solution, say $f_{\max}$. *We set* $f_{\max} := 0$ *and the facility cost greater than* $f_{\max}$ *(the nonzero value in the original instance) to* ∞. *Let us denote this new instance as* $\mathscr{I}^{(1)}$. *For the instance* $\mathscr{I}^{(1)}$, *we run the following steps.*

Step 1 *Let us introduce time* t. *The algorithm starts at time* $t = 0$. *Initially all the dual variables are zero, all the facilities are closed, and all clients are unfrozen. In the process of the algorithm, client* j *becomes frozen when the dual variable* α_j *stops increasing. Let* $\tilde{\mathscr{F}}$ *denote the temporarily open facility set,* U *denote the unfrozen client set,* $\tilde{\mathscr{P}}$ *denote the temporarily penalized client set, and* $\tilde{\mathscr{O}}$ *denote the outlier set. For each* $i \in \mathscr{F}$, *denote* N_i^{wit} *to be the set of the clients whose connecting witness is facility* i *(we will explain the connecting witness at Step 2.2). At the beginning of the algorithm, set* $\tilde{\mathscr{F}} := \emptyset$, $U := \mathscr{C}$, $\tilde{\mathscr{P}} := \emptyset$, $\tilde{\mathscr{O}} := \emptyset$, $N_i^{wit} := \emptyset$ *for all* $i \in \mathscr{F}$.

Step 2 *Constructing a dual feasible solution* (α, β, θ).
For the unfrozen client $j \in U$, *we increase* α_j *at the same rate with time* t.

Step 2.1 *If* $|U| > q$, *go to Step 2.2. Otherwise, freeze* $j \in U$, *let* $\tilde{\mathscr{O}} := U$ *and* $U := \emptyset$. *We denote this time by* t_q. *Let* $\theta := t_q$. *Go to Step 3.*

Step 2.2 *As time goes on, the following events will happen. If several events happen simultaneously, we execute the algorithm in arbitrary order.*

Event 1 *There is a client* $j \in U$ *and a facility* $i \in \mathscr{F}$, *such that* $\alpha_j = c_{ij}$.

Event 1.1 *If the facility* $i \in \tilde{\mathscr{F}}$, *we say client* j touches *the facility* $i \in \tilde{\mathscr{F}}$. *Set* $i(j) := i$ *and call* $i(j)$ *the* connecting witness *of client* j. *Freeze* j, *and update* $N_i^{wit} := N_i^{wit} \cup \{j\}$, $U := U \setminus \{j\}$.

Event 1.2 *If the facility* $i \in \mathscr{F} \setminus \tilde{\mathscr{F}}$, *we increase the corresponding dual variable* β_{ij}.

Event 2 *There is a facility* $i \in \mathscr{F} \setminus \tilde{\mathscr{F}}$, *such that* $\sum_{j \in \mathscr{C}} \beta_{ij} = f_i$. *We say that facility* i *is* fully paid, *and it can be* temporarily opened, *record this time by* $t(i)$. *Update* $\tilde{\mathscr{F}} := \tilde{\mathscr{F}} \cup \{i\}$, *and define* $N_i^{con} = \{j \in \mathscr{C} | \beta_{ij} > 0\}$ *to be the neighbor of facility* i, *i.e., the set of the clients contributing to facility* i. *For each* $j \in U \cap N_i^{con}$, *set* $i(j) := i$ *and call* $i(j)$ *the* connecting witness *of client* j. *Freeze* $j \in U \cap N_i^{con}$, *and update* $N_i^{wit} := N_i^{wit} \cup (U \cap N_i^{con})$, $U := U \setminus N_i^{con}$.

Event 3 *There is a client* $j \in U$, *such that* $\alpha_j = p_j$. *Freeze* j, *and update* $\tilde{\mathscr{P}} := \tilde{\mathscr{P}} \cup \{j\}$ *and* $U := U \setminus \{j\}$.

Step 3 *Constructing a primal integer feasible solution* $(\hat{x}, \hat{y}, \hat{z}, \hat{r})$.
Let $\hat{\mathscr{F}}$ *denote the finally open facility set, i.e., the facility set opened in the final integer solution,* $\hat{\mathscr{P}}$ *denote the penalty client set, and* $\hat{\mathscr{O}}$ *denote the outlier set.*

Step 3.1 *Determine outliers. If $|\tilde{\mathscr{O}}| = q$, set $\hat{\mathscr{O}} := \tilde{\mathscr{O}}$. Otherwise, there must be a facility i_q gets fully paid at time t_q (If not, Event 1 or Event 3 happens, this implies $|U| \geq q$). Choose the clients in $N_{i_q}^{wit}$ with the maximum $q - |\tilde{\mathscr{O}}|$ connection cost from i_q and add these clients to the set $\tilde{\mathscr{O}}$. Let us denote this set as $\hat{\mathscr{O}}$.*

Step 3.2 *Determine open facilities. Consider each facility $i \in \tilde{\mathscr{F}}$. If there is a facility $i' \in \tilde{\mathscr{F}}$, $i' \neq i$, such that $N_i^{con} \cap N_{i'}^{con} \neq \emptyset$, we say that facility i and i' are* relevant *to each other. We choose any maximal independent subset $\hat{\mathscr{F}} \subseteq \tilde{\mathscr{F}}$, open all facilities in $\hat{\mathscr{F}}$.*

Step 3.3 *Determine penalty clients. Let $\hat{\mathscr{P}} := \tilde{\mathscr{P}} \setminus \bigcup_{i \in \hat{\mathscr{F}}} N_i^{con}$.*

Step 3.4 *Connect each client in $\mathscr{C}^{(1)} := \mathscr{C} \setminus (\hat{\mathscr{P}} \cup \hat{\mathscr{O}})$ to its closest open facility in $\hat{\mathscr{F}}$ respectively.*

We declare that the dual solution obtained by Step 2 denoted by (α, β, θ) is feasible. First, the dual ascending process guarantees that the first three constraints in (DP) are established. Second, $\theta := t_q$ implies $\alpha_j \leq \theta$ for all clients. The feasibility of the solution $(\hat{x}, \hat{y}, \hat{z}, \hat{r})$ is clearly visible. Note that the new instance $\mathscr{I}^{(1)}$ just changes part of the facility cost, so $(\hat{x}, \hat{y}, \hat{z}, \hat{r})$ and (α, β, θ) are also feasible to the original instance.

3.2 Analysis

In this subsection, we analyze the approximation factor of Algorithm 1, i.e., analyze the relationship between the cost of the solution obtained from Algorithm 1 and the cost of the optimal solution denoted by OPT. Denote $OPT^{(1)}$ be the optimal solution cost of the instance $\mathscr{I}^{(1)}$. We have $OPT \geq f_{\max} + OPT^{(1)}$. At the same time, for intuitional analysis, we introduce $F^{(1)}$, $C^{(1)}$, and $P^{(1)}$, which indicate the opening cost, connection cost, and penalty cost of the solution $(\hat{x}, \hat{y}, \hat{z}, \hat{r})$, respectively. Furthermore, let $F_q^{(1)}$ denote the facility cost of $\hat{\mathscr{F}} \setminus \{i_q\}$.

In order to bound the total cost of the solution $(\hat{x}, \hat{y}, \hat{z}, \hat{r})$, we provide the following lemmas to bound $F_q^{(1)}$, $C^{(1)}$, and $P^{(1)}$ by the cost of the dual solution, respectively. The proofs of the following lemmas and theorem are deferred to the journal version.

Lemma 1.

$$F_q^{(1)} = \sum_{i \in \hat{\mathscr{F}}} \sum_{j \in N_i^{con} \setminus (N_{i_q}^{con} \cap \hat{\mathscr{O}})} \beta_{ij}.$$

For convenience, let us denote $\mathscr{C}^{con} := \bigcup_{i \in \hat{\mathscr{F}}} N_i^{con} \setminus (N_{i_q}^{con} \cap \hat{\mathscr{O}}), \mathscr{C}^{tou} := \bigcup_{i \in \hat{\mathscr{F}}} (N_i^{wit} \setminus N_i^{con}), \mathscr{C}^{clo} := \mathscr{C}^{(1)} \setminus \bigcup_{i \in \hat{\mathscr{F}}} (N_i^{wit} \cup N_i^{con})$. Note that $\mathscr{C} = \mathscr{C}^{con} \cup$

$\mathscr{C}^{tou} \cup \mathscr{C}^{clo} \cup \hat{\mathscr{P}} \cup \hat{\mathscr{O}}$. The clients in $\mathscr{C}^{con}$ contribute to some finally open facilities, the clients in $\mathscr{C}^{tou}$ touch some finally open facilities, and the connecting witnesses of the clients in $\mathscr{C}^{clo}$ are closed by some finally open facilities.

We bound the connection cost in the following lemma.

Lemma 2.

$$C^{(1)} \le \sum_{i \in \hat{\mathscr{F}}} \sum_{j \in N_i^{con} \setminus (N_{iq}^{con} \cap \hat{\mathscr{O}})} c_{ij} + \sum_{j \in \mathscr{C}^{tou}} \alpha_j + 3 \sum_{j \in \mathscr{C}^{clo}} \alpha_j .$$

For the penalty cost of our algorithm, we obtain the following lemma.

Lemma 3.

$$P^{(1)} = \sum_{j \in \hat{\mathscr{P}}} \alpha_j .$$

Now we are ready to give the main result of this subsection as the following theorem.

Theorem 5. *Algorithm 1 is a 3-approximation algorithm for the RFLPWP.*

4 Discussions

Combining with the greedy augmentation technique in [3,5], we can further improve the approximation ratio to 2. We omit the detail of this improvement and defer it to the journal version. Recall that the best known approximation factor for the FLPWP and UFLP are 1.514 and 1.488, respectively. As a variation of the above two problems, it will be interesting to further improve the approximation factor of 2 for the RFLPWP.

Acknowledgements The research of the second author is supported by Scientific Research Common Program of Beijing Municipal Commission of Education (No. KM201210005033) and China Scholarship Council. The third author's research supported by NSF of China (No.11371001).

References

1. Cornuejols, G., Nemhauser, G.L., Wolsey, L.A.: The uncapacitated facility location problem. In: Mirchandani, P.B., Francis, R.L. (eds.) Discrete Location Theory, pp. 119–171. Wiley, New York (1990)
2. Kuehn, A.A., Hamburger, M.J.: A heuristic program for locating warehouses. Manag. Sci. **9**, 643–666 (1963)

3. Charikar, M., Guha, S.: Improved combinatorial algorithms for facility location problems. SIAM J. Comput. **34**, 803–824 (2005).
4. Chudak, F.A., Shmoys, D.B.: Improved approximation algorithms for the uncapacitated facility location problem. SIAM J. Comput. **33**, 1–25 (2003)
5. Guha, S., Khuller, S.: Greedy strikes back: improved facility location algorithms. J. Algorithms **31**, 228–248 (1999)
6. Jain, K., Mahdian, M., Markakis, E., Saberi, A., Vazirani, V.V.: Greedy facility location algorithms analyzed using dual fitting with factor-revealing LP. J. ACM **50**, 795–824 (2003)
7. Jain, K., Vazirani, V.V.: Approximation algorithms for metric facility location and k-median problems using the primal-dual schema and Lagrangian relaxation. J. ACM **48**, 274–296 (2001)
8. Li, S.: A 1.488-approximation algorithm for the uncapacitated facility location problem. In: Proceedings of ICALP, Part II, pp. 77–88 (2011)
9. Mahdian, M., Ye, Y., Zhang, J.: Improved approximation algorithms for metric facility location problems. In: Proceedings of APPROX, pp. 229–242 (2002)
10. Shmoys, D.B., Tardös, E., Aardal, K.I.: Approximation algorithms for facility location problems. In: Proceedings of STOC, pp. 265–274 (1997)
11. Sviridenko, M.: An improved approximation algorithm for the metric uncapacitated facility location problem. In: Proceedings of IPCO, pp. 240–257 (2002)
12. Ageev, A., Ye, Y., Zhang, J.: Improved combinatorial approximation algorithms for the k-level facility location problem. SIAM J. Discrete Math. **18**, 207–217 (2003)
13. Chen, X., Chen, B.: Approximation algorithms for soft-capacitated facility location in capacitated network design. Algorithmica **53**, 263–297 (2007)
14. Charikar, M., Khuller, S., Mount, D.M., Naraasimban, G.: Algorithms for facility location problems with outliers. In: Proceedings of SODA, pp. 642–651 (2001)
15. Du, D., Lu, R., Xu, D.: A primal-dual approximation algorithm for the facility location problem with submodular penalties. Algorithmica **63**, 191–200 (2012)
16. Shu, J.: An efficient greedy heuristic for warehouse-retailer network design optimization. Transp. Sci. **44**, 183–192 (2010)
17. Xu, G., Xu, J.: An LP rounding algorithm for approximating uncapacitated facility location problem with penalties. Inf. Process. Lett. **94**, 119–123 (2005)
18. Xu, G., Xu, J.: An improved approximation algorithm for uncapacitated facility location problem with penalties. J. Comb. Optim. **17**, 424–436 (2008)
19. Ye, Y., Zhang, J.: An approximation algorithm for the dynamic facility location problem. In: Combinatorial Optimization in Communication Networks, pp. 623–637. Kluwer Academic, Dordrecht (2005)
20. Zhang, J.: Approximating the two-level facility location problem via a quasi-greedy approach. Math. Program. **108**, 159–176 (2006)
21. Zhang, J., Chen, B., Ye, Y.: A multiexchange local search algorithm for the capacitated facility location problem. Math. Oper. Res. **30**, 389–403 (2005)
22. Zhang, P.: A new approximation algorithm for the k-facility location problem. Theor. Comput. Sci. **384**, 126–135 (2007)
23. Li, Y., Du, D., Xiu, N., Xu, D.: Improved approximation algorithms for the facility location problems with linear/submodular penalty. In: Proceedings of COCOON, pp. 292–303 (2013)

A Discrete State Transition Algorithm for Generalized Traveling Salesman Problem

Xiaolin Tang, Chunhua Yang, Xiaojun Zhou, and Weihua Gui

Abstract Generalized traveling salesman problem (GTSP) is an extension of classical traveling salesman problem (TSP), which is a combinatorial optimization problem and an NP-hard problem. In this paper, an efficient discrete state transition algorithm (DSTA) for GTSP is proposed, where a new local search operator named *K-circle*, directed by neighborhood information in space, has been introduced to DSTA to shrink search space and strengthen search ability. A novel robust update mechanism, restore in probability and risk in probability (Double R-Probability), is used in our work to escape from local minima. The proposed algorithm is tested on a set of GTSP instances. Compared with other heuristics, experimental results have demonstrated the effectiveness and strong adaptability of DSTA and also show that DSTA has better search ability than its competitors.

Keywords Generalized traveling salesman problem • Discrete state transition algorithm • K-circle • Double R-probability

1 Introduction

Generally speaking, GTSP can be described as follows: given a completely undirected graph $G = \{V, E\}$, where V is a set of n vertices and has been partitioned into m clusters $V = \{V_1, V_2, \cdots, V_m\}$, E is a set of m edges, and the goal of GTSP is to find a tour visiting each cluster exactly once while minimizing the sum of the route costs. In this paper, the symmetric GTSP is concerned, that is to say, $c_{i,j} = c_{j,i}$, here, the associated cost $c_{i,j}$ for each pair of vertices (i, j) represents the distance from one vertex in V_i to another vertex in V_j. Since each cluster has at least one vertex and each vertex can only belong to one cluster, we have $m \leq n$. If $m = n$, GTSP is restored to TSP and both of them are NP-hard problems [1]. The sole task of dealing with TSP is to optimize the sequence of the clusters. While in the process

X. Tang • C. Yang (✉) • X. Zhou • W. Gui
School of Information Science and Engineering, Central South University, Changsha 410083, People's Republic of China
e-mail: xiaolin5789@126.com; ychh@csu.edu.cn; tiezhongyu2005@126.com; gwh@csu.edu.cn

D. Gao et al. (eds.), *Advances in Global Optimization*, Springer Proceedings in Mathematics & Statistics 95, DOI 10.1007/978-3-319-08377-3_15

of solving GTSP, it requires determining the sequence of the clusters and a vertex to be visited in each cluster simultaneously, which indicates that GTSP is more complex than TSP. Nonetheless, GTSP is extensively used in many applications, such as task scheduling, airport selection, and postal routing, etc [2, 3].

According to the characteristics of GTSP, the process of solving GTSP can be decomposed into two phases. One is to determine the visiting order of all the clusters, which is similar to TSP; the other is to find the optimal vertex in each cluster in a given order. Many reputed heuristic searching algorithms, like genetic algorithm (GA) [4], particle swarm optimization (PSO) [5], simulated annealing (SA) [6], ant colony optimization (ACO) [7], have been varied into discrete versions to solve GTSP. Though these algorithms have their own mechanisms to deal with continuous optimization problems, they have to adapt themselves to GTSP with some classic operators, such as *swap* and *insert*. These search operators change the visiting order of clusters in particular ways. Lin–Kernighan (L–K) is a well-known method to solve TSP and GTSP [8], which focuses on changing the edges instead of the visiting order of clusters. The number of edges that L–K impacts in a single operation is unknown; as result, the depth of L–K is usually limited within a constant [9]. The majority of these methods focus on finding an optimal sequence of clusters, while to solve GTSP, it still has to choose a vertex from each cluster to make the minimal cost simultaneously. This is a well-known shortest path problem in operations research which is also called cluster optimization (CO) in GTSP. The most common method to deal with this problem is dynamic programming that can give us a definitively best result which is named as layer network method in other literatures [10].

State transition algorithm is a new optimization algorithm, according to the control theory and state transition [11]. The efficiency of STA in application to continuous optimization problems has been proved [12]. In [13], discrete version of state transition algorithm has been introduced to solve a series of discrete optimization problems such as TSP and boolean integer programming. In this study, we will extend DSTA to solve the GTSP.

In Sect. 2, we give a brief description of DSTA and some transformation operators. Section 3 introduces relevancy and correlation index to describe K-Neighbor. In Sect. 4, a DSTA is presented to solve GTSP with a new updating mechanism. Some experimental results are given in Sect. 5, and the final part is the conclusion.

2 Discrete State Transition Algorithm

2.1 Description of DSTA

State transition algorithm comes from control theory. It regards a solution to an optimization problem as a state and updating of the solution as state transition. The unified form of discrete state transition algorithm is given as follows:

$$\begin{cases} \boldsymbol{x}_{k+1} = A_k(\boldsymbol{x}_k) \bigoplus B_k(\boldsymbol{u}_k) \\ y_{k+1} = f(\boldsymbol{x}_{k+1}) \end{cases}, \tag{1}$$

where $\boldsymbol{x}_k \in \mathbb{R}^n$ denotes a current state, corresponding to current solution of an optimization problem; $\boldsymbol{u}_k \in \mathbb{R}^n$ is a function of $\boldsymbol{x}_k$ and historical states; both A_k, $B_k \in \mathbb{R}^{n\times n}$ are transition operators which are usually state transition matrixes; $\bigoplus$ is an operation, which is admissible to operate on two states; f is the cost function or evaluation function.

In general, the solution to discrete optimization problem is a sequence, which means a new state $\boldsymbol{x}_{k+1}$ should also be a sequence after transformation by A_k or B_k. For the TSP, only a state transition matrix is considered, avoiding the complexity of adding one sequence to another. So the form of DSTA for TSP is simplified as follows:

$$\begin{cases} \boldsymbol{x}_{k+1} = G_k \boldsymbol{x}_k \\ y_{k+1} = f(\boldsymbol{x}_{k+1}) \end{cases} \tag{2}$$

where $\boldsymbol{x}_k = [x_{1,k}, x_{2,k}, \cdots, x_{m,k}]^T, x_{i,k} \in \{1, 2, \cdots, m\}$; G_k is the state transition matrix which is created by transformation operators. State transition matrixes are variants of identity matrix with only position value 1 in each column and each row. Multiplying a state transition matrix by a current state will get a new state which is still a sequence and the process is like this:

Current state $\boldsymbol{x}_k$: $[1\ 2\ 3\ 4\ 5]^T$
State transition matrix G_k:

$$\begin{pmatrix} 1&0&0&0&0 \\ 0&1&0&0&0 \\ 0&0&0&0&1 \\ 0&0&0&1&0 \\ 0&0&1&0&0 \end{pmatrix}$$

$$\begin{pmatrix} 1&0&0&0&0 \\ 0&1&0&0&0 \\ 0&0&0&0&1 \\ 0&0&0&1&0 \\ 0&0&1&0&0 \end{pmatrix} \times \begin{pmatrix} 1 \\ 2 \\ 3 \\ 4 \\ 5 \end{pmatrix}$$

New state:$[1\ 2\ 5\ 4\ 3]^T \leftarrow G_k \times \boldsymbol{x}_k$

2.2 State Transformation Operators

DSTA solves TSP with three efficient operators, which is the foundation of study on GTSP. All of the three operators, swap, shift, symmetry, belong to G_k .

(1) Swap

$$(x_1, x_2, x_3, \cdots, x_{i-1}, x_i, x_{i+1}, \cdots, x_{m-1}, x_m)$$

$$\rightarrow (x_1, x_i, x_3, \cdots, x_{i-1}, x_2, x_{i+1}, \cdots, x_{m-1}, x_m)$$

This is an operator to exchange several vertices in the tour and the number of the vertices to be changed is limited by a parameter m_a. With this operator, the number of edges to be changed is twice as that of vertices to be exchanged.

(2) Shift

$$(x_1, x_2, x_3, \cdots, x_{i-1}, x_i, x_{i+1}, \cdots, x_{m-1}, x_m)$$

$$\rightarrow (x_1, x_3, \cdots, x_{i-1}, x_i, x_2, x_{i+1}, \cdots, x_{m-1}, x_m)$$

This operator first removes a segment of sequence from a given tour and then inserts this segment into a random position of the remaining sequence. The length of the removed sequence is restricted to less than m_b. Three edges will be changed through this operator, of which two edges are adjacent and the last edge is non-adjacent to them.

(3) Symmetry

$$(x_1, x_2, \cdots, x_{i-3}, x_{i-2}, x_{x-1}, x_i, x_{i+1}, x_{x+2}, x_{x+3} \cdots, x_m)$$

$$\rightarrow (x_1, x_2, \cdots, x_{i-3}, x_{i+2}, x_{x+1}, x_i, x_{i-1}, x_{x-2}, x_{x+3} \cdots, x_m)$$

Symmetry is to choose a vertex in a given tour as center, and then mirror a small segment on the left side of the center to the opposite side and so does a small segment on the right side. The length of these small segments is restricted to m_c. For symmetric GTSP, the symmetry operator can change two edges every time.

(4) Circle

At the final stage of search process, a tour is usually locally optimal, which indicates that this tour has many similar segments to the global tour; therefore, the entire sequence can be regarded as the combination of these segments. To further optimize a tour, we have to change several conjoint vertices in some segments simultaneously. Considering that the number of changing vertices in swap or shift is no more than 2, a new operator called circle is proposed here to enhance the global search ability.

Circle consists of two steps. First, we divide a given tour into two circles randomly, and then break one of the circles and insert it into another to create a new complete tour. Effects of this operator can be summarized as follows:

1. One of the circles contains only one vertex (Fig. 1a);
2. Both of the circles contain more than one vertex, and change the connection at the interfaces of each circle (Fig. 1b);
3. Both of the circles contain more than one vertex, break one of the circles at its interface and insert it into another from a random position except the interface (Fig. 1c, d).

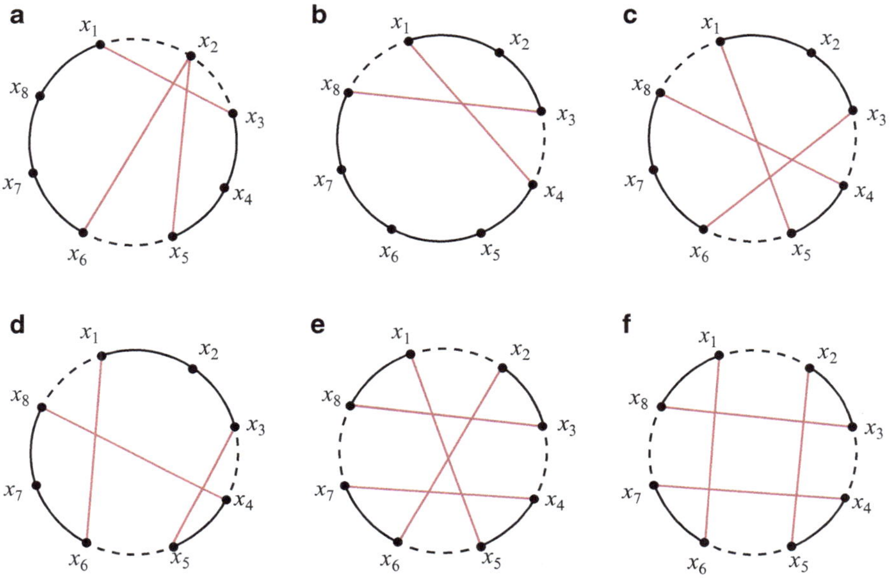

Fig. 1 Effects of circle operator in different cases

4. Both of the circles contain more than one vertex, break one of the circles and insert it into another. Both of the breaking position and inserting position are randomly chosen except two interfaces. Figure 1e, f show the results of this case.

   Obviously, circle is much more flexible than the other operators since it can gain six different kinds of cases.

(5) Cluster optimization (CO)

This is a sole operator to find the best path of given visiting order of clusters. A tour $[x_{1,k}, x_{2,k}, \cdots, x_{m,k}]$ with costs $W(\boldsymbol{x}_k)$ will be optimized into a new tour $\boldsymbol{x}_k^{'}$ after running CO, here, $W(\boldsymbol{x}_k^{'}) \leq W(\boldsymbol{x}_k)$ and $cluster(\boldsymbol{x}_k^{'}) = cluster(\boldsymbol{x}_k)$. In general, few of visiting orders of clusters will be changed from $\boldsymbol{x}_k$ to $\boldsymbol{x}_k^{'}$, thus we only need to optimize a small segment around the changed clusters.

3 K-Neighbor

To improve the global search ability for large-scale problems in limited computational time, it is necessary to avoid some potential bad search space. In this paper, the correlation index and relevancy are proposed, where correlation index is used to assess the correlation of every two clusters and the relevancy is applied to define K-Neighbor which will guide search direction as heuristic information.

Definition 1: *Let define the distance between the geometric centers of cluster i and cluster j as $d_{i,j}$, the sum of distance from the geometric centers of cluster i to the geometric centers of other clusters as d_i and denote $r_{i,j}$ as the correlation index of cluster i to cluster j:*

$$r_{i,j} = \frac{1 - \frac{d_{i,j}}{d_i}}{n-1}, \tag{3}$$

$$\sum_{j=1}^{n} r_{i,j} = 1, d_i = \sum_{j=1}^{n} d_{i,j}.$$

Definition 2: *Given $r_{i,j}$ as the correlation index of cluster i to cluster j and $r_{j,i}$ as the correlation index of cluster j to cluster i, then the relevancy $p_{i,j}$ of cluster i to cluster j can be formulated as:*

$$p_{i,j} = \frac{r_{i,j} \times r_{j,i}}{\sum_{j=1}^{n} r_{i,j} \times r_{j,i}}. \tag{4}$$

Calculating each $p_{i,j}$, we can get a relevancy matrix. The ith row of the relevancy matrix shows the relevancy of cluster i to other clusters. A big $p_{i,j}$ indicates cluster i is with high possibility connecting with cluster j. After sorting the relevancy matrix in descending order by row, the top k clusters in row i will be the K-Neighbor of cluster i. Using K-Neighbor as heuristic information, the global search ability can be improved significantly.

4 DSTA for GTSP

To adapt DSTA to GTSP and to make the algorithm more efficient, K-Neighbor is used as heuristic information to guide the search. Thus, k-shift, k-symmetry, and k-circle which are all guided by K-Neighbor are included in DSTA. The core procedure of the DSTA for GTSP can be outlined in pseudocode as follows:

repeat
    $[Best, Best^*] \leftarrow$ swap($SE, Best, Best^*, m_a$)
    $[Best, Best^*] \leftarrow$ shift($SE, Best, Best^*, m_b$)
    $[Best, Best^*] \leftarrow$ k-circle($SE, Best, Best^*, K\text{-}Neighbor$)
    $[Best, Best^*] \leftarrow$ k-symmetry($SE, Best, Best^*, K\text{-}Neighbor$)
    $[Best, Best^*] \leftarrow$ k-shift($SE, Best, Best^*, K\text{-}Neighbor$)
until the specified termination criterion is met

where SE is the search enforcement, representing the times of transformation by a certain operator; $Best$ is the best solution from the candidate state set created by transformation operators; $Best^*$ is the best solution in history. There are five operators in DSTA to optimize the sequence of clusters. A short cluster optimization which only optimizes a small segment (no more than five vertices) around the changed vertices is contained in every transformation operator to find a minimum path of a given sequence in further. To escape from local minima, a new robust update mechanism, restore and risk in probability, called double R-Probability for short, is introduced. Risk in probability is to accept a bad solution with a probability p_1. To ensure the convergence of DSTA, restore in probability p_2 is designed to recover the best solution in history.

5 Computational Results

Instances used in this paper all come from GTSPLIB [14]. The number of clusters in these instances varies from 30 to 89. Algorithms including DSTA, SA and ACO are coded in matlab and run on an Intel Core i5 3.10 GHz under Window XP environment. In order to test the performance of the proposed operators and approach, DSTA is compared with SA and ACO, and ten runs are carried out for the experiment. Some statistics are used as follows:

Opt. : the best known solution,
Best: the best solution obtained from the experiment,
Δ_{avg}: the relative error of the average solution,

$$\Delta_{avg} = \frac{mean(values) - Opt.}{Opt.} \times 100\,\%$$

t_{avg}: average time consumed.

Results of comparison among DSTA, SA, and ACO are listed in Table 1. In DSTA, we set $k = 8$, $m_a = 2, m_b = 1$. The initial temperature in SA is 5,000 and the cooling rate is 0.97. In ACO, $\alpha = 1$, $\beta = 5$, $\rho = 0.95$, where α, β are used to control the relative weight of pheromone trail and heuristic value, and ρ is the pheromone trail decay coefficient. As can be seen from Table 1, DSTA is superior to SA and ACO in both time consumption and solution quality. The Δ_{avg} of DSTA is very small, which indicates DSTA has good robustness and can obtain good solutions with high probability. In the ten runs, DSTA obtains the optimal solution at almost each run for every instance, but SA and ACO seldom find the Opt. except for 30kroB150. Δ_{avg} of SA is smaller than that of ACO because SA accepts a bad solution with probability which can help it escape from local minima.

Table 1 Results of comparison for SA, ACO, and DSTA

		SA			ACO			DSTA		
Instance	Opt.	Best	Δ_{avg}	t_{avg}	Best	Δ_{avg}	t_{avg}	Best	Δ_{avg}	t_{avg}
30kroA150	11,018	11,027	0.16	152	11,331	5.99	104	11,018	0	13
30kroB150	12,196	12,196	0.02	78	12,532	6.02	67	12,196	0.18	18
31pr152	51,576	51,584	1.12	79	51,734	1.60	69	51,576	0	25
32u159	22,664	22,916	1.90	89	24,285	8.68	75	22,664	0.74	33
39rat195	854	857	1.09	198	884	5.86	145	854	0.05	56
40d198	10,557	10,574	0.53	112	11,458	9.31	103	10,557	0.06	74
40kroA200	13,406	13,454	0.62	107	14,687	10.77	99	13,406	1.10	69
40kroB200	13,111	13,117	0.38	108	13,396	8.34	99	13,111	0.20	28
45ts225	68,340	68,401	1.57	325	70,961	5.83	223	68,340	0.66	72
45tsp225	1,612	1,618	1.77	122	1,736	8.15	119	1,612	1.35	88
46pr226	64,007	64,062	2.70	130	66,458	7.51	124	64,007	0	35
53gil262	1,013	1,047	5.24	142	1,148	15.92	148	1,013	1.30	85
53pr264	29,549	29,725	1.87	146	32,388	12.06	150	29,546	0.07	58
60pr299	22,615	23,186	7.00	165	25,296	15.97	184	22,618	2.54	114
64lin318	20,765	21,528	5.73	166	23,365	13.57	199	20,769	2.62	117
80rd400	6,361	6,920	10.36	225	8,036	21.67	299	6,361	2.52	141
84fl417	9,651	10,099	9.95	282	10,122	10.14	345	9,651	0.51	158
88pr439	60,099	66,480	13.13	276	69,271	16.14	368	60,099	2.95	156
89pcb442	21,657	23,811	11.15	253	26,233	19.48	376	21,664	3.80	163

6 Conclusions

We added a new operator and heuristic information to DSTA to solve GTSP. K-Neighbor can guide the search direction, in a way to ignore all possible connections among vertices. A flexible operator k-circle is guided by the K-Neighbor, which can change random segments freely in a tour. Double R-Possibility is helpful to escape from local minima. It accepts a bad solution with a probability p_1 and restore the history best with another probability p_2. All these strategies contribute to improving the performance of the DSTA.

Acknowledgements The work is supported by the National Science Found for Distinguished Young Scholars of China (Grant No. 61025015), Key Project of National Natural Science Funds (Grant No. 61134006), Foundation for Innovative Research Groups of the National Natural Science Foundation of China (Grant No.61321003).

References

1. Gutin, G., Yeo, A.: Assignment problem based algorithms are impractical for the generalized TSP. Australas. J. Comb. **27**, 149–153 (2003)
2. Fischetti, M., Salazar Gonzlez, J.J., Toth, P.: The symmetric generalized travelling salesman polytope. Networks **26**, 113–123 (1995)
3. Fischetti, M., Salazar, G., J.J., Toth, P.: A branch-and-cut algorithm for the symmetric generalized traveling salesman problem. Oper. Res. **45**, 378–394 (1995)
4. Gutin, G., Karapetyan, D.: A memetic algorithm for the generalized travelling salesman problem. Nat. Comput. **9**, 47–60 (2010)
5. Tasgetiren, M.F., Suganthan, P.N.: A discrete particle swarm optimization algorithm for the generalized traveling salesman problem, networks. In: 9th Annual Conference on Genetic and Evolutionary Computation, pp. 158–167 (2007)
6. Skiscim, C.C., Golden, B.L.: Optimization by simulated annealing: a preliminary computational study for the TSP. In: 15th Conference on Winter Simulation, pp. 523–535. IEEE Press, Piscataway (1983)
7. Song, X., Li, B., Yang, H.: Improved ant colony algorithm and its applications in TSP. In: 6th International Conference on Intelligent Systems Design and Applications, pp. 1145–1148, IEEE Computer Society, Washington (2006)
8. Lin, S., Kernighan, B.W.: An effective heuristic algorithm for the Traveling-Salesman Problem. Oper. Res. **21**(2), 498–516 (1973)
9. Karapetyan, D., Gutin, G.: Lin-Kernighan heuristic adaptations for the generalized traveling salesman problem. Eur. J. Oper. Res. **208**, 221–232 (2011)
10. Bondou, B., Artigues, C., Feillet, D.: A memetic algorithm with a large neighborhood crossover operator for the generalized traveling salesman problem. Comput. Oper. Res. **37**, 1844–1852 (2010)
11. Zhou, X.J., Yang, C.H., Gui, W.H.: Initial version of state transition algorithm. In: International Conference on Digital Manufacturing and Automation (ICDMA), pp. 644–647 (2011)
12. Zhou, X.J., Yang, C.H., Gui, W.H.: State transition algorithm. J. Ind. Manag. Optim. **8**(4), 1039–1056 (2012)
13. Zhou, X.J., Gao, D.Y., Yang, C.H.: Discrete state transition algorithm for unconstrained integer optimization problems. arXiv:1209.4199 [math.OC] (2012)
14. GTSP Instances Library. http://www.cs.rhul.ac.uk/home/zvero/GTSPLIB

Part III
Duality Theory

Canonical Dual Approach for Minimizing a Nonconvex Quadratic Function over a Sphere

Yi Chen and David Y. Gao

Abstract In this paper, we study global optimal solutions of minimizing a nonconvex quadratic function subject to a sphere constraint. The main challenge is to solve the problem when it has multiple global solutions on the boundary of the sphere, which is called *hard case*. By canonical duality theory, a concave maximization problem is formulated, which is one-dimensional and without duality gaps to the primal problem. Then sufficient and necessary conditions are provided to identify whether the problem is in the hard case or not. A perturbation method and associated algorithms are proposed to solve hard-case problems. Theoretical results and methods are verified by numerical examples.

Keywords Global optimization • Quadratic minimization problems • Canonical duality theory • Trust region subproblem

Mathematics Subject Classification (2010): 90C20, 90C26, 90C46

1 Introduction

In mathematical programming, the problem of minimizing a nonconvex quadratic function over a sphere constraint is known as the trust region subproblem, which arises in trust region methods [1,2]. Here, we formulate it as

$$(\mathcal{P}) \quad \min\{P(\boldsymbol{x}) = \boldsymbol{x}^T \boldsymbol{Q}\boldsymbol{x} - 2\boldsymbol{f}^T\boldsymbol{x} \mid \boldsymbol{x} \in \mathcal{X}_a\} \tag{1}$$

where the given matrix $\boldsymbol{Q} \in \mathbb{R}^{n \times n}$ is assumed to be symmetric, $\boldsymbol{f} \in \mathbb{R}^n$ is a given vector, and the feasible region is defined as

$$\mathcal{X}_a = \{\boldsymbol{x} \in \mathbb{R}^n \mid \|\boldsymbol{x}\| \leq r\}, \tag{2}$$

Y. Chen (✉) • D.Y. Gao
School of Science, Information Technology and Engineering,
Federation University Australia, Ballarat, VIC 3353, Australia
e-mail: yi.chen@federation.edu.au

D. Gao et al. (eds.), *Advances in Global Optimization*, Springer Proceedings in Mathematics & Statistics 95, DOI 10.1007/978-3-319-08377-3_16

in which r is a positive real number and $\|\boldsymbol{x}\| = \|\boldsymbol{x}\|_2$ represents ℓ_2 norm in $\mathbb{R}^n$. In literatures, two types of similar problems are also discussed: one considers a general quadratic constraint, i.e. the convexity is removed [3, 4]; the other one is equipped with a two-sided (lower and upper bound) quadratic constraint [5, 6].

It is proved that the problem $(\mathcal{P})$ possesses hidden convexity, i.e. it is actually equivalent to a convex optimization problem [5]. Thus, if the vector $\bar{\boldsymbol{x}}$ is a solution of $(\mathcal{P})$, there exists a Lagrange multiplier $\bar{\mu}$ such that besides the KKT conditions

$$(\boldsymbol{Q} + \bar{\mu}\boldsymbol{I})\bar{\boldsymbol{x}} = \boldsymbol{f}, \ \ \|\bar{\boldsymbol{x}}\| \le r, \ \ \bar{\mu} \ge 0, \ \ \bar{\mu}(\|\bar{\boldsymbol{x}}\| - r) = 0, \tag{3}$$

we also have [7]

$$\boldsymbol{Q} + \bar{\mu}\boldsymbol{I} \succeq 0, \tag{4}$$

which demonstrates the hidden convexity. If we let λ_1 be the smallest eigenvalue of the matrix $\boldsymbol{Q}$, from conditions (3) and (4), it is true that $\bar{\mu} \ge \max\{0, -\lambda_1\}$. If the problem $(\mathcal{P})$ has no solution on the boundary of $\mathcal{X}_a$, then $\boldsymbol{Q}$ must be positive definite and $\|\boldsymbol{Q}^{-1}\boldsymbol{f}\| < r$, which leads to $\bar{\mu} = 0$. While if $(\mathcal{P})$ has a solution on the boundary of $\mathcal{X}_a$ and $(\boldsymbol{Q} + \bar{\mu}\boldsymbol{I}) \succ 0$, we have $\|(\boldsymbol{Q} + \bar{\mu}\boldsymbol{I})^{-1}\boldsymbol{f}\| = r$. In this case, the multiplier $\bar{\mu}$ can be easily found. However, if the solution $\bar{\boldsymbol{x}}$ is located on the boundary of $\mathcal{X}_a$ and $\det(\boldsymbol{Q} + \bar{\mu}\boldsymbol{I}) = 0$, this situation is the so-called hard case (see [8]), which leads to numerical difficulties [9–13]. As pointed in [9, 12–14], the hard case always implies that $\boldsymbol{f}$ is perpendicular to the subspace generated by all the eigenvectors corresponding to λ_1. We will show by Theorem 3 in this paper that this condition is only a necessary condition for the hard case. Many methods have been proposed for solving this spherical constrained quadratic minimization problem, especially focusing on the hard case. They include Newton type methods [8, 15], methods recasting the problem in terms of a parameterized eigenvalue problem [12, 13], methods sequential searching Krylov subspaces [14, 16], semidefinite programming methods [9, 11], and the D.C. (difference of convex functions) method [17].

The canonical duality theory was developed from Gao and Strang's original work [18] for solving the nonconvex/nonsmooth variational problems. It is a powerful methodological theory which has been used successfully for solving a large class of difficult problems (nonconvex, nonsmooth or discrete) in global optimization (see [19, 20]) within a unified framework. This theory is mainly comprised of (1) a *canonical dual transformation*, which can be used to reformulate nonconvex/discrete problems from different systems as a unified canonical dual problem without duality gaps; (2) a *complementary-dual principle*, which provides a unified analytical solution form in terms of the canonical dual variable; and (3) a *triality theory*, which can be used to identify both global and local extrema.

The goal of this paper is to apply the canonical dual approach to find global solutions for the problem $(\mathcal{P})$, especially when it is in the hard case. We first show in the next section that the canonical dual problem is canonically (i.e., perfectly) dual to $(\mathcal{P})$ in the sense that both problems have the same set of KKT solutions. Then sufficient and necessary conditions are provided for identifying global optimal solutions. In Sect. 3, a perturbation method is proposed for problems in the hard case. Numerical results are presented in Sect. 4. The paper is ended with some conclusion remarks.

2 Canonical Dual Problem

Let $\mathcal{S}_a = \{\sigma \in \mathbb{R} \mid \sigma \geq 0,\ \det\boldsymbol{G}(\sigma) \neq 0\}$, where $\boldsymbol{G}(\sigma) = \boldsymbol{Q} + \sigma\boldsymbol{I}$. The canonical dual function $P^d : \mathcal{S}_a \to \mathbb{R}$ is defined by

$$P^d(\sigma) = -\boldsymbol{f}^T\boldsymbol{G}(\sigma)^{-1}\boldsymbol{f} - r^2\sigma, \tag{5}$$

and the stationary canonical dual problem [21] is defined by

$$\text{sta}\{P^d(\sigma) \mid \sigma \in \mathcal{S}_a\}. \tag{6}$$

Theorem 1 (Analytical Solution and Complementary-Dual Principle [20, 22]). *The problem (6) is canonically dual to the problem ($\mathcal{P}$) in the sense that if $\bar{\sigma} \in \mathcal{S}_a$ is a critical point of $P^d(\sigma)$, then $\bar{\boldsymbol{x}} = \boldsymbol{G}_a(\bar{\sigma})^{-1}\boldsymbol{f}$ is a KKT point of the primal problem ($\mathcal{P}$), and we have $P(\bar{\boldsymbol{x}}) = P^d(\bar{\sigma})$.*

Here, we focus the discussion on global optimal solutions and define the canonical dual problem to ($\mathcal{P}$) as the following maximization problem:

$$(\mathcal{P}^d) \quad \max\{P^d(\sigma) \mid \sigma \in \mathcal{S}_a^+\}, \tag{7}$$

where $\mathcal{S}_a^+ = \{\sigma \in \mathcal{S}_a \mid \boldsymbol{G}(\sigma) \succ \boldsymbol{0}\}$.

Theorem 2 (Global Optimality Condition [20, 22]). *If $\bar{\sigma} \in \mathcal{S}_a^+$ is a critical point of $P^d(\sigma)$, then $\bar{\sigma}$ is a global maximal solution of the problem ($\mathcal{P}^d$) and $\bar{\boldsymbol{x}} = \boldsymbol{G}(\bar{\sigma})^{-1}\boldsymbol{f}$ is a global minimal solution of the primal problem ($\mathcal{P}$), i.e. $P(\bar{\boldsymbol{x}}) = \min_{\boldsymbol{x}\in\mathcal{X}_a} P(\boldsymbol{x}) = \max_{\sigma\in\mathcal{S}_a^+} P^d(\sigma) = P^d(\bar{\sigma})$.*

According to the triality theorem [22, 23], the global optimality condition is called canonical min–max duality. Similar results are also discussed by Corollary 5.3 in [6] and Theorem 1 in [11].

By the symmetry of the matrix $\boldsymbol{Q}$, there exist diagonal matrix Λ and orthogonal matrix $\boldsymbol{U}$ such that $\boldsymbol{Q} = \boldsymbol{U}\Lambda\boldsymbol{U}^T$. The diagonal entities of Λ are the eigenvalues of the matrix $\boldsymbol{Q}$ and are arranged in nondecreasing order, $\lambda_1 = \cdots = \lambda_k < \lambda_{k+1} \leq \cdots \leq \lambda_n$. The columns of $\boldsymbol{U}$ are corresponding eigenvectors. Let $\hat{\boldsymbol{f}} = \boldsymbol{U}^T\boldsymbol{f}$.

Theorem 3 (Existence Conditions [24]). *Suppose that for any given symmetric matrix $\boldsymbol{Q} \in \mathbb{R}^{n\times n}$ and vector $\boldsymbol{f} \in \mathbb{R}^n$, λ_i and $\hat{f}_i$ are defined as above. The canonical dual problem ($\mathcal{P}^d$) has a unique solution $\bar{\sigma} \in (-\lambda_1, +\infty)$ if and only if either $\sum_{i=1}^k \hat{f}_i^2 \neq 0$ or $\sum_{i=k+1}^n \frac{\hat{f}_i^2}{(\lambda_i-\lambda_1)^2} > r^2$. If $\lambda_1 \leq 0$, $\bar{\boldsymbol{x}} = \boldsymbol{G}(\bar{\sigma})^{-1}\boldsymbol{f}$ is a unique* *global solution of the problem ($\mathcal{P}$). Moreover, given that ($\mathcal{P}^d$) has no solutions in $(-\lambda_1, +\infty)$, the problem ($\mathcal{P}$) has exactly two global solutions if the multiplicity of λ_1 is $k = 1$ and infinite number of solutions if $k > 1$.*

Generally speaking, the case that the canonical dual problem ($\mathcal{P}^d$) has no critical point in $\mathcal{S}_a^+$ does not imply that the problem ($\mathcal{P}$) is in the hard case. For example, if $\lambda_1 > 0$, i.e. the matrix $\boldsymbol{Q}$ is positive definite, ($\mathcal{P}^d$) may have no critical point in $\mathcal{S}_a^+ = [0, +\infty)$. But, the matrix $\boldsymbol{G}$ is not singular and therefore, the problem is not in the hard case. If $\lambda_1 \leq 0$, the hard case of ($\mathcal{P}$) is equivalent to that ($\mathcal{P}^d$) has no critical point in $\mathcal{S}_a^+ = (-\lambda_1, +\infty)$. In this case, $P^d(-\lambda_1) = \sup\{P^d(\sigma) | \sigma \in \mathcal{S}_a^+\}$. The theorem can also be stated equivalently as: *If $\lambda_1 \leq 0$, the nonconvex problem ($\mathcal{P}$) is in the hard case if and only if $\sum_{i=1}^k \hat{f}_i^2 = 0$ and $\sum_{i=k+1}^n \frac{\hat{f}_i^2}{(\lambda_i - \lambda_1)^2} \leq r^2$.* The first condition indicates that a problem could be in the hard case only when the coefficient $\boldsymbol{f}$ is perpendicular to the subspace generated by eigenvectors of the smallest eigenvalue. The second one adds that if the norm of $\boldsymbol{f}$ is relative small, comparing to the radius r and differences between the smallest eigenvalue and all other eigenvalues, the hard case would happen.

3 Perturbation Methods

In order to reinforce the existence conditions, a perturbation $\sum_{i=1}^k \alpha_i \boldsymbol{U}_i$ to $\boldsymbol{f}$ with parameters $\boldsymbol{\alpha} = \{\alpha_i\}_{i=1}^k \neq 0$ is introduced. Let $\boldsymbol{p} = \boldsymbol{f} + \sum_{i=1}^k \alpha_i \boldsymbol{U}_i$ and $\hat{\boldsymbol{p}} = \boldsymbol{U}^T \boldsymbol{p}$. The perturbed problem is

$$(\mathcal{P}_\alpha) \quad \min\{P_\alpha(\boldsymbol{x}) = \boldsymbol{x}^T \boldsymbol{Q} \boldsymbol{x} - 2\boldsymbol{p}^T \boldsymbol{x} \mid \boldsymbol{x} \in \mathcal{X}_a\}. \tag{8}$$

It is true that the existence condition, $\sum_{i=1}^k \hat{p}_i^2 \neq 0$, holds for the perturbed problem.

Theorem 4. *[24] Suppose that $\lambda_1 \leq 0$, there is no critical point of $P^d(\sigma)$ in $\mathcal{S}_a^+$, and $\bar{\boldsymbol{x}}^*$ is the optimal solution of the problem ($\mathcal{P}_\alpha$). Then, there is a global solution of the problem ($\mathcal{P}$), denoted as $\bar{\boldsymbol{x}}$, which is on the boundary of $\mathcal{X}_a$ and, for any $\varepsilon > 0$, if the parameter $\boldsymbol{\alpha}$ satisfies*

$$\|\boldsymbol{\alpha}\|^2 \leq (\lambda_2 - \lambda_1)^2 \left(r^2 - \sum_{i=k+1}^n \frac{\hat{f}_i^2}{(\lambda_i - \lambda_1)^2} \right) (1/\sqrt{2(1 - \cos(\varepsilon/r))} - 1)^{-2}, \tag{9}$$

we have $\|\bar{\boldsymbol{x}}^* - \bar{\boldsymbol{x}}\| \leq \varepsilon$.

Actually, if the perturbation parameter $\boldsymbol{\alpha}$ is properly chosen, each solution of the problem ($\mathcal{P}$) can be approximated. When the multiplicity of λ_1 is equal to one, as stated in Theorem 3, there are exactly two global solutions. In this case, $\boldsymbol{\alpha}$ becomes a scalar and has exactly two possible directions, which are mutual opposite and, respectively, lead to the two global solutions. For general cases, there may be infinite number of global solutions for the problem $(\mathcal{P})$, and we can show that between solutions of the problem $(\mathcal{P})$ and directions of $\boldsymbol{\alpha}$ there is a one-to-one correspondence [24].

4 Numerical Results

Based on the perturbation method discussed in the previous section, a canonical primal-dual algorithm is developed [24], which is matrix inverse free and the essential cost of calculation is only the matrix-vector multiplication.

One hundred examples are randomly generated, containing 50 examples of the general case and 50 examples of the hard case. Both cases have ten examples for dimensions of 500, 1,000, 2,000, 3,000, and 5,000. All elements of the coefficients, $\boldsymbol{Q}$, $\boldsymbol{f}$ and r, are integer numbers in $[-100, 100]$. For each example of the hard case, in order to make $\boldsymbol{f}$ can be easily chosen, we use a matrix $\boldsymbol{Q}$ of whom the multiplicity of the smallest eigenvalue is equal to one. Then, the vector $\boldsymbol{f}$ is constructed such that it is perpendicular to the eigenvector of the smallest eigenvalue, and a proper radius r is selected such that the existence conditions are violated.

For the hard case, a perturbation $\alpha \boldsymbol{U}_1$ is added to the vector $\boldsymbol{f}$, where $\boldsymbol{U}_1$ is the eigenvector of the smallest eigenvalue, and two values of α, 1e−3 and 1e−4, are tried. The algorithm is implemented on Matlab 7.13, which was runned in the platform with a Linux 64-bit system and a quad-core CPU.

Results are shown in Tables 1, 2, 3, and 4, and they contain the number of examples which are successfully solved (Succ.Solv.), the distance of the optimal solution to the boundary of the sphere (Dist.Boun.), the number of iterations in Algorithm: Main (Numb.Iter.) and the running time (in second) of the algorithm (Runn.Time). The values in the columns of Dist.Boun., Numb.Iter. and Runn.Time are averages of the examples successfully solved. We compare the results of the algorithm adopting "left division" and that of the algorithm adopting "quadprog" in the same table, where LD denotes "left division" and QP denotes "quadprog."

We can see that the examples are solved very accurately with error allowance being less than 1e-09. The failure in solving some examples is due to "left division"

Table 1 General case and $\alpha = 1e - 3$

	Succ.Solv.		Dist.Boun.		Numb.Iter.		Runn.Time.	
Dim	LD	QP	LD	QP	LD	QP	LD	QP
500	10	10	4.716e−09	5.245e−09	28.9	28.6	0.53	1.29
1,000	10	10	4.261e−09	3.974e−09	27.1	27.5	1.67	6.25
2,000	10	10	3.211e−09	3.822e−09	28.2	27.8	6.52	15.23
3,000	10	10	5.674e−09	5.221e−09	26.1	26.4	20.90	72.43
5,000	10	10	5.422e−09	3.873e−09	28.6	28.5	71.68	170.34

Table 2 General case and $\alpha = 1e - 4$

	Succ.Solv.		Dist.Boun.		Numb.Iter.		Runn.Time.	
Dim	LD	QP	LD	QP	LD	QP	LD	QP
500	10	10	4.532e−09	4.464e−09	28.9	28.9	0.43	1.16
1,000	10	10	3.849e−09	5.931e−09	27.4	27.1	1.47	6.08
2,000	10	10	2.648e−09	2.872e−09	27.9	28.5	6.26	15.82
3,000	10	10	5.299e−09	5.137e−09	26.2	26.2	20.15	73.60
5,000	10	10	3.188e−09	4.005e−09	28.7	28.5	65.71	171.92

Table 3 Hard case and $\alpha = 1e - 3$

	Succ.Solv.		Dist.Boun.		Numb.Iter.		Runn.Time.	
Dim	LD	QP	LD	QP	LD	QP	LD	QP
500	10	10	4.340e−09	6.297e−09	36.0	34.9	0.48	1.11
1,000	10	10	4.253e−09	4.904e−09	34.6	34.9	1.54	3.54
2,000	10	10	2.808e−09	4.255e−09	35.9	35.8	7.15	15.11
3,000	9	10	5.479e−09	4.466e−09	34.0	35.0	19.41	36.01
5,000	10	10	3.755e−09	4.705e−09	35.2	35.5	74.79	121.41

Table 4 Hard case and $\alpha = 1e - 4$

	Succ.Solv.		Dist.Boun.		Numb.Iter.		Runn.Time.	
Dim	LD	QP	LD	QP	LD	QP	LD	QP
500	7	9	2.503e−09	4.488e−09	39.6	40.6	0.51	1.36
1,000	9	9	3.148e−09	4.482e−09	37.4	38.3	1.56	3.81
2,000	5	9	8.668e−09	5.785e−09	38.6	42.6	7.36	17.95
3,000	5	10	6.003e−09	3.997e−09	38.4	40.6	20.43	41.06
5,000	8	10	4.748e−09	2.814e−09	37.8	38.8	72.72	131.51

and "quadprog" being unable to handle very nearly singular matrices. For general cases, all the examples can be solved within no more than 30 iterations, while for hard cases, the number of iterations is around 40. From the running time, we notice that our method is capable to handle very large problems in reasonable time. The algorithms using "left division" and "quadprog" have similar performances in the accuracy and the number of iterations. The one using "left division" needs much less time than that of the one using "quadprog", but "quadprog" is able to solve more examples successfully.

5 Conclusion Remarks

We have presented a detailed study on the quadratic minimization problem with a sphere constraint. By the canonical duality, this nonconvex optimization is equivalent to a concave maximization dual problem over a convex domain $\mathcal{S}_a^+$, which is true also for many other global optimization problems (see [25–31]). Based on this canonical dual problem, sufficient and necessary conditions are obtained for both general and hard cases. In order to solve hard-case problems, a perturbation method and the associated algorithm are proposed. Numerical results demonstrate that the proposed approach is able to handle the problem effectively. Combining with the trust region method, the results presented in this paper can be used to solve general global optimizations.

Acknowledgements This research is supported by US Air Force Office of Scientific Research under the grant AFOSR FA9550-10-1-0487, as well as by a grant from the Australian Government under the Collaborative Research Networks (CRN) program. The main results of this paper have been announced at the 3rd World Congress of Global Optimization, July 9–11, 2013, the Yellow Mountains, China.

References

1. Conn, A.R., Gould, N.I.M., Toint, P.L.: Trust-Region Methods. SIAM, Philadelphia (2000)
2. Powell M.J.D.: On trust region methods for unconstrained minimization without derivatives. Math. Program. **97**(3), 605–623 (2003)
3. Boyd, S.P., Vandenberghe, L.: Convex Optimization. Cambridge University Press, Cambridge (2004)
4. Xing, W.X., Fang, S.C., Gao, D.Y., Sheu, R.L., Zhang, L.: Canonical dual solutions to the quadratic programming over a quadratic constraint. ICOTA **7**, 35–36 (2007)
5. Ben-Tal, A., Teboulle, M.: Hidden convexity in some nonconvex quadratically constrained quadratic programming. Math. Program. **72**(1), 51–63 (1996)
6. Stern, R.J., Wolkowicz, H.:. Indefinite trust region subproblems and nonsymmetric eigenvalue perturbations. SIAM J. Optim. **5**(2), 286–313 (1995)
7. Sorensen, D.C.: Newton's method with a model trust region modification. SIAM J. Numer. Anal. **19**(2), 409–426 (1982)
8. Moré, J.J., Sorensen, D.C..: Computing a trust region step. SIAM J. Sci. Stat. Comput. **4**(3), 553–572 (1983)
9. Fortin, C., Wolkowicz, H.: The trust region subproblem and semidefinite programming. Optim. Method Softw. **19**(1), 41–67 (2004)
10. Jorge, N., Wright, S.J.: Numerical Optimization, vol. 2. Springer, New York (1999)
11. Rendl, F., Wolkowicz, H.: A semidefinite framework for trust region subproblems with applications to large scale minimization. Math. Program. **77**, 273–299 (1997)
12. Rojas, M., Santos, S.A., Sorensen, D.C.: A new matrix-free algorithm for the large-scale trust-region subproblem. SIAM J. Optim. **11**(3), 611–646 (2001)
13. Sorensen, D.C.: Minimization of a large-scale quadratic functionsubject to a spherical constraint. SIAM J. Optim. **7**(1), 141–161 (1997)
14. Hager, W.W.: Minimizing a quadratic over a sphere. SIAM J. Optim. **12**(1), 188–208 (2001)
15. Gay, D.M.: Computing optimal locally constrained steps. SIAM J. Sci. Stat. Comp. **2**(2), 186–197 (1981)
16. Gould, N.I.M., Lucidi, S., Roma, M., Toint, P.L.: Solving the trust-region subproblem using the lanczos method. SIAM J. Optim. **9**(2), 504–525 (1999)
17. Tao, P.D., An, L.T.H.: A dc optimization algorithm for solving the trust-region subproblem. SIAM J. Optim. **8**(2), 476–505 (1998)
18. Gao, Y., Strang, G.: Geometric nonlinearity: potential energy, complementary energy, and the gap function. Q. Appl. Math. **47**, 487–504 (1989)

19. Gao, D.Y.: Canonical duality theory: unified understanding and generalized solution for global optimization problems. Comput. Chem. Eng. **33**(12), 1964–1972 (2009)
20. Gao, D.Y., Ruan, N., Sherali, H.D.: Solutions and optimality criteria for nonconvex constrained global optimization problems with connections between canonical and Lagrangian duality. J. Glob. Optim. **45**(3), 473–497 (2009)
21. Gao, D.Y.: Canonical Duality theory and solutions to constrained nonconvex quadratic programming. J. Glob. Optim. **29**(4), 377–399 (2004)
22. Gao, D.Y.: Duality Principles in Nonconvex Systems: Theory, Methods, and Applications. Springer, Netherlands (2000)
23. Gao, D.Y., Wu, C.: On the triality theory for a quartic polynomial optimisation problem. J. Ind. Manag. Optim. **8**(1), 229–242 (2012)
24. Chen, Y., Gao, D.Y.: Global solutions to spherical constrained quadratic minimization via canonical dual approach. arXiv:1308.4450 (2013)
25. Gao, D.Y.: Perfect duality theory and complete solutions to a class of global optimization problems. Optimization **52**(4–5), 467–493 (2003)
26. Gao, D.Y.: Sufficient conditions and perfect duality in nonconvex minimization with inequality constraints. J. Ind. Manag. Optim. **1**(1), 53–63 (2005)
27. Gao, D.Y.: Complete solutions and extremality criteria to polynomial optimization problems. J. Glob. Optim. **35**(1), 131–143 (2006)
28. Gao, D.Y.: Solutions and optimality criteria to box constrained nonconvex minimization problems. J. Ind. Manag. Optim. **3**(2), 293–304 (2007)
29. Gao, D.Y., Ruan, N., Pardalos, P.M.: Canonical dual solutions to sum of fourth-order polynomials minimization problems with applications to sensor network localization. In: Pardalos, P.M., Ye, Y.Y., Boginski, V., Commander, C. (eds.) Sensors: Theory, Algorithms, and Applications, pp. 37–54. Springer, New York (2010)
30. Gao, D.Y., Ruan, N., Sherali, H.D.: Canonical dual solutions for fixed cost quadratic programs. In: Chinchuluun, A., Pardalos, P.M., Enkhbat, R., Tseveendorj, I. (eds.) Optimization and Optimal Control, pp. 139–156. Springer, New York (2010)
31. Gao, D.Y., Watson, L.T., Easterling, D.R., Thacker, W.I., Billups, S.C.: Solving the canonical dual of box- and integer-constrained nonconvex quadratic programs via a deterministic direct search algorithm. Optim. Methods Softw. **28**, 313–326 (2013)

Application of Canonical Duality Theory to Fixed Point Problem

Ning Ruan and David Yang Gao

Abstract In this paper, we study general fixed point problem. We first rewrite the original problem in the canonical framework. Then, we proposed a canonical transformation of this problem, which leads to a convex differentiable dual problem and new iteration method. An illustrative example is presented.

Keywords Fixed point problem • Double well function • Canonical duality theory

1 Problems and Motivations

The fixed point problem [1–3] is a well-established subject in the area of nonlinear analysis. There are different kinds of iterative methods for solving these kinds of problems [4, 5]. In this paper, we will discuss a new approach based on duality principle. The mathematical formulation of the general problem is

$$(\mathscr{P}_0) \quad A\mathbf{x} = F(\mathbf{x}, \lambda), \tag{1}$$

where $\mathbf{x} = \{x_i\} \in \mathbb{R}^n$ is an unknown vector, $A = \{a_{ij}\} \in \mathbb{R}^{n \times n}$ is a symmetric positive definite matrix, and λ is a given constant.

Problem $(\mathscr{P}_0)$ has many applications. In many fields, equilibria or stability are fundamental concepts that can be described in terms of fixed points. For example, in economics, a Nash equilibrium of a game is a fixed point of the game's best response correspondence. The general form of equilibrium problem was first considered by Nikaido and Isoda in 1955 as an auxiliary problem to establish existence results for Nash equilibrium points in non-cooperative games [6–9].

If $\mathbf{x} = A\mathbf{x}$, then

$$\mathbf{x}_{k+1} = F(\mathbf{x}_k, \lambda), \tag{2}$$

where k denotes the number of iteration.

N. Ruan (✉) • D.Y. Gao
School of Science, Information Technology and Engineering, Federation University Australia, Ballarat, VIC 3353, Australia
e-mail: n.ruan@federation.edu.au; d.gao@federation.edu.au

D. Gao et al. (eds.), *Advances in Global Optimization*, Springer Proceedings in Mathematics & Statistics 95, DOI 10.1007/978-3-319-08377-3_17

If we assume

$$F(\mathbf{x}, \lambda) = \nabla W(\mathbf{x}, \lambda) - \mathbf{f}, \tag{3}$$

it is equal to solve the following problem:

$$(\mathscr{P}) \quad \min \quad P(\mathbf{x}) = W(\mathbf{x}, \lambda) - \langle \mathbf{f}, \mathbf{x} \rangle - \frac{1}{2}\mathbf{x}^T A\mathbf{x}$$
$$\text{s.t.} \quad \mathbf{x} \in \mathbb{R}^n,$$

where

$$W(\mathbf{x}, \lambda) = \frac{1}{2}\alpha(\frac{1}{2}\|B\mathbf{x}\|^2 - \lambda)^2, \tag{4}$$

and $\mathbf{f} = \{f_i\} \in \mathbb{R}^n$ is a given vector, $B = \{b_{ij}\} \in \mathbb{R}^{n\times n}$ is a symmetric matrix, α is also a given constant. The traditional methods for solving these kinds of problems are mainly by iteration. Since

$$\nabla W(\mathbf{x}, \lambda) = \alpha(\frac{1}{2}\|B\mathbf{x}\|^2 - \lambda)(B^T B\mathbf{x}), \tag{5}$$

combined with (2) and (3), we have

$$\mathbf{x}_{k+1} = \alpha(\frac{1}{2}\|B\mathbf{x}_k\|^2 - \lambda)(B^T B\mathbf{x}_k) - \mathbf{f}. \tag{6}$$

Therefore, given an initial point, after several times of iteration, the convergent point is the fixed point of the function $F(\mathbf{x}, \lambda)$.

Due to the nonconvexity of function $W(\mathbf{x}, \lambda)$, during the period of iteration, it may produce the chaotic phenomenon [10, 11]. In order to deal with such problem, we introduce the canonical duality theory. This theory has been used for solving a large class of nonlinear systems. The purpose of this paper is to illustrate the application of the canonical duality theory for solving the problem $(\mathscr{P}_0)$. In the next section, we will show how to use the canonical transformation to convert the equivalent nonconvex optimization problem into a canonical dual problem. In Sect. 3, we improve the canonical dual problem and find new iteration method. In Sect. 4, an illustrative application is given to show the efficiency of the theory. Finally, in Sect. 5, concluding remarks are provided.

2 Standard Canonical Dual Problem

Following the standard procedure of the canonical dual transformation, we introduce a quadratic geometric measure

$$\xi \triangleq \Lambda(\mathbf{x}) = \frac{1}{2}\|B\mathbf{x}\|^2 \in \mathbb{R}. \tag{7}$$

For the sake of convenience, we let $W(\mathbf{x}) = W(\mathbf{x}, \lambda)$. Thus, the nonconvex function $W(\mathbf{x})$ can be written in the canonical form

$$W(\mathbf{x}) = V(\Lambda(\mathbf{x})), \tag{8}$$

where $V(\xi) = \frac{1}{2}\alpha(\xi - \lambda)^2$, i.e., the duality relation

$$\sigma = \frac{\partial V(\xi)}{\partial \xi} = \alpha(\xi - \lambda) \in \mathbb{R} \tag{9}$$

is invertible for any given $\xi \in \mathbb{R}$. Thus (ξ, σ) forms a canonical duality pair on $\mathbb{R}\times\mathbb{R}$ (see [12]) and the Legendre conjugate V^* can be uniquely defined by

$$V^*(\sigma) = \mathrm{sta}_{\xi\in\mathbb{R}}\{\xi\sigma - V(\xi)\} = \frac{1}{2\alpha}\sigma^2 + \lambda\sigma, \tag{10}$$

where sta{} denotes finding the stationary point of the statement in {}.

Replacing $W(\mathbf{x}) = V(\Lambda(\mathbf{x}))$ by $\Lambda(\mathbf{x})\sigma - V^*(\sigma)$, the *total complementary function* [12, 13] can be defined as

$$\begin{aligned}\Xi(\mathbf{x}, \sigma) &= \Lambda(\mathbf{x})\sigma - V^*(\sigma) - \langle \mathbf{f}, \mathbf{x}\rangle - \frac{1}{2}\langle \mathbf{x}, A\mathbf{x}\rangle \\ &= \frac{1}{2}\mathbf{x}^T G(\sigma)\mathbf{x} - \langle \mathbf{f}, \mathbf{x}\rangle - \lambda\sigma - \frac{1}{2\alpha}\sigma^2,\end{aligned}$$

where

$$G(\sigma) = \sigma B^T B - A, \tag{11}$$

For a fixed σ, the criticality condition $\nabla_{\mathbf{x}}\Xi(\mathbf{x}, \sigma) = 0$ leads to the following canonical equilibrium equation:

$$G(\sigma)\mathbf{x} = \mathbf{f}. \tag{12}$$

On the canonical dual feasible space $\mathscr{S}_a \subset \mathbb{R}$ defined by

$$\mathscr{S}_a^+ = \{\sigma \in \mathbb{R} |\ \sigma > -\alpha\lambda,\ G(\sigma) \succ 0\}, \tag{13}$$

the solution of the canonical equilibrium equation can be uniquely determined as $\mathbf{x} = G^{-1}(\sigma)\mathbf{f}$. Substituting this result into the total complementary function Ξ, the canonical dual problem can be finally formulated as

$$\begin{aligned}(\mathscr{P}^d) \quad &\max \ P^d(\boldsymbol{\sigma}) = -\frac{1}{2}\mathbf{f}^T G^{-1}(\sigma)\mathbf{f} - \lambda\sigma - \frac{1}{2\alpha}\sigma^2 \\ &\text{s.t.} \ \ \sigma \in \mathscr{S}_a^+.\end{aligned}$$

3 Advanced Canonical Dual Problem and New Iteration Method

Following the same idea of canonical duality theory, we develop a new iteration method. Since matrix A is positive definite, there exists a unique matrix D such that $A = D^T D$. We then introduce a quadratic geometrical measure

$$\boldsymbol{\xi} = \Lambda(\mathbf{x}) = (D\mathbf{x}, \frac{1}{2}\|B\mathbf{x}\|^2) = (\mathbf{v}, \epsilon) \in \mathbb{R}^n \times \mathbb{R}. \tag{14}$$

Thus, the nonconvex function $W(\mathbf{x}) - \frac{1}{2}\langle \mathbf{x}, A\mathbf{x}\rangle$ can be written in the canonical form

$$W(\mathbf{x}) - \frac{1}{2}\langle \mathbf{x}, A\mathbf{x}\rangle = V(\Lambda(\mathbf{x})), \tag{15}$$

where $V(\boldsymbol{\xi}) = V(\mathbf{v}, \epsilon) = \frac{1}{2}\alpha(\epsilon - \lambda)^2 - \frac{1}{2}\mathbf{v}^2$, i.e., the duality relation

$$\boldsymbol{\varsigma} = \frac{\partial V(\mathbf{v}, \epsilon)}{\partial \boldsymbol{\xi}} = (-\mathbf{v}, \alpha(\epsilon - \lambda)) = (\boldsymbol{\rho}, \sigma) \in \mathbb{R}^n \times \mathbb{R} \tag{16}$$

is invertible for any given $\boldsymbol{\xi} \in \mathbb{R}^n \times \mathbb{R}$. Thus $(\boldsymbol{\xi}, \boldsymbol{\varsigma})$ forms a canonical duality pair and the Legendre conjugate V^* can be uniquely defined by

$$\begin{aligned} V^*(\boldsymbol{\varsigma}) &= \text{sta}_{\boldsymbol{\xi}\in\mathbb{R}^n\times\mathbb{R}}\{\boldsymbol{\xi}^T\boldsymbol{\varsigma} - V(\boldsymbol{\xi})\} \\ &= \text{sta}_{\mathbf{v}\in\mathbb{R}^n}\{\mathbf{v}\boldsymbol{\rho} - (-\frac{1}{2}\mathbf{v}^2)\} + \text{sta}_{\epsilon\in\mathbb{R}}\{\epsilon\sigma - \frac{1}{2}\alpha(\epsilon-\lambda)^2\} \\ &= -\frac{1}{2}\boldsymbol{\rho}^2 + \frac{1}{2\alpha}\sigma^2 + \lambda\sigma, \end{aligned}$$

Replacing $V(\Lambda(\mathbf{x}))$ by $\Lambda(\mathbf{x})\boldsymbol{\varsigma} - V^*(\boldsymbol{\varsigma})$, the *total complementary function* can be defined as

$$\Xi(\mathbf{x}, \boldsymbol{\rho}, \sigma) = \frac{1}{2}\mathbf{x}^T\sigma B^T B\mathbf{x} - (\mathbf{f} - D^T\boldsymbol{\rho})\mathbf{x} - \frac{1}{2\alpha}\sigma^2 - \lambda\sigma + \frac{1}{2}\boldsymbol{\rho}^T\boldsymbol{\rho}.$$

For a fixed $\boldsymbol{\varsigma}$, the criticality condition $\partial_{\mathbf{x}}\Xi(\mathbf{x}, \boldsymbol{\rho}, \sigma) = 0$ leads to the following canonical equilibrium equation:

$$\sigma B^T B\mathbf{x} - (\mathbf{f} - D^T\boldsymbol{\rho}) = 0. \tag{17}$$

On the canonical dual feasible space $\mathscr{S}_a^+ \subset \mathbb{R}^n \times \mathbb{R}$, the solution of the canonical equilibrium equation can be uniquely determined as $\mathbf{x} = \frac{1}{\sigma}(B^T B)^{-1}(\mathbf{f} - D^T\boldsymbol{\rho})$. Substituting this result into the total complementary function Ξ, the canonical dual problem can be finally formulated as

$$(\mathscr{P}^d) \quad \text{Max} \;\; P^d(\rho, \sigma) = -\frac{1}{2\sigma}(\mathbf{f} - D^T\rho)^T(B^TB)^{-1}(\mathbf{f} - D^T\rho) - \frac{1}{2\alpha}\sigma^2 - \lambda\sigma + \frac{1}{2}\rho^T\rho$$

$$\text{s.t.} \;\; \varsigma \in \mathscr{S}_a^+.$$

Since $\rho = -\mathbf{v} = -D\mathbf{x}$, by $\partial_\sigma P^d = 0$, we have

$$(\frac{\sigma}{\alpha} + \lambda)\sigma^2 = \frac{1}{2}(\mathbf{f} + A\mathbf{x})^T(B^TB)^{-1}(\mathbf{f} + A\mathbf{x}).$$

Considering (2), it is easy to find the following iteration equation

$$\mathbf{x}_{k+1} = \sigma(\mathbf{x}_k)(B^TB\mathbf{x}_k) - \mathbf{f}.$$

4 Applications

We now use an example to illustrate the applications of the theory presented in this paper.

$$(\mathscr{P}^d) \quad \min \;\; P(\mathbf{x}) = \frac{1}{2}\alpha(\frac{1}{2}(b_{11}x_1^2 + b_{22}x_2^2) - \lambda)^2 - f_1x_1 - f_2x_2 - a_{11}x_1^2 - a_{22}x_2^2$$

$$\text{s.t.} \;\; \mathbf{x} \in \mathbb{R}^n.$$

On the dual feasible space

$$\mathscr{S}_a = \{\sigma \in \mathbb{R} | \; \sigma > -\alpha\lambda, \; \sigma \neq 0\},$$

the canonical dual problem has the form of

$$P^d(\sigma) = -\frac{1}{2}[f_1 \;\; f_2]\begin{bmatrix} \frac{1}{\sigma} & \\ & \frac{1}{\sigma} \end{bmatrix}\begin{bmatrix} f_1 \\ f_2 \end{bmatrix} - \lambda\sigma - \frac{1}{2}\sigma^2.$$

We let

$$A = B = \begin{bmatrix} 1 & 0 \\ 0 & 1 \end{bmatrix}, \;\; \mathbf{f} = [0.3, \;\; -0.5]^T, \alpha = 1, \lambda = 1.5, \tag{18}$$

the dual problem has three critical points (see Fig. 1):

$$\bar{\sigma}_3 = -1.47218 < \bar{\sigma}_2 = 0.723493 < \bar{\sigma}_1 = 1.24869.$$

Since $\bar{\sigma}_1 \in \mathscr{S}_a^+$, by the canonical duality theory, we know that

$$\bar{\mathbf{x}}_1 = [1.20631, -2.01052]^T$$

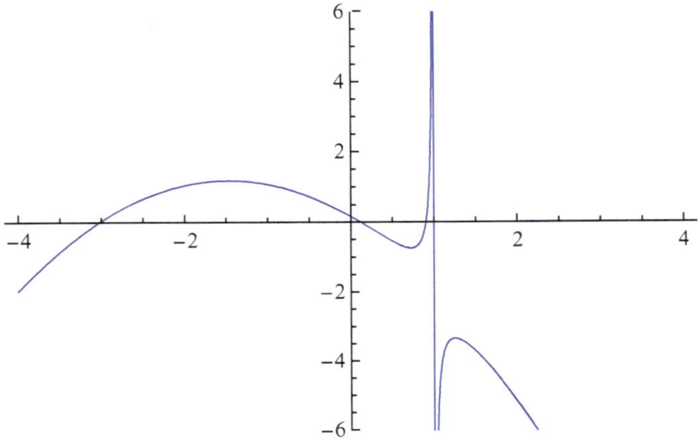

Fig. 1 Graph of $P^d(\sigma)$ for Example 1

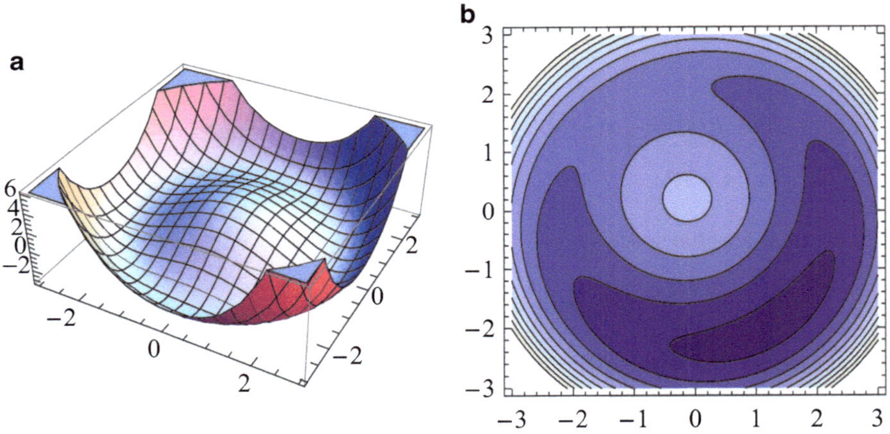

Fig. 2 Graph of $P(\mathbf{x})$ (*left*); contours of $P(\mathbf{x})$ (*right*) for Example 1

is a global minimizer (see Fig. 2). And we have

$$P(\bar{\mathbf{x}}_1) = -3.33623 = P^d(\bar{\sigma}_1).$$

5 Conclusions

We have presented a detailed application of the canonical duality theory to the general fixed point problems. This type of problems arises in many real-world applications. Using the canonical dual transformation, the canonical dual problem was formulated with zero duality gap. Application shows that the n-dimensional nonconvex problem $(\mathscr{P})$ can be reformulated as a one-dimensional concave maximization dual problem $(\mathscr{P}^d)$ on $\mathscr{S}_a^+$, which can be solved more easily.

Acknowledgements Dr. Ning Ruan was supported by a funding from the Australian Government under the Collaborative Research Networks (CRN) program.

References

1. Bierlaire, M., Crittin, F.: Solving noisy, large-scale fixed-point problems and systems of nonlinear equations. Transp. Sci. **40**, 44–63 (2006)
2. Border, K.C.: Fixed Point Theorems with Applications to Economics and Game Theory. Cambridge University Press, New York (1985)
3. Eaves, B.C.: Homotopies for computation of fixed points. Math. Program. **3**, 1–12 (1972)
4. Hirsch, M.D., Papadimitriou, C., Vavasis, S.: Exponential lower bounds for finding Brouwer fixed points. J. Complex. **5**, 379–416 (1989)
5. Huang, Z., Khachiyan, L., Sikorski, K.: Approximating fixed points of weakly contracting mappings. J. Complex. **15**, 200–213 (1999)
6. Scarf, H.: The approximation of fixed point of a continuous mapping. SIAM J. Appl. Math. **35**, 1328–1343 (1967)
7. Scarf, H.E., Hansen, T.: Computation of Economic Equilibria. Yale University Press, New Haven (1973)
8. Shellman, S., Sikorski, K.: A two-dimensional bisection envelope algorithm for fixed points. J. Complex. **2**, 641–659 (2002)
9. Shellman, S., Sikorski, K.: A recursive algorithm for the infinity-norm fixed point problem. J. Complex. **19**, 799–834 (2003)
10. Smart, D.R.: Fixed Point Theorems. Cambridge University Press, Cambridge (1980)
11. Yang, Z.: Computing Equilibria and Fixed Points: The Solution of Nonlinear Inequalities. Kluwer Academic, Dordrecht (1999)
12. Gao, D.Y.: Duality Principles in Nonconvex Systems: Theory, Methods and Applications. Kluwer Academic, Dordrecht (2000)
13. Gao, D.Y., Ruan, N., Sherali, H.D.: Solutions and optimality criteria for nonconvex constrained global optimization problems. J. Glob. Optim. **45**, 473–497 (2009)

Solving Facility Location Problem Based on Duality Approach

Ning Ruan

Abstract The facility location problem is one of the most widely studied discrete location problems, whose applications arise in a variety of settings, such as routers or servers in a communication network, warehouses or distribution centres in a supply chain, hospitals or airports in a public service system. The problem involves locating a number of facilities to minimize the sum of the fixed setup costs and the variable costs of serving the market demand from these facilities. First a dual problem is developed for the facility location problem. Then general optimality conditions are also obtained, which generate sequences globally converging to a primal and dual solutions, respectively.

Keywords Facility location • Integer programming • Canonical duality theory

1 Introduction

Many economical decision problems concern selecting and placing certain facilities to serve given demands. Examples are manufacturing plants, storage facilities, depots, warehouse, fire station, and hospital, etc. [1–4].

The most widely studied model in discrete facility location is the so-called uncapacitated facility location problem (UFLP), which involves locating an undetermined number of facilities to minimize the sum of the fixed setup costs and the variable costs of serving the market demand from these facilities [5–7].

Let I denote the set of m customer zones, indexed by i, and J denote the set of n potential facility locations, indexed by j. The (UFLP) has two sets of decision variables:

1. x_{ij}: binary variables that assume a value of 1, if assignment of customer i to facility j, and 0 otherwise;
2. y_j: binary variables that assume a value of 1, if a facility is to be established at location j, and 0 otherwise.

N. Ruan (✉)
School of Science, Information Technology and Engineering, Federation University Australia, Ballarat, VIC 3353, Australia
e-mail: n.ruan@federation.edu.au

D. Gao et al. (eds.), *Advances in Global Optimization*, Springer Proceedings in Mathematics & Statistics 95, DOI 10.1007/978-3-319-08377-3_18

The cost data is represented by the following notation:

1. c_{ij}: the total capacity, production and distribution cost of supplying all of customer zone i's demand by a facility at location j;
2. f_j: the fixed charge of establishing a facility at location j.

The variable cost c_{ij} are assumed to be linear functions of the quantities produced and shipped at each facility. The problem (UFLP) can be formulated as [8–10]:

$$(\mathscr{P}_0) \quad \min P(\mathbf{x}, \mathbf{y}) = \sum_{i \in I, j \in J} c_{ij} x_{ij} + \sum_{j \in J} f_j y_j \tag{1}$$

$$\text{s.t.} \quad \sum_{j \in J} x_{ij} = 1, \forall i \in I, \tag{2}$$

$$x_{ij} \leq y_j, \forall i \in I, j \in J, \tag{3}$$

$$x_{ij} \in \{0, 1\}, \forall i \in I, j \in J,$$

$$y_j \in \{0, 1\}, \forall j \in J.$$

The objective function (1) represents the total costs, whereas constraints (2) ensure that each customer is assigned to exactly one site. Constraints (3) guarantee that customer demand can be produced and shipped only from the locations where a facility is established, i.e., if $y_j = 1$.

The location of facilities to serve clients at minimum cost has been one of the most studied themes in the field of operations research. Before 1978, the best-known approaches for solving the UFLP were the implicit enumeration technique and the branch-and-bound algorithm, which is developed by Efroymson and Ray. Then, Thenm Khumawala developed efficient branching and separation strategies for the branch-and-bound algorithm. Erlenkotter used one kind of formulation, which is known to often produce natural integer solutions. Also Erlenkotter combined simple dual heuristics in a branch-and-bound framework. There are also a lot of works regarding approximate solutions for the problem [11, 12].

In this paper we present a generalized canonical duality theory for solving these challenging problems. Canonical duality theory [13] developed from nonconvex analysis and global optimization. It is a potentially powerful methodology, which has been used successfully for solving a large class of challenging problems in biology, network communications, and engineering. The rest of the paper is arranged as follows. In Sect. 2, we reformulated the original problem in the canonical framework. In Sect. 3, we demonstrate how to rewrite the primal problems as a dual problem by using the canonical dual transformation. In Sect. 4, we show that the obtained formulation is canonical dual to the original problems. The last section finishes with conclusions.

2 Problem Reformulation

In order to use canonical dual transformation, we firstly rewrite the **x**, **c** into the vector form:

$$\mathbf{x} = [x_{11}, \cdots, x_{1n}, \cdots, x_{m1}, \cdots, x_{mn}]^T,$$
$$\mathbf{c} = [c_{11}, \cdots, c_{1n}, \cdots, c_{m1}, \cdots, c_{mn}]^T.$$

Let

$$A = \begin{bmatrix} 1 \cdots 1 \; 0 \cdots 0 \cdots 0 \cdots 0 \\ 0 \cdots 0 \; 1 \cdots 1 \cdots 0 \cdots 0 \\ \vdots \ddots \vdots \; \vdots \ddots \vdots \ddots \vdots \ddots \vdots \\ 0 \cdots 0 \; 0 \cdots 0 \cdots 1 \cdots 1 \end{bmatrix},$$

$$D = \begin{bmatrix} I_{n\times n} \\ \vdots \\ I_{n\times n} \end{bmatrix},$$

where $I_{n\times n}$ is an identity matrix, and there are m block matrices $I_{n\times n}$ in matrix D.

$$\mathbf{e} = [1, \cdots, 1]^T$$

and matrix $B \in \mathbb{R}^{(mn)\times(mn)}$ is an identify matrix. By the fact that $\mathbf{x} \circ \mathbf{x} = \mathbf{x}$, original problem can be reformulated as

$$(\mathscr{P}) \quad \min \; P(\mathbf{x}, \mathbf{y}) = \frac{1}{2}\mathbf{x}^T \text{Diag}\,(2\mathbf{c})\mathbf{x} + \mathbf{f}^T\mathbf{y} \tag{4}$$

$$\text{s.t.} \quad A\mathbf{x} = \mathbf{e}, \tag{5}$$

$$B\mathbf{x} - D\mathbf{y} \le 0, \tag{6}$$

$$\mathbf{x} \circ (\mathbf{x} - \mathbf{e}) = 0, \tag{7}$$

$$\mathbf{y} \circ (\mathbf{y} - \mathbf{e}) = 0, \tag{8}$$

$$\mathbf{x} \in \mathbb{R}^{mn}, \mathbf{y} \in \mathbb{R}^{n}, \tag{9}$$

where the notation $\mathbf{s} \circ \mathbf{t} := (s_1t_1, s_2t_2, \cdots, s_nt_n)$ denotes the Hadamard product for any two vectors $\mathbf{s}, \mathbf{t} \in \mathbb{R}^n$. Let $\mathscr{X}$ be the primal feasible space.

$$\mathscr{X} = \{(\mathbf{x}, \mathbf{y}) \in \mathbb{R}^{mn+n} | A\mathbf{x} = \mathbf{e}, B\mathbf{x} - D\mathbf{y} \le 0, \mathbf{x} \circ (\mathbf{x} - \mathbf{e}) = 0, \mathbf{y} \circ (\mathbf{y} - \mathbf{e}) = 0\}.$$

3 Canonical Dual Transformation

In order to apply the canonical duality theory to solve this problem, we need to choose the following geometrically nonlinear operator. Define

$$\begin{aligned}\boldsymbol{\xi} &= \Lambda(\mathbf{x},\mathbf{y}) = [(A\mathbf{x}-\mathbf{e})^T, (B\mathbf{x}-D\mathbf{y})^T, (\mathbf{x}\circ(\mathbf{x}-\mathbf{e}))^T, (\mathbf{y}\circ(\mathbf{y}-\mathbf{e}))^T]^T \\ &= [(\eta)^T, (\lambda)^T, (\delta)^T, (\rho)^T]^T.\end{aligned}$$

Clearly, this is a nonlinear mapping. The canonical function associated with this geometrical operator is

$$V(\boldsymbol{\xi}) = \begin{cases} 0 & \text{if } \eta = 0, \lambda \leq 0, \delta = 0, \rho = 0, \\ +\infty & \text{otherwise.} \end{cases}$$

Let $U(\mathbf{x},\mathbf{y}) = -\frac{1}{2}\mathbf{x}^T \text{Diag}\,(2\mathbf{c})\mathbf{x} - \mathbf{f}^T\mathbf{y}$, primal problem can be rewritten in the canonical form:

$$P(\mathbf{x},\mathbf{y}) = \mathbf{V}(\Lambda(\mathbf{x},\mathbf{y})) - U(\mathbf{x},\mathbf{y}).$$

Define $\boldsymbol{\beta} = [(\boldsymbol{\sigma})^T, (\boldsymbol{\varsigma})^T, (\boldsymbol{\tau})^T, (\boldsymbol{\mu})^T]^T \in \mathbb{R}^{(2m+1)\times n+m}$ be the canonical dual variable corresponding to $\boldsymbol{\xi}$. The couple $(\boldsymbol{\xi}, \boldsymbol{\beta})$ forms a canonical duality pair with the Fenchel conjugate of the function $V^\sharp(\boldsymbol{\beta})$ defined by

$$\begin{aligned}V^\sharp(\boldsymbol{\beta}) &= \sup\{\boldsymbol{\xi}^T\boldsymbol{\beta} - V(\boldsymbol{\xi}) : \boldsymbol{\xi} \in \mathscr{X}\} \\ &= \begin{cases} 0 & \text{if } \boldsymbol{\sigma} > 0,\ \boldsymbol{\varsigma} \geq 0,\ \boldsymbol{\tau} > 0,\ \boldsymbol{\mu} > 0, \\ +\infty & \text{otherwise.} \end{cases}\end{aligned}$$

By considering that $V(\boldsymbol{\xi}) = \boldsymbol{\xi}^T\boldsymbol{\beta} - V^\sharp(\boldsymbol{\beta})$, the total complementarity function can be defined by

$$\Xi(\mathbf{x},\mathbf{y},\boldsymbol{\beta}) = \frac{1}{2}\mathbf{x}^T\mathbf{G}_1(\boldsymbol{\tau})\mathbf{x} - F_1^T(\boldsymbol{\sigma},\boldsymbol{\varsigma},\boldsymbol{\tau})\mathbf{x} + \frac{1}{2}\mathbf{y}^T\mathbf{G}_2(\boldsymbol{\mu})\mathbf{y} - F_2^T(\boldsymbol{\varsigma},\boldsymbol{\mu})\mathbf{y} - \boldsymbol{\sigma}^T\mathbf{e}.$$

By the criticality condition, we obtain

$$\mathbf{G}_1(\boldsymbol{\tau})\mathbf{x} = \mathbf{F}_1(\boldsymbol{\sigma},\boldsymbol{\varsigma},\boldsymbol{\tau}), \mathbf{G}_2(\boldsymbol{\mu})\mathbf{y} = \mathbf{F}_2(\boldsymbol{\varsigma},\boldsymbol{\mu}),$$

where

$$\begin{aligned}\mathbf{G}_1(\boldsymbol{\tau}) = 2\text{Diag}\,(\mathbf{c}+\boldsymbol{\tau}), \mathbf{G}_2(\boldsymbol{\mu}) = 2\text{Diag}\,(\boldsymbol{\mu}), \\ \mathbf{F}_1(\boldsymbol{\sigma},\boldsymbol{\varsigma},\boldsymbol{\tau}) = \boldsymbol{\tau} - B^T\boldsymbol{\varsigma} - A^T\boldsymbol{\sigma}, \mathbf{F}_2(\boldsymbol{\varsigma},\boldsymbol{\mu}) = \boldsymbol{\mu} + D^T\boldsymbol{\varsigma} - \mathbf{f}.\end{aligned}$$

Therefore,

$$\mathbf{x} = G_1^{-1}(\boldsymbol{\tau})F_1(\boldsymbol{\sigma},\boldsymbol{\varsigma},\boldsymbol{\tau}), \tag{10}$$

$$\mathbf{y} = G_2^{-1}(\boldsymbol{\mu})F_2(\boldsymbol{\varsigma},\boldsymbol{\mu}). \tag{11}$$

Thus, the canonical dual problem can be formulated as the following:

$$(\mathscr{P}^d) \quad \max \; P^d(\boldsymbol{\beta}) = P^d(\boldsymbol{\sigma}, \boldsymbol{\varsigma}, \boldsymbol{\tau}, \boldsymbol{\mu}) = -\frac{1}{2}\mathbf{F}_1^T(\boldsymbol{\sigma}, \boldsymbol{\varsigma}, \boldsymbol{\tau})\mathbf{G}_1^{-1}(\boldsymbol{\tau})\mathbf{F}_1(\boldsymbol{\sigma}, \boldsymbol{\varsigma}, \boldsymbol{\tau})$$

$$-\frac{1}{2}\mathbf{F}_2^T(\boldsymbol{\varsigma}, \boldsymbol{\mu})\mathbf{G}_2^{-1}(\boldsymbol{\mu})\mathbf{F}_2(\boldsymbol{\varsigma}, \boldsymbol{\mu}) - \boldsymbol{\sigma}^T\mathbf{e}$$

$$\text{s.t. } \boldsymbol{\beta} = (\boldsymbol{\sigma}, \boldsymbol{\varsigma}, \boldsymbol{\tau}, \boldsymbol{\mu}) \in \mathscr{S}_a,$$

where dual feasible space is

$$\mathscr{S}_a = \{\boldsymbol{\beta} = (\boldsymbol{\sigma}, \boldsymbol{\varsigma}, \boldsymbol{\tau}, \boldsymbol{\mu}) \in \mathbb{R}^{(2m+1)\times n+m} \mid \; \boldsymbol{\sigma} \neq 0, \; \boldsymbol{\varsigma} \geq 0, \; \boldsymbol{\tau} \neq 0, \; \boldsymbol{\mu} \neq 0\}.$$

4 Optimality Criterion

Theorem 1 (Complementary-Dual Principle). *The problem* $(\mathscr{P}^d)$ *is canonically dual to the primal problem* $(\mathscr{P})$ *in the sense that* $(\bar{\boldsymbol{\sigma}}, \bar{\boldsymbol{\varsigma}}, \bar{\boldsymbol{\tau}}, \bar{\boldsymbol{\mu}})$ *is a KKT point of* $P^d(\boldsymbol{\sigma}, \boldsymbol{\varsigma}, \boldsymbol{\tau}, \boldsymbol{\mu})$ *over* $\mathscr{S}_a$ *if and only if* $(\bar{\mathbf{x}}, \bar{\mathbf{y}})$ *is a KKT point of* $(\mathscr{P})$*, with*

$$\delta_{\mathbf{x}}\Xi(\mathbf{x}, \mathbf{y}, \boldsymbol{\sigma}, \boldsymbol{\varsigma}, \boldsymbol{\tau}, \boldsymbol{\mu}) = 0, \delta_{\mathbf{y}}\Xi(\mathbf{x}, \mathbf{y}, \boldsymbol{\sigma}, \boldsymbol{\varsigma}, \boldsymbol{\tau}, \boldsymbol{\mu}) = 0.$$

Furthermore, the following relation holds:

$$P(\bar{\mathbf{x}}, \bar{\mathbf{y}}) = \Xi(\bar{\mathbf{x}}, \bar{\mathbf{y}}, \bar{\boldsymbol{\sigma}}, \bar{\boldsymbol{\varsigma}}, \bar{\boldsymbol{\tau}}, \bar{\boldsymbol{\mu}}) = P^d(\bar{\boldsymbol{\sigma}}, \bar{\boldsymbol{\varsigma}}, \bar{\boldsymbol{\tau}}, \bar{\boldsymbol{\mu}}).$$

Proof. By introducing the Lagrange multiplier vectors $\eta \in \mathbb{R}^m$, $\boldsymbol{\lambda} \leq 0 \in \mathbb{R}^{mn}$, $\boldsymbol{\delta} \in \mathbb{R}^{mn}$, and $\boldsymbol{\rho} \in \mathbb{R}^n$ to relax the constraints, the Lagrangian function associated with the dual function $P^d(\boldsymbol{\sigma}, \boldsymbol{\varsigma}, \boldsymbol{\tau}, \boldsymbol{\mu})$ becomes

$$L(\eta, \boldsymbol{\lambda}, \boldsymbol{\delta}, \boldsymbol{\rho}, \boldsymbol{\sigma}, \boldsymbol{\varsigma}, \boldsymbol{\tau}, \boldsymbol{\mu}) = P^d(\boldsymbol{\sigma}, \boldsymbol{\varsigma}, \boldsymbol{\tau}, \boldsymbol{\mu}) - \eta^T\boldsymbol{\sigma} - \boldsymbol{\lambda}^T\boldsymbol{\varsigma} - \boldsymbol{\delta}^T\boldsymbol{\tau} - \boldsymbol{\rho}^T\boldsymbol{\mu}.$$

Then, in terms of $\mathbf{x} = G_1^{-1}(\boldsymbol{\tau})F_1(\boldsymbol{\sigma}, \boldsymbol{\varsigma}, \boldsymbol{\tau})$ and $\mathbf{y} = G_2^{-1}(\boldsymbol{\mu})F_2(\boldsymbol{\varsigma}, \boldsymbol{\mu})$, the KKT conditions of the dual problem become

$$\frac{\partial L(\eta, \boldsymbol{\lambda}, \boldsymbol{\delta}, \boldsymbol{\rho}, \boldsymbol{\sigma}, \boldsymbol{\varsigma}, \boldsymbol{\tau}, \boldsymbol{\mu})}{\partial \boldsymbol{\sigma}} = A\mathbf{x} - \mathbf{e} - \eta = 0,$$

$$\frac{\partial L(\eta, \boldsymbol{\lambda}, \boldsymbol{\delta}, \boldsymbol{\rho}, \boldsymbol{\sigma}, \boldsymbol{\varsigma}, \boldsymbol{\tau}, \boldsymbol{\mu})}{\partial \boldsymbol{\varsigma}} = B\mathbf{x} - D\mathbf{y} - \boldsymbol{\lambda} = 0,$$

$$\frac{\partial L(\eta, \boldsymbol{\lambda}, \boldsymbol{\delta}, \boldsymbol{\rho}, \boldsymbol{\sigma}, \boldsymbol{\varsigma}, \boldsymbol{\tau}, \boldsymbol{\mu})}{\partial \boldsymbol{\tau}} = \mathbf{x} \circ (\mathbf{x} - \mathbf{e}) - \boldsymbol{\delta} = 0,$$

$$\frac{\partial L(\eta, \boldsymbol{\lambda}, \boldsymbol{\delta}, \boldsymbol{\rho}, \boldsymbol{\sigma}, \boldsymbol{\varsigma}, \boldsymbol{\tau}, \boldsymbol{\mu})}{\partial \boldsymbol{\mu}} = \mathbf{y} \circ (\mathbf{y} - \mathbf{e}) - \boldsymbol{\rho} = 0,$$

$$\boldsymbol{\varsigma} \geq 0, \; \boldsymbol{\lambda} \leq 0, \boldsymbol{\varsigma}^T\boldsymbol{\lambda} = 0; \boldsymbol{\sigma} \neq 0, \; \eta = 0; \boldsymbol{\tau} \neq 0, \; \boldsymbol{\delta} = 0; \boldsymbol{\mu} \neq 0, \; \boldsymbol{\rho} = 0.$$

They can be written as:

$$A\mathbf{x} - \mathbf{e} = 0, \tag{12}$$

$$B\mathbf{x} - D\mathbf{y} \leq 0, \tag{13}$$

$$\mathbf{x} \circ (\mathbf{x} - \mathbf{e}) = 0, \tag{14}$$

$$\mathbf{y} \circ (\mathbf{y} - \mathbf{e}) = 0, \tag{15}$$

$$\boldsymbol{\sigma} \neq 0, \quad A\mathbf{x} - \mathbf{e} = 0, \tag{16}$$

$$\boldsymbol{\varsigma} \geq 0, \quad B\mathbf{x} - D\mathbf{y} \leq 0, \tag{17}$$

$$\boldsymbol{\tau} \neq 0, \quad \mathbf{x} \circ (\mathbf{x} - \mathbf{e}) = 0, \tag{18}$$

$$\boldsymbol{\mu} \neq 0, \quad \mathbf{y} \circ (\mathbf{y} - \mathbf{e}) = 0. \tag{19}$$

This proves that if $(\bar{\boldsymbol{\sigma}}, \bar{\boldsymbol{\varsigma}}, \bar{\boldsymbol{\tau}}, \bar{\boldsymbol{\mu}})$ is a KKT solution of $(\mathscr{P}^d)$, then (12)–(15) are the so-called primal feasibility conditions, while (16) and (17) are the so-called dual feasibility conditions and dual complementary conditions. Therefore, the vector $(\bar{\mathbf{x}}, \bar{\mathbf{y}})$ is a KKT solution of Problem $(\mathscr{P})$.

Again, by the complementary conditions and (10) and (11), we have

$$\begin{aligned} P^d(\boldsymbol{\sigma}, \boldsymbol{\varsigma}, \boldsymbol{\tau}, \boldsymbol{\mu}) &= -\frac{1}{2}\mathbf{F}_1^T(\boldsymbol{\sigma}, \boldsymbol{\varsigma}, \boldsymbol{\tau})\mathbf{G}_1^{-1}(\boldsymbol{\tau})\mathbf{F}_1(\boldsymbol{\sigma}, \boldsymbol{\varsigma}, \boldsymbol{\tau}) \\ &\quad -\frac{1}{2}\mathbf{F}_2^T(\boldsymbol{\varsigma}, \boldsymbol{\mu})\mathbf{G}_2^{-1}(\boldsymbol{\mu})\mathbf{F}_2(\boldsymbol{\varsigma}, \boldsymbol{\mu}) - \boldsymbol{\sigma}^T\mathbf{e} \\ &= \frac{1}{2}\mathbf{x}^T \text{Diag}\,(2\mathbf{c})\mathbf{x} + \mathbf{f}^T\mathbf{y} + \boldsymbol{\sigma}^T(A\mathbf{x} - \mathbf{e}) + \boldsymbol{\varsigma}^T(B\mathbf{x} - D\mathbf{y}) \\ &\quad +\boldsymbol{\tau}^T(\mathbf{x} \circ (\mathbf{x} - \mathbf{e})) + \boldsymbol{\mu}^T(\mathbf{y} \circ (\mathbf{y} - \mathbf{e})) \\ &= \frac{1}{2}\mathbf{x}^T \text{Diag}\,(2\mathbf{c})\mathbf{x} + \mathbf{f}^T\mathbf{y} = P(\mathbf{x}, \mathbf{y}). \end{aligned}$$

Therefore, the theorem is proved. □

Theorem 1 shows that if $(\bar{\mathbf{x}}, \bar{\mathbf{y}})$ is a KKT point of the primal problem $(\mathscr{P})$ if and only if the associated $(\bar{\boldsymbol{\sigma}}, \bar{\boldsymbol{\varsigma}}, \bar{\boldsymbol{\tau}}, \bar{\boldsymbol{\mu}})$ is a KKT point of its canonical dual. Furthermore, they have the same optimal function value. Thus, there is no duality gap between the primal problem $(\mathscr{P})$ and its canonical dual $(\mathscr{P}^d)$.

In order to identify the global minimizer of $(\mathscr{P})$, we introduce

$$\mathscr{S}_a^+ = \{\boldsymbol{\beta} = (\boldsymbol{\sigma}, \boldsymbol{\varsigma}, \boldsymbol{\tau}, \boldsymbol{\mu}) \in \mathscr{S}_a \mid \mathbf{G}_1(\boldsymbol{\tau}) \succeq 0, \mathbf{G}_2(\boldsymbol{\mu}) \succeq 0\},$$

then, we have the following theorem.

Theorem 2 (Global Optimality Condition). *Suppose that* $(\bar{\boldsymbol{\sigma}}, \bar{\boldsymbol{\varsigma}}, \bar{\boldsymbol{\tau}}, \bar{\boldsymbol{\mu}})$ *is a critical point of* $P^d(\boldsymbol{\sigma}, \boldsymbol{\varsigma}, \boldsymbol{\tau}, \boldsymbol{\mu})$*. If* $(\bar{\boldsymbol{\sigma}}, \bar{\boldsymbol{\varsigma}}, \bar{\boldsymbol{\tau}}, \bar{\boldsymbol{\mu}}) \in \mathscr{S}_a^+$*, then* $(\bar{\boldsymbol{\sigma}}, \bar{\boldsymbol{\varsigma}}, \bar{\boldsymbol{\tau}}, \bar{\boldsymbol{\mu}})$ *is a global maximizer of* $P^d(\boldsymbol{\sigma}, \boldsymbol{\varsigma}, \boldsymbol{\tau}, \boldsymbol{\mu})$ *and* $(\bar{\mathbf{x}}, \bar{\mathbf{y}})$ *defined by*

$$\bar{\mathbf{x}} = \mathbf{G}_1^{-1}(\bar{\boldsymbol{\tau}})\mathbf{F}_1(\bar{\boldsymbol{\sigma}}, \bar{\boldsymbol{\varsigma}}, \bar{\boldsymbol{\tau}}),$$
$$\bar{\mathbf{y}} = \mathbf{G}_2^{-1}(\bar{\boldsymbol{\mu}})\mathbf{F}_2(\bar{\boldsymbol{\varsigma}}, \bar{\boldsymbol{\mu}})$$

is a global minimizer of $P(\mathbf{x}, \mathbf{y})$ *on* $\mathscr{X}$, *i.e.,*

$$P(\bar{\mathbf{x}}, \bar{\mathbf{y}}) = \min_{(\mathbf{x},\mathbf{y})\in\mathscr{X}} P(\mathbf{x}, \mathbf{y}) = \max_{(\boldsymbol{\sigma},\boldsymbol{\varsigma},\boldsymbol{\tau},\boldsymbol{\mu})\in\mathscr{S}_a^+} P^d(\boldsymbol{\sigma}, \boldsymbol{\varsigma}, \boldsymbol{\tau}, \boldsymbol{\mu}) = P^d(\bar{\boldsymbol{\sigma}}, \bar{\boldsymbol{\varsigma}}, \bar{\boldsymbol{\tau}}, \bar{\boldsymbol{\mu}}).$$

Proof. On $\mathscr{S}_a^+$, both $\mathbf{G}_1(\boldsymbol{\tau})$ and $\mathbf{G}_2(\boldsymbol{\mu})$ are positive semi-definite. Their inverse are also positive semi-definite. Therefore, the canonical dual function $P^d(\boldsymbol{\sigma}, \boldsymbol{\varsigma}, \boldsymbol{\tau}, \boldsymbol{\mu})$ is concave on $\mathscr{S}_a^+$. Thus, a KKT point $(\boldsymbol{\sigma}, \boldsymbol{\varsigma}, \boldsymbol{\tau}, \boldsymbol{\mu}) \in \mathscr{S}_a^+$ must be a global maximizer of $P^d(\boldsymbol{\sigma}, \boldsymbol{\varsigma}, \boldsymbol{\tau}, \boldsymbol{\mu})$ on $\mathscr{S}_a^+$. For any given $(\boldsymbol{\sigma}, \boldsymbol{\varsigma}, \boldsymbol{\tau}, \boldsymbol{\mu}) \in \mathscr{S}_a^+$, the complementary function $\Xi(\mathbf{x}, \mathbf{y}, \boldsymbol{\sigma}, \boldsymbol{\varsigma}, \boldsymbol{\tau}, \boldsymbol{\mu})$ is convex in $\mathbf{x}, \mathbf{y}$ and concave in $(\boldsymbol{\sigma}, \boldsymbol{\varsigma}, \boldsymbol{\tau}, \boldsymbol{\mu})$, the critical point $(\mathbf{x}, \mathbf{y}, \boldsymbol{\sigma}, \boldsymbol{\varsigma}, \boldsymbol{\tau}, \boldsymbol{\mu})$ is a saddle point of the complementary function. More specifically, we have

$$
\begin{aligned}
& P^d(\bar{\beta}) = \max_{\boldsymbol{\beta}\in\mathscr{S}_a^+} P^d(\boldsymbol{\beta}) \\
&= \max_{\boldsymbol{\beta}\in\mathscr{S}_a^+} \min_{(\mathbf{x},\mathbf{y})\in\mathscr{X}} \Xi(\mathbf{x}, \mathbf{y}, \boldsymbol{\beta}) = \min_{(\mathbf{x},\mathbf{y})\in\mathscr{X}} \max_{\boldsymbol{\beta}\in\mathscr{S}_a^+} \Xi(\mathbf{x}, \mathbf{y}, \boldsymbol{\beta}) \\
&= \min_{(\mathbf{x},\mathbf{y})\in\mathscr{X}} \max_{\boldsymbol{\beta}\in\mathscr{S}_a^+} \left\{ \frac{1}{2}\mathbf{x}^T\mathbf{G}_1(\boldsymbol{\tau})\mathbf{x} - F_1^T(\boldsymbol{\sigma}, \boldsymbol{\varsigma}, \boldsymbol{\tau})\mathbf{x} + \frac{1}{2}\mathbf{y}^T\mathbf{G}_2(\boldsymbol{\mu})\mathbf{y} - F_2^T(\boldsymbol{\varsigma}, \boldsymbol{\mu})\mathbf{y} - \boldsymbol{\sigma}^T\mathbf{e} \right\} \\
&= \min_{(\mathbf{x},\mathbf{y})\in\mathscr{X}} \max_{\boldsymbol{\beta}\in\mathscr{S}_a^+} \left\{ \frac{1}{2}\mathbf{x}^T \mathrm{Diag}\,(2\mathbf{c})\mathbf{x} + \mathbf{f}^T\mathbf{y} + \boldsymbol{\sigma}^T(A\mathbf{x} - \mathbf{e}) + \boldsymbol{\varsigma}^T(B\mathbf{x} - D\mathbf{y}) \right. \\
&\quad \left. + \boldsymbol{\tau}^T(\mathbf{x} \circ (\mathbf{x} - \mathbf{e})) + \boldsymbol{\mu}^T(\mathbf{y} \circ (\mathbf{y} - \mathbf{e})) \right\} \\
&= \min_{(\mathbf{x},\mathbf{y})\in\mathscr{X}} \max_{\boldsymbol{\beta}\in\mathscr{S}_a^+} \{ \frac{1}{2}\mathbf{x}^T \mathrm{Diag}\,(2\mathbf{c})\mathbf{x} + \mathbf{f}^T\mathbf{y} + \boldsymbol{\beta}^T\boldsymbol{\xi} \} \\
&= \min_{(\mathbf{x},\mathbf{y})\in\mathscr{X}} \max_{\boldsymbol{\beta}\in\mathscr{S}_a^+} \{ \frac{1}{2}\mathbf{x}^T \mathrm{Diag}\,(2\mathbf{c})\mathbf{x} + \mathbf{f}^T\mathbf{y} + \boldsymbol{\beta}^T\boldsymbol{\xi} - V^\sharp(\boldsymbol{\beta}) \} \\
&= \min_{(\mathbf{x},\mathbf{y})\in\mathscr{X}} \{ \frac{1}{2}\mathbf{x}^T \mathrm{Diag}\,(2\mathbf{c})\mathbf{x} + \mathbf{f}^T\mathbf{y} \} + \max_{\boldsymbol{\beta}\in\mathscr{S}_a^+} \{ \boldsymbol{\xi}^T\boldsymbol{\beta} - V^\sharp(\boldsymbol{\beta}) \} \\
&= \min_{(\mathbf{x},\mathbf{y})\in\mathscr{X}} \{ \frac{1}{2}\mathbf{x}^T \mathrm{Diag}\,(2\mathbf{c})\mathbf{x} + \mathbf{f}^T\mathbf{y} \} \\
&= \min_{(\mathbf{x},\mathbf{y})\in\mathscr{X}} P(\mathbf{x}, \mathbf{y}) = P(\bar{\mathbf{x}}, \bar{\mathbf{y}}).
\end{aligned}
$$

This completes the proof. □

Theorem 2 provides a sufficient condition for a global minimizer of the primal problem ($\mathscr{P}$).

5 Conclusion

Facility location problems have received much interest in the past 40 years. Most of them are motivated from various practical applications. In this paper, we concentrated on simple model and try to solve it in the elegant way. By using the canonical dual transformation, the integer programming problem can be converted into a continuous canonical dual problem with zero duality gap. The analytical solutions are also obtained.

Acknowledgements Dr. Ning Ruan was supported by a funding from the Australian Government under the Collaborative Research Networks (CRN) program.

References

1. Aardal, K.: Capacitated facility location: separation algorithms and computational experience. Math. Program. **81**, 149–175 (1998)
2. Aardal, K., Chudak, F.A., Shmoys, D.B.: A 3-approximation algorithm for the k-level uncapacitated facility location problem. Inf. Process. Lett. **72**, 161–167 (1999)
3. Ageev, A.A.: Improved approximation algorithms for multilevel facility location problems. Oper. Res. Lett. **30**, 327–332 (2002)
4. Ageev, A., Ye, Y., Zhang, J.: Improved combinatorial approximation algorithms for the k-level facility location problem. SIAM J. Discrete Math. **18**, 207–217 (2005)
5. Balinski, M.L.: Integer programming: methods, uses, computation. Manag. Sci. **12**, 253–313 (1965)
6. Chudak, F.A., Shmoys, D.B.: Improved approximation algorithms for the uncapacitated facility location problem. SIAM J. Comput. **33**, 1–25 (2003)
7. Erlenkotter, D.: A dual-based procedure for uncapacitated facility location. Oper. Res. **1991**, 280–297 (1991)
8. Guha, S., Khuller, S.: Greedy strikes back: improved facility location algorithms. J. Algorithms **31**, 228–248 (1999)
9. Jain, K., Mahdian, M., Markakis, E., Saberi, A., Vazirani, V.V.: Greedy facility location algorithms analyzed using dual fitting with factor revealing LP. J. ACM **50**, 795–824 (2003)
10. Jain, K., Vazirani, V.V.: Approximation algorithms for metric facility location and k-median problems using the primal-dual schema and Lagrangian relaxation. J. ACM **48**, 274–296 (2001)
11. Arya, V., Garg, N., Khandekar, R., Meyerson, A., Munagala, K., Pandit, V.: Local search heuristics for k-median and facility location problems. SIAM J. Comput. **33**, 544–562 (2004)
12. Korupolu, M., Plaxton, C., Rajaraman, R.: Analysis of a local search heuristic for facility location problems. J. Algorithms **37**, 146–188 (2000)
13. Gao, D.Y. : Duality Principles in Nonconvex Systems: Theory, Methods and Applications. Kluwer Academic, Dordrecht (2000)

Duality Method in the Exact Controllability of Hyperbolic Electromagnetic Equations

Xiaojun Lu, Ziheng Tu, and Xiaoxing Liu

Abstract In this paper, we address the exact controllability problem for the hyperbolic electromagnetic equation, which plays an important role in the research of quantum mechanics. Typical techniques such as Hamiltonian induced Hilbert spaces and pseudodifferential operators are introduced. By choosing appropriate multipliers, we proved the observability inequality with sharp constants. In particular, a genuine compactness-uniqueness argument is applied to obtain the minimal time. In the final analysis, a suitable boundary control is constructed by the systematic Hilbert Uniqueness Method introduced by J.L. Lions. Compared with the micro-local discussion in Bardos et al. (SIAM J Control Optim 30:1024–1065, 1992), we do not require the coefficients belong to C^{∞}. Actually, C^1 is already sufficient for the vector potential of the hyperbolic electromagnetic equation.

Keywords Hamiltonian operator • Pseudodifferential operators • Trace theorem • Energy conservation law • Observability inequality • Hilbert uniqueness method • Unique continuation theorem • Compactness-uniqueness argument

---

X. Lu (✉)
Department of Mathematics, Southeast University, 211189 Nanjing, China

School of Economics and Management, Southeast University, 211189 Nanjing, China

BCAM, Alameda de Mazarredo 14, 48009 Bilbao, Bizkaia, Spain
e-mail: lvxiaojun1119@hotmail.de

Z. Tu
School of Mathematics and Statistics, Zhejiang University of Finance and Economics, 310018 Hangzhou, China
e-mail: tuziheng@zufe.edu.cn

X. Liu
School of Economics and Management, Southeast University, 211189 Nanjing, China
e-mail: starsunmoon198@163.com

D. Gao et al. (eds.), *Advances in Global Optimization*, Springer Proceedings in Mathematics & Statistics 95, DOI 10.1007/978-3-319-08377-3_19

1 Introduction to Hyperbolic Electromagnetic Equations and Exact Controllability

In the study of quantum mechanics, a magnetic field is defined as a vector field produced by electric fields varying in time, spinning of the elementary particles, or moving electric charges, etc. For instance, as is known to all, the earth's magnetic field is a consequence of the movement of convection currents in the outer ferromagnetic liquid of the core. Nowadays, with the fast development of modern technology, electromagnetic theory is widely utilized in medical research of organs' biomagnetism, studying the vortex in the superconductor which carries quantized magnetic flux, and predicting geographical cataclysms, such as earthquakes, volcanic eruptions, geomagnetic reversal, etc.

Mathematically speaking, the magnetic field **B** is a solenoidal vector field whose field line either forms a closed curve or extends to infinity. In contrast, a field line of the electric field **E** starts at a positive charge and ends at a negative charge.

Let $\mathbf{A}(x)$ be the vector potential of $\mathbf{B}$, which does not depend on time, $\mathbf{B} = \nabla\times\mathbf{A}$. Clearly, $\nabla \cdot \mathbf{B} = \operatorname{div}\operatorname{rot}\mathbf{A} = 0$. We deduce from the Maxwell's equation (μ is the magnetic permeability) $\nabla \times \mathbf{E} = -\mu\frac{\partial \mathbf{B}}{\partial t} = 0$ that $\mathbf{E} = -\nabla\phi$, where the scalar ϕ represents the electric potential. Next we choose an appropriate Lagrangian for the charged particle in the electromagnetic field (q is the electric charge of the particle, and $\mathbf{v}$ is its velocity, m is mass) $\mathscr{L} = \frac{m\mathbf{v}^2}{2} - q\phi + q\mathbf{v}\cdot\mathbf{A}$. The canonical momentum is specified by the equation $\mathbf{p} = \nabla_{\mathbf{v}}\mathscr{L} = m\mathbf{v} + q\mathbf{A}$. Then we define the classical Hamiltonian by Legendre transform,

$$\mathscr{H} \triangleq \mathbf{p}\cdot\mathbf{v} - \mathscr{L} = \frac{(\mathbf{p} - q\mathbf{A})^2}{2m} + q\phi.$$

In quantum mechanics, we replace $\mathbf{p}$ by $-i\hbar\nabla$, ($\hbar$ is the Planck constant) and we have

$$\mathscr{H} = \frac{(i\hbar\nabla + q\mathbf{A})^2}{2m} + q\phi.$$

When we do not consider the influence from the electric field **E**, then the above Hamiltonian can be simplified as the differential operator $\mathscr{H}_{\mathbf{A}}^2 \triangleq (i\nabla + \mathbf{A})^2 : \mathscr{H} \to \mathscr{H}^*$. $\mathscr{H}$ and $\mathscr{H}^*$ are concerned function spaces. This Hamiltonian operator phenomenologically describes a quantity of behaviors discovered in superconductors and quantum electrodynamics (QED). Ginzburg–Landau equations, Schrödinger equations, Dirac equations, and the matrix Pauli operator are famous examples in this respect. Interesting readers please refer to [1–6] for more details.

We are interested in addressing the following variational problem in a suitable function space $\mathscr{U}$, which will be explained later,

$$\min_{u\in\mathscr{U}}\left\{\frac{1}{2}\int_{\Omega}\left(|u_t|^2 - |(i\nabla + \mathbf{A}(x))u|^2 - \phi(x)|u|^2\right)dx\right\}.$$

Let the Lagrangian be

$$\mathscr{L}(t, x_1, \cdots, x_n, u, \bar{u}, u_t, \bar{u}_t, \nabla u, \nabla \bar{u}) \triangleq |u_t|^2 - |(i\nabla + \mathbf{A}(x))u|^2 - \phi(x)|u|^2.$$

The Euler–Lagrangian equation for $\mathscr{L}$ is of the form

$$\frac{\partial \mathscr{L}}{\partial u} - \frac{\partial}{\partial t}\Big(\frac{\partial \mathscr{L}}{\partial u_t}\Big) - \sum_{i=1}^{n} \frac{\partial}{\partial x_i}\Big(\frac{\partial \mathscr{L}}{\partial u_{x_i}}\Big) = 0.$$

In fact, simple calculation leads to

$$\frac{\partial \mathscr{L}}{\partial u} = i\mathbf{A}\cdot\nabla\bar{u} - \mathbf{A}^2\bar{u} - \phi(x)\bar{u}, \quad \frac{\partial \mathscr{L}}{\partial u_t} = \bar{u}_t, \quad \frac{\partial \mathscr{L}}{\partial u_{x_i}} = -\bar{u}_{x_i} - ia_i\bar{u}.$$

Consequently, one has

$$u_{tt} + \mathscr{H}_{\mathbf{A}}^2 u + \phi(x)u = 0.$$

Let $\Omega \subset \mathbb{R}^N$ be a bounded open set with a time-independent vector potential $\mathbf{A}(x)$. In this work, we mainly discuss the exact controllability problem of the hyperbolic electromagnetic equation in the following form,

$$\begin{cases} u_{tt} + \mathscr{H}_{\mathbf{A}}^2 u = 0 & (t,x) \in (0,T)\times\Omega \\ u = \psi & (t,x) \in (0,T)\times\Gamma \\ u(0,x) = u_0(x), u_t(0,x) = u_1(x) \; x \in \Omega. \end{cases} \tag{1}$$

We are interested in the property, i.e. for every initial data (u_0, u_1) and every target (u_0^T, u_1^T) in appropriate function spaces, whether there exists a boundary control ψ such that the solution u of (1) satisfies

$$(u(T,x), u_t(T,x)) = (u_0^T, u_1^T) \text{ fora.e. } x \in \Omega. \tag{2}$$

Due to the time-reversibility for hyperbolic operators, we can decompose the linear problem (1) into two parts, u^a and u^b. u^a is solution of the homogeneous Dirichlet problem

$$\begin{cases} u^a_{tt} + \mathscr{H}_{\mathbf{A}}^2 u^a = 0 & (t,x) \in (0,T)\times\Omega \\ u^a = 0 & (t,x) \in (0,T)\times\Gamma \\ u^a(T,x) = u_0^T(x), u^a_t(T,x) = u_1^T(x) \;\; x \in \Omega. \end{cases}$$

Assume that there exists a function ψ such that the solution u^b of the problem

$$\begin{cases} u^b_{tt} + \mathscr{H}_{\mathbf{A}}^2 u^b = 0 & (t,x) \in (0,T)\times\Omega \\ u^b = \psi & (t,x) \in (0,T)\times\Gamma \\ u^b(0,x) = u_0(x) - u^a(0,x), u^b_t(0,x) = u_1(x) - u^a_t(0,x) \;\; x \in \Omega \end{cases}$$

satisfies

$$u^b(T,x) = u_t^b(T,x) = 0.$$

It is evident that $u = u^a + u^b$ is the solution of (1) and it satisfies (2). In view of this, it is sufficient to consider the null controllability of (1). Henceforth, we shall assume the target $u_0^T = u_1^T = 0$. Partial boundary control can be introduced similarly. More details are explained in Sect. 2.

2 Main Results and Conclusion

In this section, as an example, we mainly consider the whole boundary control problem (1). Partial boundary control problems can be treated in a similar manner. And once the boundary control problem is solved, one can even treat the inner control problem with the techniques in [7]. Of course, the regularity of the vector potential and boundary must be sufficiently high.

Theorem 2.1. *Assume that* $A \in (C^1(\overline{\Omega}))^N$, *and the boundary* $\Gamma \in C^2$. *When* $T > 2\max_\Omega \|x\|_2$, *then for any initial data* $(u_0, u_1) \in L^2 \times H^{-1}$, *we can find a boundary control* $\psi \in L^2([0,T]; L^2(\Gamma))$ *such that the hyperbolic electromagnetic problem (1) is exactly controllable.*

Sketch of proof: First we describe the $\mathscr{H}_{\mathbf{A}}^2$-induced function spaces $\mathscr{H}_0^1$, $\mathscr{H}^{-1}$ and introduce $\mathscr{H}_{\mathbf{A}}^2$-pseudodifferential operators in the distributional sense. And the usual compactness-uniqueness argument is applied to demonstrate a generalized Poincaré's inequality, which helps to establish the equivalence between $\mathscr{H}_0^1$ and the classical Sobolev space H_0^1. With these tools, we use the idea of HUM (Hilbert Uniqueness Method) and apply suitable multipliers to obtain the hidden regularity inequality and observability inequality of the adjoint problem of (1), respectively. Readers can also refer to [8–13] for the philosophy of HUM. In order to prove the minimal time for exact controllability, one uses a genuine compactness-uniqueness argument introduced in [7, 14]. In addition, a rigorous proof of the unique continuation theorem for the elliptic operator $\mathscr{L} = \mathscr{H}_{\mathbf{A}}^2 - \phi$ is given in [15]. The most crucial part is how to prove the hidden regularity and observability inequality for the adjoint problem,

$$\begin{cases} w_{tt} + \mathscr{H}_{\mathbf{A}}^2 w = 0 & (t,x) \in (L,S) \times \Omega \\ w(t,x) = 0 & (t,x) \in (L,S) \times \Gamma \\ w(L,x) = w^L(x), w_t(L,x) = w_t^L(x) \; x \in \Omega. \end{cases} \tag{3}$$

Step 1

Lemma 2.2. *(Hidden Regularity Inequality) For any given initial data* $(w^L, w_t^L) \in \mathscr{H}_0^1 \times L^2$ *and* $S \in \mathbb{R}$, *the homogeneous problem (3) has a unique solution such that*

$$w \in C([L,S];\mathscr{H}_0^1)\bigcap C^1([L,S];L^2)\bigcap C^2([L,S];\mathscr{H}^{-1}).$$

Moreover, the outward normal derivative satisfies

$$\frac{\partial w}{\partial \nu} \in L^2(L,S;L^2(\Gamma))$$

and

$$\int_L^S \int_\Gamma |\frac{\partial w}{\partial \nu}|^2 d\Gamma dt \le C(1+(S-L))\mathbb{E}(w)(L),$$

where

$$\mathbb{E}(w)(L) \triangleq \|w_t(L,\cdot)\|_{L^2}^2 + \|w(L,\cdot)\|_{\mathscr{H}_0^1}^2.$$

In particular, for the one-dimensional case, let $\Omega = (-1,1)$, then the above inequality can be rewritten as

$$\int_L^S (|w_x(t,-1)|^2 + |w_x(t,1)|^2)dt \le C(1+(S-L))\mathbb{E}(w)(L).$$

Sketch of proof: It can be proved by applying the multiplier $\mathbf{H}(x)\cdot\mathscr{H}_{\mathbf{A}}w$ to (3), where $\mathbf{H} \in C^1(\overline{\Omega})$ and $\mathbf{H} = \nu$ on Γ. Indeed, since $\overline{\Omega}$ is compact, then it can be covered with a finite number of neighborhoods $\mathscr{O}_k \subset \mathbb{R}^N$, $k = 1,2,\cdots,m$, in which there is a unique $\mathscr{O}_1$ satisfying $\mathscr{O}_1 \cap \Gamma = \varnothing$. Next, we choose $\zeta_1 = 0$ in $\mathscr{O}_1$ and $\zeta_k \in C^1(\overline{\mathscr{O}_k})$ such that $\zeta_k = \nu$ on $\mathscr{O}_k \cap \Gamma$ for $k = 2,\cdots,m$. Let $\{\theta_k \in C_0^2(\mathscr{O}_k)\}_k$ be a partition of unity, corresponding to the covering $\{\mathscr{O}_k\}_k$. Then we define $\mathbf{H}(x) \triangleq \left(\sum_k \theta_k\zeta_k\right)\Big|_{\overline{\Omega}}$. Q. E. D.

Step 2

Lemma 2.3. *(Observability Inequality) Let $T^* \triangleq 2\max_\Omega \|x\|_2$. For any given $L,S \in \mathbb{R}$ and initial data $(w^L, w_t^L) \in \mathscr{H}_0^1 \times L^2$, then when $S-L > T^*$, the outward normal derivative satisfies*

$$((S-L)-T^*)\mathbb{E}(w)(L) \le C(\Omega)\int_L^S \int_\Gamma |\frac{\partial w}{\partial \nu}|^2 d\Gamma dt.$$

In particular, in one-dimensional case, let $\Omega = (-1,1)$, the above inequality can be rewritten as

$$((S-L)-T^*)\mathbb{E}(w)(L) \le C(\Omega)\int_L^S (|w_x(t,-1)|^2 + |w_x(t,1)|^2)dt.$$

Sketch of proof: For the multidimensional case, by applying the multipliers $x \cdot \mathscr{H}_{\mathbf{A}} w$ and $i\frac{N-1}{2}w$ to the adjoint equation $w_{tt} + \mathscr{H}_{\mathbf{A}}^2 w = 0$, respectively, one has the following identity as a result,

$$\frac{1}{2}\int_L^S \int_\Gamma \left|\frac{\partial w}{\partial \nu}\right|^2 \cdot \Big(x \cdot \nu\Big) d\Gamma dt$$

$$= \underbrace{-\operatorname{Im}(w_t, x \cdot \mathscr{H}_{\mathbf{A}} w)_{L^2}\Big|_L^S}_{(I)} \underbrace{-\operatorname{Im}(w_t, \frac{N-1}{2} i w)_{L^2}\Big|_L^S}_{(II)}$$

$$\underbrace{+\operatorname{Re}\int_L^S \int_\Omega \overline{w}\mathscr{H}_{\mathbf{A}} w \times \Xi_{\mathbf{A}} \times \mathbf{H}^T dx dt}_{(III)}$$

$$+(S-L)\mathbb{E}(w)(L).$$

Here $\Xi_{\mathbf{A}}$ is the compatibility matrix, which serves as a test matrix for the magnetic field, i.e.

$$\Xi_{\mathbf{A}} \triangleq \begin{pmatrix} \xi_{11} & \xi_{12}, & \cdots & \xi_{1N} \\ \xi_{21} & \xi_{22} & \cdots & \xi_{2N} \\ \vdots & \vdots & \cdots & \vdots \\ \xi_{N1} & \xi_{N2} & \cdots & \xi_{NN} \end{pmatrix}$$

where

$$\xi_{jk} \triangleq \begin{vmatrix} \nabla_j & \nabla_k \\ a_j & a_k \end{vmatrix}.$$

Actually,

$$|(I) + (II)| \le 2 \max_\Omega \|x\|_2 \mathbb{E}(w)(L).$$

As to (III), apply Schwartz's inequality and one has

$$|(III)| \le \epsilon \max_x (\|\Xi_{\mathbf{A}}\|_F \|\mathbf{H}\|_2)(S-L)\mathbb{E}(w)(L)$$

$$+\frac{1}{4\epsilon} \max_x (\|\Xi_{\mathbf{A}}\|_F \|\mathbf{H}\|_2) \int_L^S \int_\Omega |w|^2 dx dt,$$

where $\|\Xi_{\mathbf{A}}\|_F$ is the Frobenius norm of $\Xi_{\mathbf{A}}$. Afterwards, one applies the usual compactness-uniqueness argument introduced in [7, 10, 14]. With this method, the lower order terms can be absorbed by the boundary integral, i.e. there exists a positive constant $C(\epsilon, \Omega)$ such that

$$\int_L^S \int_\Omega |w|^2 dx dt \le C(\epsilon, \Omega) \int_L^S \int_\Gamma \left|\frac{\partial w}{\partial \nu}\right|^2 d\Gamma dt.$$

This concludes our proof. Q.E.D.

Step 3

Lemmas 2.2 and 2.3 together indicate, for $T > 2\max_\Omega \|x\|_2$,

$$\int_0^T \int_\Gamma |\frac{\partial w}{\partial \nu}|^2 d\Gamma dt$$

defines an equivalent norm for $\mathcal{H}_0^1 \times L^2$. While Lemma 2.2 also demonstrates that, for any given $T \in \mathbb{R}$ and initial data $(w(0,x), w_t(0,x)) \in \mathcal{H}_0^1 \times L^2$, there exists a unique solution

$$w \in C([0,T]; \mathcal{H}_0^1) \bigcap C^1([0,T]; L^2) \bigcap C^2([0,T]; \mathcal{H}^{-1})$$

for the homogeneous problem (3). The outward normal derivative satisfies

$$\frac{\partial w}{\partial \nu} \in L^2([0,T]; L^2(\Gamma))$$

and it is continuous with respect to the initial data. If we choose

$$\psi \triangleq \frac{\partial w}{\partial \nu} \in L^2([0,T]; L^2(\Gamma))$$

and consider problem (1) with the initial data $(u(T,x), u_t(T,x)) = 0$, then problem (1) has a unique solution satisfying

$$(u_0, u_1) \triangleq (u(0,x), u_t(0,x)) \in L^2 \times \mathcal{H}^{-1}.$$

And the linear mapping

$$(u(T,x), u_t(T,x), \psi) \longmapsto (u(0,x), u_t(0,x))$$

is continuous from $L^2 \times \mathcal{H}^{-1} \times L^2([L,S]; L^2(\Gamma))$ into $L^2 \times \mathcal{H}^{-1}$ with respect to these topologies. Let $(w_0, w_1) = (w(0,x), w_t(0,x))$ be the initial data of (3). Thus, in a unique fashion, one can define a linear and bounded mapping

$$\mathcal{S} : \mathcal{H}_0^1 \times L^2(\Omega) \longrightarrow \mathcal{H}^{-1} \times L^2(\Omega),$$

$$(w_0, w_1) \mapsto (u_1, -u_0).$$

By applying the Lions–Lax–Milgram lemma in [16], one knows that $\mathscr{S}$ is surjective. Then $\psi \triangleq \frac{\partial w}{\partial \nu}$ is an appropriate boundary control which drives $(u_0, u_1) \in L^2 \times \mathscr{H}^{-1}$ to rest. Q.E.D.

In the above theorem, one applies a control on the whole boundary Γ. Next we consider the partial boundary control problem. For fixed $x^0 \in \mathbb{R}^N$, let

$$\Gamma_+ \triangleq \{x \in \Gamma : (x - x^0) \cdot \nu(x) > 0\},$$

$$\Gamma_- \triangleq \{x \in \Gamma : (x - x^0) \cdot \nu(x) \leq 0\}.$$

And our control problem is stated in the following form,

$$\begin{cases} u_{tt} + \mathscr{H}_{\mathbf{A}}^2 u = 0 & (t, x) \in (0, T) \times \Omega \\ u = \psi & (t, x) \in (0, T) \times \Gamma_+ \\ u = 0 & (t, x) \in (0, T) \times \Gamma_- \\ u(0, x) = u_0(x), u_t(0, x) = u_1(x) \ x \in \Omega. \end{cases} \tag{4}$$

Apply the same techniques as in Theorem 2.1, and one proves the partial boundary control problem (4).

Theorem 2.4. *Assume that $A \in (C^1(\overline{\Omega}))^N$, and the boundary $\Gamma \in C^2$. When $T > 2\max_\Omega \|x - x^0\|_2$, then for any initial data $(u_0, u_1) \in L^2 \times H^{-1}$, we can find a partial boundary control $\psi \in L^2([0, T]; L^2(\Gamma_+))$ such that the hyperbolic electromagnetic problem (4) is exactly controllable.*

When we consider the influence from the electric field **E**, e.g. replacing $\mathscr{H}_A^2$ in (4) by $(i\nabla + A(x))^2 + \phi(x)$, actually, by applying the same multipliers and compactness-uniqueness argument, one is able to prove the following fact.

Theorem 2.5. *Assume that $A \in (C^1(\overline{\Omega}))^N$, $\phi \in L^\infty(\Omega)$ is a nonnegative real function, and the boundary $\Gamma \in C^2$. When $T > 2\max_\Omega \|x - x^0\|_2$, then for any initial data $(u_0, u_1) \in L^2 \times H^{-1}$, we can find a partial boundary control $\psi \in L^2([0, T]; L^2(\Gamma_+))$ such that the hyperbolic electromagnetic problem with $\mathscr{H}_A^2$ in (4) replaced by $(i\nabla + A(x))^2 + \phi(x)$ is exactly controllable.*

Remark 2.6. *Compared with the classical multiplier $H(x) \cdot \nabla w$, the new multiplier $H(x) \cdot \mathscr{H}_A w$ has several advantages. On the one hand, it allows to utilize the special quantum structure of the Hamiltonian, such as the magnetic energy conservation law, the test matrix for magnetic field Ξ_A, Coulomb gauge condition, etc. However, $H(x) \cdot \nabla w$ will destroy the particular physical structure. On the other hand, it helps to obtain the optimal minimal control time. In contrast, if we use the multiplier $H(x) \cdot \nabla w$, by the compactness-uniqueness argument, there exist several remainder terms which can only be estimated by uncertain constants, such as Poincaré constant, etc., which keep us from getting the optimal minimal control time $2\|x - x^0\|_2$. Moreover, this new multiplier can be successfully applied to the discussion of exact controllability problems for magnetic Schrödinger equations.*

Remark 2.7. *During the proof, one finds that, in a high magnetic field with large* $\|\Xi_A\|_F$, *it is difficult to impose some exterior force on the boundary to influence the interior activity. Astronomically speaking, when the solar wind with coronal mass ejections encounters Earth's magnetosphere, most of the radioactive particles are deflected around the earth instead of impacting the atmosphere or the earth's surface, although some leakage occurs, resulting in auroras and Van Allen belts.*

Remark 2.8. *Actually, by applying the transmutation method introduced in [17, 18], one can check the null controllability results for magnetic heat equations and magnetic Schrödinger equations. It is really challenging to investigate the control theory in the field of quantum mechanics. Many interesting problems, such as the exact controllability of Maxwell's equations, nonlinear Ginzburg–Landau equations, etc., are to be addressed. More references are to be found in [19, 20].*

Acknowledgements This project was suggested by Prof. Enrique Zuazua during the first author's post-doctoral research in BCAM from 2010–2011. This project is supported by Natural Science Foundation of Jiangsu Province, BK20130598, SBK201342035, NSFC 71273048, 70973028, 11001049, Fundamental Research Funds for the Central Universities No. 3207012208 (2012) and No. 3207013210 (2013) in Southeast University. This project is also supported by Grant MTM2008-03541 and MTM2011-29306 of the MICINN (Spain), the ERC Advanced Grant FP7-246775 NUMERIWAVES, the ESF Research Networking Programme OPTPDE and the Grant PI2010-04 of the Basque Government. Moreover, the authors also express their deep gratitude to the referees for their careful reading and useful remarks.

References

1. Cazanave, T.: Semilinear Schrödinger Equations. Courant Lecture Notes. AMS, Providence (2003)
2. Fanelli, L.: Electromagnetic Schrödinger flow: multiplier methods for dispersion. In: Proceedings Journées EDP, Port D'Albret, GDR 2434 (CNRS) (2010)
3. Goldberg, M.: Strichartz estimates for Schrödinger operators with a non-smooth magnetic potential. Discrete Contin. Dyn. Syst. **31**, 109–118 (2011)
4. Kovarik, H.: Large time behavior of the heat kernel of two-dimensional magnetic Schrödinger operators. Preprint (2010)
5. Peskin, M.E., Schröder, D.V.: An Introduction to Quantum Field Theory. Westview Press, Boulder (1995)
6. Reed, M., Simon, B.: Methods of Modern Mathematical Physics, I–IV. Academic, London (1978)
7. Bardos, C., Lebeau, G., Rauch, J.: Sharp sufficient conditions for the observation, control, and stabilization of waves from the boundary. SIAM J. Control Optim. **30**, 1024–1065 (1992)
8. Fernández-Cara, E., Zuazua, E.: On the null controllability of the one-dimensional heat equation with non-smooth coefficients. Preprint (2000)
9. Komornik, V.: Contrôlabilité exacte en un temps minimal. C. R. Acad. Sci. Paris, Série I, **304**, 223–225(1987).
10. Komornik, V.: A new method of exact controllability in short time and applications. Annales Faculté des Sciences de Toulouse Série 5 **10**, 415–464 (1989)
11. Komornik, V.: Exact Controllability and Stabilization, The Multiplier Method. Masson, Paris (1994)

12. Lions, J.L.: Contrôlabilité Exacte et Stabilisation de Systèmes Distribués, vol. 1. Masson, Paris (1988)
13. Vancostenoble, J., Zuazua E.: Hardy inequalities, observability, and control for the wave and Schrödinger equations with singular potentials. SIAM J. Math. Anal. **41**, 1508–1532 (2009)
14. Zuazua, E.: Controllability of the linear system of thermoelasticity. J. Math. Pures Appl. **74**, 291–315 (1995)
15. Lu, X.: Unique continuation for magnetic Schrödinger operators. Preprint (2013)
16. Lions, J.L., Magenes, E.: Problèmes aux limites non homogènes et applications I–III. Dunod, Paris (1968–1970)
17. Miller, L.: Geometric bounds on the growth rate of null-controllability cost for the heat equation in small time. J. Differ. Equ. **204**, 202–226 (2004)
18. Miller, L.: Controllability cost of conservative systems: resolvent condition and transmutation. J. Funct. Anal. **218**, 425–444 (2005)
19. Komornik, V.: Boundary stabilization, observation and control of Maxwell's equations. Panam. Math. J. **4**, 47–61 (1994)
20. Komornik, V.: Rapid boundary stabilization of Maxwell's equations. Équations aux Dérivées Partielles et Applications, Articles dédiés à Jacques-Louis Lions, pp. 611–622. Gauthier-Villars, Paris (1998)

Conceptual Study of Inter-Duality Optimization

Shaokun Chen

Abstract This paper intends to reveal the inter-duality nature of a system and the intrinsic structure of a general system. Furthermore, the inter-duality theory will be introduced and employed to analyse the nature of optimization and optimal behaviour relative to management systems.

Keywords General system • Systematics inter-duality • Optimization management

1 Introduction

The birth of *Systems Science* has had epoch-making significance, which, in terms of philosophy, indicates that the human scientific process has converted from the mechanical reductionism era, with the feature of $1 + 1 = 2$, into a new age, the systematic thinking era, holding the characteristic of $1 + 1 \neq 2$. With the revolution of this new thinking pattern, the optimization behaviour has eventually come into our consciousness and has formed a scientific group with the help of various theories and diverse applications according to different fields of human activities.

Meanwhile, a new epoch of *Human Sciences* has been switched into *Post-positivism*, i.e. a theory can be established on the basis of a set of conjectures (or axioms) instead of finding entirely available evidence (just accepting the logical verification and striving to gain wholly solid evidence). Seriously, almost all modern theories have these post-positivism characteristics, such as mathematics, micro-physics, cosmology, social sciences, and even include theology and philosophy, etc. For instance, Newtonian mechanics was built on three postulates; Einstein's Theory of Relativity was initially constructed by two axioms, which had not been verified at that time. This feature also includes the postulates of economics and laws and the regulations and rules in management activities. We integrate the *axiomatization* and *formalization* into systematic insights and axiomatic thought, which is naturally

S. Chen (✉)
Institute for Systematics [www.interduality.com], Attached No. 1, No. 155, West 1st Section of First Ring Road, Qing Yang District, Chengdu City, Sichuan Province, China
e-mail: chenshaokun1989@gmail.com

D. Gao et al. (eds.), *Advances in Global Optimization*, Springer Proceedings in Mathematics & Statistics 95, DOI 10.1007/978-3-319-08377-3_20

employed as a new thinking style for recognizing and modelling the objective objects—systems. This axiomatic thought should absolutely be introduced into management science, social science, theology, philosophy and systems science, etc.

Human Sciences have evolved for thousands of years, and we have created a range of thoughts, methods and models to make sense of the laws and principles of physical reality, the behaviour of living systems, the activities of humanity and the phenomena of abstract existence, etc. Actually, from the viewpoint of *General System Theory* [1], extracted from sophisticated phenomena and complex objects and intricate human activities and proposed explicitly by *Ludwig Von Bertalanffy*, it is evidently reasonable that existence or reality or entity, in terms of generalized category, can be accurately described as a system. Therefore, it is unquestionable that we can employ revealing the nature and intrinsic structure and evolution of a general system to represent the study of essence, laws and principles of objects involved in an *Ultimate Objective World* [2].

As an overview of the research characteristics with regard to systems science all over the world, we can briefly conclude that it focuses on material systems in Europe, concentrates on living systems in the America, and social systems are main focus in China. In summary, the structure of systems science basically involves four fields: the systematic philosophy, the systematic theories, the systematic engineering and systematics (or systematology), as is shown in Fig. 1.

In this paper, we concentrate on the conceptual study of optimization based on theory of systematics, which is the fundamental theory of systems science [3]. There are four parts to our research of the nature, the intrinsic structure of the general system, and the optimization of cognition of the general system. Later we will employ a management system as a representative example. A brief instruction is as follows.

1. Concept study: introducing definitions of the concept of a general system.
2. Structure study: revealing the nature and the intrinsic structure of a general system and proposing the *Inter-duality Theory* by introducing the inter-duality concept, and then concisely studying inter-dual systems.

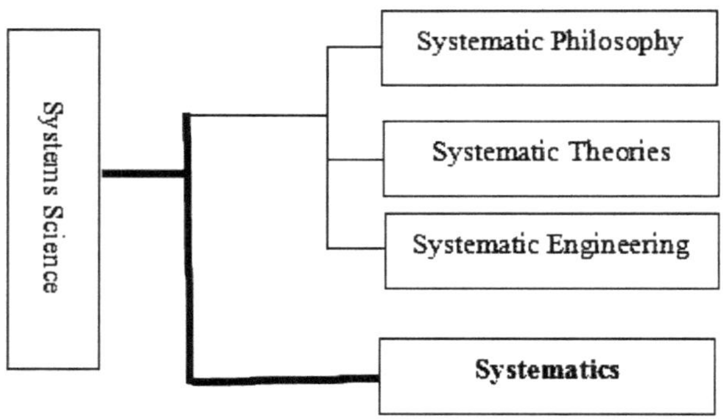

Fig. 1 The structure of systems science

3. Optimization study: studying the optimization conceptually by means of inter-duality thought.
4. Application study: applying inter-duality thought and optimization cognition to management systems.

2 Concept Study

Any independent science has its fundamental theory [3, 4].

2.1 Concept of Systematics

Systematics or systematology, like mathematics, physics, sociology, politics, etc., is a fundamental theory corresponding to the associated independent science. There are three common features.

- To explore the general problems
- To explain the conjunct phenomena
- To reveal and establish a fundamental logic system and to have higher abstraction and more profundity

Ultimately, a fundamental theory must deal with basic concepts and answer the question of being or reveal the nature of associated objects instead of modelling or merely defining and explaining these objective objects. This is the distinct difference between reality and approximation or essential otherness between *conception* and *definition*, shown as follows.

- Conception is the real reflection of objective reality through consciousness behaviour; its characteristics are *objectivity*, *uniqueness*, *being* and *abstraction*, i.e., one recognizes the general system from revealing *what it is actually being*.
- Definition is the description and explanation of a concept by concise words which depend on profundity of cognition and deliberation for the purpose of understanding, communicating and convenient expression; its features are *subjectivity*, *non-uniqueness*, *approximation* and *conciseness*, i.e., one employs definition to express and share a concept that as close as possible strives to answer *what it is really like.*

Roughly speaking, the definition is divided into *descriptive definition*[1] and *axiomatic definition* [2][5].

---

[1]Such as the definition of society, mathematics, etc.

[2]Such as the definition of differential, legal regulations, etc.

Accordingly, a definition of systematics is provided.

Definition 1. Systematics *is a basic theoretical branch of systems science, which satisfies the following axioms.*

- In allusion to the most widely objective objects involved in systems science, i.e., the general system;
- The research objects are a complete system which is based on *Ultimately Objective World View* [2, 3, 5];
- The research objectives are revealing the nature of the system and its intrinsic structure, the fundamental laws and principles of the general system, i.e., the *being cognition* and *complete space* investigation of the general system;
- To create and to apply the most profound and powerful methodology;

2.2 Concept of General System

As the concept of system has been proposed for almost 80 years, and the complex system study has been involved in this colourful and intricate world more widely in the recent decades, there must be many definitions from different fields with various descriptions [6]. Therefore, in this paper, it is reasonable that we can define a system as follows:

Definition 2. *An object is a system if it is treated as something that contains contents and details.*

By explanation, we usually say a system consists of elements. Through the above definition, we know that this sentence in quotation marks is not a system, while it actually is a system, once one focuses on its contents and details. It is true that from the viewpoint of nowadays a system is made up of elements, yet we concentrate more on this constitution with respect to the system. That is, we focus on firstly, what is the result of this constitution; and secondly, what is the constitution itself. It is reasonable from definitions as follows that the constitution means *forming a system through relationship*. Therefore, it is of remarkable significance to explore the relationship of a system, and we will understand that the optimization of a system is intimately connected with this inner relationship.

Definition 3. *A system is a triple group.*

This definition means that a system is a triple structure that consists of its Objective Space, Elements Space, and Relationship Space. Let a system be S, then it can be defined as a set form:

$$S = (Y; X, F) \tag{1}$$

Here Y represents the objective space of a system; X expresses the elements space of a system; F indicates the relationship space of a system. Further, we understand

Definition 4. *A system can be equally represented by its objective spaces.*

Let a system be S, then we get

$$S = (Y; X, F) = \{y : y = f(x, a), y \in Y, x \in X, (f, a) \in F\} \quad (2)$$

Proposition 1. *A system can be embodied by subsystems of different hierarchies.*

Let a system be S, and its subsystems are $\{s\}$, then for s, it has its own objective space and we can denote it as:

$$s = y = f(x, a), y \in Y, x \in X, (f, a) \in F \quad (3)$$

Therefore, $S = \{y\} = \{s\}$

In principle, an arbitrary system can be an element of its higher hierarchical system; meanwhile, it also occupies its lower hierarchical systems or subsystems as its elements. So is the view of the objective spaces of a system.

Proposition 2. *A general system is an open system.*

Generally speaking, a system exists in an environment or takes a higher hierarchical system as its circumstance; furthermore, a system has the capacity to exchange energy and information and substances with its environment. Therefore, it is true that general system must be an open system.

In this part, the definitions of the concept of a general system have been constructed. The next parts will reveal the nature of a general system and do a profound study on the intrinsic structure of a system.

3 Structure Study

Although we have established definitions of the concept of a general system, we have not yet revealed the nature and intrinsic structure and inner properties of general system. This section will explain these definitions.

3.1 Inter-Dual System

Under the description in 2, we have to ask further whether the notion of a system could be excavated deeply and examined more profoundly? Can a system have intrinsic and fundamental structure? If so, how can we reveal these essential elements? We understand that the best method of study would be to explore and investigate a system from its abstract and virtual hierarchies. Indeed, we can analyse a system from its spatial hierarchies. Meanwhile, if we use spatial consciousness to rethink the characteristics of profoundly investigating systems, it is obvious that the methods of analysis of a system are usually manifested as a flat thinking style, such as row and column, item and piece, far and near, inside and outside, domain and range, scope and region, etc. Sometimes these styles mention the hierarchy, gradation, and vertical, jump, etc., but it is merely through subconsciousness, e.g., from the meaning of daily life and the sense of experienced thinking. Therefore, we need the breakthrough of spatial consciousness, especially when studying the *complete space* of a system and *inter-duality* of a general system.

From Definitions 3 and 4, Proposition 1, we can easily demonstrate that if a system is denoted by S, and all its subspaces the system occupies are denoted by $\{s\}$, we get:

$$S = \{s_{ij}\} = \{s_{ij}\}_{i,j=1}^{m,n} \tag{4}$$

Here i denotes spatial hierarchies of subsystems and j denotes the number of subsystems in the same level.

Proposition 3. *A system encompasses the completely spatial hierarchies of the system or complete space of the system.*

From (4) and Proposition 1, we can verify succinctly that the completely spatial hierarchies determined by all the subsystems of a general system integrate into complete space of the objective system. In this way, we understand clearly and demonstrate vividly that the complete space of a general system can be an underlying notion to recognize the general system visually and analyse the objective system profoundly.

Moreover, for $\forall S_{i_1}, S_{i_2}$, if S_{i_1} is a higher hierarchical system related to S_{i_1}, we can get: $S_{i_1} \neq S_{i_2}$, and S_{i_2} has a higher spatial level than S_{i_1}. In this case, we denoted S_{i_2} as the abstraction, functional or promotion of S_{i_1}. That is: for S_{i_1} and S_{i_2}, there is a mapping:

$$\wp_i : s_{i_1} \longrightarrow s_{i_2} \tag{5}$$

Therefore, considering the different hierarchical subsystems of a system, there are distinct relationships between them of this system. In depth, if a subsystem S_2 is the abstraction, promotion, functional or a higher spatial hierarchy with respect to a subsystem S_1 of a system, we denote S_1 as real-like (or real space), while S_2 is imaginary-like (or imaginary space); furthermore, we denote the *real-like* and *imaginary-like* duality as *inter-duality*. In this sense, the real spaces and imaginary spaces involving in a general system integrated into a whole can be a *complete space* of the system.

Accordingly, the terms real-like and imaginary-like are relative concepts and we can only distinguish them from conceptual or ideational meaning. Furthermore, the real-like and imaginary-like concepts in a general system are non-uniqueness, for instance, when thinking about the pair of an entity and its attributes or the elements and relationship duality of this entity, etc. Now we can build the concept of an inter-dual system naturally, an underlying way to reveal the nature and intrinsic structure of general system.

Definition 5. *A system is an inter-dual system if it comprises inseparable real-like and imaginary-like dualities that are inextricably intertwined.*

Let a system be an inter-dual system, denoted as S, and let real-like be X, while imaginary-like is X^*, then:

$$S = (X^*, X) \tag{6}$$

The inseparable duality of real-like and imaginary-like in a system, such as a concrete image and an abstract image of a system, a real image and an imaginary image of a system, an actual image and a virtual image of a system, or a hard part and a soft part of a system, etc., can be denoted as *inter-duality*.

As is shown above, we introduce a mapping $\wp$ to present the connection and distinction between real-like X (real space) and imaginary-like X^* (imaginary space) with respect to a system S. That is:

$$\wp : X \longrightarrow X^* or \wp_i : X \longrightarrow x^*, x^* \in X^* \tag{7}$$

It is obvious that real-like and imaginary-like duality must be essentially different, i.e., an element of imaginary space, at least, is the abstraction, functional, promotion and global mapping with respect to the real space, i.e., $x^* = \wp_i(X)$.

Proposition 4. *An inter-dual system is a complete system.*

From Proposition 2, we know that a system contains its entirely spatial hierarchies that can be denoted as real-space and imaginary-space, which are integrated as a complete space of the system. It can be expressed as $S = (X^*, X) = \{s_{ij}\}$, i.e., an arbitrary system must be a complete system with an intrinsic structure of inter-duality. Actually, inter-duality is just one intrinsic structure of general system, but it is at least, from my point of view, an underlying way to reveal the nature of a general system and, furthermore, it is the intrinsic mechanism to generate or to create the inner power or inherent impetus of the configuration or formation of a system. Thirdly, inter-duality is the fundamental structure that drives a system evolution and development or destruction. Essentially, from complete space thinking, we can understand the entire structure and whole evolution of an arbitrary system, and this fundamental thought is the basis for being cognizant of optimization with respect to a system.

Proposition 5. *Prove* $(Y; X, F) = (X^*, X)$*, for a system* S.

Prove: for a system S, the elements and real-like are the same, i.e.,$X = X$; from Definition 4, $S = Y, Y = \{y\}$, for $\forall y$, it satisfies $\{y\} = \{y : y = f(x, a), y \in Y, x \in X, (f, a) \in F\}$; for X and X^*, there exists $\wp : X \longrightarrow X^*$; then $X^* = (Y; F)$. Therefore $(Y; X, F) = (X^*, X)$.

This proposition is also a demonstration that the objective space and relationship space of an arbitrary system must stand at the imaginary-like or imaginary space of the system. From this viewpoint, it is powerful and profound that the inner nature and global behaviour of objective space and relationship space of general systems are revealed based on inter-duality thought and the breakthrough of spatial consciousness.

3.2 Inter-Duality Theory

As we have introduced and constructed the concept of an inter-dual system, it is reasonable that the properties and principles should be established in an underlying and universal way. It is better to be presented these ideas using theorems:

Theorem 1. *An arbitrary system inherently has an inter-dual structure, and the real-like and the imaginary-like elements of the system must be concurrently emerging and simultaneously vanishing—that is to say, they are inextricably intertwined and cannot be divorced from each other.*

It can be easily proven using Definition 6. Essentially, the relationship between real-like and imaginary-like must be non-linear and none of them can be zero, i.e., inter-duality is the underlying structure and essential law of a general system.

Theorem 2. *The real-like and the imaginary-like elements of an inter-dual system must exist with an essentially spatial hierarchical difference.*

According to Definitions 3 and 4 and from Proposition 4, $(Y; X, F) = (X^*, X)$, it is obvious that the objective space and relationship space of a general system must be substantial otherness and intrinsic difference with respect to its elements space. Therefore, it is true that the imaginary-like X^* and the real-like X of an inter-dual system have an essentially spatial hierarchical difference.

Moreover, from mapping $\wp : X \longrightarrow X^*$, we denote X^* as the dual space with respect to real space X. Mathematically, in terms of linear functional, a dual space is defined as follows:

Let X be a linear space, if there is another space X^*, and they satisfy the inner product (denoted as $\langle \bullet, \bullet \rangle$), i.e., for $x_1, x_2 \in X; x_1, x_2 \in X^*; a, b \in R$, then

1. $\langle x_1^*, ax_1 + bx_2 \rangle = a\langle x_1, x_1 \rangle + \langle x_1^*, x_2 \rangle, or \langle ax_1^* + bx_2^*, x_1 \rangle = a\langle x_1^*, x_1 \rangle + b\langle x_2^*, x_1 \rangle$
2. for $x_0 \in X$, if for $\forall x^* \in X^*$, there is $\langle x_0, x^* \rangle = 0$, then $x_0 = 0$, or for $x_0^* \in X^*$, if for $\forall x \in X$, there is $\langle x_0^*, x \rangle = 0$, then $x_0^* = 0$;

And here, X^* is the dual space of X, denoted as (X^*, X).

Therefore, the inter-duality, also denoted as (X^*, X), of a general system is generalized from linear functional, and from the definition of the concept of dual space, X^* is the set of all functional, denoted as $X^* = \wp(X) = \{\wp_i(X)\}$, and $\wp_i : X \longrightarrow R$, then $\wp(X, R) = X^*$. Obviously, the imaginary-like X^* and the real-like X must exist spatial hierarchical difference.

Theorem 3. *The real-like and imaginary-like elements of an inter-dual system must possess an intrinsic duality relationship or inter-duality relationship.*

Let real-like be X and imaginary-like be X^*, due to Definition 6 and the description of Theorem 2, if the real space is X, then imaginary space X^* can be:

$$X^* = \wp(X) = \{\wp_i(X)\} \tag{8}$$

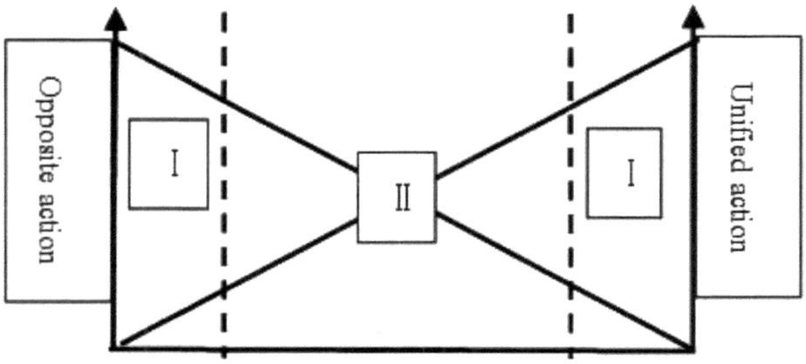

Fig. 2 The inter-duality relationship

It is reasonable that $\wp$ can be an operator between real-space X and imaginary-space X^*, i.e.

$$\wp_i(X) = x^*, x^* \in X^* \tag{9}$$

According to (9), we can also know the difference between the real-like and imaginary-like, and we can unquestionably define the difference as *opposite action* in an inter-dual system. Meanwhile, an inter-dual system must be unified underlying this inextricably and inseparably opposite operation, therefore, we denote this unification behaviour as *unified action* according to the general system. In conclusion, the simultaneous *opposite and unified action* can be denoted as an inter-duality relationship (or intrinsic duality relationship), as shown in Fig. 2.

The opposite action and unified action relationship of inter-duality are intrinsic duality relationship and inextricably interaction. This inter-duality relationship of an arbitrary system is concurrently emerging and simultaneously vanishing as is consistent with the real-like and imaginary-like of an inter-dual system.

Corollary 1. *Opposite action generates dynamic or impetus, while unified action creates progress or advance.*

Because of the intrinsic difference of an inter-dual system and this essential difference is the potential of the system, the real-like and imaginary-like can inherently evolve. This is the reason of why a system moving and evolving. Of course, due to this impetus, a system can reach and achieve its systematic equilibrium, and this is the progress or advance that results from the unified action.

Theorem 4. *The real-like and imaginary-like elements of an inter-dual system must have an inherently interactive relationship or interaction of inter-duality.*

Let $X = x$, $X^* = \{x^*\} = \{\wp_i(X)\}$. If there is any change in real-like X or imaginary-like X^*, the other must be inherently changing synchronously to satisfy its inner equilibrium state.

We can also present the change as $X' = X + \Delta_i$, then $\wp(X') = x^* + \Delta_i^*$, that is, X^* can also be changed to Δ_i^* in this action, and vice versa. This reciprocally intrinsic change can be defined as an interactive relationship of an inter-dual system, or denoted as interaction of inter-duality.

Theorem 5. *The real-like and imaginary-like elements of an inter-dual system must exist in a proportional relationship or flexibility of inter-duality.*

Let real-space be X, and imaginary-space be X^*; there must exist a mapping $F_i : X \longrightarrow R$; $F_j : X^* \longrightarrow R$, then $F_i : F_j = \alpha \in (\delta - \varepsilon, \delta + \varepsilon)$, and $(\delta - \varepsilon, \delta + \varepsilon)$ is denoted as the domain or interval of flexibility of inter-duality.

This is the underlying principle which distinguishes the domain of flexibility of an arbitrary system, in which the behaviour of equilibrium, optimization and security would be basically revealed. Furthermore, with these criteria, destructive or dangerous behaviours would also be taken into consideration.

Corollary 2. *The inter-dual structure of a general system is a multi-layered structure of fractal modality.*

In general, the different hierarchical real-like and imaginary-like elements themselves can be an inter-dual system. Philosophically, the inter-dual structure of any general system could consist of infinite hierarchies. Therefore, a general system could be the same structural formation in a fractal pathway or iteration way.

Corollary 3. *A general system must be a complex system.*

In principle, an arbitrary system can consist of different hierarchies of subsystems and these subsystems can also be regarded as inter-dual structures. Moreover, the relationship between subsystems, generally speaking, are non-linear or irreversible or uncertain or any combination of these three properties. Essentially, the real-like and imaginary-like elements of a system must be non-linear. Therefore, a general system should be treated as a complex system. In other words, the *Inter-duality Theory* is a theory of complex systems.

Inter-Duality Theory. When a system is investigated using inter-duality thought, it is an inter-dual system, and the theory of revealing the inter-dual structure, inter-dual behaviour and inter-dual mechanism etc., and establishing the laws, principles and methods of inter-dual systems, is denoted as the *Systematics Inter-duality Theory* or called *Inter-duality Theory*.

Actually, the Inter-duality Theory is a basic branch of *systematics*, and this theory reveals the fundamental reality and intrinsic structure and inner mechanism of a general system involved in an *Ultimate Objective World*. Moreover, this very theory is, historically, based on the investigation of *Ultimate Reality*, and naturally, on the basis of *Meta-space Conjecture* that underlies the *Ultimate Objective World View*. All the original work has been proposed and studied firstly by Professor *Longchang Gao* dating back to the end of the twentieth century.

The Inter-duality Theory, roughly speaking, would be applied to many fields, such as mathematics, physics, politics, sociology, management science and systematics, etc. For instance, from the theory of inter-duality, we can reveal the nature of a mathematical model, especially the parameter space in various models [2,3,5,7] and reveal the nature of space–time structure, the origin and essence of logic, the essence of humanity, the essential structure of particles, and the nature of consciousness, etc. [2, 3, 5, 7–9]. Further, we have investigated the nature of society, the principles and essence of generalized measurement, and the nature and principles of management, etc. [2,4,10].

4 Optimization Study

When it comes to optimization in daily life, we could say there is a natural need to realize the most efficient and sufficient utility, benefit or value, etc., with regard to associated systems or relative objects. But why can this happen? And how do we reach this prospect? In other words, we should reveal the nature and mechanism of optimization behaviour or explore the Tao of optimization. In terms of the goal of this paper, it is modestly discussed in a conceptual way.

Lemma 1. *There is no optimization if a system has a nature of* $1 + 1 = 2$*, i.e., it cannot be an objective reality with* $1 + 1 = 2$*.*

This lemma is somewhat awful and distressing because from this basic view, one will essentially prove that if a system is formed by a $1 + 1 = 2$ property, it must be true imagination and cannot exist in this objective world and likewise will not be a reality in an ultimately objective world. This is a revolutionary point that, in terms of philosophy, there is no completely common existence between two arbitrary certain realities in this world. Correspondingly, one can prove that $1 + 1 \neq 2$ represents the natural essence of an arbitrary existence, of an actual reality and of an entity, underlying the essential cognition of inter-duality.

Essentially, $1+1 = 2$ is a purely imaginary formula and cannot be a real equation of reality, but this law is the fundamental way to reality. Perhaps this is the greatest paradox in this world and the most beautiful and profound paradox existing in an objective world. From the view of inter-duality, any system is an inter-dual existence of the nature of $1 + 1 \neq 2$ between the real-like and imaginary-like.

Theorem 6. *An inter-dual system is a reality of optimization, and optimization behaviour becomes a reality when considering the underlying mechanism of interaction of inter-duality.*

Firstly, as is shown in the inter-duality theory, an inter-dual system is a complete system, denoted as $S = (X, X^*)$. It is the optimization of itself, or the perfect mapping by itself completely. It is reasonably obvious that an optimization behaviour must be an equilibrium by its complete space. Furthermore, if it happens in two systems, there must be a global optimization through their equilibrium of interaction between the two complete spaces, and eventually, they integrate into another inter-dual system, as existed in an inter-dual system.

It is evidently proven that optimization behaviour exists universally with respect to the associated objective systems relative to other systems as reference. Optimization behaviour must happen, at least, basically between two systems, such as the optimization of a management system relative to the objective system, even though it is created by humans, and this objective system is a reality when it is generated and exists in reality. Therefore, let the optimization operator be $\wp$, and one system S_1, and another S_2; then S_1 and S_2 are the optimization system, respectively. When there is an optimization behaviour between the two systems, denoted as $S = \wp(S_1, S_2) = S_1 \wp S_2$, one can finally gain another inter-dual system S, a new optimization state in terms of S_1 or S_2.

Corollary 4. *Optimization can be achieved through the equilibrium underlying the interaction of inter-duality.*

As is shown by Theorems 4 and 6, an inter-dual equilibrium is an inter-dual optimization under the interaction of inter-duality. It is obvious that this equilibrium must be a global equilibrium when it takes the complete space of inter-duality into account.

4.1 Equilibrium Remarks

Equilibrium is the most hotly pursued notion by people, by organizations, by society, by governments, by our whole world, and even by any system. Its existence is widely universal. There are many analogues of equilibrium, including equality, balance, average, fair, impartial, justice, optimization, optimum, harmony, singularity, stationary point or extreme point, etc. From the viewpoint of inter-duality, the antithesis concepts of equilibrium are extensive as well, such as break, fracture, burst, rupture, damage, accident, calamity, disaster, and ruin with negative connotations and breakthrough, penetration, innovation, insight, etc., which indicate positive meanings. From systematic space, the domain of medium banding between the equilibrium state and its dual state, i.e., the normal or general state, is the basic status or normal condition. Furthermore, equilibrium, geometrically, is a series or set of equilibrium points, which is hyper-dimensional, strictly speaking. Frequently, equilibrium looks like a local concept, yet it must be a global evolution inherently linking the associated complete space. Mathematics and mechanics have revealed a host of perfect results, and in this paper, there is no need to report them again [11, 12]. But for an appreciation, there is a result demonstration in [12], as shown in Fig. 3.

It is obvious that these equilibrium points are primary types, and I want to introduce some advanced equilibriums:

1. Higher order equilibrium points, a comprehensive, synthetic and superimposing form of primary equilibrium points;

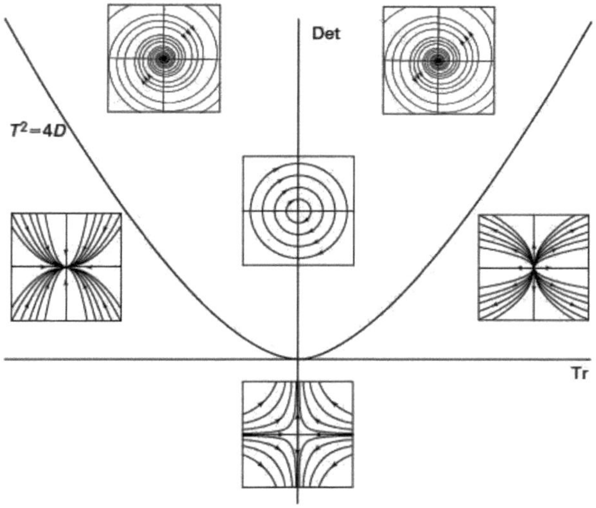

Fig. 3 A brief illustration of equilibrium neighbourhood and trajectory characteristics

2. Senior equilibrium, which belongs to psychological satisfaction, involved in the hierarchy of spiritual space, such as utility, benefit, and function, etc., mathematically, representing the derivative of corresponding non-linear function being zero;
3. Trajectory equilibrium, isolated senior equilibrium points or sets representing the derivative of corresponding functional being zero, such as the optimal path;
4. Equilibrium growth, the equilibrium points or sets are a function of time, and these trajectory equilibrium processes are equilibrium growth, using variational method to get these patterns;
5. Space equilibrium, equilibrium sets which are high dimensional space constrained by the complete space of a system, such as the Pareto equilibrium state, with high dimensional manifolds;
6. System equilibrium, or complete space equilibrium, an inter-dual optimization, which are not equilibrium points or sets, but all in all, must be equilibrium underlying the interaction of inter-duality;

In conclusion, primary equilibrium is merely an optimization in the same spatial hierarchy, holding the $1 + 1 = 2$ behaviour, and we have shown that this optimization is inter-duality breaking, strictly speaking, and is not actual optimization, while, the other equilibriums must be optimization in complete space involved in real-like and imaginary-like elements of an inter-dual system.

From this way, we can reveal the generalize d equilibrium or global optimization of a complex system. By explanation, the optimization behaviour or equilibrium process underlying the mechanism of interaction of inter-duality can be illustrated in Fig. 4.

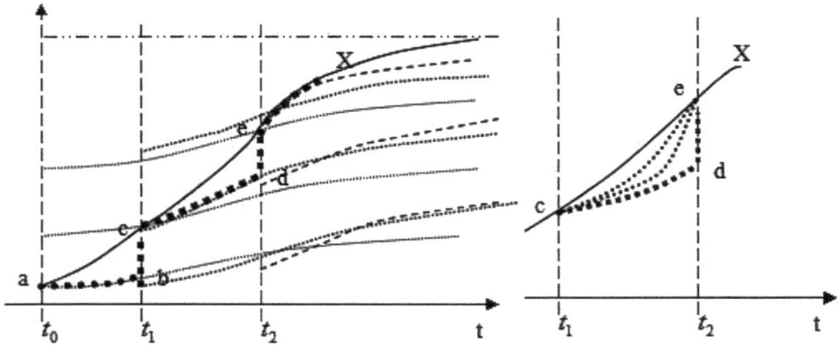

Fig. 4 A brief illustration of equilibrium process

For a general system, the curved line $X(t)$ is its optimization path. Through this way, it dissipates the least energy (including resources, materials, information, etc.) and has the highest function. The other pathways, like $cd - de$ or the ways between $cd - de$ and $X(t)$, are non-optimization, and the least effective way is surely $cd - de$ pattern, such as $ab - bc$.

5 Application Study

As an application, let's consider the representative inter-dual system, the management system, and we will reveal its nature and operational law, and explore how a management system can be optimized.

In order to study the nature and law of management systems, it is effectively to denote a management system as manager system M^* and managed system M.

For this analysis, we merely concentrate on the standard management process, a management system that holds its exact objective space O, managed space M and the relationship space F between O and M; at the same time, it involves one periodicity of management action. From the view of the inter-duality theory, the O and F are imaginary-like M^* (manager system), while M is real-like (managed system), in terms of an inter-dual management system S_M.

Then, we denote the standard management process as a mapping process:$\wp_i$: $M_i \longrightarrow m_i^*$ and $\wp_j^*$: $m_i^* \longrightarrow M_j$, i.e., the first operation $\wp$ is the management with the drive from the managed system M, while the second management action $\wp^*$ is motivated by the manager system M^*. We define this continuous management process as the standard period of management based on interaction of inter-duality.

Lemma 2. *If a management system is in its optimization state, or in an inter-dual equilibrium state, there is no need to run a management behaviour.*

According to our hypothesis, there is no change in imaginary-like or real-like. By contrast, if there is any change in this inter-dual system, there must exist a management behaviour to adjust and adapt the system to a new equilibrium state or new optimization. Therefore, we propose the nature of management:

Theorem 7. *The nature of management is to achieve the optimization of the real-like from the operation of the imaginary-like in a management system.*

Corollary 5. *The management behaviour is to maintain the equilibrium by the means of the mechanism of interaction of inter-duality in the management system.*

Therefore, a management system can be denoted as its objective space or its imaginary-like, such as management is decision, or management is control, etc. It is without doubt that a management system is

$$S_M = (M^*, M) = (O; M, F) \tag{10}$$

Operator $\wp$ represents the first management action:

$$\wp_i : M_i \longrightarrow m_i^*, m_i^* \in M^* \tag{11}$$

where $\wp$ denotes an operation process based on M_i. The inverse operator $\wp^*$, as the second management operation, satisfies:

$$\wp_j^*(m_i^*) = M_j \tag{12}$$

Further, we can conclude that

$$\{M_{ij}\} = \wp_j^*(\wp_i(M)) = \wp^*(\wp(M)) \tag{13}$$

According to the description of a standard management process, denoted as M_{ij}, it is clearly indicated that a management behaviour is always a series of M_{ij}. Further, M_{ij} is an inter-dual dynamical system, due to the interaction of inter-duality.

For the dynamical process of an inter-dual management system, as in a management mathematical application, it is useful for an inter-dual dynamical system of management to be established at this proper position:

The current management system can be presented as:

$$O_i = F_i(M_i, \varepsilon_i) \tag{14}$$

where ε_i is the change of imaginary-like with respect to the operation of $\wp_i$, i.e., the manager system makes a valuation and decision based on the first process of management, which is driven by the inner difference of managed system relative to its objective space or the optimization state. When it comes to the action from a manager system to a managed system, there should be an inter-dual intrinsic change ε_j^* of real-like result from the change of imaginary-like ε_i and the operation $\wp^*$.

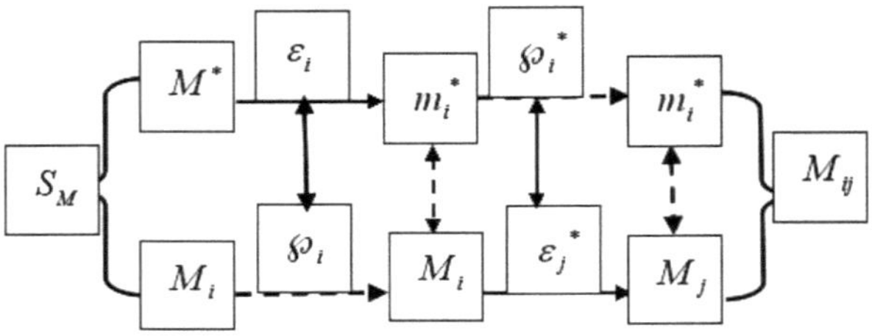

Fig. 5 The brief illustration of interaction of inter-duality (for the same strategic structure)

Essentially, ε_i and ε_j^* are merely the fine adjustments underlying the same strategic structure, i.e., the macro-structure is not changed in this management dynamical process. Therefore, we get:

$$O_i = F_i(M_i, \varepsilon_j^*) \tag{15}$$

According to the description, we can naturally reveal the interaction of the inter-dual fluctuation, i.e., an inherent impetus determined by an inter-dual management system drives any perturbation into an equilibrium state. We can illustrate this in Fig. 5.

This dynamic behaviour of an inter-dual system is also defined as self-organization. Therefore, we reveal another definition of the nature of management.

Definition 6. *Management is a specific self-organization process.*

As we can see, there are many essential and powerful laws and conclusions which are aided by the inter-duality theory. Furthermore, there are many mathematical analysis methods which reflect these management natural processes, such as the Fibre Bundles [13], the Functional Theories [14], Dynamical System Theories [15], Inter-dual analysis methods [4, 10, 16], etc.

Due to the aims and limitations of this paper, other essential problems and fundamental principles, such as the paradox of management, the relationship between the manager system and the managed system, etc., cannot be completely studied. We will reveal them and analyse them in another convenient place.

6 Conclusion

This paper concentrates closely on the conceptual study of revealing the nature and intrinsic structure of optimization from the viewpoint of an inter-duality theory. Therefore, there is limited introduction to the corresponding laws, principles and methods of systems, for instance, the rigorous proof of each theorem, corollary, lemma and propositions. Also, there is less research regarding the relationship between two systems, compound inter-dual systems, or multi-layered systems, etc.

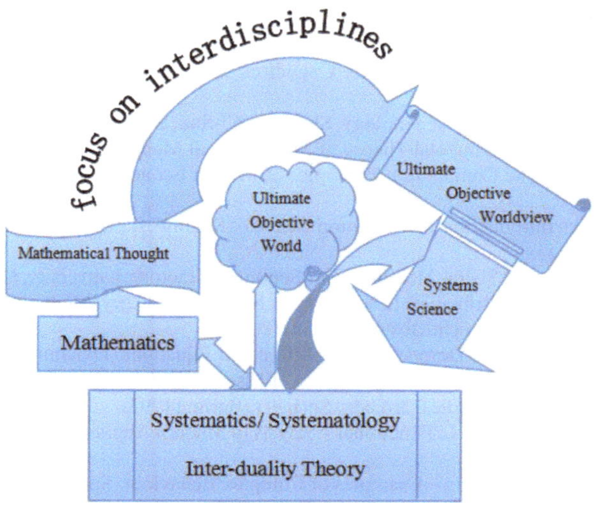

Fig. 6 Brief illustration of inter-duality theory

Moreover, we do not study the laws of evolution and development or destruction in an inter-dual system which involves profound optimization behaviour, such as social systems, political systems, ecological systems, physical systems and living systems, etc. and the environment or neighbourhood investigations, etc. Ultimately, in this paper we study the inter-dual system as a bold attempt in a manner that leads us to additional structures and further methodology on system theories. Essentially, we initiated and proposed the underlying road that we should take to further study an ultimately objective world in a Being way and using inter-duality thought, not merely the Modelling methods. Although the Inter-duality Theory was proposed originally at the end of the twentieth century by Professor *Longchang Gao*, it has not been popularized and has not been widely received. Therefore, we want to present a brief illustration of the development of this theory, as shown in Fig. 6.

Acknowledgements This paper is supported by the *Institute for Systematics*. The author would like to thank Professor *Longchang Gao* for his detailed guidance and suggestions. The author is grateful for *Cheng Yu* and Professor *Karen Mosley* with their carefully editing and corrections.

References

1. Von Bertalanffy, L.: General System Theory—Foundations, Development, Applications. George Braziller, New York (1973)
2. Gao, L.: Complexity Principles in Ultimately Objective World. Science, Beijing (2004)
3. Gao, L.: The Principles of Systematics, 2nd edn. Science, Beijing (2010)

4. Gao, L., Xu, F.: Inter-dual Space and Management Inter-duality Theory—An Fundamental Exploration of Management Sciences. Science, Beijing (2005)
5. Gao, L., Li, W.: Mathematics and Its Cognition, 2ed edn. Southwest Jiaotong University, Chengdu (2011)
6. Piao, C.: Introduction to Systematology. Shanghai Ci-Shu, Shanghai (2005)
7. Gao, L., Yang, Y.: Fundamental Theory of Mathematical Modeling. Science, Beijing (2007)
8. Gao, L., Tao, R.: The Competition Principles of Market Economy. Railway of China, Beijing (2002)
9. Gao, L., et al.: Introduction of Cognition Science. Southwest Jiaotong University, Chengdu (2004)
10. Gao, L.: Principles of Social Measurement. Southwest Jiaotong University, Chengdu (2000)
11. Yang Gao, D.: Duality Principles in Non-convex Systems—Theory, Methods and Applications. Kluwer Academic, Dordrecht/Boston/London (2000)
12. Hirsch, M.W., Smale, S., Devaney, R.L.: Differential Equations, Dynamical Systems, and An Introduction to Chaos. Academic, an imprint of Elsevier (2004)
13. Husemoller, D.: Fibre Bundles, 3rd edn. Springer, Berlin (1993)
14. Curtain, R.F., Pritchard, A.J.: Functional Analysis in Modern Applied Mathematics. Academic, New York (1977)
15. Jost, J.: Dynamical Systems—Examples of Complex Behaviour. Springer, Berlin (2005)
16. Gao, L., et al.: Systematics Inter-duality Theory—Theories and Methods. Science, Beijing (2013)

Part IV
Topology Optimization

The Interval Uncertain Optimization Strategy Based on Chebyshev Meta-model

Jinglai Wu, Zhen Luo, Nong Zhang, and Yunqing Zhang

Abstract This paper proposes a new design optimization method for structures subject to uncertainty. Interval model is used to account for uncertainties of uncertain-but-bounded parameters. It only requires the determination of lower and upper bounds of an uncertain parameter, without necessarily knowing its precise probability distribution. The interval uncertain optimization problem containing interval design variables and/or interval parameters will be formulated as a nested double-loop procedure, in which the outer loop optimization updates the midpoint of interval variables while the inner loop optimization calculates the bounds of objective and constraints. However, the nested double-loop optimization strategy will be computationally prohibitive, and it may be trapped into some local optimal solutions. To reduce the computational cost, the interval arithmetic is applied to the inner loop to directly evaluate the bounds of interval functions, so as to eliminate the optimization of the inner loop. The Taylor interval inclusion function is introduced to control the overestimation induced by the intrinsic wrapping effect of interval arithmetic. Since it is hard to evaluate the high-order coefficients in the Taylor inclusion function, a Chebyshev meta-model is proposed to approximate the Taylor inclusion function. Two numerical examples are used to demonstrate the effectiveness of the proposed method in the uncertain design optimization.

1 Introduction

In engineering, there are many uncertain factors inevitably related to material properties, geometry dimensions, loads and tolerance in the whole life cycle of design, manufacturing, service, and aging of the structure [1], due to the inherent

J. Wu • Z. Luo (✉) • N. Zhang
School of Electrical, Mechanical and Mechatronic Systems, The University of Technology, Sydney, NSW 2007, Australia
e-mail: jinglai.wu@student.uts.edu.au; zhen.luo@uts.edu.au; nong.zhang@uts.edu.au

Y. Zhang
National Engineering Research Center for CAD, Huazhong University of Science and Technology, Wuhan, Hubei 430074, China
e-mail: zhangyq@hust.edu.cn

D. Gao et al. (eds.), *Advances in Global Optimization*, Springer Proceedings in Mathematics & Statistics 95, DOI 10.1007/978-3-319-08377-3_21

uncertain nature of the real-world systems. The design under the deterministic assumption may not satisfy the expected design goal or even lies in the unfeasible region. Hence, there is an increasing demand to consider the impact of uncertainties quantitatively in the optimization of structures in spite of unavoidable variability and uncertainty, to enhance system safety and avoid failure in extreme working conditions. To incorporate uncertainties in the design optimization, the deterministic design problem should be suitably modified and enhanced.

There have been many different methods that can be applied to model uncertainties, among which the reliable-based optimization (RBO) [2] and the robust design optimization (RDO) [3] represent two major paradigms. RDO aims at determining a robust design to optimize the deterministic performance about a mean value, while making it insensitivity with respect to uncertain variations by minimizing the performance variance. RBO focuses on a risk-based solution taking into account the feasibility of design target at expected probabilistic levels, in which the failure probabilities and expected values are used to quantitatively express the effects of uncertainties.

In fact, RDO and RBO can be represented in the uniform theory framework. For instance, Du et al. [4] proposed an integrated framework for the design optimization under uncertainty, which took both the robust of the design objective and the probability of the constraints into account. In RDO and RBO methods, uncertain parameters are mostly treated as random variables, with precise probability distributions to be predefined based on the availability of complete information. However, it is generally a time-consuming and even an impossible process to achieve sufficient uncertain information to determine probability distributions, due to the complexity of practical problems [5]. Hence, probabilistic methods may experience difficulty for engineering problems. To this end, some non-probabilistic methods have emerged as beneficial supplements to the conventional probabilistic methods. In engineering, there are a large number of design problems involved uncertain-but-bounded parameters.

The uncertainties induced by the bounded parameters can be treated with interval parameters [6]. In particular, the interval model has attracted much attention recently in the uncertain optimization [7]. In interval models, the interval number is used to measure the uncertainty, because the representation of intervals only requires bounds of uncertain variables. The determination of lower and upper bounds of an interval is relatively easier, compared to a precise probability distribution. The interval model has been successfully applied to the optimization problems involving uncertain-but-bounded parameters [8]. For instance, Luo et al. [9] studied a new mathematical definition of non-probabilistic reliability index using the ellipsoid convex model. It can be found that most convex models involve a nested double-loop procedure. A nested double-loop optimization method using ellipsoid models was proposed to structural optimization [10], which included the method of moving asymptotes in the outer loop and a sequential quadratic programming (SQP) in the inner loop. Although the nested double-loop optimization is applicable, the computational cost is still prohibitive, as each individual outer loop consists of an inner loop minimization.

To reduce the computation cost and avoid trapping into local solution, the interval arithmetic [11], which defines the fundamental arithmetic operators, is introduced to replace the inner optimization to evaluate the maximum and minimum values of an interval function, as the interval arithmetic can easily obtain the bounds of a design function with interval parameters. However, the range of an interval function will be enlarged in the numerical implementation, due to the inherent wrapping effect of the interval arithmetic [11]. To control the overestimation, the Taylor inclusion function [11] with the high-order Taylor series is utilized to approximate the original function as a polynomial function, and then the interval arithmetic is used to calculate the range of the polynomial function. However, the coefficients, a set of high-order derivatives, in the polynomial function are hard to be obtained even for functions with explicit expressions. To this end, the Chebyshev series [6] are used to approximate coefficients of the Taylor inclusion (polynomial) function, so as to develop a Chebyshev meta-model. This meta-model can be constructed by evaluating function values at some specified interpolation points rather than the high-order derivatives, to improve computational efficiency [6]. After obtaining the Chebyshev approximation, the interval arithmetic can be used to calculate the bounds of the Taylor inclusion function in the inner loop.

2 Interval Uncertainty Optimization Model

This section proposes a new uncertain optimization model, in which both the design variables and parameters are considered as interval numbers. The uncertainty of both the objective and constraints induced by interval numbers is calculated, which may be similar to the concept of traditional RDO and RBO, respectively.

A general deterministic optimization model for the design of structures is given by

$$\begin{cases} \min\limits_{\mathbf{x}} & f(\mathbf{x}, \mathbf{y}) \\ \text{s.t.} & g_i(\mathbf{x}, \mathbf{y}) \le 0, \quad i = 1, 2, \ldots, n \\ & \mathbf{x}^l \le \mathbf{x} \le \mathbf{x}^u \end{cases} \tag{1}$$

The above mathematical model is used to minimize the objective f subject to constraints g_i. $\mathbf{x} \in R^k$ is the vector including deterministic design variables, and $\mathbf{y} \in R^q$ is the vector of consisting of deterministic parameters. To describe uncertainties in the design, interval numbers [11] are introduced to express the variations induced by the uncertainty. Any interval $[x]$ can be expressed as

$$[x] = \left[\underline{x}, \overline{x}\right] = x_c + [\Delta x] \tag{2}$$

where $\underline{x}$ and $\overline{x}$ denotes the lower and upper bounds of $[x]$, respectively, $x_c = \left(\overline{x} + \underline{x}\right)/2$ denotes the midpoint of $[x]$, and $[\Delta x]$ denotes the symmetric interval of $[x]$, which is defined by

$$[\Delta x] = [\text{-rad}([x]), \text{rad}([x])], \text{ where rad}([x]) = \left(\overline{x}\text{-}\underline{x}\right)/2 \quad (3)$$

where the radius rad($[x]$) reflects the uncertain degree of $[\mathbf{x}]$.

Consider the uncertainties, the deterministic optimization model (1) can be re-defined as follows:

$$\begin{cases} \min\limits_{[\mathrm{x}]} & f([\mathbf{x}], [\mathbf{y}]) \\ \text{s.t.} & g_i([\mathbf{x}], [\mathbf{y}]) \leq 0, \ i = 1, 2, \ldots, n \\ & \mathbf{x}^l \leq [\mathbf{x}] \leq \mathbf{x}^u \end{cases} \quad (4)$$

Here, the ranges for the interval parameters $[\mathbf{y}]$ will in general be pre-determined. Since the radius of an interval design variable $[\mathbf{x}]$ is also pre-given as $\boldsymbol{\xi}$, any interval design variable can be expressed as

$$[\mathbf{x}] = \mathbf{x}_c + [-\boldsymbol{\xi}, \ \boldsymbol{\xi}] \quad (5)$$

The responses of the objective and constraints would also be interval numbers, denoted by $[f]$ and $[g]$, respectively, because the design variables and parameters are interval vectors. This minimization problem is to minimize both the average value and the radius of the uncertain objective function, to ensure the "robustness" of the design. The minimization of the radius will lead to the decrease of the variance of the objective function, to make the uncertain objective function insensitive to the variation due to the uncertainty. It is noted that the midpoint value and radius are functionally similar to the probabilistic counterparts in the conventional RDO [3], which is a standard technique to minimize both the mean value and the standard deviation of the objective function.

To optimize the objective, both the midpoint and radius of the objective should be minimized, which can actually be regarded as a type of robust designs [12]. Thus, the new objective f_{obj} can be specified as

$$f_{obj} = \alpha f_c + \beta \text{rad}([f]) \quad (6)$$

where α and β denotes the weighting coefficients, and we set both α and β as 1 in this study. Then the objective can be re-defined as follows:

$$f_{obj} = f_c + \text{rad}([f]) = \overline{f} \quad (7)$$

Then the objective would be the upper bound of interval $[f]$, which is the maximum value of f under the uncertainty. For the interval constraints, there are three cases in the design space: $0 \leq \underline{g}$, $\underline{g} \leq 0 \leq \overline{g}$, and $\underline{g} \leq 0$. The first case violates the constraint, and the second case contains the possibility of violating the constraint. Only the last case can guarantee the design points in the feasible region, which denotes a 100 % reliability index. So the upper bounds should be used to meet the constraints

$$\overline{g}_i\left([\mathbf{x}], [\mathbf{y}]\right) \leq 0, \quad i = 1, 2, \ldots, n \tag{8}$$

The upper bounds of the objective and constraints can be calculated through maximizing the value in the range of uncertainty. Consider Eqs. (5), (7), and (8), the optimization model can be finally expressed as

$$\begin{cases} \min\limits_{\mathbf{x}_c} & \max\limits_{\mathbf{x}\in[\mathbf{x}],\mathbf{y}\in[\mathbf{y}]} f\left(\mathbf{x}, \mathbf{y}\right) \\ \text{s.t.} & \max\limits_{\mathbf{x}\in[\mathbf{x}],\mathbf{y}\in[\mathbf{y}]} g_i\left(\mathbf{x}, \mathbf{y}\right) \leq 0, \ i = 1, 2, \ldots, n \\ & [\mathbf{x}] = \mathbf{x}_c + [\text{-}\boldsymbol{\xi}, \boldsymbol{\xi}], \ [\mathbf{y}] = \left[\underline{\mathbf{y}}, \overline{\mathbf{y}}\right] \\ & \mathbf{x}^l + \boldsymbol{\xi} \leq \mathbf{x}_c \leq \mathbf{x}^u\text{-}\boldsymbol{\xi} \end{cases} \tag{9}$$

3 Nested Double-Loop Optimization

The optimization model in Eq. (9) involves a nested double-loop optimization process. The outer loop searches the optimal midpoint of interval design variables to minimize the objective, while the inner loop finds the maximum values (or minimum values) of the objective and constraints within the ranges of interval variables and parameters. The two optimization loops should use different optimization algorithms to balance numerical accuracy and computational efficiency, as the different characteristics are possessed by the optimization models at two different layers.

The design space of outer loop optimization is relatively large, so it may contain multiple local optimal points. To seek the global optimal solution and avoid multiple local minima, the heuristic techniques with strong global searching ability may be used. This study employs the Multi-Island Genetic Algorithm (MIGA) [13] to solve the outer loop optimization problem. MIGA divides each population of individuals into several sub-populations called "islands." In the evolution, some individuals are selected from each island and migrated to different islands periodically, which may make the solution converge to the global optimal solution faster than conventional genetic algorithms. To improve the efficiency, the SQP is included to search the optimal point after MIGA, which means the optimal point of MIGA is used as the initial point of SQP. In this case, the number of generations in MIGA can be reduced, because only a limited number of points near the global optimal solution are required, which will greatly decrease the calculation time of MIGA.

The design space of the inner loop optimization is relatively narrow, and the inner loop optimization can be searched by using many conventional optimization algorithms. This paper employs the Active Set Optimization (ASO) method in MATLAB. The idea of ASO is to define a working set as the active set in terms of a set of constraints at each step. The working set is chosen to be a subset of the constraints that are actually active at the current point, and hence the current point is feasible for the working set. The algorithm then proceeds to move on the surface defined by the working set of constraints to an improved point. At this new point the working set may be changed. An ASO consists of the following components: (1) determination of a current working set that is a subset of the current active constraints, and (2) movement on the surface defined by the working set to an improved point.

The flowchart for the nested double-loop optimization is shown in Fig. 1, where the outer loop is implemented by a combination of MIGA and SQP, in which the initial point of SQP is the optimal point obtained by MIGA. The midpoints of the interval variables are updated in each step of the outer loop, and the bounds of the interval design functions are calculated in the inner loop using the ASO algorithm.

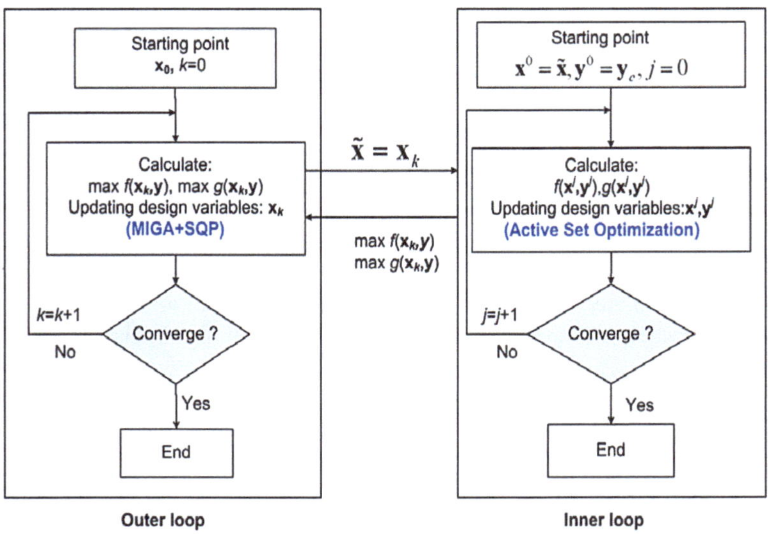

Fig. 1 Flowchart of double-loop process

4 Interval Optimization Strategy Based on Chebyshev Meta-models

In the nested double-loop optimization, the computational efficiency of the design problem is a key issue to be considered. The computational cost for the double-loop optimization will be computationally prohibitive. To improve the computational efficiency, the interval arithmetic is used to replace the inner loop optimization process. In this section, the interval arithmetic is introduced to calculate the bounds of interval functions, to eliminate the inner optimization, and the Taylor inclusion function is used to reduce the overestimation triggered by the interval arithmetic. However, higher-order derivatives are involved as coefficients in the calculation of the Taylor inclusion function, which again weights the computational cost, so a Chebyshev meta-model is proposed to approximate the Taylor inclusion function, to improve the efficiency and accuracy.

4.1 Taylor Inclusion Function in the Interval Arithmetic

The notation of interval numbers has been introduced in Sect. 2. The interval arithmetic defines some basic arithmetic operations between two different interval numbers. Consider two interval variables $[x]$ and $[y]$, and the basic arithmetic operations [11] between them can be defined as follows:

$$\begin{cases} [x]+[y]=\left[\underline{x}+\underline{y},\overline{x}+\overline{y}\right], \\ [x]-[y]=\left[\underline{x}-\overline{y},\overline{x}-\underline{y}\right], \\ [x]\times[y]=\left[\min\left(\underline{x}\,\underline{y},\underline{x}\,\overline{y},\overline{x}\,\underline{y},\overline{x}\,\overline{y}\right),\max\left(\underline{x}\,\underline{y},\underline{x}\,\overline{y},\overline{x}\,\underline{y},\overline{x}\,\overline{y}\right)\right], \\ [x]\div[y]=\left[\min\left(\underline{x}/\underline{y},\underline{x}/\overline{y},\overline{x}/\underline{y},\overline{x}/\overline{y}\right),\max\left(\underline{x}/\underline{y},\underline{x}/\overline{y},\overline{x}/\underline{y},\overline{x}/\overline{y}\right)\right], \quad if\ 0\notin[y] \end{cases} \tag{10}$$

From Eq. (10), it can be found that the interval arithmetic only depends on the bound of interval variables, which can obviously improve the computational efficiency of the optimization problem. However, interval arithmetic will lead to a large overestimation in the optimization, because of the dependence between interval variables.

Although the interval arithmetic can be used to eliminate the inner loop in interval uncertain optimizations, the wrapping effect of the interval arithmetic is required to be well monitored. Here, the Taylor inclusion function with high-order series is introduced to control the overestimation for general interval functions. For a function, which is $(n+1)$ times partially differentiable with respect to the vector $\mathbf{x}=[x_1,\ldots,x_k]^{\mathrm{T}}$ can be expanded with the nth order Taylor series at the midpoint $\mathbf{x}_c$ as follows:

$$f(\mathbf{x}) = f(\mathbf{x}_c) + \sum_{\sum_{i=1}^{k} p_{1,i}} = 1 \left(\frac{\partial f}{\partial x_1^{p_{1,1}} \ldots \partial x_k^{p_{1,k}}} \right) \Bigg|_{\mathbf{x}_c} (x_1\text{-}x_{1c})^{p_{1,1}} \ldots (x_k\text{-}x_{kc})^{p_{1,k}}$$
$$+ \cdots + \frac{1}{p_{n,1}! \ldots p_{n,k}!} \sum_{\sum_{i=1}^{k} p_{n,i}=n} \left(\frac{\partial^n f_c}{\partial x_1^{p_{n,1}} \ldots \partial x_k^{p_{n,k}}} \right) \Bigg|_{\mathbf{x}_c}$$
$$\times (x_1\text{-}x_{1c})^{p_{n,1}} \ldots (x_k\text{-}x_{kc})^{p_{n,k}} + R_{n+1} \tag{11}$$

where $p_{i,j}$ are the non-negative integers ($p_{n,i} = 0, 1, 2 \ldots, n$) and R_{n+1} is the series remainder.

Changing the real variable $\mathbf{x}$ to an interval variable $[\mathbf{x}]$ can lead to the Taylor inclusion function as:

$$[f]([\mathbf{x}]) = \sum_{0 \le i_1 + \cdots + i_k \le n} \beta_{i_1 \ldots i_k} [\Delta x_1]^{i_1} \ldots [\Delta x_k]^{i_k} + [R_{n+1}] \tag{12}$$

$[R_{n+1}]$ in Eq. (13) denotes the high-order remainder. In engineering, this term is usually neglected to achieve the truncated Taylor series. From Eq. (13), the Taylor inclusion function transforms the original interval function $f[(x)]$ to a *nth* order polynomial with respect to the symmetric interval $[\Delta \mathbf{x}]$ as

$$[f]([\mathbf{x}]) \approx \sum_{0 \le i_1 + \cdots + i_k \le n} \beta_{i_1 \ldots i_k} [\Delta x_1]^{i_1} \ldots [\Delta x_k]^{i_k} \tag{13}$$

where $\beta_{i_1 \ldots i_k}$ denote the coefficients which are related with the partial derivatives of f with respect to $\mathbf{x}$, and the total number of coefficients is $N_T = (n + k)!/n!k!$.

In most cases, the Taylor inclusion function will produce a narrower interval than the interval calculated directly by interval arithmetic. However, a major problem of the higher-order Taylor inclusion function is that a set of high-order partial derivatives, acting as the coefficients of the evaluation function, are required to be calculated. Since the high-order derivatives are hard to calculate, another numerical method will be applied to evaluate these coefficients, which will lead to a meta-model for the approximation of the high-order Taylor inclusion function.

4.2 Chebyshev Meta-model

Similar to the Taylor series, the Chebyshev series can also be used to expand the continuous function, with the Chebyshev polynomials to replace the power function in the Taylor expansion. Wu et al. [6] has shown that the Chebyshev polynomials have higher approximation accuracy than the Taylor polynomials under the same orders.

To simplify the problem but without losing any generality, we consider a variable $\mathbf{x} \in [-1, 1]^k$. The continuous function $f(\mathbf{x})$ can be approximated by

$$f(\mathbf{x}) \approx \sum_{i_1=0}^{n} \cdots \sum_{i_k=0}^{n} \left(\frac{1}{2}\right)^p f_{i_1 \ldots i_k} C_{i_1 \ldots i_k}(\mathbf{x}) \tag{14}$$

where p denotes the total number of zero(s) to be occurred in the subscripts $i_1, \ldots, i_k$. $C_{i_1 \ldots i_k}(\mathbf{x})$ is the k-dimensional Chebyshev polynomials [6], $f_{i_1 \cdots i_k}$ is a kth-order tensor with $(n+1)^k$ elements. Each coefficient of the Chebyshev polynomials can be calculated using the following integral formula [6]:

$$\begin{aligned} f_{i_1 \ldots i_k} &= \left(\frac{2}{\pi}\right)^k \int_{-1}^{1} \cdots \int_{-1}^{1} \frac{f(\mathbf{x})\, C_{i_1 \ldots i_k}(\mathbf{x})}{\sqrt{1-x_1^2} \cdots \sqrt{1-x_k^2}} dx_1 \ldots dx_k \\ &\approx \left(\frac{2}{m}\right)^k \sum_{j_1=1}^{m} \cdots \sum_{j_k=1}^{m} f\left(x_{j_1}, \ldots, x_{j_k}\right) C_{i_1 \ldots i_k}\left(x_{j_1}, \ldots, x_{j_k}\right) \end{aligned} \tag{15}$$

where m denotes the order of numerical integral formula ($m = n + 1$ in this study), x_j are the interpolation points of numerical integral formula. The interpolation points in each dimension are the zeros of $(n+1)th$ order Chebyshev polynomial, to be determined by

$$x_j = \cos\theta_j, \text{ where } \theta_j = \frac{2j-1}{n+1}\frac{\pi}{2}, \ j = 1, 2, \ldots, n+1 \tag{16}$$

Thus, the number of interpolation points for a k-dimensional problem would be $N_s = (n+1)^k$.

From Eqs. (14) to (16), it can be found that the process of constructing the Chebyshev approximant is similar to the response surface methodology (RSM), which obtains the data at sampling points (or interpolation points in this study) and then produces the coefficients based on these data. Equation (14) can be transformed to a polynomial based on the power function

$$f(\mathbf{x}) \approx \sum_{i_1=0}^{n} \cdots \sum_{i_k=0}^{n} \left(\frac{1}{2}\right)^p f_{i_1 \ldots i_k} C_{i_1 \ldots i_k}(\mathbf{x}) = \sum_{i_1=0}^{n} \cdots \sum_{i_k=0}^{n} F_{i_1 \ldots i_k} x_1^{i_1} \ldots x_k^{i_k} \tag{17}$$

where $F_{i_1 \ldots i_k}$ denotes the coefficients after the transformation. In Eq. (17), replacing the variable $\mathbf{x}$ with the interval variable $[\mathbf{x}]$, we can obtain the Taylor inclusion function. However, Eq. (17) contains $(n+1)^k$ terms, while the number of items in the Taylor inclusion function [Eq. (13)] is $N_T = (n+k)!/n!\,k!$ which is usually smaller than $(n+1)^k$.

Thus, if the Chebyshev polynomials are used to approximate the Taylor inclusion function, some higher-order items will not be necessary. At the same time, the number of interpolation points for constructing Chebyshev polynomial equals to the number of items in Eq. (14), which is still computationally expensive, especially for the high dimensional problems. To further save the computational cost, only a part of the interpolation points will be used to build the Chebyshev polynomials, which is termed Chebyshev meta-model. Removing the items with orders higher than n in Eq. (17), the meta-model can be expressed by

$$f(\mathbf{x}) \approx \sum_{0 \leq i_1 + \cdots + i_k \leq n} \left(\frac{1}{2}\right)^p f_{i_1 \ldots i_k} C_{i_1 \ldots i_k}(\mathbf{x}) = \sum_{0 \leq i_1 + \cdots + i_k \leq n} F_{i_1 \ldots i_k} x_1^{i_1} \ldots x_k^{i_k} \quad (18)$$

Since only a part of interpolation points are used to construct the Chebyshev meta-model, the Eq. (15) cannot be used to calculate the coefficients. To reduce the error between the meta-model and evaluation function, the least squares method (LSM) can be employed to produce the coefficients. The number of coefficients in Eq. (18) is N_T, so the number of sampling points from the interpolation should not be less than N_T. At the same time, the number of interpolation points is $N_s = (n+1)^k$, which is larger than N_T when $k > 1$. Therefore, the number of sampling points can be chosen as any number in the interval $[N_T, N_s]$. The larger number of sampling points, the smaller error of the approximation, but lower efficiency. Some studies [14] show there will be a good balance between the accuracy and efficiency, when the number of the sampling points is twice of the number of the coefficients. Thus, when $N_s > 2N_T$, the $2N_T$ interpolation points are chosen as the sampling points randomly. Otherwise, all the interpolation points are chosen as the sampling points. After the set of sampling data is obtained, the LSM is used to calculate the coefficients and establish the meta-model.

It should be noted that the coefficients $F_{i_1,\ldots,i_k}$ in Eq. (1) may be different from the coefficients $\beta_{i_1,\ldots,i_k}$ in Eq. (14), because $F_{i_1,\ldots,i_k}$ can be calculated via LSM while $\beta_{i_1,\ldots,i_k}$ are calculated via derivatives. In most cases, the Chebyshev meta-model in Eq. (19) can provide higher approximation accuracy than the truncated Taylor series given in Eq. (13). The Chebyshev meta-model can then be combined with the outer loop optimization (MIGA + SQP) to implement the uncertain optimization. The major advantage of the interval arithmetic is that the maximum and minimum values of a function are contained in the interval results, which provide rigorous constraints for the outer loop to guarantee the outer loop optimal solution is in the feasible region. The optimal design of the interval arithmetic may be more conservative than that of the double-loop optimization, but it is more reliable than the double-loop optimization.

The flowchart in Fig. 2 illustrates the numerical process of the proposed interval optimization strategy. The first step is to define some initialization parameters, where $\mathbf{x}_c$ denotes the nominal value of design variables, and n denotes the order of the Chebyshev meta-model. The second step is to calculate the objective $f_{\max}(\mathbf{x}_c, \mathbf{y})$, and constraints $g_{\max}(\mathbf{x}_c, \mathbf{y})$ in the outer loop using the interval arithmetic. The second step contains several sub-steps to build the Chebyshev meta-model. In this stage, we produce several interpolation points through Eq. (16), and then choose some

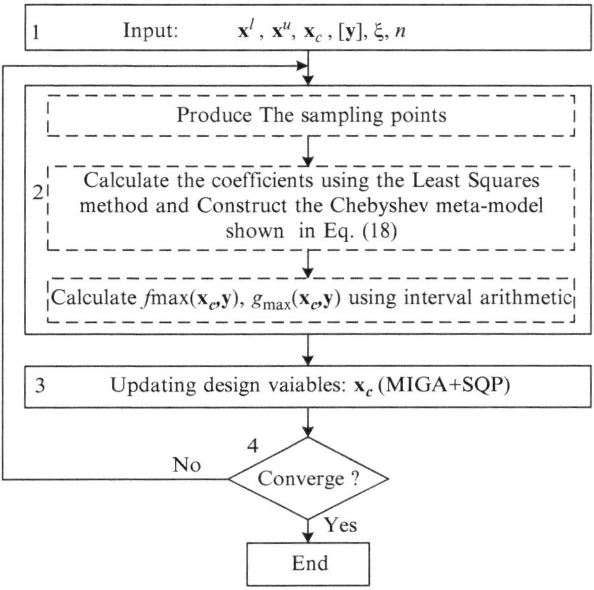

Fig. 2 The flowchart of interval optimization

interpolation points as the sampling points to calculate the values of the evaluation function at these sampling points. Then, the Least Square method is used to calculate the coefficients of the Chebyshev meta-model and construct the Chebyshev approximation model. Based on the Chebyshev meta-model, the interval arithmetic is used to evaluate the objective $f_{\max}(\mathbf{x}_c,\mathbf{y})$ and constraints $g_{\max}(\mathbf{x}_c,\mathbf{y})$. Hence, the third step is to update the nominal values of design variables based on the outer loop optimization algorithm (MIGA + SQP). If the result satisfies the convergence condition, the algorithm will be ended; otherwise the algorithm will go to the step 2.

5 Numerical Examples

In this section, the optimization of the 18-bar cantilever planar truss with interval uncertainty is provided to validate the accuracy and efficiency of the proposed interval uncertain optimization strategy. Figure 3 shows the 18-bar cantilever planar truss. The objective is to minimize the total weight of the truss subject to the stress limitations of ±20,000 lb/in.^{2} and Euler buckling compressive stress limitation [15]

$$b^{\sigma_i} = -\frac{KEA_i}{L_i^2} \tag{19}$$

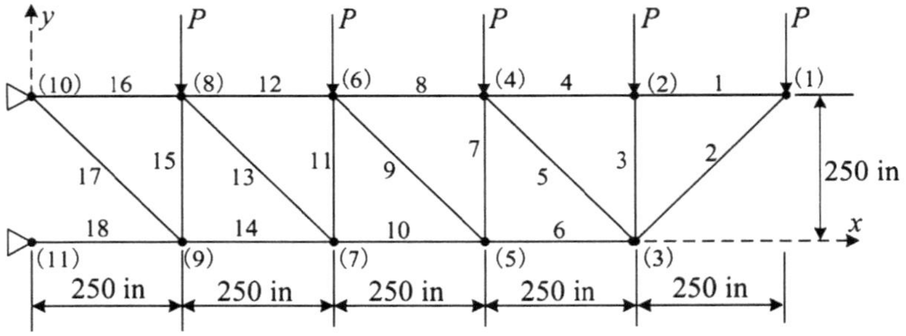

Fig. 3 18-Bar planar truss structure

where $K = 4$ denotes a constant determined by the cross-sectional geometry, $E = 10^7$ lb/in.^{2} is the modulus of elasticity, L_i is the ith member length, and A_i denotes the cross-sectional area of the ith member. The minimum cross-sectional area of members is 0.1 in.^{2}, and the maximum value is 50 in.^{2}

The members can be categorized into different groups, according to the cross-sectional areas (design variables): $x_1 = A_1 = A_4 = A_8 = A_{12} = A_{16}$, $x_2 = A_2 = A_6 = A_{10} = A_{14} = A_{18}$, $x_3 = A_3 = A_7 = A_{11} = A_{15}$, and $x_4 = A_5 = A_9 = A_{13} = A_{17}$. Here the design variables are considered as interval variables that has the interval width of 0.2 in.^{2} The material density is $\rho = 0.1$ lb/in.^{3} The vertical loads $P = 20{,}000$ lb are applied at the upper side of the truss.

The uncertain optimization model can be defined as follows:

$$\begin{aligned}
&\min_{\mathbf{x}_c} \quad \max_{\mathbf{x}\in[\mathbf{x}]} w = \sum_{i=1}^{18} A_i L_i \rho \\
&\text{s.t.} \quad g_1 = \max_{i=1,\dots,18}\left(\max_{\mathbf{x}\in[\mathbf{x}]}(|\sigma_i|)\right) \le 20{,}000 \\
&\qquad\; g_2 = \max_{i=1,\dots,18}\left(\max_{\mathbf{x}\in[\mathbf{x}]}(\sigma_i / b^{\sigma_i})\right) \le 1 \\
&\qquad\; [\mathbf{x}] = \mathbf{x}_c + [-\boldsymbol{\xi}, \boldsymbol{\xi}] \quad \boldsymbol{\xi} = \left[0.2 \dots 0.2\right]^{\mathrm{T}}_{1\times 4} \\
&\qquad\; \left[0.1 \dots 0.1\right]^{\mathrm{T}}_{1\times 4} + \boldsymbol{\xi} \le \mathbf{x}_c \le \left[50 \dots 50\right]^{\mathrm{T}}_{1\times 4} - \boldsymbol{\xi}
\end{aligned} \tag{20}$$

where σ_i denotes the stress of ith member, c_1 and c_2 denotes the stress constraints and Euler buckling compressive stress limitation, respectively.

The results, obtained with the three different methods: the nested optimization, linear model, and the proposed interval strategy, are shown in Table 1.

The objective and constraint shown in the bracket denotes the validated value, which is obtained through the scanning method in the uncertain range around the design point. The results show that the double-loop optimization method is 6,623.95 lb and the interval solution is 6,650.27 lb. However, the second constraint is violated for the double-loop optimization solution shown in italics. The interval

Table 1 The optimization results

	x_1 (in.^{2})	x_2 (in.^{2})	x_3 (in.^{2})	x_4 (in.^{2})	g_1 (lb/in.^{2})	g_2	w (lb)	Time (s)
Nested optimization	10.2024	21.8514	12.6160	7.2761	19,995 (19,995)	1.000 (*1.014*)	6,623.95	288
Interval optimization	10.2000	21.8517	12.8889	7.2711	20,000 (20,000)	1.000 (1.000)	6,650.27	139

method can ensure the satisfaction of both constraints, and so we can say that the interval method is able to provide more reliable optimization results. For the calculation time, the interval method takes 139 s, which is much less than the double-loop optimization method.

6 Conclusions

This research has proposed a new uncertain optimization method for the problem involving uncertain-but-bounded parameters. The interval model is used to describe uncertainties of the bounded parameters, which only requires the lower and upper bounds of an interval number. The identification of bounds for an interval parameter is easier than the determination of a precise probability distribution in the probability theory. The proposed interval uncertain optimization model has the characteristics of both the robust design and reliability based optimization. The interval optimization commonly leads to the nested double-loop process, but it is computationally prohibitive. Besides, the inner loop optimization may be trapped into local optimal solution, because only local optimization algorithm is used in common.

To save the expensive computational cost of the double-loop optimization, the interval arithmetic has been introduced into the inner loop to directly evaluate the bounds of interval design functions, so as to eliminate the inner optimization. Furthermore, to reduce the overestimation in the interval arithmetic, the high-order Taylor inclusion function is utilized to calculate the bounds of the interval design functions. However, the calculation of the high-order derivatives in the inclusion function is not easy. Hence, the Chebyshev meta-model is incorporated in the inclusion function to approximate the high-order derivatives, so that a Chebyshev model is developed, which can provide higher approximation accuracy than the truncated Taylor series. Typical structure optimization example is used to demonstrate the effectiveness of the proposed interval optimization methodology. In contrast to the nested double-loop process, this method can effectively improve computational efficiency and accuracy.

Acknowledgments This research was partially supported by the National Natural-Science-Foundation-of-China (No. 11172108), and also supported by the Chancellor's Research Fellowship (2032062), the University of Technology, Sydney (UTS).

References

1. Schuëller, G.I., Jensen, H.A.: Computational methods in optimization considering uncertainties – an overview. Comput. Methods Appl. Mech. Eng. **198**, 2–13 (2008)
2. Valdebenito, M.A., Schuëller, G.I.: A survey on approaches for reliability-based optimization. Struct. Multidiscip. Optim. **42**, 645–663 (2010)
3. Beyer, H.G., Sendhoff, B.: Robust optimization – a comprehensive survey. Comput. Methods Appl. Mech. Eng. **196**, 3190–3218 (2007)
4. Du, X., Sudjianto, A., Chen, W.: An integrated framework for optimization under uncertainty using inverse reliability strategy. J. Mech. Des. **126**, 562 (2004) [Design. 1, 47–66 (2001)]
5. Ben-Haim, Y., Elishakoff, I.: Convex models of uncertainties in applied mechanics. Elsevier, Amsterdam (1990)
6. Wu, J., Zhang, Y., et al.: Interval method with Chebyshev series for dynamic response of nonlinear systems. Appl. Math. Model. **37**, 4578–4591 (2013)
7. Qiu, Z., Elishakoff, I.: Anti-optimization of structures with large uncertain-but-non-random parameters via interval analysis. Comput. Methods Appl. Mech. Eng. **152**, 361–372 (1998)
8. Li, F., Luo, Z., et al.: Interval multi-objective optimisation of structures using adaptive Kriging approximations. Comput. Struct. **119**, 68–84 (2013)
9. Luo, Y., Kang, Z., et al.: Continuum topology optimization with non-probabilistic reliability constraints based on multi-ellipsoid convex model. Struct. Multidiscip. Optim. **39**, 297–310 (2008)
10. Ishibuchi, H., Tanaka, H.: Multiobjective programming in optimization of the interval objective function. Eur. J. Oper. Res. **48**, 219–225 (1990)
11. Jaulin, L.: Applied interval analysis: with examples in parameter and state estimation, robust control and robotics. Springer, New York (2001)
12. Doltsinis, I., Kang, Z.: Robust design of structures using optimization methods. Comput. Methods Appl. Mech. Eng. **193**, 2221–2237 (2004)
13. Whitley, D., Rana, S., Heckendorn, R.B.: The island model genetic algorithm: on separability, population size and convergence. J. Comput. Inf. Technol. **7**, 33–47 (1998)
14. Isukapalli, S.S.: Uncertainty Analysis of Transport-Transformation Models. The State University of New Jersey, New Brunswick (1999)
15. Gao, W.: Natural frequency and mode shape analysis of structures with uncertainty. Mech. Syst. Signal Process. **21**, 24–39 (2007)

An Element-Free Galerkin Method for Topology Optimization of Micro Compliant Mechanisms

Yu Wang, Zhen Luo, and Nong Zhang

Abstract This paper proposes an alternative topology optimization approach for the design of the large displacement compliant mechanisms with geometrical non-linearity by using the Element-free Galerkin (EFG) Method. In this study, because of its non-negative and range-bounded properties, Shepard function method, as a density filter, is used to generate a non-local nodal density field with enriched smoothness over the design domain. Besides, the Shepard function method is employed to build a point-wise density interpolation, the numerical implementation to calculate the artificial densities at all Gauss points. The moving least squares (MLS) method is then used to construct shape functions with compactly supported weight functions, to assemble the meshless approximations of system state equations. A typical large deformation compliant mechanism is presented to demonstrate the effectiveness of the proposed method.

Keywords Topology optimization • Shepard function • EFG method • Geometrical non-linearity • Micro compliant mechanisms

1 Introduction

Topology optimization is a mathematical approach to determine the best distribution of material within a design space, under a given set of loads and boundary conditions, so that the resulting layout meets a prescribed set of performance targets. In the past two decades, structural topology optimization as a new approach in structural optimization has experienced considerable development with many new contributions to theory, computational methods and applications in a wide range of engineering disciplines [1]. By now, various schemes have been developed, such as the homogenization method [2, 3], the SIMP method [4–6], the ESO method [7] and the level set-based method [8–11]. In recent years, topology optimization

Y. Wang • Z. Luo (✉) • N. Zhang
School of Electrical, Mechanical and Mechatronic Systems, The University of Technology, Sydney, NSW 2007, Australia
e-mail: zhen.luo@uts.edu.au

D. Gao et al. (eds.), *Advances in Global Optimization*, Springer Proceedings in Mathematics & Statistics 95, DOI 10.1007/978-3-319-08377-3_22

methods had been employed by a variety of applications successfully. In particular, the design of multi-physics compliant mechanism has received much concern from researchers.

In regards to the topology optimization design of compliant mechanism, it is extremely crucial to include geometrically non-linear problems in the numerical analysis. By far, most of the research works numerically processed the topology optimization of compliant mechanisms with geometrically non-linearity based on the standard finite element method (FEM) [12, 13]. Most recently, there have emerged several alternative methods to perform numerical analysis for topology optimization of structures based on nodal design variables of finite elements [14–19]. However, there are only a few research works can be found in topology optimization of compliant mechanism via the meshless method [20, 21]. Liew et al. [20] developed a thermo-mechanical constitutive model for analysing the behaviours of shape memory alloy and demonstrate the viability and advantages of meshless methods for modelling large deformation problems with geometrically non-linearity. Du et al. [21] applied the element-free Galerkin (EFG) method to implement the geometrical non-linear thermo-mechanical compliant mechanisms and showed that the meshless method can overcome the convergence difficulty in standard FEM.

Since the meshless method is in some cases more capable of modelling the large displacement mechanisms with the geometrical non-linearity, this paper attempts to propose a meshless topology optimization method based on EFG method for the topology optimization of micro complaint mechanism. In this method, the nodal densities are considered as the design variables are uniformly described based on a set of scattered field nodes inside the design domain. Firstly, in terms of the original set of density field, the Shepard function method worked as a density filter is applied to generate a non-local nodal density field with enriched smoothness over the design domain. Secondly, instead of using the Moving Least Square (MLS) approximants to formulate both the shape function and approximate the densities on computational points, the Shepard function method is employed to approximate the densities on computational points while MLS approximants is used to formulate the trial function. Because Shepard function method possesses non-negative and range-bounded properties, it can ensure a physically meaningful approximation of topology optimization design. Final, the MLS-shape function together with the Galerkin global weak-form is applied to develop the meshless approximation. Considering that the shape function founded by MLS approximants does not satisfy the KroneckerDelta criterion [22], a penalty method is then used to enforce essential boundary conditions. As to the simulation of the deformation of compliant mechanisms undergoing large-displacements, it is not always feasible to apply the linear elastic assumption, thus, the output displacement maximization has been used as an objective function to appropriately capture the behaviour of compliant mechanisms in this study.

2 Non-local Nodal Density Approximation Using Shepard function

It is known that the Shepard function has a mechanism similar to the smoothing effect of the density filtering schemes [23–25]. Meanwhile, the approximated values via the Shepard function are bounded between lower and upper values of the sampling points. This is the essential property for a physically meaningful density field approximant in topology optimization. The Shepard function is originally defined as a global interpolation. To improve the computational efficiency while keeping reasonable approximation accuracy, this study approximates the density at any node in terms of the nodal density variables within a compactly supported influence domain. With the Shepard function method, any nodal density variable relative can be given as

$$\overline{\rho}(x) = \sum_{i=1}^{n_H} \Theta_i(x)\rho_i \tag{1}$$

where any nodal density variable $\overline{\rho}(x)$ can be obtained by searching the total number of surrounding nodal variables ρ_i within the influence domain of the node x, and $\overline{\rho}(x)$ is density at the concerned node to be approximated by the Shepard function. n_H is the number of nodes within the influence domain. In this study, when the Shepard function used as the nodal density approximant, the weight function is chosen as

$$\omega_i\,(x - x_i) = \frac{3}{\pi r^2} \max\left(0,\ 1 - \frac{D_i(x)}{r}\right) \tag{2}$$

where $D_i(x) = x - x_i = \sqrt{(X - X_i)^2 + (Y - Y_i)^2}$.

This weight function is a radially linear "hat" function defined by [23]. It means that only nearby points are considered in computing any approximated value. In this way, the cost of computation is greatly saved by eliminating calculations with distant data points. It is straightforward that the Shepard function can meet the following necessary conditions for a physically meaningful density approximant in topology optimization:

(1) $0 \leq \overline{\rho}(x) \leq 1$
(2) $\partial\overline{\rho}(x)/\partial\rho_i \geq 0$

With the density field approximant, the Young's modulus at the node x can be defined by

$$E(x) = \overline{\rho}^p(x)E_0 = \left(\sum_{i=1}^{n_H} \Theta_i(x)\rho_i\right)^p E_0 \tag{3}$$

where E_0 represents the full-solid state material property over all nodes. The design variable ρ_i acts as the intrinsic nodal density allowing intermediate values between 0 and 1.

3 Point-Wise Density Interpolation Using Shepard Function

In this study, the Shepard function is utilized to construct an interpolation scheme for evaluating point-wise densities over computational points inside the design domain, according to the previously obtained nodal densities. For implementing Gauss quadrature of the system stiffness matrix, the background virtual cells are required, which are independent of the set of field nodes. Here, 4 × 4 Gauss quadrature is used to numerically calculate the uniform integration cells according to the location of the computational points. The densities on the computational points (Gauss points) are interpolated via the Shepard function method, which can be given as

$$\overline{\rho}_{gp} = \sum_{i=1}^{\mathrm{n_s}} \theta_i \left(x_{gp}\right) \overline{\rho}(x) = \sum_{i=1}^{\mathrm{n_s}} \frac{\omega_i (x)}{\sum_{\mathrm{j}=1}^{\mathrm{n_s}} \omega_j (x)} \overline{\rho}(x) \tag{4}$$

To make the Shepard function $\theta_i(x)$ satisfy the interpolation condition $\theta_i(x_j) = \delta_{ij}$, where $i, j = 1, 2, \ldots, \mathrm{n_s}$, a point-wise density field over the computational points can be constructed via the interpolation of the Shepard function with the following weight functions, which expressed as

$$\omega_i(x) = \omega(D) = \begin{cases} \frac{2}{3} - 4D^2 + 4D^3 & D \leq 1/2 \\ \frac{4}{3} - 4D + 4D^2 - \frac{4D^3}{3} & 1/2 < D \leq 1 \\ 0 & D > 1 \end{cases} \tag{5}$$

where $D = x_{gp} - x = \frac{\sqrt{(X_{gp}-X)^2+(Y_{gp}-Y)^2}}{r}$.

4 Meshless Approximations Using Moving Least Squares-Shape Function

The moving least squares (MLS) technique is used to construct the meshless approximations of the system state equation, and the MLS approximation for a general function $u(x)$ at x can be described as below[22]:

$$u^h(x) = \sum_{j=1}^{m} p_j(x) a_j(x) = \mathbf{p}^{\mathrm{T}}(\mathbf{x})\, \mathbf{a}(\mathbf{x}) \tag{6}$$

where $\mathbf{p}(\mathbf{x})$ is a complete polynomial of order, m acting as the basis at x, and $\mathbf{a}(\mathbf{x})$ is the vector consisting of unknown coefficients. $a_j(x)$ $(j = 1, \ldots, m)$ are the unknown parameters related to given points, which can be determined by minimizing a weighted discrete $\mathbf{L}_2$ norm over all nodes in terms of the pre-known parameters u_I.

$$J = \sum_{I=1}^{n} \tilde{w}\left(x - x_I\right) \left(\sum_{j=1}^{m} p_j\left(x_I\right) a_j(x) - u_I \right)^2 \tag{7}$$

where n is the number of nodes within the local support of x where the weight function $\tilde{w}\left(x - x_I\right) \neq 0$. u_I is the nodal parameter of u at $x = x_I$. The minimization of J with respect to the coefficients $\mathbf{a}(\mathbf{x})$ results in a set of linear equations as

$$\frac{\partial J}{\partial a_j(x)} = 2\sum_{I=1}^{n} \tilde{w}\left(x - x_I\right) \left(\sum_{j=1}^{m} p_j\left(x_I\right) a_j(x) - u_I \right) p_j\left(x_I\right) = 0 \tag{8}$$

The compact form for the above equation is given by

$$\mathbf{A}(x)\mathbf{a}(x) = \mathbf{B}(x)\mathbf{u} \tag{9}$$

Here $\mathbf{u}$ is the vector consisting of the nodal parameters for all nodes inside the support domain, and $\mathbf{u}^{\mathrm{T}} = [u_1, u_2, \ldots, u_n]$. Solving Eq. (16) for $\mathbf{a}(x)$ leads to

$$\mathbf{a}(\boldsymbol{x}) = \mathbf{A}^{-1}(\boldsymbol{x})\,\mathbf{B}(\boldsymbol{x})\,\mathbf{u} \tag{10}$$

Substituting the above equation into Eq. (13), we have the following MLS approximant

$$u^h(x) = \sum_{I=1}^{n} \phi_I(x) u_I = \boldsymbol{\Phi}(\boldsymbol{x})\,\mathbf{u} \tag{11}$$

where $\mathbf{N}(x)$ is the vector of MLS-shape functions related to the n nodes in the local support domain of $\mathbf{x}$. The shape function $N_I(x)$ associated with node I at point x can be written as

$$\boldsymbol{\Phi}(\boldsymbol{x}) = \mathbf{p}^{\mathrm{T}}(\boldsymbol{x})\,(\mathbf{A}(\boldsymbol{x}))^{-1}\mathbf{B}(\boldsymbol{x}) \tag{12}$$

5 Formulation of Topology Optimization

There are many different objective functions that can be founded for the design of compliant mechanisms. It is noted that many objective functions are originally designed for compliant mechanisms under the assumption of linear elasticity, which may not be suitable for the problem with geometrically non-linearity. It has shown that the displacement output can be used as the objective function to model the large displacement compliant mechanisms [11, 13]. Thus, the optimization problem based on mesh-free method can be established as

$$\begin{cases} \underset{\rho}{Maximize}: & u_{out} \\ Subject\ to: & \sum_{j=1}^{n} \rho_j V_j - \overline{V} = 0, \\ & \rho_j^{min} \leq \rho_j \leq 1, (j = 1, 2, \ldots, n) \\ & \mathbf{a}(u, \delta u) = \mathbf{l}(\delta u),\ u|_{\Gamma_D} = \overline{u}, \forall \delta u \in \mathbf{H}^1 \end{cases} \tag{13}$$

where

$$f(u, \delta u) = \frac{1}{2} \varepsilon_{ij}(u) D_{ijkl}(\rho(x)) \varepsilon_{kl}(\delta u) \tag{14}$$

As aforementioned, u is the displacement field, and δu is the virtual displacement field belonging to $\mathbf{H}^1$. $\overline{u}$ is the prescribed displacement on the admissible Dirichlet boundary Γ_D. ρ is the design variable, which is the nodal density in this study. V_j is the discrete material volume and $\overline{V}$ is the total material constrain. n is the number of the discrete design variables in the design space, and ρ_j^{min} is the lower bound of the design variables that is determined as 0.0001 to avoid the numerical singularity when computing the global stiffness matrix.

6 Numerical Example

As a numerical example of the micro compliant mechanism design problem, the displacement inverter is chosen as shown in Fig. 1. The example is solved via using the proposed non-linear modelling of EFG method modelling, and to demonstrate the effectiveness of the proposed approach; the results were compared with the results obtained by using non-linear modelling of FEM and linear modelling of EFG method, respectively. As seen in the Fig. 1, the design domain of displacement inverter is $400 \times 400\ \mu m^2$. On the input port, the input actuator is modelled by a linear spring with stiffness K_{in} and a force F_{in}. The goal of the optimization problem is to maximize the displacement U_{out} performed on the workpiece, which is modelled by a spring with stiffness K_{out}. For simplifying the computing progress, the design domain is discretized uniformly by 41×41 nodes as the design variables

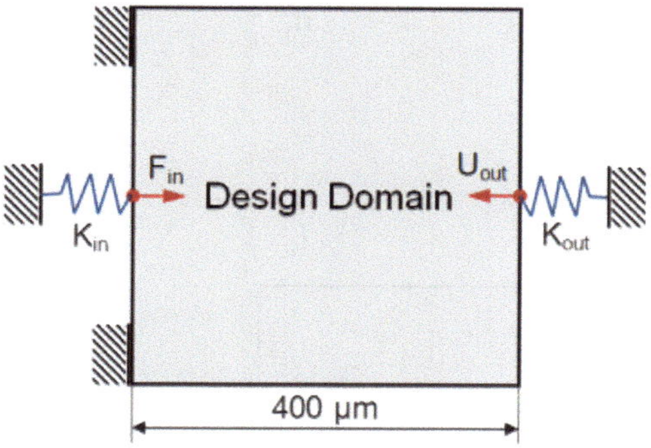

Fig. 1 The displacement inverter design problem

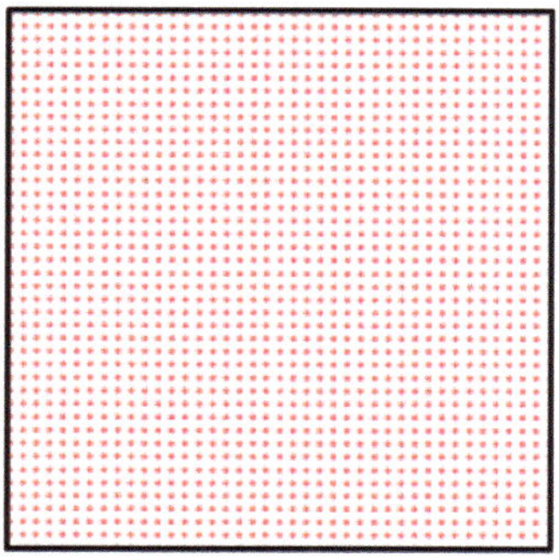

Fig. 2 Design variables in design domain

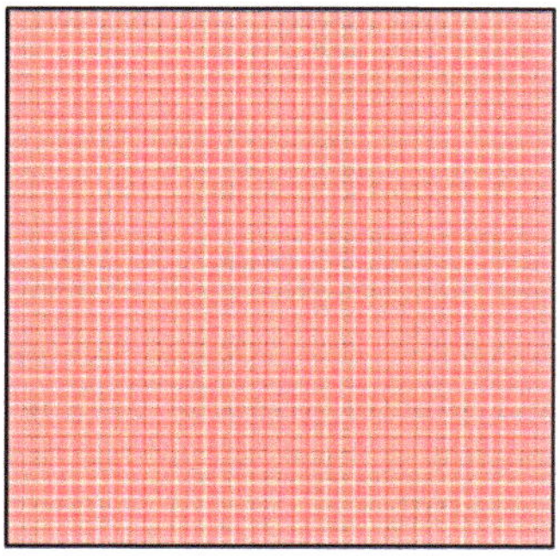

Fig. 3 Computational points in design domain

as shown in Fig. 2, and 40×40 integration cells are used based on the location of the nodes, inside with the 4×4 Guass points are used as the computational points, which has been shown in Fig. 3.

The topologies under different iteration are shown in Fig. 6. In this case, Young's modulus is 3 GPa, Poisson's Ratio is 0.3. An input force $F_{in} = 1$ N, and an artificial spring with stiffness $K_{in} = 5 \times 10^4$ N/m is attached to the input port to

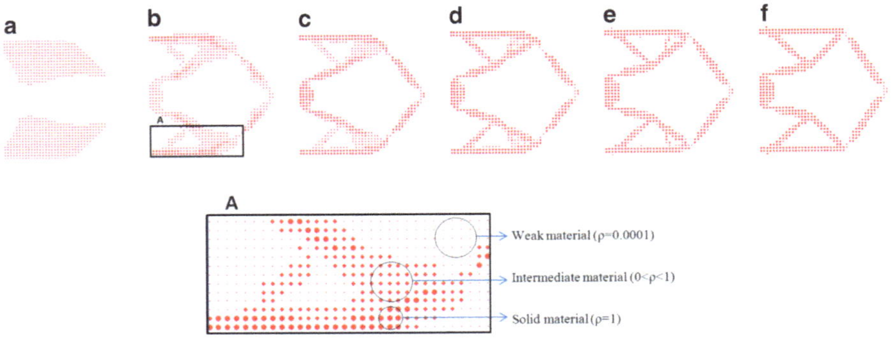

Fig. 4 Topology plots of point-wise nodal material densities: (**a–e**) intermediate results and (**f**) optimal design

simulate the input work. On the output point, an artificial spring with stiffness $K_{out} = 0.1 \times 10^4$ N/m is used to simulate the resistance from a workpiece. The material usage is limited to 30 %.

As shown in Fig. 4, the set of nodal densities serve as the design variables of the topology optimization to map the material distribution, and the design gradually moves towards the lower limit 0.0001 (weak material phase) and upper limit 1 (solid material phase) during the optimization. So we can deduce that the topology optimization is actually an iterative process to re-distribute a number of density points in the design space until the design approaches towards a so-called "0–1" distribution. It can be seen that the optimal topology does not have the discontinuously scattered nodes. In addition, the boundary of the optimal topology is acceptable smooth for providing curves and distinct material interface, which is beneficial to manufacturing procedure.

The output displacement of the optimal design is 37.06 μm. Figure 5 shows curves of the objective function and the volume constraint over the iterations. The convergence history of the optimization process using the proposed EFG method is convergent after 103 iterations. According to the curve of constraint, the proposed method is well mass conservative. The optimal design obtained by the proposed non-linear modelling of EFG method is similar to those reported in [1]. Therefore, we can deduce that the proposed method can be applied to design the micro compliant mechanisms successfully.

7 Conclusions

This paper proposes an alternative EFG method for topology design of large displacement compliant mechanism with geometrical non-linearity. The Shepard function method is applied to generate a non-local nodal density field with enriched smoothness over the design domain, so that there is no other filter scheme required

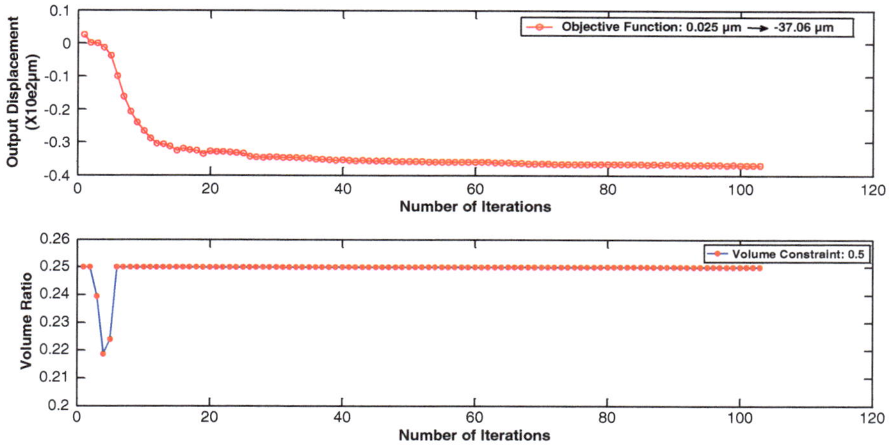

Fig. 5 Iteration histories of objective function and volume constraint

during the numerical analysis. Furthermore, the Shepard function method is used again to interpolate the densities at all computational points. In this way, a physically meaningful material density representation is obtained based on a set of design variables located on the field nodes. To implement the meshless approximations of system state equations, the MLS method is used to construct shape functions with compactly supported weight functions. The numerical example has demonstrated that the proposed method is more capable to handle the large deformation compliant mechanism and can avoid the undesirable mesh distortion caused by large deformation and convergence problem. With the development of the non-linear analysis of EFG method, it is straightforward to extend the proposed topology optimization method to more advanced mechanics problems.

References

1. Bendsøe, M.P., Sigmund, O.: Topology Optimization: Theory, Methods and Applications. Springer, New York (2003)
2. Bendsøe, M.P., Kikuchi, N.: Generating optimal topologies in structural design using a homogenization method. Comput. Methods Appl. Mech. Eng. **71**(2), 197–224 (1988)
3. Guedes, J.M., Kikuchi, N.: Preprocessing and postprocessing for materials based on the homogenization method with adaptive finite element methods. Comput. Methods Appl. Mech. Eng. **83**(2), 143–198 (1990)
4. Zhou, M., Rozvany, G.I.N.: The COC algorithm, part II: topological, geometrical and generalized shape optimization. Comput. Methods Appl. Mech. Eng. **89**(1), 309–336 (1991)
5. Mlejnek, H.P.: Some aspects of the genesis of structures. Struct. Multidiscip. Optim. **5**(1), 64–69 (1992)
6. Bendsøe, M.P., Sigmund, O.: Material interpolation schemes in topology optimization. Arch. Appl. Mech. **69**(9–10), 635–654 (1999)

7. Xie, Y.M., Steven, G.P.: A simple evolutionary procedure for structural optimization. Comput. Struct. **49**(5), 885–896 (1993)
8. Sethian, J.A., Wiegmann, A.: Structural boundary design via level set and immersed interface methods. J. Comput. Phys. **163**(2), 489–528 (2000)
9. Wang, M.Y., Wang, X., et al.: A level set method for structural topology optimization. Comput. Methods Appl. Mech. Eng. **192**(1), 227–246 (2003)
10. Allaire, G., Jouve, F., et al.: Structural optimization using sensitivity analysis and a level-set method. J. Comput. Phys. **194**(1), 363–393 (2004)
11. Luo, Z., Tong, L.: A level set method for shape and topology optimization of large-displacement compliant mechanisms. Int. J. Numer. Methods Eng. **76**(6), 862–892 (2008)
12. Sigmund, O.: On the design of compliant mechanisms using topology optimization*. J. Struct. Mech. **25**(4), 493–524 (1997)
13. Pedersen, C.B.W., Buhl, T., et al.: Topology synthesis of large-displacement compliant mechanisms. Int. J. Numer. Methods Eng. **50**(12), 2683–2705 (2001)
14. Guest, J.K., Prévost, J.H., et al.: Achieving minimum length scale in topology optimization using nodal design variables and projection functions. Int. J. Numer. Methods Eng. **61**(2), 238–254 (2004)
15. Matsui, K., Terada, K.: Continuous approximation of material distribution for topology optimization. Int. J. Numer. Methods Eng. **59**(14), 1925–1944 (2004)
16. Rahmatalla, S.F., Swan, C.C.: A Q4/Q4 continuum structural topology optimization implementation. Struct. Multidiscip. Optim. **27**(1), 130–135 (2004)
17. Paulino, G.H., Le, C.H.: A modified Q4/Q4 element for topology optimization. Struct. Multidiscip. Optim. **37**(3), 255–264 (2009)
18. Kang, Z., Wang, Y.: Structural topology optimization based on non-local Shepard interpolation of density field. Comput. Methods Appl. Mech. Eng. **200**(49), 3515–3525 (2011)
19. Wang, Y., Luo, Z., et al.: Topological optimization of structures using a multilevel nodal density-based approximant. Comput. Model. Eng. Sci. **84**(3), 229 (2012)
20. Liew, K.M., Ren, J., et al.: Numerical simulation of thermomechanical behaviours of shape memory alloys via a non-linear mesh-free Galerkin formulation. Int. J. Numer. Methods Eng. **63**(7), 1014–1040 (2005)
21. Du, Y., Luo, Z., et al.: Topology optimization for thermo-mechanical compliant actuators using mesh-free methods. Eng. Optim. **41**(8), 753–772 (2009)
22. Belytschko, T., Lu, Y.Y., et al.: Element-free Galerkin methods. Int. J. Numer. Methods Eng. **37**(2), 229–256 (1994)
23. Bourdin, B.: Filters in topology optimization. Int. J. Numer. Methods Eng. **50**(9), 2143–2158 (2001)
24. Luo, Z., Chen, L., et al.: Compliant mechanism design using multi-objective topology optimization scheme of continuum structures. Struct. Multidiscip. Optim. **30**(2), 142–154 (2005)
25. Luo, Z., Zhang, N., et al.: Structural shape and topology optimization using a meshless Galerkin level set method. Int. J. Numer. Methods Eng. **90**, 369–389 (2012)

A Level Set Based Method for the Optimization of 3D Structures with the Extrusion Constraint

Hao Li, Liang Gao, Peigen Li, and Tao Wu

Abstract An extrudable structure is formed via the extrusion process in which materials are squeezed through a die with pre-specified shape. Then, we obtain the long and straight metal parts with the fixed cross sections. The key of implementing the extrusion process is that the cross sections are perpendicular to the specified direction should be kept constant. To this end, this paper proposes a level set based optimization method for 3D structures with the extrusion constraint. The compactly supported radial basis functions (CS-RBFs) are introduced to convert the conventional level set method to a parametric one. The cross section projection strategy is applied to reduce the design variables and satisfy the extrusion constraint. Several 3D numerical examples are also provided.

Keywords Structure optimization • Level set • Manufacturability • Extrusion constraint

1 Introduction

Optimization plays an important role for improving the performance of structures. Many methods have been established for solving the structure optimal design problem, such as the homogenization method [1], SIMP (solid isotropic material with penalization) scheme [2], level-set based method [3–5], ESO (evolutionary structural optimization) approach [6], and the genetic algorithm [7]. While when traditional structural optimization methods are applied to these design problems, another critical factor of a structure—manufacturability—also needs to be intentionally considered.

The extrusion constraint discussed in this paper ensures a structure to be fabricated by the extrusion process. Generally, in the extrusion manufacturing process, materials are squeezed through an orifice of the required shape in a die by using the pressure from a ram, which is illustrated in Fig. 1. This technique

H. Li • L. Gao (✉) • P. Li • T. Wu
Huazhong University of Science and Technology, Wuhan, China
e-mail: helloleehao@gmail.com; gaoliang@mail.hust.edu.cn; lipg@mail.hust.edu.cn; wutaooptal@126.com

D. Gao et al. (eds.), *Advances in Global Optimization*, Springer Proceedings in Mathematics & Statistics 95, DOI 10.1007/978-3-319-08377-3_23

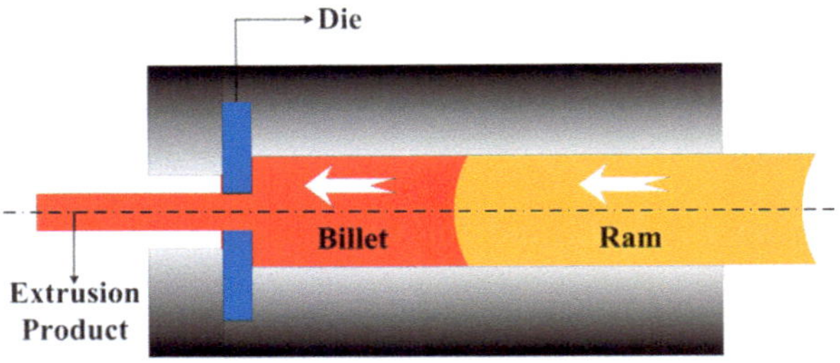

Fig. 1 Illustration of the extrusion process

is capable of generating compressive and shear force in the product with a lower manufacturing cost. It is important to maintain the same shape of cross section along an extruding path to guarantee the parts can be fabricated properly. Therefore, the design of an extrudable product should not only optimize the structural performance but also should satisfy the extrusion constraint.

Due to the asymmetric or inconsonant boundary conditions, the three-dimensional structure optimal design problem cannot be always simplified to two dimensions. Thus, some alternatives have been made for incorporating the extrusion constraint into the structure optimization issues. Kim and Kim [8] are among the earliest researchers who studied the topology optimization of beam cross section. Zhou et al. proposed the mathematical formulation for the topology optimization with extrusion constraint, which is embedded in the software Optistruct [9]. Ishii and Aomura [10] utilized the homogenization method to solve the extrusion-based structural optimization problem. Liu et al. solved the optimization problems for the beam cross section considering warping of sections and coupling among deformations by using the SIMP-based approach [11]. Zuberi et al. investigated the influence of different configuration and location of the load and boundary conditions on the optimal results for the extrudable designs [12].

In this paper, we propose a level set-based method for the extrusion-based optimization problem of 3D structures via the cross section projection strategy, by which the aim of obtaining the constant cross section along the extrusion axis is achieved.

2 Parameterization for Structural Optimization

2.1 The Level Set Based Method in Structural Optimization

In the level set-based structural optimization framework, the free boundary of a design is defined as the zero level set, which is embedded implicitly in a one-higher dimensional scalar function, i.e., the level set function $\Phi(\mathbf{x}, t)$. Then, the propagation of structural interface can be driven by the iteratively expanding of a specified speed

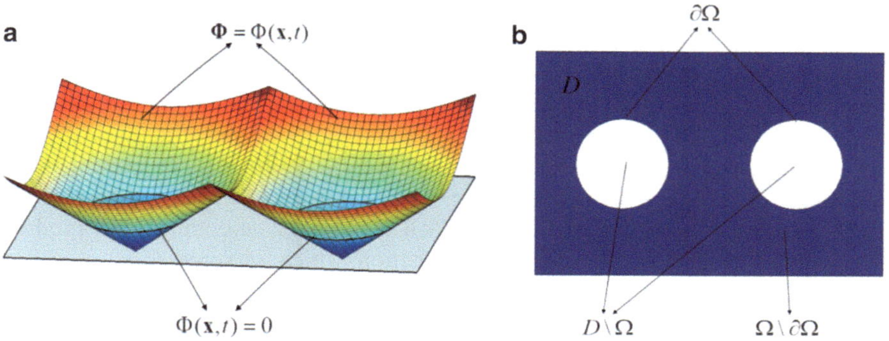

Fig. 2 The level set function and its zero isosurface

field. This process is achieved by solving a hyperbolic PDE called Hamilton–Jacobi equation on the fixed Eulerian grids.

Supposed that all the admissible shapes Ω are varying within a given design domain $D(D \subset R^d, d = 2or3)$, the Lipschitz-continuous level set function, as well as its zero isosurface, which can be shown in Fig. 2, is defined as:

$$\begin{cases} \Phi(\mathbf{x},t) > 0, & \forall \mathbf{x} \in \Omega \backslash \partial\Omega \quad (solid) \\ \Phi(\mathbf{x},t) = 0, & \forall \mathbf{x} \in \partial\Omega \cap D \quad (boundary) \\ \Phi(\mathbf{x},t) < 0, & \forall \mathbf{x} \in D \backslash \Omega \quad (hole) \end{cases} \tag{1}$$

Then, by introducing the pseudo-time t, the Hamilton–Jacobi equation can be specified as:

$$\frac{\partial \Phi(\mathbf{x},t)}{\partial t} + \mathbf{v_n} |\nabla \Phi| = 0, \quad \Phi(\mathbf{x},0) = \Phi_0(\mathbf{x}) \tag{2}$$

where $\mathbf{v_n} = \mathbf{v} \cdot (\nabla \Phi / |\nabla \Phi|)$ is the normal velocity. $\Phi_0(\mathbf{x})$ represents the initial level set function.

Though the implicit boundary expression contains several favorable features [3], moving the structural surface towards optimum is not an easy task when associated with the conventional level set based method.

2.2 RBF-Based Parametric Model

To address the numeric drawbacks that hamper the implement of classical level set method, several RBF-based parameterization approaches have been derived [13–15]. In this paper, a CS-RBF-based parametric model is employed to solve the optimization problem.

RBFs are radially symmetric functions centered at a particular point, which are widely utilized to interpolate or approximate a function from a given set of scattered samples. Unlike the globally supported RBFs, which require fully dense matrix and an elaborate shape function to guarantee a desired accuracy, the CS-RBF can be employed as an alternative kernel to circumvent these disadvantages. Due to the strictly positive definiteness and sparseness, CS-RBFs achieve a competitive advantage when incorporated into the level set based structural optimization method. As it is examined in [13], the CS-RBF with C2 smoothness can produce favorable approximation for the level set function, which is expressed as:

$$\varphi(r) = \max\left\{0, (1-r)^4\right\}(4r+1) \quad \text{(Wendland-C2)} \tag{3}$$

With regard to the 3D structural optimization, the radius of support is defined in a three-dimensional Euclidean space:

$$r = \frac{d_I}{R} = \frac{\sqrt{(x-x_i)^2 + (y-y_i)^2 + (z-z_i)^2}}{R} \tag{4}$$

where the parameter R indicates the radius of influences from the other adjacent knots. The radius of support r should be chosen properly, since it strongly affects the non-singularity of interpolation and the efficiency of approximation.

By introducing the CS-RBF, the originally coupled time and space variables have been separated naturally:

$$\Phi(\mathbf{x}, t) = \phi(\mathbf{x})^T \cdot \boldsymbol{\alpha}(t) = \sum_{i=1}^{N} \varphi_i(\mathbf{x}) \cdot \alpha_i(t) \tag{5}$$

where $\phi(\mathbf{x}) = [\varphi_1(\mathbf{x}), \varphi_2(\mathbf{x}), \ldots, \varphi_N(\mathbf{x})]^T \in R^N$ is the univariate, radially symmetric kernel, and $\boldsymbol{\alpha} = [\alpha_1, \alpha_2, \ldots, \alpha_N]^T \in R^N$ is the expansion coefficient vector, which is only time dependent.

Substituting Eq. (5) into the Hamilton–Jacobi equation (2), it yields:

$$\phi^{\mathbf{T}}(\mathbf{x}) \frac{d\boldsymbol{\alpha}(t)}{dt} + \mathbf{v_n} \left| (\nabla\phi(\mathbf{x}))^T \boldsymbol{\alpha}(t) \right| = 0 \tag{6}$$

Then, the normal velocity $\mathbf{v_n}$ can be rewritten as:

$$\mathbf{v_n} = -\frac{\phi(\mathbf{x})^T}{\left|(\nabla\phi)^T \boldsymbol{\alpha}(t)\right|} \cdot \frac{d\boldsymbol{\alpha}(t)}{dt} \tag{7}$$

Hereto, the classical level set method has been transformed into a parametric form for solving the structural optimization problems. Thus, the time-consuming process of handling Hamilton–Jacobi PDE is replaced by solving the rather convenient ODEs [13]. It should be noted that all the fixed nodes in the Eulerian type approach are now regarding as the CS-RBF interpolation knots, which means the level set propagation is now extending to the entire design domain rather than merely the front, making it possible to nucleate new holes during iteration.

3 A Cross Section Projection Strategy

An extrudable design means to create a constant cross-sectional product, which is seldom discussed in the conventional structural optimization techniques. Regarding the proposed cross section projection method, the surfaces of the 3D structure that are perpendicular to the specified axis, namely extrusion direction, are projected onto a uniformly meshed plane in 2D. In this circumstance, the original three-dimensional design problem is converted to an issue of identifying the optimal material distribution within a 2D space with much less number of design variables.

For the sake of simplicity but without losing generality, in this paper, we consider the widely used compliance design problem to illustrate the developed approach. In general, the structural compliance optimization with extrusion constraint can be formulated as:

$$\begin{aligned} &\underset{(u,\Phi)}{Minimize} : J(u,\Phi) = \int_D f(u) H(\Phi)\, d\Omega \\ &Subject\ \ to : a(u,v,\Phi) = l(v,\Phi),\ \forall v \in U,\ u|_{\partial\Omega} = u_0 \\ &\qquad G(u,\Phi) = \int_D H(\Phi)\, d\Omega - V_{\max} \le 0 \\ &\qquad \left(\Phi_i = \Phi_j = \cdots = \Phi_{ne}\right)_k,\ \ k = 1,2,\ldots,K \end{aligned} \tag{8}$$

where J is the objective function and G is the global volume constraint with a upper bound of $V_{\max}$, respectively. H is the Heaviside function associated with the implicit level set Φ. K denotes the number of elements in a single cross section, and ne represents the number of elements along the path of the extrusion. We see that numerous extra constraints are considered for an extrudable design in Eq. (8).

To implement the cross section projection method, the fundamental is to appropriately handle individual elements within the discrete 3D finite element (FE) model. Suppose that E_i is an arbitrary element in the discrete design space, of which the corresponding domain is denoted by Ω_{Ei}. Generally, there are two types of elements in the FE model that influence E_i, i.e. adjacent elements and parallel elements, which are shown in Fig. 3a, b. The adjacent elements indicate the elements in the fixed range of neighborhood of E_i, and the parallel elements refer to the elements on the same extrusion axis of E_i. The data of every finite element is aggregated onto a uniform-meshed 2D projection plane, which can be seen in Fig. 3c. It should be noticed that the structural design that demands fixed cross section is still in need of performing a 3D finite element analysis. After that, the data in the original FE model are mapped to the relevant finite elements on the projection plane via two specified operations.

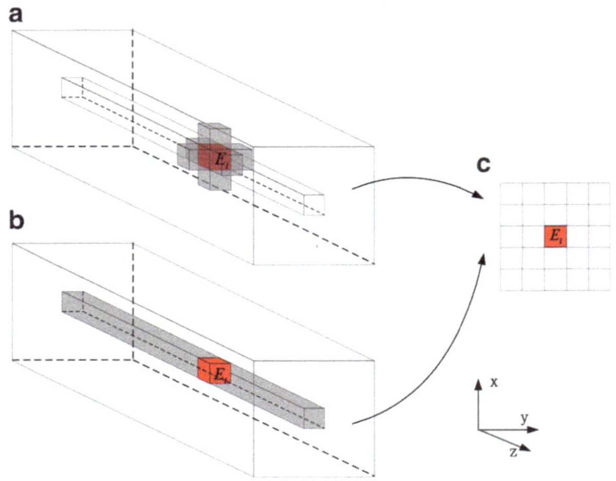

Fig. 3 Description of the cross section projection strategy. (**a**) the adjacent elements; (**b**) the parallel elements; (**c**) the projection plane

In the proposed method, the bilinear functional associated with 2D domain is established as:

$$a_P(u, v, \Phi) = \sum_{k=1}^{K} \left\{ \sum_{i=1}^{ne} \overline{a}(u, v, \Phi)_i / ne \right\}_k \tag{9}$$

$$\overline{a}(u, v, \Phi)_i = \frac{1}{\sum_{j=1}^{nae} w(i, j)} \sum_{j=1}^{nae} \left[w(i, j) \left(\int_{D_{Ei}} (\varepsilon(v_i))^T C(E_i) \varepsilon(u_i) H(\Phi_i) \, d\Omega_{Ei} \right) \right],$$
$$w(i, j) = r - dist(i, j) \tag{10}$$

where Eq. (9) is used to average the influences from the elements in the identical axis with E_i, and Eq. (10) is implemented to aggregate the influences from the neighbor elements of E_i. The bilinear functional $\overline{a}(u, v, \Phi)_i$ in a sub-domain Ω_{Ei} is given in the so-called weak form. *w(i, j)* is the weight coefficient that can be calculated by the radius of neighborhood and the distance between the relatively adjacent elements, i.e. *r* and *dist(i, j)*. *nae* represents the number of adjacent elements within radius *r*.

Similarly, the loading functional is derived as:

$$l_P(v, \Phi) = \sum_{k=1}^{K} \left\{ \sum_{i=1}^{ne} \overline{l}(v, \Phi)_i / ne \right\}_k \tag{11}$$

$$\bar{l}(v,\Phi)_i = \frac{1}{\sum_{j=1}^{nae} w(i,j)} \sum_{j=1}^{nae} \left[w(i,j) \left(\int_{D_{Ei}} p v H(\Phi_i)\, d\Omega_{Ei} + \int_{\Gamma_{Ei}} \tau v H d\Gamma_{Ei} \right) \right] \tag{12}$$

Then, after projection, we obtain a new formulation which is described in a two-dimensional space Ω_P:

$$\begin{aligned} &\underset{(u,\Phi)}{Minimize} : J_P(u,\Phi) = \int_{D_P} f(u) H(\Phi)\, d\Omega_P \\ &Subject\ \ to : a_P(u,v,\Phi) = l_P(v,\Phi),\ \forall v \in U_P,\ u|_{\partial\Omega_P} = u_0 \\ &\qquad\qquad G_P(u,\Phi) = \int_{D_P} H(\Phi)\, d\Omega_P - V_{\max} \le 0 \end{aligned} \tag{13}$$

In the approach specified above, no extra manufacturing constraints are explicitly existed when Eq. (13) can be applied to ensure an optimum design is able to be fabricated by extrusion. The reason is that each design variable in the original extrusion-based structural optimization is mapped to a relevant design variable associated with the 2D projection plane, which simultaneously produces a reduced set of design variables and guarantees an extrudable structure along the pre-specified path. Furthermore, the computational effort for the RBF interpolation is greatly decreased in that the identical grids in both RBF and FEA with regard to the parametric level set method reduce the number of interpolation kernels and corresponding summation operators. It should be pointed out that even a structure is meshed with irregular grids, it can be projected onto the uniform-meshed 2D domain by using the shape function.

4 Sensitivity Analysis and Optimization Algorithm

In Sect. 3 we obtained the optimization model for the structural optimization problem, which offers a tool to ensure a structure have constant cross section geometry during iterations so that it can be manufactured by the low-cost extrusion technique. To improve the structure's performance, a design sensitivity analysis is needed to conduct. In this section, we introduce the shape sensitivity analysis [8] to establish the relationship between the objective function and design variables.

For the compliance minimization of an extrusion product, the Lagrangian function with a positive Lagrange multiplier Λ can be established as:

$$L_P(u,\Phi) = J_P(u,\Phi) + l_P(v,\Phi) - a_P(u,v,\Phi) + \Lambda G_P(u,\Phi) \tag{14}$$

The shape derivative of the Lagrangian function can be derived as:

$$\frac{dL_P(u,\Phi)}{dt} = \int_{D_P} [\beta(u,v,\Phi) + \Lambda] |\nabla\Phi| \mathbf{v_n} \delta(\Phi) d\Omega_P = \int_{\Gamma_P} [\beta(u,v,\Phi) + \Lambda] \mathbf{v_n} d\Gamma_P \tag{15}$$

where the shape gradient density is obtained by the two particular operators specified in Sect. 3:

$$\beta(u,v,\Phi) = \sum_{i=1}^{ne} \overline{\beta}(u,v,\Phi)_i / ne \tag{16}$$

$$\overline{\beta}(u,v,\Phi)_i = \frac{1}{\sum_{j=1}^{nae} w(i,j)} \sum_{j=1}^{nae} \Big\{ w(i,j) \Big(f(u) - (\varepsilon(v))^T C(E_i) \varepsilon(u) + [pv + \nabla(\tau v) \cdot \mathbf{n} + \nabla \cdot \mathbf{n}(\tau v)] \Big) \Big\} + \Lambda \tag{17}$$

On the one hand, we substitute Eq. (7) into Eq. (15) to rewrite the shape derivative in term of expansion coefficients α_i; and on the other hand, we calculate the corresponding derivative by utilizing the chain rule. Then, the design sensitivities of the objective function and the volume constraint can be derived by comparing the results from the two different ways:

$$\frac{dJ_P}{d\alpha_i} = -\int_{\Gamma_P} J_P(u,\Phi) \frac{\phi_i}{\left|(\nabla\phi)^T \boldsymbol{\alpha}(t)\right|} d\Gamma_P \tag{18}$$

$$\frac{dG_P}{d\alpha_i} = \int_{\Gamma_P} \frac{\phi_i}{\left|(\nabla\phi)^T \boldsymbol{\alpha}(t)\right|} d\Gamma_P \tag{19}$$

To solve the structural optimization of 3D structure under extrusion constraint, some of the well-established techniques can be used, such as the steepest descent method, the method of moving asymptote (MMA), and the optimality criteria (OC). Owing to easy implement and high efficiency in dealing with large-scale topology optimization problems with single global constraint, the OC-based method [2] is employed to handle the optimization problem in this paper. It should be pointed out that the expansion coefficients of the CS-RBF interpolation are selected as the design objectives and are proceed with the OC approach iteratively.

5 Numerical Examples

In this section, the benchmark of the 3D cantilever beam is used as the numerical example to illustrate the characteristics of the proposed method. The elastic material is assumed with a Young's modulus of $E = 180$ GPa and the Poisson ratio of 0.3. The weak material has a Young's modulus of 0.001 and the same Poisson ratio.

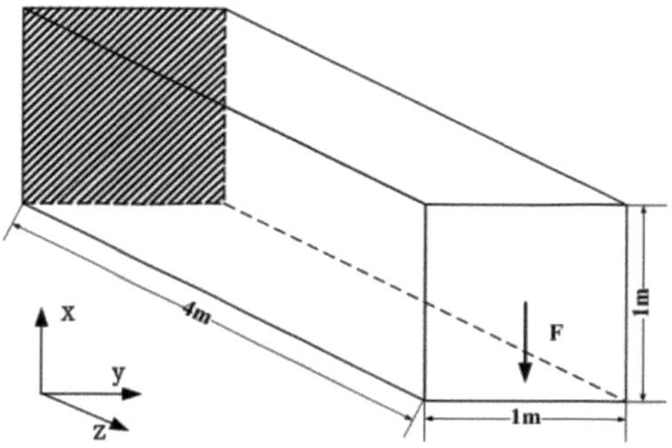

Fig. 4 The design domain

The initial design, which can be shown in Fig. 4, is a beam-like structure of size $1 \times 1 \times 4$ m with one end being fixed as the Dirichlet boundary. While the other end with a concentrated force of $F = 50$ kN is vertically loaded at the center point of the surface is treated as a non-homogeneous Neumann boundary. The design domain is discretized by 10-by-10-by-40 tri-linear 8-node cube elements for FEA. The level set surface is evolved on a 10-by-10-by-40 fixed Eulerian grid, accordingly. The objective functional is to minimize the structural strain energy with a material usage of 30 % in volume fraction.

To illustrate the method, solutions from the following two scenarios are compared: the design problem without and with considering the extrusion constraint. In the first example, we do not impose the manufacturing constraint on the optimization model. Although the parametric level set method can nucleate new holes within the design domain, we still observe that several regular voids are added factitiously into the rectangular solid in order to speed up the optimization process. The evolution process can be seen as in Fig. 5a–d. It is obviously that the structure changed dramatically during the initial 20 iterations, and the optimal topology has been almost formed. The subsequent iterations are carried out to perform the shape optimization to get a uniform distribution of strain energy densities at the boundary. From the observation of the final result that is shown in Fig. 5d, the cross sections along each direction are absolutely different and the optimal structure is unable to be fabricated with the low-cost extrusion process.

In the second example, we consider the extrusion constraint. We assume that the extrusion path is along the Z axis, and it means, in other word, the cross sections that are perpendicular to the pre-defined path should be kept constant to ensure the application of extrusion technique. It should be pointed out that the initial design is different from the first example which contains several holes penetrating the entire structure along the extrusion direction to maintain the uniform cross section from the very beginning. From Fig. 5e–h, we see clearly that the motion of structural boundary is restricted to be the same within every section along the Z axis during the optimization process. Such restriction is done by unifying the value of level set function for the corresponding nodes on each cross section via the proposed

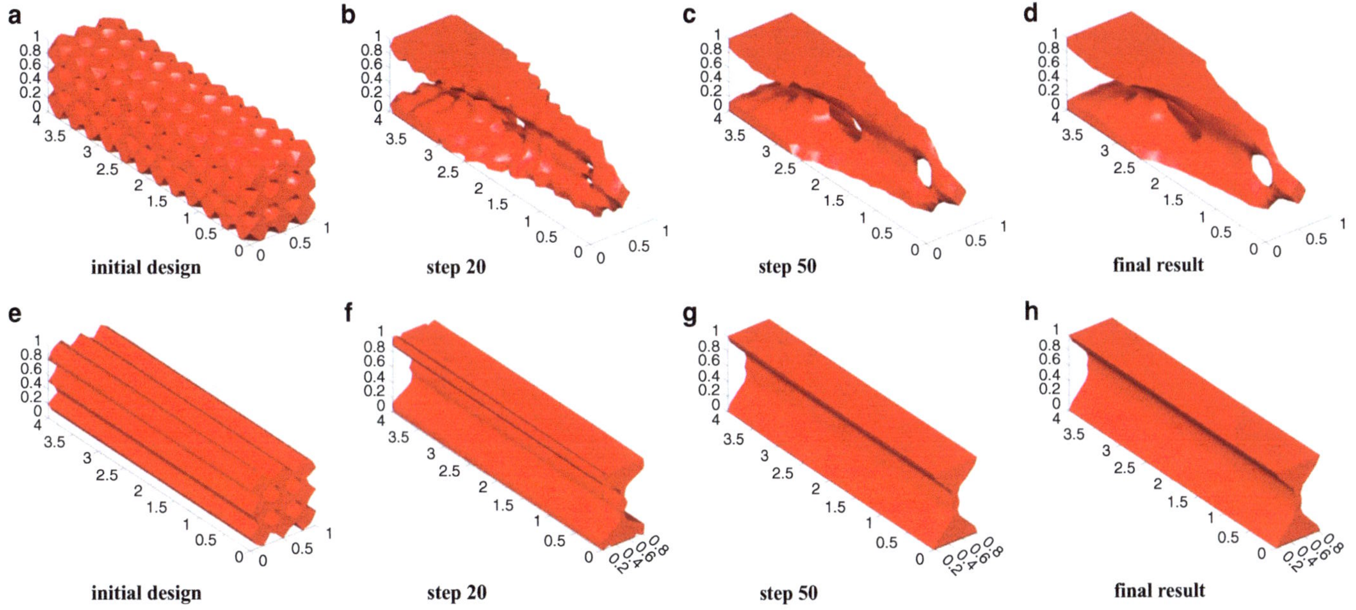

Fig. 5 Process of the optimization. (**a–d**) optimization without the extrusion constraint; (**e–h**) optimization with the extrusion constraint

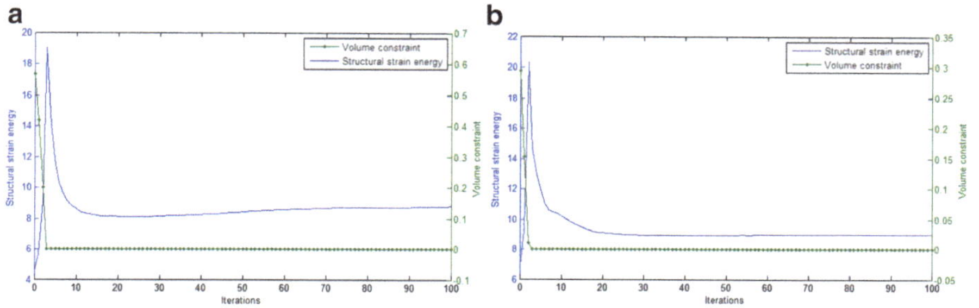

Fig. 6 The convergence history. (**a**) without the extrusion constraint, the structural strain energy is 8.8005; (**b**) with the extrusion constraint, the structural strain energy is 9.0104

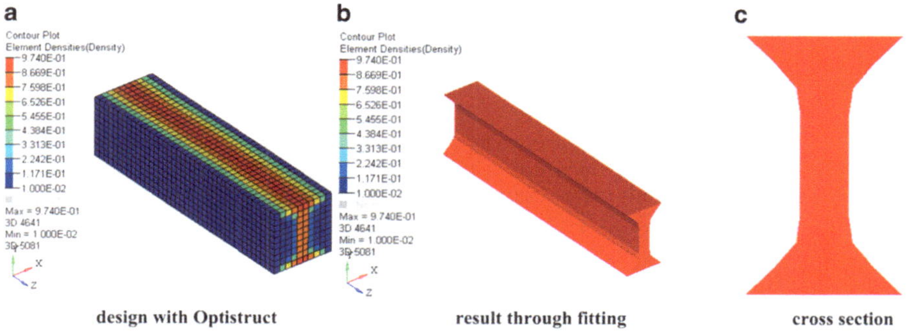

Fig. 7 Optimal results with Hyperworks

projection method. It is straightforward to observe that the optimal result, which is plotted in Fig. 5h, can be manufactured with the extrusion process.

The convergent histories of the structural strain energy and the volume constraint over iterations are in Fig. 6.

In addition, we further compare the result in the second example with the topologies obtained from the software Hyperworks, which contains the optimization module—Optistruct—that can solve the topology optimization problem considering several manufacturing constraints via the well-established SIMP interpolation model. The extrudable result from Optistruct is shown in Fig. 7a. The zigzag structural boundary can be observed, owing to the element-wise approach. Through the surface fitting we can get the smooth boundary, as shown in Fig. 7b, c. By comparing the result with that in the second example, we see that the final topologies in both cases are approximately the same. Nevertheless, the result from the SIMP-based method cannot be directly used without the fitting process. In our proposed method, the structural boundary is always kept smooth, and the final result can be imported straightforwardly into the CAD software for further detailed design.

6 Conclusions

A new level set-based method has been proposed in this study for 3D structures with extrusion constraint. The CS-RBFs are introduced to interpolate the level set function on the fixed Eulerian grids, which transform the original PDE-based approach into the easily solved parameterization one. The cross section projection strategy is applied to convert the extrusion-based 3D structure optimization to the 2D issue with much less number of design variables. The extrudable design is achieved with the constant cross sections along the pre-defined path. Several numerical examples are provided.

The presented method only deals with the extrusion-based design with simple loading case. The future work would extend the method to solve different optimization problems, such as dynamics issues or considering warping deformation conditions. Also, the future work should address the issue of a structure that is meshed with irregular grids.

Acknowledgments This work is supported by the National Natural Science Foundation of China (Grant No. 51175197) and the Open Research Fund Program (No. 31115020) of the State Key Laboratory of Advanced Design and Manufacturing for Vehicle Body, Hunan University, China.

References

1. Bendsøe, M.P., Kikuchi, N.: Generating optimal topology in structural design using a homogenization method. Comput. Methods Appl. Mech. Eng. **71**, 197–224 (1988)
2. Zhou, M., Rozvany, G.I.N.: The COC algorithm, part II: topological, geometry and generalized shape optimization. Comput. Methods Appl. Mech. Eng. **89**, 309–336 (1991)
3. Wang, M.Y., Wang, X., Guo, D.: A level set method for structural topology optimization. Comput. Methods Appl. Mech. Eng. **192**, 227–246 (2003)
4. Allaire, G., Jouve, F., Toader, A.M.: Structural optimization using sensitivity analysis and a level-set method. J. Comput. Phys. **194**, 363–393 (2004)
5. Luo, Z., Tong, L.Y., Kang, Z.: A level set method for structural shape and topology optimization using radial basis functions. Comput. Struct. **84**, 127–140 (2009)
6. Xie, Y.M., Steven, G.: A simple evolutionary procedure for structural optimization. Comput. Struct. **49**, 885–896 (1993)
7. Wang, S.Y., Tai, K., Wang, M.Y.: An enhanced genetic algorithm for structural topology optimization. Int. J. Numer. Methods Eng. **65**, 18–44 (2006)
8. Kim, Y.Y., Kim, T.S.: Topology optimization of beam cross sections. Int. J. Solids Struct. **37**, 477–493 (2000)
9. Zhou, M., Fleury, R., Shyy, Y.K., Thomas, H.L., Brennan, J.M.: Progress in topology optimization with manufacturing constraints. In: 9th AIAA/ISSMO Symposium on Multidisciplinary Analysis and Optimization, Atlanta, 2002
10. Ishii, K., Aomura, S.: Topology optimization for the extruded three dimensional structure with constant cross section. JSME Int. J., Ser. A **47**, 198–206 (2004)
11. Liu, S.T., An, X.M., Jia, H.P.: Topology optimization of beam cross-section considering warping deformation. Struct. Multidiscip. Optim. **35**, 403–411 (2008)

12. Zuberi, R.H., Zuo, Z.X., Long, K.: Topological optimization of beam cross section by employing extrusion constraint. In: Proceedings of ISCMII and EPMESCXII, pp. 964–969. Hong Kong-Macau, 2009
13. Luo, Z., Wang, M.Y., Wang, S.Y., Wei, P.: A level set-based parameterization method for structural shape and topology optimization. Int. J. Numer. Methods Eng. **76**, 1–26 (2008)
14. Belytschko, T., Xiao, S.P., Parimi, C.: Topology optimization with implicit functions and regularization. Int. J. Numer. Methods Eng. **57**, 1177–1196 (2003)
15. Wang, S.Y., Wang, M.Y.: Radial basis functions and level set method for structural topology optimization. Int. J. Numer. Methods Eng. **65**, 2060–2090 (2006)

Topology Optimization of Structures Using an Adaptive Element-Free Galerkin Method

Yixian Du, Shuangqiao Yan, De Chen, Qingping Long, and Xiang Li

Abstract This paper proposes a topology optimization method based on an adaptive element-free Galerkin (EFG) method. Since there are a large number of discrete design variables in the meshless based topology optimization, the adaptive EFG method is included to improve the computational efficiency. The topology optimization problem is formulated as a mean compliance design of structures, in which the nodal densities are regarded as design variable. In the proposed method, the criteria of adding node, the scheme of adaptively inserting nodes, and the method of updating the variable for any new node are discussed. A typical example is used to demonstrate the effectiveness of the proposed topology optimization method using the adaptive EFG method.

Keywords Topology optimization • Adaptive EFG method • Nodal encryption criteria

1 Introduction

Topology optimization of structures has experienced considerable development over the past two decades [1]. The typical methods for topology optimization of structures include the homogenization and SIMP method [1], the nodal density description method (Kang and Wang 2012) [2], and the level set method (Wang et al. 2003) [3, 4]. However, most current topology optimization methods are based on finite element methods, and will be easily subjected to numerical instabilities [5]. With the development of the meshless method in the numerical simulation, alternative topology optimization methods based on element-free methods have gradually

Y. Du (✉) • S. Yan • D. Chen • Q. Long • X. Li
College of Mechanical and Material Engineering, China Three Gorges University, Yichang 443002, China

Hubei Key Laboratory of Hydroelectric Machinery Design & Maintenance, China Three Gorges University, Yichang 443002, China

Collaborative Innovation Center for Energy Equipment of Three Gorges Region, China Three Gorges University, Yichang 443002, China
e-mail: duyixian@aliyun.com

D. Gao et al. (eds.), *Advances in Global Optimization*, Springer Proceedings in Mathematics & Statistics 95, DOI 10.1007/978-3-319-08377-3_24

become a research focus, such as the EFG method and RKPM (reproducing kernel particle method) have been applied to topology optimization of structures [6] (Luo et al. 2012) has become a research focus in the field. These methods have shown their advantages in overcoming the typical numerical instabilities. However, the major shortcoming of the meshless topology optimization methods is the expensive computational cost. Hence, how to improve the numerical efficiency is an interesting topic in meshless topology optimization.

In finite element based topology optimization, Xu and Cheng [7] studied the topology optimization of structures based on an adaptive mesh refinement scheme in the process of optimization. According to the information of the node density field, the adaptive encryption for the elemental mesh was completed and the shape function interpolation method was used to get the new design space and the next-stage topology optimization is performed, to reduce the dimension of design variable space and amount of iterations. Lin and Chou [8] used a smaller number of finite elements to increase the chance of obtaining a good topology under a limited computational cost. In term of meshless method, the gauss point density values were used as design variables. However, this method cannot construct a continuous density field, and will easily lead to numerical instability due to separate point-clouds. Many researchers began using the nodal relatively density as the design variable to avoid this phenomenon. Wu and Gong [9] treated the nodal densities as the design variables, which initially used sparse nodes in the process of topology optimization and then added nodes gradually via an adaptive node refinement scheme until the distance between the adjacent nodes is less than the fixed value to improve the efficiency.

Therefore, the paper will study an adaptive topology optimization using EFG method, which takes the nodal relative densities as the design variables, combining the adaptive numerical technology with EFG method to reduce the number of design variables and to improve the efficiency.

2 Topology Optimization Based on Adaptive Element-Free Galerkin Method

2.1 Element-Free Galerkin Method

The element-free Galerkin (EFG) method is one of the most promising meshless methods in the area of computational solid mechanics [10, 11]. The key concept in the EFG method is to employ the moving least-square (MLS) technique to enable the numerical approximation of the discrete govern equation.

$$u^h(x,\overline{x}) = \Phi(x,\overline{x})\,u \tag{1}$$

where $\Phi\,(x, \overline{x})$ is the shape function, $u^h\,(x, \overline{x})$ is local approximation, and u are nodal parameters. The EFG shape function consists of polynomial $p(x)$ and a cubic spline weight function [12]. The governing equation for two-dimensional elastostatics problems on the domain Ω bounded by Γ can be given as follows:

$$\begin{cases} \nabla \cdot \boldsymbol{\sigma} + \mathbf{b} = 0 & \text{in } \Omega \\ \mathbf{u} = \bar{\mathbf{u}} & \text{on } \Gamma_u \\ \boldsymbol{\sigma} \cdot \mathbf{n} = \bar{\mathbf{t}} & \text{on } \Gamma_t \end{cases} \tag{2}$$

where $\mathbf{b}$ is a body force vector, $\bar{\mathbf{t}}$ the prescribed traction on the natural boundary, and $\mathbf{n}$ the unit normal outward to the domain boundary.

It is noted that the MLS shape function $\Phi\,(x, \overline{x})$ does not possess the Kronecker delta δ function properties. As a result, the essential boundary cannot be directly applied. In the paper the Lagrange multiplier method is applied to enforce the essential boundary. The final discrete equations can be obtained by

$$\begin{bmatrix} K & G \\ G^T & 0 \end{bmatrix} \begin{Bmatrix} u \\ \lambda \end{Bmatrix} = \begin{Bmatrix} F \\ q \end{Bmatrix} \quad \text{or} \quad \overline{K}d = \overline{F} \tag{3}$$

where the details of the above equation can be referred to [12]. In this paper the Gauss integration method is used to solve the Eq. (3) to obtain nodal displacements.

2.2 Topology Optimization Based on the Adaptive EFG Method

In the paper, the nodal relative densities ρ_i are used as the design variables. So every nodal density in the domain can be obtained through the following shape function interpolation:

$$\rho^* = \sum_{i=1}^{N} \psi_i \rho_i \tag{4}$$

where N is the number of nodes inside the support domain, and ψ is the shape function.

The topology optimization formulation of the mean compliance design of structures is developed using the SIMP method [13]. SIMP is currently experiencing popularity because of its conceptual simplicity and numerical easiness. The key concept of the SIMP method is to relax the original discrete topology optimization problem to allow the intermediate densities of the design variables, so as to enable the gradient-based optimization methods (e.g., SLP, SQP, and SCP optimization algorithms) and avoid the "N–P" hard phenomena [14, 15]. This work employs the nodal relative densities as the design variables and the mathematical model of the optimization can be written as

$$\begin{cases} \min : \ C(\rho) = \overline{F}^T u \\ s.t. : \ V(\rho) - f V_0 \le 0 \\ \qquad \overline{K} d = \overline{F} \\ \qquad 0 < \rho_{\min} \le \rho \le 1 \end{cases} \tag{5}$$

where $C(\rho)$ is the objective, ρ is design variable, $V(\rho)$ and V_0 is the material volume and design domain volume, respectively, and f is the prescribed volume fraction, $\overline{F}$ is the force vector, and $\rho_{\min}$ is a vector of minimum relatives densities to avoid numerical singularity in numerical analysis. For the simplicity the standard OC method is used as the optimizer [16].

2.3 Local-Encryption of the Adaptive EFG Method

(1) The criterion of nodes adaptive encryption

In processing the topology optimization, the nodal relative densities are used as the design variables taking the values 0 or 1. A critical density value will be selected as the criteria of adding node in the region.

(2) The method of nodal encryption

According to the adaptive EFGM, the new nodes can be inserted in its local domains (Fig. 1).

For the special node I, we need to insert new nodes surrounding its domain influence by following the method: $(x_I, y_I + d_{Im}/2)$, $(x_I, y_I - d_{Im}/2)$, $(x_I + d_{Im}/2, y_I)$, $(x_I - d_{Im}/2, y_I)$, $(x_I + d_{Im}/2, y_I + d_{Im}/2)$, $(x_I - d_{Im}/2, y_I + d_{Im}/2)$, $(x_I + d_{Im}/2, y_I - d_{Im}/2)$, $(x_I - d_{Im}/2, y_I - d_{Im}/2)$, where the d_{Im} is the distance of the recent point to I.

(3) The criteria of new additional nodes update design variables

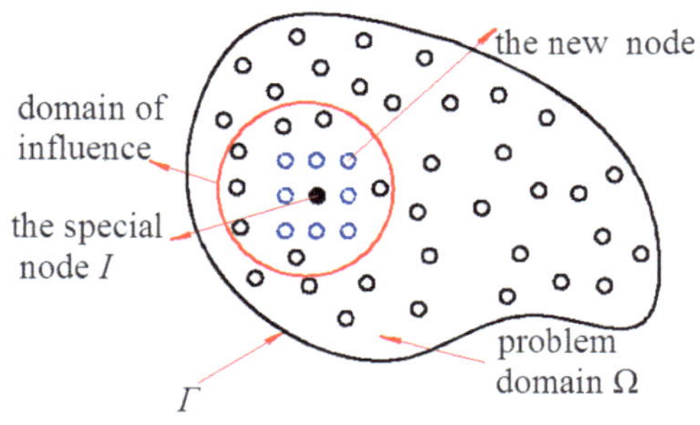

Fig. 1 Local nodal adding method

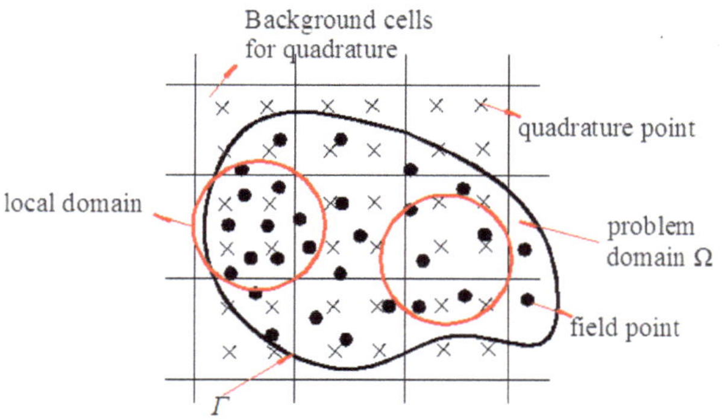

Fig. 2 The radius of adaptive search

After nodes local refinement, the new additional nodes need to be added into the original design variable to prepare for the next topology optimization. As nodal relative density values closed to other nodes in its local domain, the new additional nodal density value may be produced by the interpolation on the nodal relative densities in the latest topology optimization loop

$$\rho'' = \sum_{i=1}^{N} \xi(x_i)\rho_i' \tag{6}$$

where $\xi(x_i)$ is the interpolation shape function, ρ' is the relative density value of the last topology optimization, N is the number of nodes in the local domain, ρ'' is the relative density value of the new additional node.

(4) Determination of the radius on the adaptive method (Fig. 2)

Following the adaptive analysis, the nodes could be very arbitrary. So it is necessary to determine the size of the influence locally. The scaling factor c_I need to be assigned to each node to reflect local nodal densities. It is evaluated from the influence area by using the following form:

$$c_I' = \sqrt{\frac{2}{N} A_k} \tag{7}$$

where N is the number of nodes in domain, and the A_k the area of the k_{th} domain.

$$d_{mI}' = d_{\max}' c_I' \tag{8}$$

where $d_{mI}' \geq 2$ is often used for the static analysis.

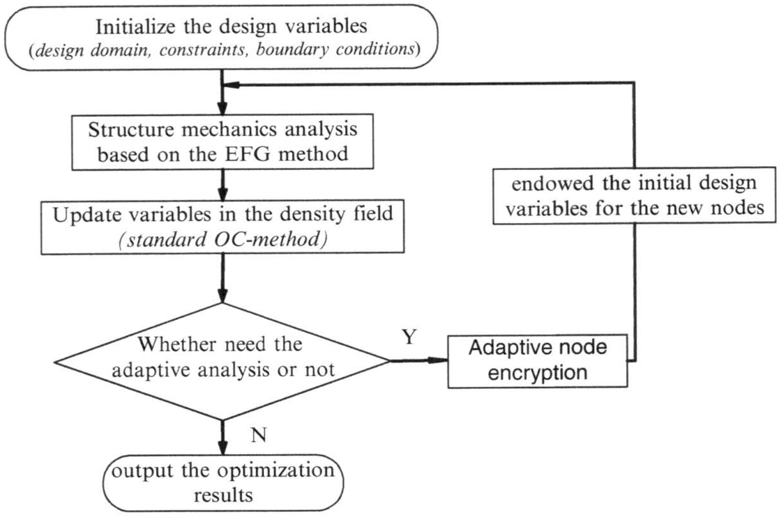

Fig. 3 Flowchart of topology optimization based on the adaptive EFG method

2.4 Flow Diagram of Adaptive EFGM Topology Optimization

The flowchart of topology optimization based on the adaptive EFG method shown in Fig. 3.

Step 1. Initialize the design variables: the design domain, constraints, and boundary conditions.

Step 2. Calculate the displacement at each node using the EFG method in those nodes distribution.

Step 3. Solve the optimization problem (5) using the OC method to update the design variables.

Step 4. Following the distribution of nodal relative density to judge whether the next adaptive analysis is to be performed or not.

Step 5. According to the criterion of the refinement $\rho_i > \rho_o$, select those nodes to be refined and insert nodes surrounding them.

Step 6. According to the new node distribution to update the design variables and return to *Step 2*.

3 Numerical Example for Adaptive Topology Optimization

The first example is a cantilever beam using the proposed adaptive topology optimization. The parameters for the design domain are $L = 20$, $D = 10$, and $P = 1{,}000$, prescribed in Fig. 4. The material properties: Young's Modulus $E = 3.0 \times 10^7$

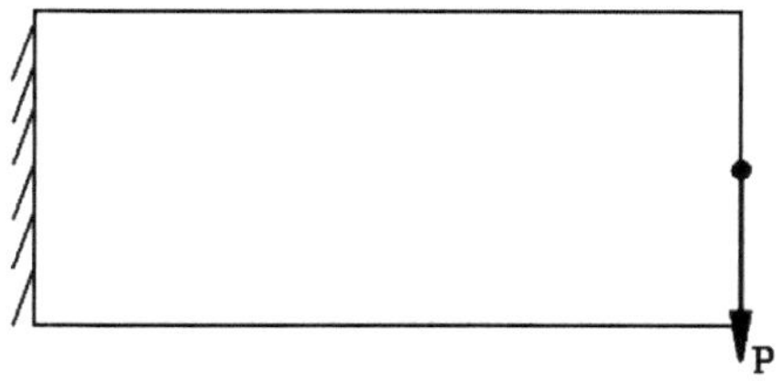

Fig. 4 Geometric model of the first numerical example

and Poisson's Ratio $\mu = 0.3$, the adaptive parameters $d_{max} = 2.0$, C = 2.5, and the volume constraint is limited to 30 %. The nodal distribution, the number of the design variables, and the results of the adaptive topology optimization are shown in Table 1.

The objective is to determine the optimal layout of the problem illustrated in Fig. 4. Every adaptive results of the topology optimization are illustrated in Table 1. Notice that the original domain is distributed with relatively sparse nodes (231 design variables), the optimal topology is clear but it not inadequate continuous and integrity. With the proposed adaptive topology optimization method, a better topology of the structure can be obtained after the first and second adaptive encryption. After the first encryption, the number of the design variables was increased to 554, compared to a number of 861, which has the same node distribution but the node number is reduced by 35.66 %. In the second nodal encryption, the number of the design variables was 1,090, compared to the same node distribution numbers of 3,321, which denotes a reduction of 42.52 %. From the analysis, it can be found that the adaptive meshless topology optimization method can get a better optimal topology and reduce the total number of the design variables, so as to improve the computational efficiency of the optimization.

4 Conclusions

This paper has proposed an adaptive meshless topology optimization method for the design of continuum structures. The numerical example shows that the proposed method can improve the computational efficiency of the optimization. A better topology description of the structure can be obtained with relatively a small number of field nodes. The proposed topology optimization can be further extended to more advanced mechanics problems.

Acknowledgements This research was partially funded by National Natural Science Foundation of China (51105229; 51305232), Natural Science Foundation of Hubei Province of China (2010CDB10805), and National Science Foundation for Distinguished Young Scholars of Hubei Province of China (2013CFA022).

Table 1 The results of topology optimization for the cantilever beam

Adaptive steps		The initial node map	Results of the topology optimization	Numbers of the design variables
0		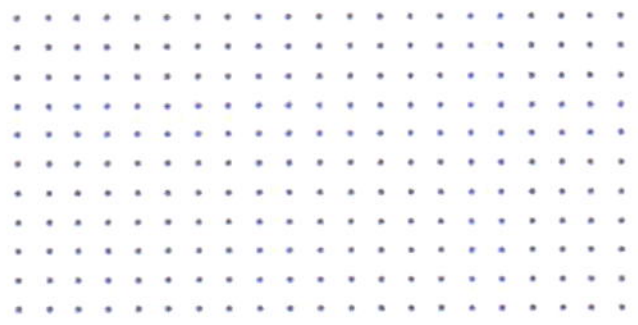		231
1	Don't use Adaptive technology	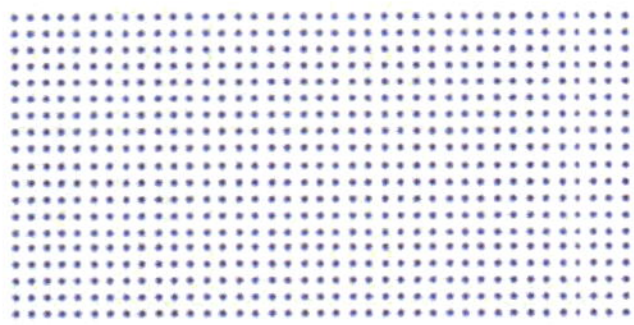		741
	using Adaptive technology			554
2	Don't use Adaptive technology	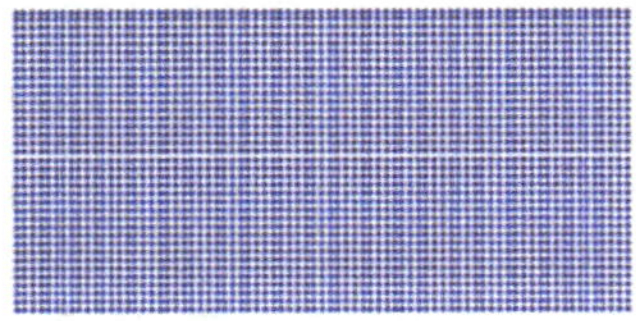		2849
	using Adaptive technology	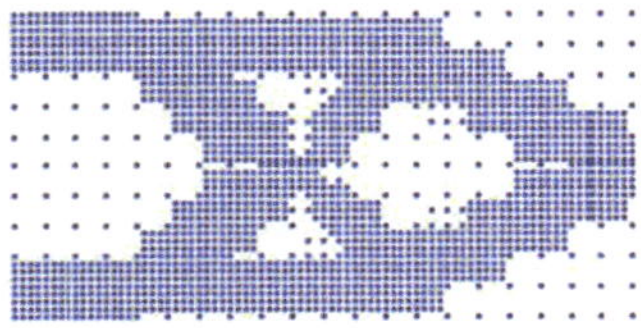		1909

References

1. Bendsøe, M.P., Sigmund, O.: Topology Optimization: Theory, Methods, and Applications. Springer, Berlin (2003)
2. Luo, Z., Zhang, N., Wang, Y., Gao, W.: Topology optimization of structures using meshless density variable approximants. Int. J. Numer. Methods Eng. **93**, 443–464 (2013)
3. Allaire, G., Jouve, F., Toader, A.M.: Structural optimization using sensitivity analysis and a level-set method. J. Comput. Phys. **194**, 363–393 (2004)

4. Luo, Z., Wang, M.Y., Wang, S., Wei, P.: A level set-based parameterization method for structural shape and topology optimization. Int. J. Numer. Methods Eng. **76**, 1–16 (2008)
5. Sigmund, O., Petersson, J.: Numerical instabilities in topology optimization: a survey on procedures dealing with checkerboards, mesh-dependencies and local minima. Struct. Multidiscip. Optim. **16**, 68–75 (1998)
6. Du, Y., Luo, Z., Tian, Q., Chen, L.: Topology optimization for thermomechanical compliant actuators using mesh-free methods. Eng. Optim. **41**, 753–772 (2009)
7. Xu, S., Cheng, G.D.: Structural topology optimization based on adaptive mesh. J. Dalian Univ. Technol. **49**, 469–475 (2009)
8. Lin, C.Y., Chou, J.N.: A two-stage approach for structural topology optimization. Adv. Eng. Softw. **30**, 261–271 (1999)
9. Wu, X., Gong, S.: Research of EFGM structure topology optimization based on adaptive scheme. Ph.D. Thesis, Xiangtan University (2011)
10. Belytschko, T., Lu, Y.Y., Gu, L.: Element-free Galerkin methods. Int. J. Numer. Methods Eng. **37**, 229–256 (1994)
11. Liu, G.R., Gu, Y.T.: An Introduction to Meshfree Methods and Their Programming. Springer, Berlin (2005)
12. Dolbow, J., Belytschko, T.: An introduction to programming the meshless element free Galerkin method. Arch. Comput. Methods Eng. **5**(3), 207–241 (1998)
13. Bendsøe, M.P., Sigmund, O.: Material interpolation schemes in topology optimization. Arch. Appl. Mech. **69**, 635–654 (1999)
14. Gao, D.Y.: Solutions and optimality criteria to box constrained nonconvex minimization problems. J. Ind. Manag. Optim. **3**, 293–304 (2007)
15. Gao, D.Y., Ruan, N.: Solutions to quadratic minimization problems with box and integer constraints. J. Glob. Optim. **47**, 463–484 (2012)
16. Rozvany, G.I.N., Kirch, U., Bendsøe, M.P., Sigmund, O.: Layout optimization of structures. Appl. Mech. Rev. **48**, 41–119 (1995)
17. Zhou, J.X., Zou, W.: Meshless approximation combined with implicit topology description for optimization of continua. Struct. Multidiscip. Optim. **36**, 347–353 (2008)
18. Liu, G.R., Tu, Z.H.: An adaptive procedure based on background cells for meshless methods. Comput. Methods Appl. Mech. Eng. **191**, 1923–1943 (2002)
19. Feng, T., Liu, X.: Study on adaptive meshfree method in solid mechanics. Ph.D. Thesis, Zhejiang University (2007)

Modeling and Multi-Objective Optimization of Double Suction Centrifugal Pump Based on Kriging Meta-models

Yu Zhang, Sanbao Hu, Jinglai Wu, Yunqing Zhang, and Liping Chen

Abstract This paper presents the multi-objective optimization problem of double suction centrifugal pump using Kriging meta-model. A set of double suction centrifugal pumps with various blade shapes were numerically simulated in the CFD software. Efficiency η and required net positive suction head (NPSHr) were investigated through these numerical simulations. Kriging meta-models were built to approximate the pump characteristic performance functions η and NPSHr using the design variables related to the blade geometrical shape. The objectives are to maximize η, as well as to minimize the NPSHr, which are two important indicates of centrifugal pump. Non-dominated Sorting Genetic Algorithm II (NSGA II) is used as the optimization algorithm. A tradeoff optimal point was selected in the Pareto-optimal solution set by means of robust design based on Monte Carlo simulations, and then the simulation result of the optimal solution was compared with experiment result, which shows that the proposed optimal solution coincides with the experiment well.

Keywords Centrifugal pump • Multi-objective optimization • Kriging meta-model • Computational fluid dynamics

1 Introduction

Centrifugal pumps are the group of turbomachines used for transporting liquids by raising a specified volume flow to a specified pressure level [1]. Anagnostopoulos [2] investigated the characteristic curves centrifugal pump by using CFD software, and the objective was to find the impeller geometry that maximizes the best efficiency value of the pump among a set of blade angle. Derakhshan et al. [3]

Y. Zhang • J. Wu • Y. Zhang (✉) • L. Chen
Center for Computer-Aided Design, School of Mechanical Science & Engineering, Huazhong University of Science & Technology, Wuhan, Hubei 430074, PR China
e-mail: zhangyq@hust.edu.cn

S. Hu
Hubei Key Laboratory of Advanced Technology of Automobile Parts, Wuhan University of Technology, Wuhan, Hubei 430074, PR China

D. Gao et al. (eds.), *Advances in Global Optimization*, Springer Proceedings in Mathematics & Statistics 95, DOI 10.1007/978-3-319-08377-3_25

used incomplete sensitivities and genetic algorithms to obtain a higher efficiency by re-designing the shape of impeller blades. These researches only deal with the design of centrifugal pumps as a single objective optimization problem. Actually, optimization of double suction centrifugal pumps is indeed a multi-objective optimization problem rather than a single objective optimization problem that has been considered so far in the literature [4]. Safikhani et al. [5] presented a multi-objective optimization progress on centrifugal pumps and proposed the Pareto-optimal solutions using genetic algorithm based on neural network meta-model. Nourbakhsh et al. [4] used particle swarm optimization (PSO) and Non-dominated Sorting Genetic Algorithm II (NSGA II) algorithm with neural network meta-model to find the Pareto front of two conflict objectives: efficiency and required net positive suction head (NPSHr). In contrast with these applications mentioned above, Kriging meta-models scarcely appear in the literatures relating to centrifugal pump, not to mention double suction type.

Multi-objective optimization is a process of optimizing a collection of objective functions systematically and simultaneously [6]. Modem design and real-world engineering problems often relate to multiple objectives, and they are thus treated as multi-objective optimization problems. Liu et al. [7] proposed a multi-objective optimization method based on an approximation model, and used the trust region move limits updating strategy to guarantee the accuracy of the approximation. Lin et al. [8] combined solid isotropic material with penalization (SIMP) with physical programming (PP) to achieve multi-objective topology optimization.

Kriging meta-model was originally proposed by Kleijnen [9] the South African mining engineer Daniel Gerhard Krige. It may give an optimal unbiased prediction for the unknown response points [10], which is a remarkable trait different from other meta-models. Compared to traditionally RSM, Kriging meta-models have the advantage in high dimensional nonlinear problems and prediction accuracy. Zakerifar et al. [11] constructed meta-models respectively by RSM and Kriging method. The comparison of the two methods showed that the Kriging meta-models are superior to the RSM, especially in the case of a multiple-objective optimization problem.

η and NPSHr are both important characteristic parameters for a double suction centrifugal pump. They are obtained either by experiments or by CFD simulations. The objective η needs to be maximized while the second objective NPSHr is required to be minimized. Actually, there is a conflict between the two objectives.

This paper studies multi-objective optimization for the design of double suction centrifugal pumps. η and NPSHr have been considered as two conflicting objectives. The well-established NSGA II is employed to perform the multi-objective optimization. The present work has been successful in applying Kriging Meta-models for multi-objective optimization of double suction centrifugal pump, and a robust design method based on Monte Carlo simulation is used to determine a tradeoff point from the Pareto frontier.

2 Modeling and CFD Simulation of Centrifugal Pumps

2.1 Illustration of Objective Functions

Centrifugal pump's efficiency is the ratio of useful power to input power, defined as follows.

$$\eta = P_u / P_{in} = \rho g H Q / P_{in} \tag{1}$$

Where P_u is the useful power, P_{in} is the input power, ρ denotes the fluid density, and g denotes the standard acceleration of gravity, H is the pump Head, Q is the volume flow rate.

NPSHr, which is a significant cavitation parameter for pumps, relates to the minimum pressure required at the suction port of the pump to keep the pump from cavitation. It can be depicted with the following equation

$$NPSHr = P_1 - P_{\min}/\rho g + u_1^2/2g \tag{2}$$

Where $P_{\min}$ is the minimum pressure of the impeller blade.

2.2 Definition of Design Variable

Impeller's meridional section is shown in Fig. 1. The shape of double suction centrifugal pump impeller is determined by the solid line as shown in Fig. 1, and it is parameterized by using quartic Bezier curve with five control points. There are four design variables: α_1, r_1, α_2, and r_2. r is the relative position in the line segment. Take line segment AC for example, r_1 is the ratio of the line length AB to AC.

2.3 Numerical Simulation of Centrifugal Pumps

The fluid flowing through the centrifugal pump is assumed as incompressibility, and the governing equations for conservation of mass and momentum are given as

$$\frac{\partial U_i}{\partial x_i} = 0 \tag{3}$$

$$\frac{\partial U_i}{\partial t} = -\frac{1}{\rho}\frac{\partial P}{\partial x_i} - U_j \frac{\partial U_i}{\partial x_i} + \frac{1}{\rho}\frac{\partial}{\partial x_j}\left(2\mu S_{ji} - \tau_{ji}\right) \tag{4}$$

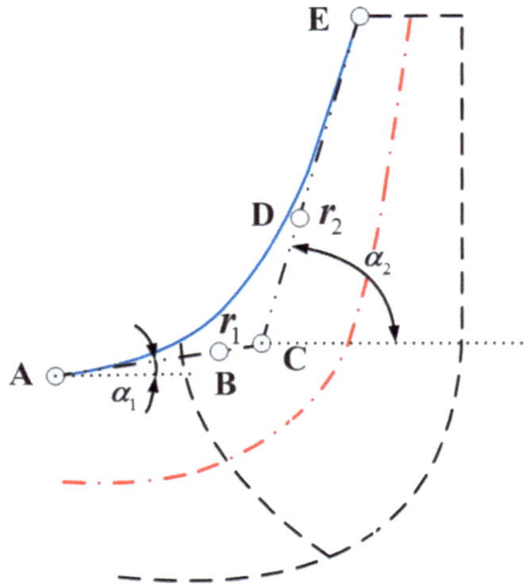

Fig. 1 Meridional section of the impeller

Where U_i denotes the velocity in tensor notation, x_i is the position vector in tensor notation, t denotes the time, ρ is the fluid density, P is the pressure, μ is molecular viscosity, Sji is the strain-rate tensor, and τji is Reynolds-stress tensor.

The standard $k-\varepsilon$ turbulence models are adopted and their equations mainly consist of turbulence kinetic energy and dissipation rate equations, expressed as

$$\frac{\partial k}{\partial t} = \frac{1}{\rho}\tau_{ij}\frac{\partial U_i}{\partial x_i} - U_j\frac{\partial k}{\partial x_i} - \varepsilon + \frac{1}{\rho}\frac{\partial}{\partial x_i}\left[(\mu + \mu_t/\sigma_k)\frac{\partial k}{\partial x_j}\right] \tag{5}$$

$$\frac{\partial \varepsilon}{\partial t} = \frac{1}{\rho}C_{\varepsilon 1}\frac{\varepsilon}{k}\tau_{ij}\frac{\partial U_i}{\partial x_j} - U_j\frac{\partial \varepsilon}{\partial x_j} - C_{\varepsilon 2}\frac{\varepsilon^2}{k} + \frac{1}{\rho}\frac{\partial}{\partial x_j}\left[(\mu + \mu_t/\sigma_\varepsilon)\frac{\partial \varepsilon}{\partial x_j}\right] \tag{6}$$

Where k is the kinetic energy of turbulent fluctuations per unit mass, μ_t is the eddy viscosity and $\mu_t = \rho C_\mu k^2/\varepsilon$, and ε is the dissipation per unit mass. $C_{\varepsilon 1}$, $C_{\varepsilon 2}$, C_μ, σ_k, and σ_ε are empirical constants.

The boundary conditions are mass flow inlet and pressure outlet, the flow rate is 2,000 m^3/h, and pressure is 1 *atm* respectively. The rotational speed is 1,400 *rpm*, flowing medium is water, and the residual error is set as 1e−5 to judge whether the calculation has converged.

The space-filling design method named Latin Hypercube Designs (LHD), which homogeneously covers the entire design domains, was employed to obtain 50 sample points. Table 1 shows the simulation results of the sample points, which are used to built the Kriging meta-models.

Table 1 The results of CFD simulations

	Design variable				Objective parameter	
Serial number	α_1 (deg)	r_1	α_2 (deg)	r_2	η (%)	NPSHr (m)
1	0.00	0.921	76.94	0.921	86.45	4.25
2	0.61	0.647	89.59	0.745	90.69	4.67
3	1.22	0.667	74.90	0.608	83.21	4.19
4	1.84	0.843	77.35	0.510	84.32	4.35
5	2.45	0.392	79.80	0.862	84.77	4.27
6	3.06	0.098	84.29	0.020	86.92	4.53
...	...	...	...	...	...	...
49	29.39	0.020	85.51	0.255	81.42	4.09
50	30.00	0.490	70.00	0.843	71.77	3.37

3 Modeling the Kriging Meta-models of Centrifugal Pumps

3.1 Kriging Model for Approximations

Kriging meta-models have global performance rather than local characteristics, used to predict the unknown values based on the given values. A Kriging model is described by the combination of a known polynomial and departures of the form

$$y(x) = \mathbf{f}(x)^{\mathrm{T}}\boldsymbol{\beta} + z(x) \tag{7}$$

Where $y(x)$ denotes the response function, $\mathbf{f}(x)^{\mathrm{T}}\boldsymbol{\beta}$ is the regression model, $\mathbf{f}(x)$ is the regression basis function, $\boldsymbol{\beta}$ is the regression coefficient, and $z(x)$ is assumed as an independent Gaussian random process which is characterized by zero mean and covariance

$$Cov\left[z(\omega), z(x)\right] = \sigma^2 R(\theta, \omega, x) \tag{8}$$

Where σ^2 is the process variance, θ is the unknown correlation parameter, $R(\theta, \omega, x)$ is the correlation function between the points ω and x.

The estimate value at point x and the prediction error are respectively given as

$$\widehat{y}(x) = \mathbf{C}^T(x)\mathbf{Y} \tag{9}$$

$$\widehat{y}(x) - y(x) = \mathbf{C}^T\mathbf{Y} - y(x) = \mathbf{C}^T\mathbf{Z} - z + \left(\mathbf{F}^T\mathbf{C} - \mathbf{f}(x)\right)^T\boldsymbol{\beta} \tag{10}$$

Where $\boldsymbol{C}$ is the unknown weight vector which is selected to minimize the mean square error (MSE), $\boldsymbol{Y}$ is the response of the sample points, $\mathbf{Z} = [z_1, \ldots, z_m]^T$, m is the number of the sample points, and $\mathbf{F}$ is a vector which consists of the value of $\mathbf{f}(x)$ at each sample points.

According to the unbiased condition, $(\mathbf{F}^T\mathbf{C}-\mathbf{f}(x))^T=0$ or $\boldsymbol{F}^T\boldsymbol{C}(x)=\boldsymbol{f}(x)$ is demanded. Without loss of generality, the MSE of the predictor's values under this condition is given by

$$s(x) = E\left[(\widehat{y}(x) - y(x)^2\right] = \sigma^2\left(1 + \mathbf{C}^T\mathbf{R}\mathbf{C} - 2\mathbf{C}^T\mathbf{r}\right) \tag{11}$$

where $\boldsymbol{r}^T(x)=[R(\theta,x,x_1)\cdots R(\theta,x,x_m)]$, which is a vector of an unknown point and all known sample points, $\boldsymbol{R}$ is an $m\times m$ symmetric correlation matrix.

Under the unbiased condition, the Lagrangian function is introduced to minimize the value of MSE with respect to $\boldsymbol{C}$. The unknown parameters $\boldsymbol{\beta}$ and σ^2 can be estimated by

$$\widehat{\boldsymbol{\beta}} = \left(\mathbf{F}^T\mathbf{R}^{-1}\mathbf{F}\right)^{-1}\mathbf{F}^T\mathbf{R}^{-1}\mathbf{Y} \tag{12}$$

$$\widehat{\sigma}^2 = \left(\mathbf{Y} - \mathbf{F}\widehat{\boldsymbol{\beta}}\right)^T\mathbf{R}^{-1}\left(\mathbf{Y} - \mathbf{F}\widehat{\boldsymbol{\beta}}\right)/m \tag{13}$$

The Kriging model and Eq. (13) can be given further as

$$\widehat{y}(x) = \mathbf{f}^T(x)\widehat{\boldsymbol{\beta}} + \mathbf{r}^T(x)\boldsymbol{R}^{-1}\left(\boldsymbol{Y} - \boldsymbol{F}\widehat{\boldsymbol{\beta}}\right) \tag{14}$$

$$s(x) = \sigma^2\left(1 + \mathbf{u}^T\left(\mathbf{F}^T\mathbf{R}^{-1}\mathbf{F}\right)^{-1}\mathbf{u} - \mathbf{r}(x)^T\mathbf{R}^{-1}\mathbf{r}(x)\right) \tag{15}$$

Where $\mathbf{u}=\mathbf{F}^T\mathbf{R}^{-1}\mathbf{r}(x)-\mathbf{f}(x)$.

3.2 Kriging Models Verification

In order to validate the predictor's ability of the Kriging meta-models, an additional set of 30 points obtained by LHD are used as test points. The root mean square error (RMSE) between the estimated values (Kriging meta-models) and the actual values obtained by CFD simulation is defined as following equation, and we use the RSME as it indicates to validate that if the meta-models satisfy the requirement or not.

$$RMSE = \sqrt{\sum_i^N d_i{}^2/N} \tag{16}$$

Where N is the total number of the test points, and $d_i=Y_{i\,(Kriging)}-Y_{i\,(CFD)}$. RMSE is used to measure the average error between values predicted by a model and the actual values. The predictive value and CFD value of RMSE are calculated according to Eqs. (15) and (16), respectively. Table 2 shows that the prediction

Table 2 The values of RMSE

Meta-models for input data-η		Meta-models for input data-NPSHr	
Prediction	CFD	Prediction	CFD
0.9361	0.9202	0.0473	0.0435

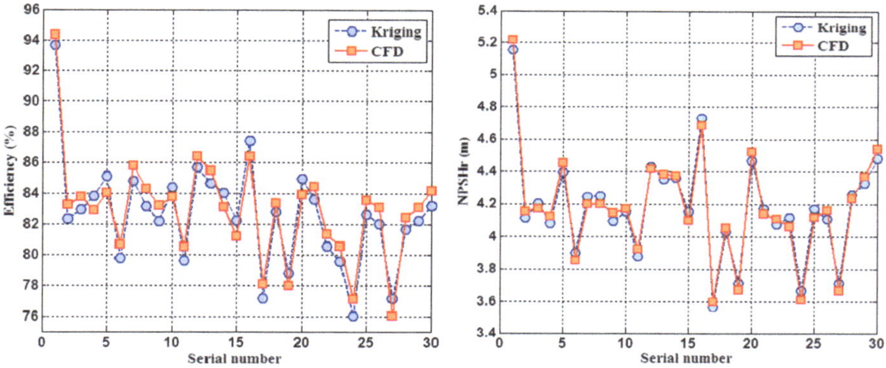

Fig. 2 Kriging predictor vs. CFD simulation

RMSE of the Kriging meta-models is coincided with the CFD results well. The values of the RMSE indicate that the prediction accuracy of the Kriging meta-models is pretty reliable.

Figure 2 shows the comparison of calculation results between the Kriging predication values and CFD simulation values. From the figures, we find that the estimated values of Kriging meta-models are coincided with the CFD simulation values well.

4 Multi-Objective Optimization of Centrifugal Pumps Using NSGA II

4.1 Problem Definition and Optimization Results

The problem in this paper can be described as follows

$$\begin{cases} Find & \boldsymbol{X} = [\alpha_1, r_1, \alpha_2, r_2]^T \\ Maximize & \eta = f_1(\alpha_1, r_1, \alpha_2, r_2) \\ Minimize & NPSHr = f_2(\alpha_1, r_1, \alpha_2, r_2) \\ Subject\ to & 0^\circ \leq \alpha_1 \leq 30^\circ;\ 0.02 \leq r_1 \leq 0.98; \\ & 70^\circ \leq \alpha_2 \leq 90^\circ; 0.02 \leq r_2 \leq 0.98 \end{cases} \tag{17}$$

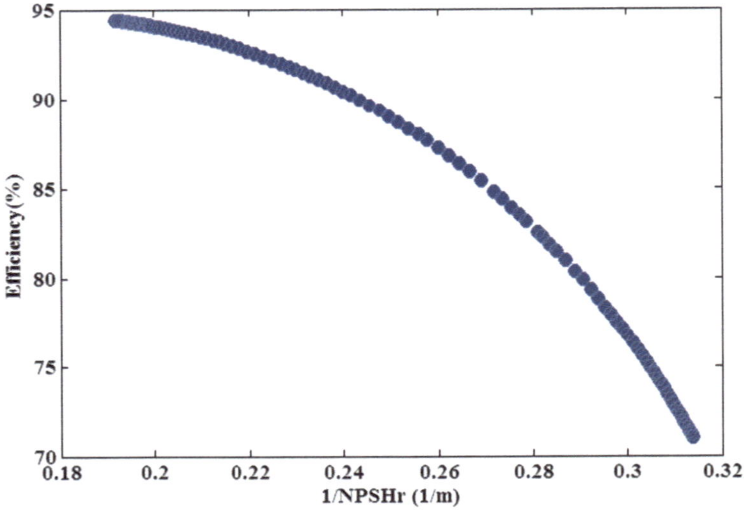

Fig. 3 Pareto-optimal front of efficiency and 1/*NPSHr* using NSGA II

Where f_1 and f_2 is the optimization objectives. NSGA II is one of the most popular multi-objective optimization algorithms. It is characterized by fast non-dominated sorting approach and simple crowded comparison operator [12]. NSGA II is usually a good choice to deal with multi-objective problem involving multiple conflicting objectives for its good performance of global searching. SBX and polynomial mutation have been used as operators for crossover and mutation, respectively. The distribution indexes for both operators are $\eta_c = 20$ and $\eta_m = 20$, respectively. The crossover probability is set to 0.9 and the mutation probability of $1/n$, where n is the number of decision variables. The NSGA II was running using the population size of 100 and the maximum generation's number of 200.

Figure 3 shows the Pareto-optimal front of the two conflict objectives η and 1/*NPSHr*. As shown in Fig. 3, when a beneficial choice leans to one objective, the other will go worse, and vice versa. These non-dominated optimal points obviously present tradeoffs between objective functions η and NPSHr.

Figure 4 show the objectives efficiency η and 1/NPSHr with respect to design variables in the Pareto front, respectively. Both figures are divided into four regions by five dashed lines. The four regions are respectively AB, BC, CD, and DE as shown in Fig. 4a. On the contrary, the order of these regions in Fig. 4b reverses to the Fig. 4a, which are due to the conflict between efficiency and 1/NPSHr as is shown in Fig. 3. There are some particular phenomena in Fig. 4. For example, in Fig. 4a, each region from AB to DE, the design variables r_2, α_2, r_1, and α_1 change in turn while the other three design variables are approximated to keep constant in the corresponding region. Additionally, there are two design variables which change in region CD, and it is different from other regions with only one design variable variation.

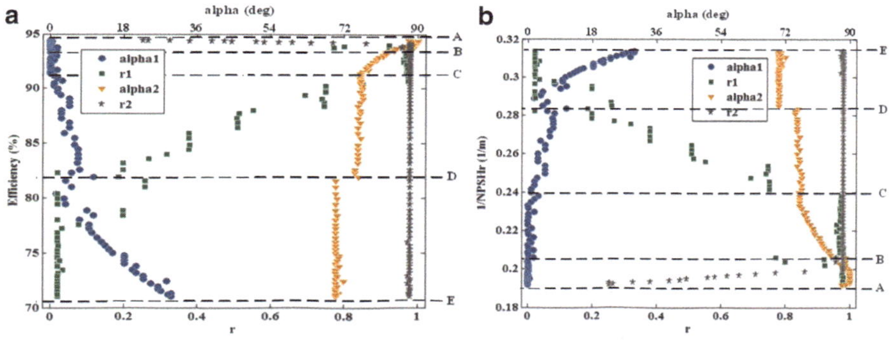

Fig. 4 Design variables vs. objective functions. (**a**) Design variables vs. η. (**b**) Design variables vs. 1/NPSHr

4.2 Optimal Point Decision and Physical Prototype Verification

For a centrifugal pump designer, a tradeoff point should be chosen among the Pareto-optimal solutions as the final pump design point. In order to guide the tradeoff selection, the robust design based on Monte Carlo simulation is proposed to determine the best point.

Monte Carlo simulation can obtain the stochastic characteristics containing mean value and variance of an objective variable by evaluating a set of random design alternative. Robust design is the method that improves quality and reliability by reducing the functional variation of a system without removing the causes of variation [13]. Given the objective functions efficiency and 1/NPSHr in Fig. 4 are f_1 and f_2 for convenience, respectively. The index function is defined as

$$F = w_1\left[w_2(\mu_1 - f_1)^2 + w_3\sigma_1^2\right] + w_4\left[w_5(\mu_2 - f_2)^2 + w_6\sigma_2^2\right] \tag{18}$$

Where w_1, w_2, w_3, w_4, w_5, and w_6 are the weight coefficients, μ_1 and μ_2 are the mean values of f_1 and f_2 in each point of the Pareto-optimal solutions, respectively, σ_1^2 and σ_2^2 are the variance of f_1 and f_2 in each point of the Pareto-optimal solutions, respectively. The design variables α_1, r_1, α_2, and r_2 are assumed to obey normal distribution. In order to obtain the stochastic characteristics μ_1, μ_2, σ_1^2, and σ_2^2 of each points in the Pareto-optimal front, Monte Carlo method is carried out by 1,000 simulations with five samples per design variable.

The tradeoff optimal point is the one that makes the value of index function F minimum, which indicates the lowest functional variation. The values of design variables in the optimal point are shown in Table 3. The physical prototype with respect to the tradeoff point is shown in Fig. 5. Figure 5a shows the impeller with corresponding design parameters of the selected point. Figure 5b shows the experiment of the double suction centrifugal pump. The pump is fitted on the test bench, and the electric machinery drives the impeller. The water is transmitted

Table 3 Values of design variables in the optimal point

Design variable			
α_1 (deg)	r_1	α_2 (deg)	r_2
9.231	0.2169	76.41	0.07

Fig. 5 The physical prototype with respect to the tradeoff point. (**a**) Impeller, (**b**) Pump

Table 4 Comparison among CFD simulation, predicted, and test values

	CFD	Predictor	Test
η (%)	88.12	87.65	86.53
NPSHr (m)	3.81	3.88	4.01

circularly in the test pool. Table 4 shows the CFD simulation, predicted, and test values of the tradeoff optimal point. The result indicates that the values of the tradeoff point predicted by the CFD simulation, Kriging meta-models, and experimental results agree with each other well.

5 Conclusion

In this paper, the multi-objective optimization is used to double suction centrifugal pumps. The Kriging meta-models of pump performance parameters η and NPSHr have been built and agree with CFD numerical simulation well in the unknown points. The Pareto-optimal solutions have been obtained using NSGA II, which reveals the relationship between the design variables and objective functions. The CFD simulation results indicate that the design variables have significant influence on the centrifugal pump performance such as efficiency and NPSHr. A tradeoff point has been selected using the robust design based on Monte Carlo simulations, and the objectives of the optimal point have been compared with corresponding physical prototype, and the results present that the optimization result is believable.

References

1. Gülich, J.: Pump Types and Performance Data. Centrifugal Pumps, pp. 39–68. Springer, Berlin (2010)
2. Anagnostopoulos, J.S.: A fast numerical method for flow analysis and blade design in centrifugal pump impellers. Comput. Fluids **38**(2), 284–289 (2009)
3. Derakhshan, S., Mohammadi, B., et al.: The comparison of incomplete sensitivities and Genetic algorithms applications in 3D radial turbomachinery blade optimization. Comput. Fluids **39**(10), 2022–2029 (2010)
4. Nourbakhsh, A., Safikhani, H., et al.: The comparison of multi-objective particle swarm optimization and NSGA II algorithm: applications in centrifugal pumps. Eng. Optim. **43**(10), 1095–1113 (2011)
5. Safikhani, H., Khalkhali, A., et al.: Pareto based multi-objective optimization of centrifugal pumps using CFD, neural networks and genetic algorithms. Eng. Appl. Comput. Fluid Mech. **5**(1), 37–48 (2011)
6. Deb, K., Kalyanmoy, D.: Multi-objective Optimization Using Evolutionary Algorithms. Wiley, Chichester (2001)
7. Liu, G.P., Han, X., et al.: A novel multi-objective optimization method based on an approximation model management technique. Comput. Methods Appl. Mech. Eng. **197**(33–40), 2719–2731 (2008)
8. Lin, J., Luo, Z., et al.: A new multi-objective programming scheme for topology optimization of compliant mechanisms. Struct. Multidiscip. Optim. **40**(1–6), 241–255 (2010)
9. Kleijnen, J.P.C.: Kriging metamodeling in simulation: a review. Eur. J. Oper. Res. **192**(3), 707–716 (2009)
10. Martin, J.D.: Computational improvements to estimating Kriging metamodel parameters. J. Mech. Des. **131**(8), 084501 (2009)
11. Zakerifar, M., Biles, W.E., et al.: Kriging metamodeling in multiple-objective simulation optimization. Simulation **87**(10), 843–856 (2011)
12. Deb, K., Pratap, A., et al.: A fast and elitist multiobjective genetic algorithm: NSGA-II. IEEE Trans. Evol. Comput. **6**(2), 182–197 (2002)
13. Chen, S.K., Chen, W., et al.: Level set based robust shape and topology optimization under random field uncertainties. Struct. Multidiscip. Optim. **41**(4), 507–524 (2010)

Topology Optimization for Human Proximal Femur Considering Bi-modulus Behavior of Cortical Bones

Kun Cai, Zhen Luo, and Yu Wang

Abstract The material in the human proximal femur is considered as bi-modulus material and the density distribution is predicted by topology optimization method. To reduce the computational cost, the bi-modulus material is replaced with two isotropic materials in simulation. The selection of local material modulus is determined by the previous local stress state. Compared with density prediction results by traditional isotropic material in proximal femur, the bi-modulus material layouts are different obviously. The results also demonstrate that the bi-modulus material model is better than the isotropic material model in simulation of density prediction in femur bone.

Keywords Topology optimization • Proximal femur • Bone remodeling • Multiple load cases • Bi-modulus material

1 Introduction

Osteoporosis, a kind of bone illness, happens widely in aged people. Serious osteoporosis leads to fracture of bone happening easily. To give an efficient therapy of osteoporosis (e.g., improving the speed of bone apposition), the mechanical properties of (both of cortical and cancellous) bone should be investigated and some results may act an important role in therapy of osteoporosis. In research of bone mechanics, the Wolff's law [1], which states that bone micro-structure (Fig. 1) and local stiffness tend to align with the principal stress directions according to its mechanical environments, is actually a core concept. In the viewpoints of some researchers, the bone is an optimal structure with "maximum" mechanical

K. Cai
College of Water Resources and Architectural Engineering, Northwest A&F University, Yangling, Xianyang 712100, PR China
e-mail: kuncai99@163.com

Z. Luo (✉) • Y. Wang
School of Mechanical and Mechatronic Engineering, The University of Technology, Sydney, NSW 2007, Australia
e-mail: zhen.luo@uts.edu.au; Yu.Wang-11@student.uts.edu.au

D. Gao et al. (eds.), *Advances in Global Optimization*, Springer Proceedings in Mathematics & Statistics 95, DOI 10.1007/978-3-319-08377-3_26

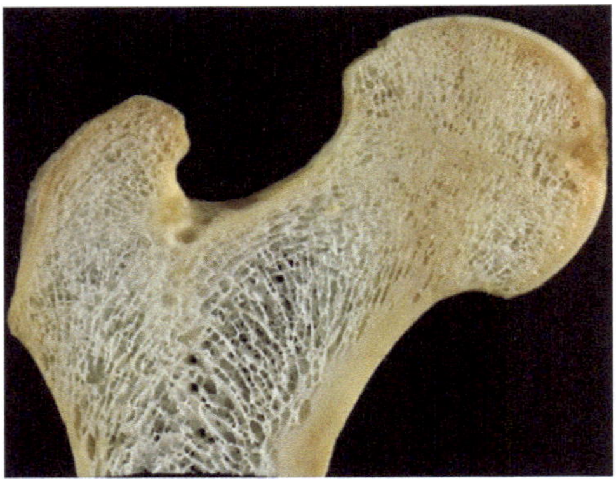

Fig. 1 Photograph and radiograph of human proximal femur [2]

efficiency but "minimum" mass according to the Wolff's law. During the last 40 years experimental equipments and computational techniques have been developed rapidly. Nowadays, the quantitative study of the Wolff's law can be carried out both of experimentally and numerically.

During the last four decades, the Wolff's law has been validated by many investigations. Much attention has been paid on the relations between the loading environments and the micro-structure of bone [3–8].

It is interesting that bone modeling process is often considered as an optimization process. Hollister et al. [9] tried to answer the question: whether bone remodeling can approach a globally optimized structure. Confined by computational conditions, the conclusion was not positive. Later, Jang and Kim [10] quantitatively investigated the validity of Wolff's law by topology optimization method. They claimed that topology optimization (with minimal structural compliance) and the bone remodeling (SED distributing uniformly) are equivalent. Fernandes et al.[11] an analytical parametric micro-structural model for trabecular bone in proximal femur was proposed according to the homogenization theory, and optimal densities and orientations were obtained by topology optimization. Similar work was given by Kowalczyk [12]. Design space optimization (DSO) method by Kim and Kwak [13] was adopted for micro-structure prediction of proximal femur by their group [10, 14].

Zhu et al. [6] tested the tensile and compressive modulus of Takin femoral cortical bone. They found that the compressive modulus of bone is 5–6 times of the tensile modulus. The difference also exists in other bones [7, 8]. However, till now, no work gives the prediction of cancellous bone in proximal femur with consideration of the bi-modulus behavior of bone. In this study, a material replacement method [15] is used to investigate the optimal material layout in proximal femur.

2 Material Properties

2.1 Bi-modulus Behavior of Elastic Material

Experiments [6] show a cortical bone has different elastic behaviors under tension and compression load. Therefore, bone is a typical bi-modulus material. Figure 2 gives the stress–strain curve of a bi-modulus material. Tangent α gives the tension modulus of material, i.e., $E_T = \tan\alpha$; and the tangent β is the compression modulus, i.e., $E_C = \tan\beta$. The ratio between them is marked with R, e.g., $R = E_T/E_C$. The material becomes an isotropic material when $\alpha = \beta$. Clearly, bi-modulus material properties are stress-dependent. Usually, in structural analysis, the piecewise linear material has to be treated as nonlinear and approximated with differentiable curve in structural reanalysis [16]. But in the present work, the character of the curve, i.e., piecewise linear, is adopted by material replacement operation [15], i.e., the bi-modulus material is replaced with two isotropic material and one of them will be used for an element in finite element analysis according to the previous stress state of the element.

2.2 Stiffness of Porous Material

For porous material (e.g., the mth material sample or finite element), the relationship between the stiffness tensor and the volume fraction (relative density) is expressed as

$$D_{m,ijkl} = \rho_m^p D_{0,ijkl} \tag{1}$$

where $D_{m,\ ijkl}$ is the elastic tensor of porous material, and the relative density $\rho_m \in [0\ 1.0]$. According to experiments data of bone [17], the power coefficient p is set to be 2 in the present study. $D_{0,\ ijkl}$ is the elastic tensor of fully solid material.

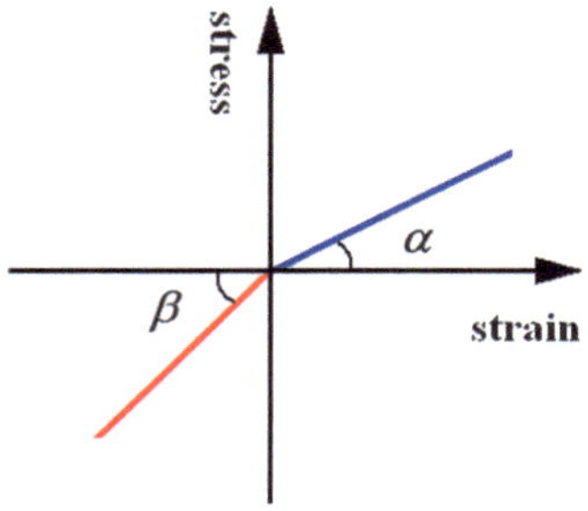

Fig. 2 The stress–strain curve of a bi-modulus material, $\alpha \neq \beta$

3 Optimization Model

3.1 Formulations of Topology Optimization Problem

Continuum topology optimization method became a hot point in structural optimization since the homogenization design method (HDM) was presented by Bendsøe and Kikuchi [18]. A large number of efforts have been paid on the development of continuum topology optimization theories during last 25 years. Besides the HDM, the most popular methods are the solid isotropic micro-structures with penalization (SIMP) method by Rozvany et al. [19]; the evolutionary structural optimization (ESO) method by Xie and Steven [20]; and the level set method by Wang et al. [21].

In the present work, the final density distribution of bone in proximal femur is obtained by modified SIMP method, rather than bone remodeling model. The volume constrained optimization of a structure with minimum of the structural mean compliance under multiple load cases is expressed as

$$\begin{aligned} \min_{\{\rho_m\}} c &= \sum_{i}^{N} w_i \mathbf{U}_i^T \overline{\mathbf{K}_i} \mathbf{U}_i = \sum_{i}^{N} w_i \sum_{m=1}^{M} \left(\mathbf{u}_m^T \overline{\mathbf{k}}_m \mathbf{u}_m \right)_i \\ \text{s.t.} \quad & \sum_{i}^{N} \sum_{m=1}^{M} (\rho_m v_m)_i = f_v \cdot V_0 \\ & \mathbf{K}_i \mathbf{U}_i = \mathbf{P}_i \,, \quad i = 1,\ 2, \ldots,\ N \\ & 0 < \rho_{\min} \le \rho_m \le 1.0 \,, \quad m \in \Omega \end{aligned} \tag{2}$$

where the objective function c is the sum of the structural mean compliance. N is the number of load cases the structure subjected to. w_i is the weighted coefficient for the ith load case. M is the total number of elements. $\{\rho_m\}$ is the set of relative densities of elements. $\mathbf{U}_i$ and $\mathbf{P}_i$ are the global nodal displacement and force vectors in the ith load case, respectively. $\overline{\mathbf{k}}_m$ is the modified matrix of $\mathbf{k}_m$ (the stiffness matrix of the mth element with isotropic material). The global stiffness matrix of structure $\mathbf{K}_i$ is assembled with $\{\mathbf{k}_m\}_i$. $\overline{\mathbf{K}}_i$ is assembled with $\left\{\overline{\mathbf{k}}_m\right\}_i$. $\mathbf{u}_m$ is the nodal displacement vector of the mth element. v_m is the volume of the mth element. f_v is the critical volume ratio of the final structure. V_0 is the total volume of solid design domain.

3.2 Selection Criterion of Elastic Modulus of Material

After being replaced with two isotropic materials, the original bi-modulus material in an element will be considered as one of the two isotropic materials for structural analysis. The elastic modulus of an element in structure under MLC is determined by the following equation:

$$E_m = \begin{cases} E_T, & \text{if} \quad (\text{TSED} > \text{CSED})_m \\ E_C, & \text{if} \quad (\text{TSED} < \text{CSED})_m \\ \max(E_T, E_C), & \text{others} \end{cases} \tag{3}$$

where the TSED (tension strain energy density) and CSED (compression strain energy density) can be calculated by the following formulations:

$$\begin{aligned} \text{TSED}_m &= \frac{1}{4}\sum_{i=1}^{N}\sum_{j=1}^{3}\left(\sigma_{j,i} + \left|\sigma_{j,i}\right|\right)\cdot\varepsilon_{j,i} \\ \text{CSED}_m &= \frac{1}{4}\sum_{i=1}^{N}\sum_{j=1}^{3}\left(\sigma_{j,i} - \left|\sigma_{j,i}\right|\right)\cdot\varepsilon_{j,i} \end{aligned} \tag{4}$$

where $\sigma_{j,i}$ and $\varepsilon_{j,i}$ are the mean principal stress and strain of element, respectively.

4 Results and Discussions

The deformation analysis in topology optimization of proximal femur is accomplished with the commercial software ANSYS [22] and 4-node plane stress element is used.

4.1 Finite Element Model for Proximal Femur

The upper part of proximal femur of human (see Fig. 3a) is discretized into 9,240 elements. The bottom is simply supported. In this study, the cortical bone with tensile elastic modulus of 17.0 GPa (the mean value) and Poisson's ratio of 0.3 [23]. The relative density of bone varies within the interval [0.05 1.0]. The modulus of bone under tension is not greater than that under compression. And the ratios between tension and compression moduli are as follows: $R = 1.0$ and 0.5, respectively. As R equals 1.0, the material shows isotropic. At the same time, two cases on considering the volume ratio of proximal femur are discussed, e.g., 35 and 46 %. From the data in Table 1, the three weighted coefficients are 0.6, 0.2, and 0.2 for three cases, respectively.

4.2 Optimal Material Distributions in Proximal Femur

Figure 3b, c give the final isotropic material distributions in proximal femur with different volume ratios. The amount of material supporting the first load case (one-legged stance) increases obviously. It reflects the real loading activity of one proximal femur.

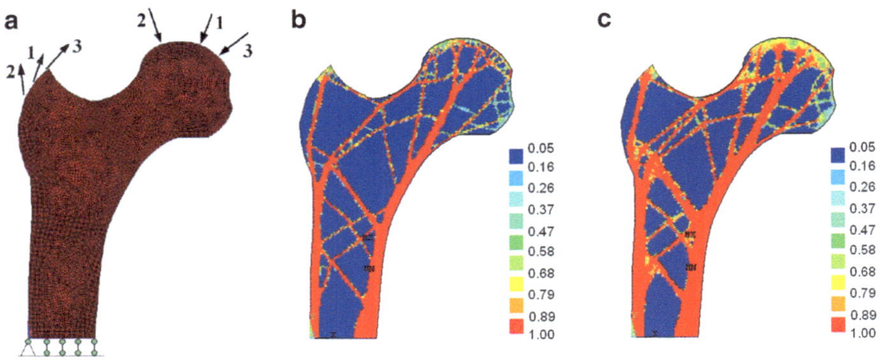

Fig. 3 Finite element of proximal femur (**a**) and density distributions of isotropic material ($R = 1$). (**a**) Three load cases. (**b**) Density plot as f_v=35 %. (**c**) Density plot as f_v=46 %

Table 1 Forces in three load cases on the proximal femur model

		Abductor reaction		Joint reaction	
Load case	Cycles/day	Magnitude (N)	Orientation (°)	Magnitude (N)	Orientation (°)
1 (one-legged stance)	6,000	703	28	−2,317	24
2 (abduction)	2,000	351	−8	−1,158	−15
3 (adduction)	2,000	468	35	−1,548	56

Orientations are given according to vertical (*negative* for *left* and *positive* for *right*). Negative force means compression and positive means tension [10]

Figure 4 gives the final bi-modulus material ($R = 0.5$) distributions in proximal femur with different amount of residual material. The amounts of material with mid-density are greater than those in Fig. 3b, c, respectively. Comparing with Fig. 3b, c, the result in Figure 4b gives a better prediction of cancellous distribution in femur bone (Fig. 1).

5 Conclusions

In simulation of density distribution of the human proximal femur, the material constitutive model is essential to the results. From the results given above, the bi-modulus material model matches the real architecture better than the isotropic material model. As the finer finite element mesh is adopted, a perfect detailed architecture of trabecular in femur bone can be found.

Acknowledgments The financial supports of the National Natural Science Foundation of China (Grant No. 50908190, 51179164) and the Fundamental Research Foundation of Northwest A&F University (Grant No. QN2011125) are greatly acknowledged.

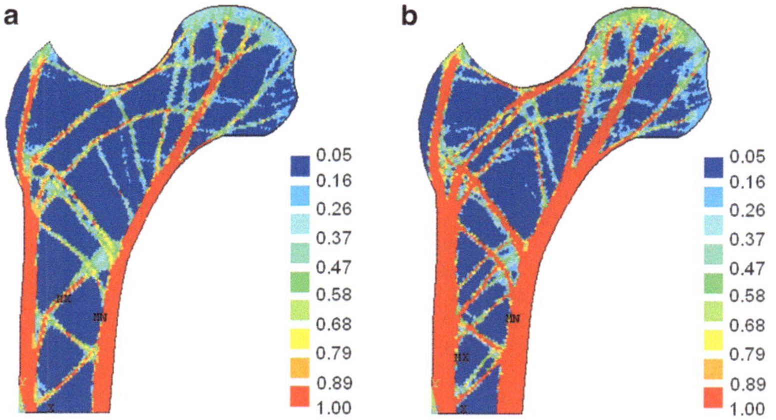

Fig. 4 Density distributions of isotropic material ($\boldsymbol{R} = 0.5$) in proximal femur. (**a**) Density plot as f_v=35 %. (**b**) Density plot as f_v=46 %

References

1. Wolff, J.: Das Gesetz Der Transformation Der Knochen. Hirschwald, Berlin (1892)
2. Skedros, J.G., Baucom, S.L.: Mathematical analysis of trabecular 'trajectories' in apparent trajectorial structures: the unfortunate historical emphasis on the human proximal femur. J. Theor. Biol. **244**, 15–45 (2007)
3. Whitehouse, W.J., Dyson, E.D.: Scanning electron microscope studies of trabecular bone in the proximal end of the human femur. J. Anat. **118**(3), 417–444 (1974)
4. Cowin, S.C.: Wolff's law of trabecular architecture at remodeling equilibrium. J. Biomech. Eng. **108**, 83–88 (1986)
5. Odgaard, A., Kabel, J., van Rietbergen, B., Dalstra, M., Huiskes, R.: Fabric and elastic principal directions of cancellous bone are closely related. J. Biomech. **30**, 487–495 (1997)
6. Zhu, L., Zhao, H.P., Song, Y.L., Feng, X.Q.: Experimental investigation of the mechanical properties of Takin femoral cortical bone. J. Tsinghua Univ. **46**(2), 301–304 (2006)
7. Cory, E., Nazarian, A., Entezari, V., Vartanians, V., Muller, R., Snyder, B.D.: Compressive axial mechanical properties of rat bone as functions of bone volume fraction, apparent density and micro-ct based mineral density. J. Biomech. **43**, 953–960 (2010)
8. Nazarian, A., Araiza Arroyo, F.J., Rosso, C., Aran, S., Snyder, B.D.: Tensile properties of rat femoral bone as functions of bone volume fraction, apparent density and volumetric bone mineral density. J. Biomech. **44**, 2482–2488 (2011)
9. Hollister, S.J., Kikuchi, N., Goldstein, S.A.: Do bone ingrowth processes produce a globally optimized structure? J. Biomech. **26**(4–5), 391–407 (1993)
10. Jang, I.G., Kim, I.Y.: Computational study of Wolff's law with trabecular architecture in the human proximal femur using topology optimization. J. Biomech. **41**, 2353–2361 (2008)
11. Fernandes, P., Rodrigues, H., Jacobs, C.: A model of bone adaptation using a global optimization criterion based on the trajectorial theory of Wolff. Comput. Methods Biomech. Biomed. Eng. **2**(2), 125–138 (1999)
12. Kowalczyk, P.: Simulation of orthotropic micro-structure remodelling of cancellous bone. J. Biomech. **43**, 563–569 (2010)
13. Kim, I.Y., Kwak, B.M.: Design space optimization using a numerical design continuation method. Int. J. Numer. Methods Eng. **53**, 1979–2002 (2002)

14. Kim, H.A., Clement, P.J., Cunningham, J.L.: Investigation of cancellous bone architecture using structural optimisation. J. Biomech. **41**, 629–635 (2008)
15. Cai, K., Gao, Z.L., Shi, J.: Compliance optimization of a continuum with bi-modulus material under multiple load cases. Comput. Aided Des. **45**, 195–203 (2013)
16. Medri, G.: A nonlinear elastic model for isotropic materials with different behavior in tension and compression. Trans. Am. Soc. Mech. Eng. **104**(1), 26–28 (1982)
17. Buchler, P., Ramaniraka, N.A., Rakotomanana, L.R., Iannotti, J.P., Farron, A.: A finite element model of the shoulder: application to the comparison of normal and osteoarthritic joints. Clin. Biomech. **17**(9–10), 630–639 (2002)
18. Bendsøe, M.P., Kikuchi, N.: Generating optimal topologies in structural design using a homogenization method. Comput. Methods Appl. Mech. Eng. **71**, 197–224 (1988)
19. Rozvany, G.I.N., Zhou, M., Birker, T.: Generalized shape optimization without homogenization. Struct. Optim. **4**, 250–252 (1992)
20. Xie, Y.M., Steven, G.P.: A simple evolutionary procedure for structural optimization. Comput. Struct. **49**, 885–896 (1993)
21. Wang, M.Y., Wang, X., Guo, D.: A level set method for structural topology optimization. Comput. Methods Appl. Mech. Eng. **192**, 227–46 (2003)
22. Ansys, Inc.: Ansys, 2013. http://www.ansys.com (2013)
23. Gupta, S., vander Helm, F.C., van Keulen, F.: The possibilities of uncemented glenoid component–a finite element study. Clin. Biomech. **19**(3), 292–302 (2004)

Topology Optimization of Microstructures for Multi-Functional Graded Composites

A. Radman, X. Huang, and Y.M. Xie

Abstract Functionally graded materials (FGMs) are inhomogeneous composites which are characterized by gradual variation in their physical properties. This study proposes a computational approach based on the bi-directional evolutionary structural optimization (BESO) for topologically designing microstructures of such materials with multi-functional properties, e.g. bulk modulus and thermal conductivity. It is assumed that the base cells are composed of two constituents. The smooth transition between adjacent base cells is realized by considering three base cells at each stage of the optimization. Effectiveness and efficiency of the proposed approach has been demonstrated by several numerical examples.

Keywords Topology optimization • Bi-directional evolutionary structural optimization (BESO) • Functionally graded materials (FGMs) • Base cell

1 Introduction

Functionally graded materials (FGMs) are inhomogeneous composites which are characterized by gradual changes in physical properties. According to this definition many biological materials can be categorized as FGMs. The idea for application of FGMs for industrial purposes has been first discussed in the 1970s [1, 2]. However, its primary application dates back to the mid-1980s when the concept was used for controlling the thermal response of materials in aerospace industry [3, 4]. In recent years, FGMs have been extensively studied in terms of developing new manufacturing methodologies and modelling techniques.

Figure 1 schematically illustrates the idea behind the development of primary FGM as a heat-shielding structural material, made up of ceramic and metallic phases [3, 5]. In Fig. 1, the ceramic phase acts as a thermal barrier protecting metallic phase from corrosion and oxidation. The metallic phase with high thermal

A. Radman • X. Huang (✉) • Y.M. Xie
Centre for Innovative Structures and Materials, School of Civil, Environmental and Chemical Engineering, RMIT University, GPO Box 2476, Melbourne 3001, VIC, Australia
e-mail: arash.radman@rmit.edu.au; huang.xiaodong@rmit.edu.au; xiaodong.huang@rmit.edu.au; mike.xie@rmit.edu.au

D. Gao et al. (eds.), *Advances in Global Optimization*, Springer Proceedings in Mathematics & Statistics 95, DOI 10.1007/978-3-319-08377-3_27

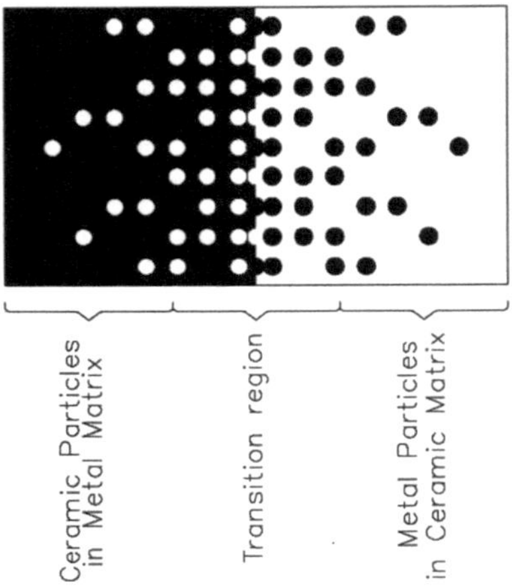

Fig. 1 Example of FGM composition and functional properties

conductivity, and high strength in low temperatures, but low corrosion resistance in high temperatures, strengthens the composite. Spherical or near spherical particles of one material are randomly dispersed within the matrix of the other material with varying proportion. As a result, inhomogeneous thermo-physical characteristics are developed into the material solely by varying the volume fractions of constituent phases [6, 7].

The inverse homogenization proposed by Sigmund [8] provided a systematic approach for topology optimization of the microstructures of composites with the goal of materials properties at macro-scale to be improved or tailored for desired functional properties. Since its primary introduction, some studies have been conducted to extend the methodology for topology optimization of microstructures with multi-functional characteristics such as materials with prescribed combinations of stiffness and thermal conductivity [9], heat and electricity transport [10], stiffness and permeability [11], and so forth [12].

The common approach for the design of multi-functional materials in these studies is to extremize a linear combination of materials functional properties [12]. The optimization problem is formulated with a single objective function formed by the weighted sum of functional properties. Different properties of FGMs could be attained by varying the weighted factors in the optimization [10–13]. However as indicated in [13] by applying different weighting factors, a proportional variation in the objective functions could not be anticipated. One reason for such a phenomenon is the possible non-linear cross-properties of the objective functions especially when these functional properties are selected from different physical natures. Most importantly the existence of the local optima may cause the topology optimization algorithms to get stuck in a nearby solution when the fixed weighting factors are used. Therefore, the optimization objective defined by the given weighting factors is inappropriate where the accurate control over the performance of the material is needed.

In this paper, a computational algorithm for the topological design of microstructures for FGMs with gradation in two functional properties is proposed. It is assumed that an FGM consists of two constituents and the design objective is to attain the FGM with variations in bulk modulus and the thermal conductivity. The microstructure of FGMs is modelled as a series of the connected base cells in which the optimal distribution of constituent phases is achieved by utilizing the BESO technique [14, 15]. Instead of using a weighting factor to individual objective function, an optimization problem statement is defined by maximizing the bulk modulus while satisfying the given variation of the thermal conductivity. To improve the connectivity of adjacent base cells, FGMs are optimized progressively by considering three base cells at each stage. Numerical examples indicate that the approach provides with an accurate control over the thermal conductivity with the smooth connectivity of neighbouring cells and low computational cost.

2 Problem Statement and Sensitivity Analysis

By assuming that the FGM totally consists of N base cells (as shown in Fig. 2) along the gradation direction, the mathematical statement of topology optimization of the jth base cell for maximum bulk modulus and prescribed thermal conductivity and volume fraction can be expressed as:

$$\begin{aligned}
&\text{Maximize} \quad K^j \\
&\text{Subject to:} \quad {k_c}^j = {k_c}^{j*} \\
&\qquad V^j = \sum_{i=1}^{M} x_i^j V_i^j \quad x_i^j = x_{min} \quad \text{or} \quad 1 \\
&\qquad (i = 1,\ 2, \cdots, M) \text{ and } (j = 1, 2, \cdots, N)
\end{aligned} \tag{1}$$

where M is the total number of elements within each base cell; V^j denotes the volume (or weight) of material 1 in the jth base cell; K^j is the bulk modulus of the jth base cell; k_c^j and k_c^{j*} are the effective thermal conductivity and its prescribed value of the jth base cell, respectively. The effective bulk modulus and thermal conductivity of material can be correlated with the elements of stiffness and conductivity matrix as:

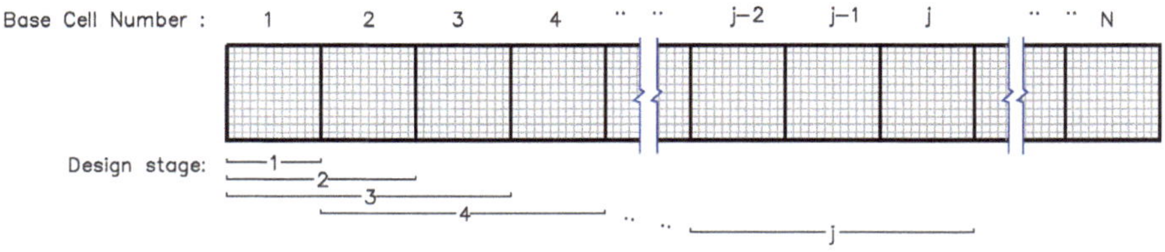

Fig. 2 FGM base cells numbering and design stages

$$K = \frac{1}{\zeta^2} \sum_{m,n=1}^{\varsigma} D_{mn}^H \tag{2}$$

$$k_c = \frac{1}{\zeta} \sum_{m}^{\varsigma} k_{mm}^H \tag{3}$$

where ζ is the material model dimension which is either 2 or 3 for 2D and 3D problems, respectively.

The design variable x_i^j of the ith element within the jth base cell can take a binary values of either 1 or a small value (i.e. $x_i = 0.001$) for the elements with constituent phases 1 or 2, respectively. In the following section, all parameters are operated within the jth unit cell by omitting the superscript j. The local material of an element within the PBC is assumed to be isotropic, with the physical property that varies between the properties of the two constituent phases. For the challenging cases where the two constituent phases have ill-ordered properties (i.e. $E^1 > E^2$ and $k^1 < k^2$), the following [16–18] interpolation scheme is applied:

$$\mathbf{D} = x_i^p \mathbf{D}^1 + \left(1 - x_i^p\right) \mathbf{D}^2 \tag{4}$$

$$\frac{1}{\mathbf{k}} = \frac{x_i}{\mathbf{k^1}} + \frac{(1 - x_i)}{\mathbf{k^2}} \tag{5}$$

in which **D** and **k** are the stiffness and thermal conductivity matrices, respectively, and the superscripts indicate the material numbers; p is the penalty exponent ($p = 3$ is used in this paper).

For a material with periodic microstructures, the effective properties such as bulk modulus and thermal conductivity could be calculated by using the homogenization theory [19–22]. With the help of the abovementioned interpolation schemes, the sensitivity of the effective elasticity matrix property $\mathbf{D}^H$ which estimates its variation due to the change of the design variable x_i can be found by

$$\frac{\partial \mathbf{D}^H}{\partial x_i} = \frac{1}{|Y|} \int_Y (\mathbf{I} - \mathbf{Bu})^T \frac{\partial \mathbf{D}}{\partial x_i} (\mathbf{I} - \mathbf{Bu})\, dY \tag{6}$$

in which **u** denotes the resulted displacement field when the PBC analysed under periodic boundary conditions and unit strains fields (e.g. $\{1,\ 0,\ 0\}^T$, $\{0,\ 1,\ 0\}^T$ and $\{0,\ 0,\ 1\}^T$ in 2D cases); $|Y|$ is the total area or volume of the base cell; **I** is the unit matrix and **B** is the strain–displacement matrix. Similarly, the sensitivity of elements with respect to the thermal conductivity can be expressed as [19, 20, 23]:

$$\frac{\partial \mathbf{k}^H}{\partial x_i} = \frac{1}{|Y|} \int_Y (\mathbf{I} - \boldsymbol{\mu})^T \frac{\partial \mathbf{k}}{\partial x_i} (\mathbf{I} - \boldsymbol{\mu})\, dY \tag{7}$$

in which $\boldsymbol{\mu}$ is the induced temperature field resulting from thermal analysis of the base cell under the periodical boundary conditions and unit heat fluxes (e.g. $\{1,\ 0, 0\}^T$, $\{0,\ 1, 0\}^T$ and $\{0,\ 0, 1\}^T$ in 3D problems).

3 Numerical Implementation

For solving the optimization problem (1) the original objective function is modified by adding the constraint function through introducing a Lagrangian multiplier. The modified objective function is expressed as [14, 24]:

$$f(\mathbf{x}) = (1 - \lambda)\, K^j + \lambda \left(k_c^j - k_c^{j*}\right) \tag{8}$$

where the Lagrangian multiplier λ $(0 \le \lambda \le 1)$ can take 0, when the prescribed thermal conductivity is attained. Otherwise, it is determined in such a way that the thermal conductivity (k_c^j) tends to its prescribed value k_c^{j*} in the subsequent iterations. To determine the Lagrangian multiplier λ, the variation of the thermal conductivity due to the changes of the design variables is estimated by the following equation:

$$\left(k_c{}^j\right)_{\eta+1} = \left(k_c{}^j\right)_\eta + \sum_i \frac{\partial k_c}{\partial x_i} \Delta x_i \tag{9}$$

where the subscript η denotes the current iteration number. Thus, the Lagrangian multiplier can be determined by using a bi-section method, as exemplified in [14, 24–26].

The elemental sensitivity numbers of the modified objective function [Eq. (8)] is expressed by

$$\alpha_i = \frac{\partial f(\mathbf{x})}{\partial x_i} = (1 - \lambda)\, \alpha_{1i} + \lambda \alpha_{2i} \tag{10}$$

where α_{1i} and α_{2i} are

$$\alpha_{1i} = \frac{1}{\zeta^2} \sum_{i,j=1}^{\eta} \frac{\partial D_{ij}^H}{\partial x_i} \tag{11}$$

$$\alpha_{2i} = \frac{1}{\zeta} \sum_i^{\eta} \frac{\partial k_{ii}^H}{\partial x_i} \tag{12}$$

Another important issue that should be accounted for is taking measures for smooth transition between topologies of adjacent cells. Based on simultaneous design of cells, Zhou and Li proposed different approaches for preserving the connectivity of PBCs [23, 27]. In this study, the procedure proposed in [26] is utilized where PBCs are progressively optimized by filtering the elemental sensitivities of three adjacent base cells at each stage of design [14, 28]. While designing the jth base cell, the connection between cells j and $j-1$ and between $j-1$ and $j-2$ is maintained by filtering their sensitivities together (see Fig. 1). The cells j, $j-1$ and $j-2$ are treated differently at each stage of design; while the cells j and $j-1$ are optimized based on the optimization statement (1), the topology of the cell $j-2$ is kept unchanged. As the result of filtering, the topology of the middle cell ($j-1$) gradually evolves and provides a smooth transition between the other two adjacent cells. It has been shown that this method is capable of providing a computationally more efficient and low cost solution in comparison with simultaneous design of base cells, particularly in 3D cases [25].

The filtering scheme is used to circumvent the numerical instabilities such as checkerboard pattern and mesh-dependency and is formulated as [28]:

$$\overline{\alpha}_i = \frac{\sum_{j=1}^{N} w_{ij}\alpha_i}{\sum_{j=1}^{N} w_{ij}} \tag{13}$$

in which $\overline{\alpha}_i$ is the filtered sensitivity number of the ith element and N represents the total number of elements within the cells j, $j-1$ and $j-2$. The weight factor w_{ij} is defined by

$$w_{ij} = \begin{cases} r_{min} - r_{ij} & \text{if } r_{ij} < r_{min} \\ 0 & \text{otherwise} \end{cases} \tag{14}$$

where r_{ij} is the distance between the centres of elements i and j; the filter radius r_{min} identifies the neighbouring elements that affect the sensitivity of element i.

4 Results and Discussion

Effectiveness of the approach is demonstrated by two examples. In both examples the FGM is divided into ten base cells which are composed of two ill-ordered constituent phases. Non-dimensional Young's modulus and thermal conductivity of constituents are defined with $E^1 = 3$ and $k_c^1 = 1$ for material 1, and $E^2 = 1$ and $k_c^2 = 3$ for materials 2.

The objective of the first example is to obtain a 2D FGM with maximum stiffness (bulk modulus) while the materials thermal conductivity linearly varies from Hashin–Shtrikman (HS) lower bound corresponding to 60 % volume fraction

Fig. 3 Three rows of 2D base cells, designed for the FGM with maximum bulk modulus and variation in thermal conductivity and volume fraction (material 1 is shown in *dark blue*)

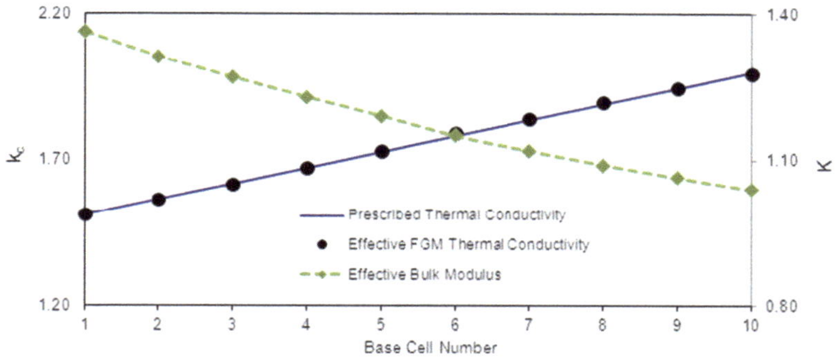

Fig. 4 Variation of thermal conductivity and bulk modulus along the gradation direction (k_c thermal conductivity, K bulk modulus)

of material 1, to the upper bound corresponding to 40 % of material 1 [29]. Each base cell is discretized into 80×80, 4-node quadrilateral elements. To initialize the optimization procedure, four elements at the centre of first base cell are assigned as material 2 while other elements are material 1.

The resulting topology is given in Fig. 3, which shows the gradual transition of cells topologies from the left to the right. Figure 4 shows the variations of the bulk modulus and thermal conductivity. It can be seen that the bulk modulus gradually decreases from the left to the right while the thermal conductivity gradually increases. These property variations are partly because the volume fraction of stiffer phase becomes less and less (from 60 to 40 %). Another important reason comes from the changes of topological patterns. It is noted that the resulting thermal conductivity agrees well with the prescribed values, with a deviation of less than 0.9 %.

To exemplify the algorithm for 3D cases an FGM is modelled with the gradation in thermal conductivity from HS lower bound to the upper bound while the corresponding prescribed volume fraction of material 1 varies from 80 to 50 %. Each base cell is discretized into $29 \times 29 \times 29$, 8-node cubic elements.

The final topology is shown in Fig. 5, which demonstrates a smooth transition between cells. Figure 6 shows the variations of the bulk modulus and thermal conductivity. The FGM thermal conductivity agrees to the prescribed values with

Fig. 5 (**a**) 3D cells for the FGM with variation in thermal conductivity and volume fraction of materials; (**b**) longitudinal sections of (**a**) (material 1 is shown in *dark blue*)

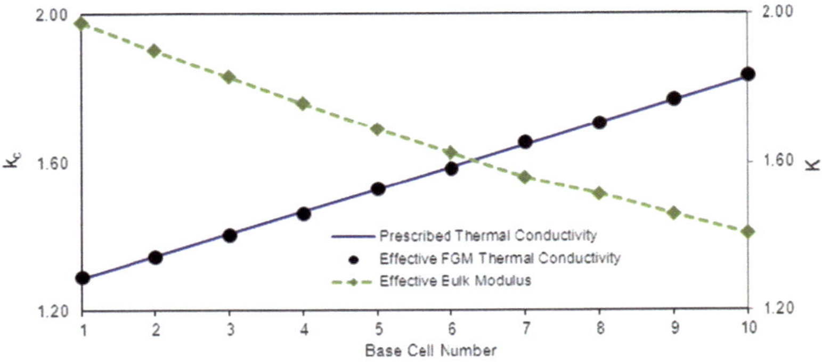

Fig. 6 Variation of thermal conductivity and bulk modulus along the gradation direction of 3D FGM (k_c thermal conductivity, K bulk modulus)

less than 0.5 % deviation. It should be mentioned that simultaneous design of such 3D base cells is almost impossible with ordinary computers.

5 Conclusions

Based on the BESO, this paper proposed an optimization technique for the topological design of a series of base cells for functionally graded composites with multi-functional properties. The variations of functional properties of the composite along a certain direction are realized by gradual changes of the volume fraction and spatial distribution of constituent phases within the base cells. To save the computational cost and obtain the smooth transition between adjacent base cells, a progressive design approach is used by considering topology optimization of three adjacent base cells at each stage. The numerical examples demonstrate the effectiveness of the approach which yields precise control over the variations of multi-functional properties. The proposed procedure can also be extended for topology optimization of other multi-functional properties of FGMs.

References

1. Bever, M.B., Duwez, P.E.: Gradients in composite materials. Mater. Sci. Eng. **10**, 1–8 (1972)
2. Shen, M., Bever, M.B.: Gradients in polymeric materials. J. Mater. Sci. **7**, 741–746 (1972)
3. Koizumi, M.: FGM activities in Japan. Compos. Part B **28**(1–2), 1–4 (1997). doi:10.1016/s1359-8368(96)00016-9
4. Hirai, T., Chen, L.: Recent and prospective development of functionally graded materials in Japan. Mater. Sci. Forum **308–311**, 509–514 (1999). doi:10.4028/www.scientific.net/MSF.308-311.509
5. Birman, V., Byrd, L.W.: Modeling and analysis of functionally graded materials and structures. Appl. Mech. Rev. **60**(5), 195–216 (2007). doi:10.1115/1.2777164
6. Lin, C.-Y., Hsiao, C.-C., Chen, P.-Q., Hollister, S.J.: Interbody fusion cage design using integrated global layout and local microstructure topology optimization. Spine **29**(16), 1747–1754 (2004)
7. Chen, K.-Z., Feng, X.-A.: CAD modeling for the components made of multi heterogeneous materials and smart materials. Comput. Aided Des. **36**(1), 51–63 (2004). doi:10.1016/s0010-4485(03)00077-0
8. Sigmund, O.: Materials with prescribed constitutive parameters: an inverse homogenization problem. Int. J. Solids Struct. **31**(17), 2313–2329 (1994). doi:10.1016/0020-7683(94)90154-6
9. Challis, V.J., Roberts, A.P., Wilkins, A.H.: Design of three dimensional isotropic microstructures for maximized stiffness and conductivity. Int. J. Solids Struct. **45**(14–15), 4130–4146 (2008). http://dx.doi.org/10.1016/j.ijsolstr.2008.02.025
10. Torquato, S., Hyun, S., Donev, A.: Optimal design of manufacturable three-dimensional composites with multifunctional characteristics. J. Appl. Phys. **94**(9), 5748–5755 (2003). doi:10.1063/1.1611631
11. Guest, J.K., Prévost, J.H.: Optimizing multifunctional materials: design of microstructures for maximized stiffness and fluid permeability. Int. J. Solids Struct. **43**(22–23), 7028–7047 (2006). http://dx.doi.org/10.1016/j.ijsolstr.2006.03.001
12. Cadman, J., Zhou, S., Chen, Y., Li, Q.: On design of multi-functional microstructural materials. J. Mater. Sci. **48**(1), 51–66 (2012). doi:10.1007/s10853-012-6643-4
13. de Kruijf, N., Zhou, S., Li, Q., Mai, Y.-W.: Topological design of structures and composite materials with multiobjectives. Int. J. Solids Struct. **44**(22–23), 7092–7109 (2007). http://dx.doi.org/10.1016/j.ijsolstr.2007.03.028
14. Huang, X., Xie, Y.M.: Evolutionary Topology Optimization of Continuum Structures: Methods and Applications. John Wiley & Sons, Chichester (2010)
15. Huang, X., Radman, A., Xie, Y.M.: Topological design of microstructures of cellular materials for maximum bulk or shear modulus. Comput. Mater. Sci. **50**(6), 1861–1870 (2011). doi:10.1016/j.commatsci.2011.01.030
16. Rozvany, G.I.N., Zhou, M., Birker, T.: Generalized shape optimization without homogenization. Struct. Multidiscip. Optim. **4**(3), 250–252 (1992). doi:10.1007/bf01742754
17. Bendsøe, M.P., Sigmund, O.: Topology Optimization: Theory, Methods and Application. Springer, Berlin (2003)
18. Huang, X., Xie, Y., Jia, B., Li, Q., Zhou, S.: Evolutionary topology optimization of periodic composites for extremal magnetic permeability and electrical permittivity. Struct. Multidiscip. Optim. **46**(3), 385–398 (2012). doi:10.1007/s00158-012-0766-8
19. Hassani, B., Hinton, E.: A review of homogenization and topology optimization \hbox{I—homogenization} theory for media with periodic structure. Comput. Struct. **69**(6), 707–717 (1998). doi:10.1016/s0045-7949(98)00131-x
20. Hassani, B., Hinton, E.: A review of homogenization and topology opimization II—analytical and numerical solution of homogenization equations. Comput. Struct. **69**(6), 719–738 (1998). doi:10.1016/s0045-7949(98)00132-1

21. Bendsøe, M.P., Kikuchi, N.: Generating optimal topologies in structural design using a homogenization method. Comput. Methods Appl. Mech. Eng. **71**(2), 197–224 (1988). http://dx.doi.org/10.1016/0045-7825(88)90086-2
22. Haug, E.J., Choi, K.K., Komkov, V.: Design Sensitivity Analysis of Structural Systems. Academic, Orlando (1986)
23. Zhou, S., Li, Q.: Computational design of multi-phase microstructural materials for extremal conductivity. Comput. Mater. Sci. **43**(3), 549–564 (2008). http://dx.doi.org/10.1016/j.commatsci.2007.12.021
24. Huang, X., Xie, Y.M.: Evolutionary topology optimization of continuum structures with an additional displacement constraint. Struct. Multidiscip. Optim. **40**(1–6), 409–416 (2010). doi:10.1007/s00158-009-0382-4
25. Radman, A., Huang, X., Xie, Y.M.: Topology optimization of functionally graded cellular materials. J. Mater. Sci. **48**(4), 1503–1510 (2013). doi:10.1007/s10853-012-6905-1
26. Radman, A., Huang, X., Xie, Y.M.: Topological optimization for the design of microstructures of isotropic cellular materials. Eng. Optim. **45**(11), 1331–1348 (2013). doi:10.1080/0305215x.2012.737781
27. Zhou, S., Li, Q.: Microstructural design of connective base cells for functionally graded materials. Mater. Lett. **62**, 4022–4024 (2008). doi:10.1016/j.matlet.2008.05.058
28. Huang, X., Xie, Y.M.: Convergent and mesh-independent solutions for the bi-directional evolutionary structural optimization method. Finite Elem. Anal. Des. **43**(14), 1039–1049 (2007). doi:10.1016/j.finel.2007.06.006
29. Hashin, Z., Shtrikman, S.: A variational approach to the theory of the elastic behaviour of multiphase materials. J. Mech. Phys. Solids **11**(2), 127–140 (1963). doi:10.1016/0022-5096(63)90060-7

Part V
Variational Inequalities and Complementarity Problems

Evolution Inclusions in Nonsmooth Systems with Applications for Earth Data Processing

Uniform Trajectory Attractors for Nonautonomous Evolution Inclusions Solutions with Pointwise Pseudomonotone Mappings

Michael Z. Zgurovsky and Pavlo O. Kasyanov

Abstract For the class of nonautonomous evolution inclusions with pointwise pseudomonotone multi-valued mappings the dynamics as $t \to +\infty$ of all global weak solutions defined on $[0, +\infty)$ is studied. The existence of a compact uniform trajectory attractor is proved. The results obtained allow one to study the dynamics of solutions for new classes of evolution inclusions related to nonlinear mathematical models of controlled geophysical and socioeconomic processes and for fields with interaction functions of pseudomonotone type satisfying the condition of polynomial growth and the standard sign condition.

Keywords Evolution inclusion • Pseudomonotone map • Uniform trajectory attractor • Feedback control

1 Introduction and Setting of the Problem

For evolution triple $(V; H; V^*)$[1] and multi-valued map $A : \mathbb{R}_+ \times V \rightrightarrows V^*$ we consider a problem of long-time behavior of all globally defined weak solutions for nonautonomous evolution inclusion

$$y'(t) + A(t, y(t)) \ni \bar{0}, \tag{1}$$

[1]That is V is a real reflexive separable Banach space continuously and densely embedded into a real Hilbert space H, H is identified with its topologically conjugated space H^*, V^* is a dual space to V. So, there is a chain of continuous and dense embeddings: $V \subset H \equiv H^* \subset V^*$ (see, for example, Gajewski et al. [1, Chap. I]).

M.Z. Zgurovsky (✉) • P.O. Kasyanov
Institute for Applied System Analysis, National Technical University of Ukraine "Kyiv Politechnic Institute", Peremogy Ave. 37, Build. 35, 03056 Kyiv, Ukraine
e-mail: zgurovsm@hotmail.com; kasyanov@i.ua

D. Gao et al. (eds.), *Advances in Global Optimization*, Springer Proceedings in Mathematics & Statistics 95, DOI 10.1007/978-3-319-08377-3_28

as $t \to +\infty$. Let $\langle \cdot, \cdot \rangle_V : V^* \times V \to \mathbb{R}$ be the pairing in $V^* \times V$ that coincides on $H \times V$ with the inner product $(\cdot, \cdot)$ in the Hilbert space H.

Note that Problem (1) arises in many important models for distributed parameter control problems and that large class of identification problems enter this formulation. Let us indicate a problem which is one of motivations for the study of the nonautonomous evolution inclusion (1) (see, for example, Migórski and Ochal [2]; Zgurovsky et al. [3] and references therein). In a subset Ω of $\mathbb{R}^3$, we consider the nonstationary heat conduction equation

$$\frac{\partial y}{\partial t} - \Delta y = f \text{ in } \Omega \times (0, +\infty)$$

with initial conditions and suitable boundary ones. Here $y = y(x,t)$ represents the temperature at the point $x \in \Omega$ and time $t > 0$. It is supposed that $f = f_1 + f_2$, where f_2 is given and f_1 is a known function of the temperature of the form

$$-f_1(x,t) \in \partial j(x,t,y(x,t)) \text{ a.e. } (x,t) \in \Omega \times (0, +\infty).$$

Here $\partial j(x,t,\xi)$ denotes generalized gradient of Clarke (see Clarke [4]) with respect to the last variable of a function $j : \Omega \times \mathbb{R} \to \mathbb{R}$ which is assumed to be locally Lipschitz in ξ (cf. Migórski and Ochal [2] and references therein). The multivalued function $\partial j(x,t,\cdot) : \mathbb{R} \to 2^{\mathbb{R}}$ is generally nonmonotone and it includes the vertical jumps. In a physicist's language it means that the law is characterized by the generalized gradient of a nonsmooth potential j (cf. Panagiotopoulos [5]). Models of physical interest include also the next (see, for example, Balibrea et al. [6] and references therein): a model of combustion in porous media; a model of conduction of electrical impulses in nerve axons; a climate energy balance model; etc.

To introduce the assumptions on parameters of Problem (1) we need to present some additional constructions. A function $\varphi \in L_1^{\mathrm{loc}}(\mathbb{R}_+)$ is called *translation-compact (tr.-c.)* in $L_1^{\mathrm{loc}}(\mathbb{R}_+)$, if the set $\{\varphi(\cdot + h) : h \geq 0\}$ is precompact in $L_1^{\mathrm{loc}}(\mathbb{R}_+)$; cf. Chepyzhov and Vishik [7, p. 917]. Here $L_1^{\mathrm{loc}}(\mathbb{R}_+)$ is the space of locally integrable on $\mathbb{R}_+$ functions; see, for example, Chepyzhov and Vishik [7, p. 919] and references therein. Note that a function $\varphi \in L_1^{\mathrm{loc}}(\mathbb{R}_+)$ is tr.-c. in $L_1^{\mathrm{loc}}(\mathbb{R}_+)$ iff two conditions hold: (a) $\sup\limits_{t\geq 0} \int_t^{t+h} |\varphi(s)|ds < +\infty$ for any $h > 0$; (b) there exists a function $\psi(s)$, $\psi(s) \to 0+$ as $s \to 0+$ such that

$$\int_t^{t+1} |\varphi(s) - \varphi(s+h)|ds \leq \psi(|h|) \text{ for any } t, h \geq 0;$$

Chepyzhov and Vishik [7, Proposition 6.5].

Throughout this paper we suppose that the listed below assumptions hold.

Assumption I. Let $p \geq 2$, $q > 1$ are such that $\frac{1}{p} + \frac{1}{q} = 1$, and the embedding $V \subset H$ is compact one.

Assumption II (Grows Condition). There exist a tr.-comp. in $L_1^{\text{loc}}(\mathbb{R}_+)$ function $c_1 : \mathbb{R}_+ \to \mathbb{R}_+$ and a constant $c_2 > 0$ such that $\|d\|_{V^*}^q \le c_1(t) + c_2\|u\|_V^p$ for any $u \in V$, $d \in A(t,u)$, and a.e. $t > 0$.

Assumption III (Signed Assumption). There exist a constant $\alpha > 0$ and a tr.-comp. in $L_1^{\text{loc}}(\mathbb{R}_+)$ function $\beta : \mathbb{R}_+ \to \mathbb{R}_+$ such that $\langle d, u\rangle_V \ge \alpha\|u\|_V^p - \beta(t)$ for any $u \in V$, $d \in A(t,u)$, and a.e. $t > 0$.

Assumption IV (Strong Measurability). If $C \subseteq V^*$ is a closed set, then the set $\{(t,u) \in (0,+\infty) \times V \,:\, A(t,u) \cap C \ne \emptyset\}$ is a Borel subset in $(0,+\infty) \times V$.

Assumption V (Pointwise Pseudomonotonicity). Let for a.e. $t > 0$ two assumptions hold:

(a) for every $u \in V$ the set $A(t,u)$ is nonempty, convex, and weakly compact one in V^*;
(b) if a sequence $\{u_n\}_{n\ge1}$ converges weakly in V towards $u \in V$ as $n \to +\infty$, $d_n \in A(t,u_n)$ for any $n \ge 1$, and $\limsup_{n\to+\infty}\langle d_n, u_n - u\rangle_V \le 0$, then for any $\omega \in V$ there exists $d(\omega) \in A(t,u)$ such that

$$\liminf_{n\to+\infty}\langle d_n, u_n - \omega\rangle_V \ge \langle d(\omega), u - \omega\rangle_V.$$

Let $0 \le \tau < T < +\infty$. As a *weak solution* of evolution inclusion (1) on the interval $[\tau, T]$ we consider an element $u(\cdot)$ of the space $L_p(\tau,T;V)$ such that for some $d(\cdot) \in L_q(\tau,T;V^*)$ it is fulfilled:

$$-\int_\tau^T (\xi'(t), y(t))dt + \int_\tau^T \langle d(t), \xi(t)\rangle_V dt = 0 \quad \forall \xi \in C_0^\infty([\tau,T];V), \qquad (2)$$

and $d(t) \in A(t, y(t))$ for a.e. $t \in (\tau, T)$.

2 Preliminary Properties of Weak Solutions

For fixed nonnegative τ and T, $\tau < T$, let us consider

$$X_{\tau,T}=L_p(\tau,T;V), \quad X_{\tau,T}^* = L_q(\tau,T;V^*), \quad W_{\tau,T} = \{y \in X_{\tau,T} \mid y' \in X_{\tau,T}^*\},$$

$$\mathcal{A}_{\tau,T} : X_{\tau,T} \rightrightarrows X_{\tau,T}^*, \quad \mathcal{A}_{\tau,T}(y)=\{d \in X_{\tau,T}^* \mid d(t) \in A(t,y(t)) \text{ for a.e. } t \in (\tau,T)\},$$

where y' is a derivative of an element $u \in X_{\tau,T}$ in the sense of $\mathcal{D}([\tau,T];V^*)$ (see, for example, Gajewski et al. [1, Definition IV.1.10]). Note that the space $W_{\tau,T}$ is a reflexive Banach space with the graph norm of a derivative $\|u\|_{W_{\tau,T}} = \|u\|_{X_{\tau,T}} + \|u'\|_{X_{\tau,T}^*}$, $u \in W_{\tau,T}$. Let $\langle\cdot,\cdot\rangle_{X_{\tau,T}} : X_{\tau,T}^* \times X_{\tau,T} \to \mathbb{R}$ be the pairing in $X_{\tau,T}^* \times X_{\tau,T}$

that coincides on $L_2(\tau, T; H) \times X_{\tau,T}$ with the inner product in $L_2(\tau, T; H)$, i.e. $\langle u, v\rangle_{X_{\tau,T}} = \int_\tau^T (u(t), v(t))dt$ for any $u \in L_2(\tau, T; H)$ and $v \in X_{\tau,T}$. Gajewski et al. [1, Theorem IV.1.17] provide that the embedding $W_{\tau,T} \subset C([\tau, T]; H)$ is continuous and dense one. Moreover,

$$(u(T), v(T)) - (u(\tau), v(\tau)) = \int_\tau^T \Big[\langle u'(t), v(t)\rangle_V + \langle v'(t), u(t)\rangle_V\Big]dt, \qquad (3)$$

for any $u, v \in W_{\tau,T}$.

Migórski [8, Lemma 7, p. 516] (see paper and references therein) and Assumptions I–V provide that multi-valued mapping $\mathcal{A}_{\tau,T} : X_{\tau,T} \rightrightarrows X^*_{\tau,T}$ satisfies the listed below properties:

Property I. There exists a positive constant $C_1 = C_1(\tau, T)$ such that $\|d\|_{X^*_{\tau,T}} \leq C_1(1 + \|y\|^{p-1}_{X_{\tau,T}})$ for any $y \in X_{\tau,T}$ and $d \in \mathcal{A}_{\tau,T}(y)$.

Property II. There exist positive constants $C_2 = C_2(\tau, T)$ and $C_3 = C_3(\tau, T)$ such that $\langle d, y\rangle_{X_{\tau,T}} \geq C_2\|y\|^p_{X_{\tau,T}} - C_3$ for any $y \in X_{\tau,T}$ and $d \in \mathcal{A}_{\tau,T}(y)$.

Property III. The multi-valued mapping $\mathcal{A}_{\tau,T} : X_{\tau,T} \rightrightarrows X^*_{\tau,T}$ is (generalized) pseudomonotone on $W_{\tau,T}$, i.e. (a) for every $y \in X_{\tau,T}$ the set $\mathcal{A}_{\tau,T}(y)$ is a nonempty, convex, and weakly compact one in $X^*_{\tau,T}$; (b) $\mathcal{A}_{\tau,T}$ is upper semi-continuous from every finite dimensional subspace $X_{\tau,T}$ into $X^*_{\tau,T}$ endowed with the weak topology; (c) if a sequence $\{y_n, d_n\}_{n\geq 1} \subset W_{\tau,T}\times X^*_{\tau,T}$ converges weakly in $W_{\tau,T}\times X^*_{\tau,T}$ towards $(y, d) \in W_{\tau,T}\times X^*_{\tau,T}$, $d_n \in \mathcal{A}_{\tau,T}(y_n)$ for any $n \geq 1$, and $\limsup_{n\to+\infty}\langle d_n, y_n - y\rangle_{X_{\tau,T}} \leq 0$, then $d \in \mathcal{A}_{\tau,T}(y)$ and $\lim_{n\to+\infty} \langle d_n, y_n\rangle_{X_{\tau,T}} = \langle d, y\rangle_{X_{\tau,T}}$.

Formula (2) and definition of the derivative for an element from $\mathcal{D}([\tau, T]; V^*)$ yield that each weak solution $y \in X_{\tau,T}$ of Problem (1) on $[\tau, T]$ belongs to the space $W_{\tau,T}$ and $y' + \mathcal{A}_{\tau,T}(y) \ni \bar{0}$. Vice versa, if $y \in W_{\tau,T}$ satisfies the last inclusion, then y is a weak solution of Problem (1) on $[\tau, T]$.

Assumption I, Properties I–III, and Denkowski et al. [9, Theorem 1.3.73] (see also Zgurovsky et al. [10, Chap. 2] and references therein) provide the existence of a weak solution of Cauchy problem (1) with initial data $y(\tau) = y^{(\tau)}$ on the interval $[\tau, T]$, for any $y^{(\tau)} \in H$.

For fixed τ and T, such that $0 \leq \tau < T < +\infty$, we denote

$$\mathcal{D}_{\tau,T}(y^{(\tau)}) = \{y(\cdot) \mid y \text{ is a weak solution of (1) on } [\tau, T],\ y(\tau) = y^{(\tau)}\}, \quad y^{(\tau)} \in H.$$

We remark that $\mathcal{D}_{\tau,T}(y^{(\tau)}) \neq \emptyset$ and $\mathcal{D}_{\tau,T}(y^{(\tau)}) \subset W_{\tau,T}$, if $0 \leq \tau < T < +\infty$ and $y^{(\tau)} \in H$. Moreover, the concatenation of Problem (1) weak solutions is a weak solutions too, i.e. if $0 \leq \tau < t < T$, $y^{(\tau)} \in H$, $y(\cdot) \in \mathcal{D}_{\tau,t}(y^{(\tau)})$, and $v(\cdot) \in \mathcal{D}_{t,T}(y(t))$, then

$$z(s) = \begin{cases} y(s), & s \in [\tau, t], \\ v(s), & s \in [t, T], \end{cases}$$

belongs to $\mathcal{D}_{\tau,T}(y^{(\tau)})$; cf. Zgurovsky et al. [3, pp. 55–56].

Gronwall lemma provides that for any finite time interval $[\tau, T] \subset \mathbb{R}_+$ each weak solution y of Problem (1) on $[\tau, T]$ satisfies estimates

$$\|y(t)\|_H^2 - 2\int_0^t \beta(\xi)d\xi + 2\alpha \int_s^t \|y(\xi)\|_V^p d\xi \le \|y(s)\|_H^2 - 2\int_0^s \beta(\xi)d\xi, \quad (4)$$

$$\|y(t)\|_H^2 \le \|y(s)\|_H^2 e^{-2\alpha\gamma(t-s)} + 2\int_s^t (\beta(\xi) + \alpha\gamma)e^{-2\alpha\gamma(t-\xi)}d\xi, \quad (5)$$

where $t, s \in [\tau, T]$, $t \ge s$; $\gamma > 0$ is a constant such that $\gamma\|u\|_H^p \le \|u\|_V^p$ for any $u \in V$; cf. Zgurovsky et al. [3, p. 56]. In the proof of (5) we used the inequality $\|u\|_H^2 - 1 \le \|u\|_H^p$ for any $u \in H$.

Therefore, any weak solution y of Problem (1) on a finite time interval $[\tau, T] \subset \mathbb{R}_+$ can be extended to a global one, defined on $[\tau, +\infty)$. For arbitrary $\tau \ge 0$ and $y^{(\tau)} \in H$ let $\mathcal{D}_\tau(y^{(\tau)})$ be the set of all weak solutions (defined on $[\tau, +\infty)$) of Problem (1) with initial data $y(\tau) = y^{(\tau)}$. Let us consider the family $\mathcal{K}_\tau^+ = \cup_{y^{(\tau)} \in H} \mathcal{D}_\tau(y^{(\tau)})$ of all weak solutions of Problem (1) defined on the semi-infinite time interval $[\tau, +\infty)$.

3 Uniform Trajectory Attractor and Main Result

Consider the Fréchet space $C^{\mathrm{loc}}(\mathbb{R}_+; H)$. We remark that the sequence $\{f_n\}_{n\ge1}$ converges in $C^{\mathrm{loc}}(\mathbb{R}_+; H)$ towards $f \in C^{\mathrm{loc}}(\mathbb{R}_+; H)$ as $n \to +\infty$ iff the sequence $\{\Pi_{t_1,t_2} f_n\}_{n\ge1}$ converges in $C([t_1, t_2]; H)$ towards $\Pi_{t_1,t_2} f$ as $n \to +\infty$ for any finite interval $[t_1, t_2] \subset \mathbb{R}_+$, where Π_{t_1,t_2} is the restriction operator to the interval $[t_1, t_2]$; Chepyzhov and Vishik [7, p. 918]. We denote $T(h)y(\cdot) = y_h(\cdot)$, where $y_h(t) = y(t + h)$ for any $y \in C^{\mathrm{loc}}(\mathbb{R}_+; H)$ and $t, h \ge 0$.

In the autonomous case, when $A(t, y)$ does not depend on t, the long-time behavior of all globally defined weak solutions for Problem (1) is described by using trajectory and global attractors theory; Kasyanov [11, 12], Zgurovsky et al. [3, Chap. 2] and references therein; see also Balibrea [6]. In this situation the set $\mathcal{K}^+ := \mathcal{K}_0^+$ is *translation invariant*, i.e. $T(h)\mathcal{K}^+ \subseteq \mathcal{K}^+$ for any $h \ge 0$. As trajectory attractor it is considered a classical global attractor for translation semigroup $\{T(h)\}_{h\ge0}$ that acts on $\mathcal{K}^+$.

In the nonautonomous case we notice that $T(h)\mathcal{K}_0^+ \not\subseteq \mathcal{K}_0^+$. Therefore, we need to consider *united trajectory space* that includes all globally defined on any $[\tau, +\infty) \subseteq \mathbb{R}_+$ weak solutions of Problem (1) shifted to $\tau = 0$:

$$\mathcal{K}^+ = \mathrm{cl}_{C^{\mathrm{loc}}(\mathbb{R}_+;H)} \left[\bigcup_{\tau\ge0} \{y(\cdot + \tau) : y \in \mathcal{K}_\tau^+\} \right],$$

where $\mathrm{cl}_{C^{\mathrm{loc}}(\mathbb{R}_+;H)}[\,\cdot\,]$ is the closure in $C^{\mathrm{loc}}(\mathbb{R}_+;H)$. Note that $T(h)\{y(\cdot+\tau)\ :\ y\in\mathcal{K}_\tau^+\}\subseteq\{y(\cdot+\tau+h)\ :\ y\in\mathcal{K}_{\tau+h}^+\}$ for any $\tau,h\geq 0$. Moreover,

$$T(h)\mathcal{K}^+\subseteq\mathcal{K}^+ \text{ for any } h\geq 0,$$

because

$$\rho_{C^{\mathrm{loc}}(\mathbb{R}_+;H)}(T(h)u,T(h)v)\leq\rho_{C^{\mathrm{loc}}(\mathbb{R}_+;H)}(u,v) \text{ for any } u,v\in C^{\mathrm{loc}}(\mathbb{R}_+;H),$$

where $\rho_{C^{\mathrm{loc}}(\mathbb{R}_+;H)}$ is a standard metric on Fréchet space $C^{\mathrm{loc}}(\mathbb{R}_+;H)$; Chepyzhov and Vishik [7].

A set $\mathcal{P}\subset C^{\mathrm{loc}}(\mathbb{R}_+;H)\cap L_\infty(\mathbb{R}_+;H)$ is said to be a *uniformly attracting set* (cf. Chepyzhov and Vishik [7, p. 921]) for the united trajectory space $\mathcal{K}^+$ of Problem (1) in the topology of $C^{\mathrm{loc}}(\mathbb{R}_+;H)$, if for any bounded in $L_\infty(\mathbb{R}_+;H)$ set $\mathcal{B}\subseteq\mathcal{K}^+$ and any segment $[t_1,t_2]\subset\mathbb{R}_+$ the following relation holds:

$$\mathrm{dist}_{C([t_1,t_2];H)}(\Pi_{t_1,t_2}T(t)\mathcal{B},\Pi_{t_1,t_2}\mathcal{P})\to 0,\quad t\to+\infty, \tag{6}$$

where $\mathrm{dist}_{C([t_1,t_2];H)}$ is the Hausdorff semi-metric.

A set $\mathcal{U}\subset\mathcal{K}^+$ is said to be a *uniform trajectory attractor* (cf. Chepyzhov and Vishik [7, p. 921]) of the translation semigroup $\{T(t)\}_{t\geq 0}$ on $\mathcal{K}^+$ in the induced topology from $C^{\mathrm{loc}}(\mathbb{R}_+;H)$, if

(1) $\mathcal{U}$ is a compact set in $C^{\mathrm{loc}}(\mathbb{R}_+;H)$ and bounded in $L_\infty(\mathbb{R}_+;H)$;
(2) $\mathcal{U}$ is strictly invariant with respect to $\{T(h)\}_{h\geq 0}$, i.e. $T(h)\mathcal{U}=\mathcal{U}\ \forall h\geq 0$;
(3) $\mathcal{U}$ is a minimal uniformly attracting set for $\mathcal{K}^+$ in the topology of $C^{\mathrm{loc}}(\mathbb{R}_+;H)$, i.e. $\mathcal{U}$ belongs to any compact uniformly attracting set $\mathcal{P}$ of $\mathcal{K}^+$: $\mathcal{U}\subseteq\mathcal{P}$.

Note that uniform trajectory attractor of the translation semigroup $\{T(t)\}_{t\geq 0}$ on $\mathcal{K}^+$ in the induced topology from $C^{\mathrm{loc}}(\mathbb{R}_+;H)$ coincides with the classical trajectory attractor for the continuous semi-group $\{T(t)\}_{t\geq 0}$ defined on $\mathcal{K}^+$ (see, for example, Chepyzhov and Vishik [13, Definition 1.1]).

Presented construction is coordinated with the theory of uniform trajectory attractors for nonautonomous problems of the form

$$\partial_t u(t)=A_{\sigma(t)}(u(t)), \tag{7}$$

where $\sigma(s)$, $s\geq 0$, is a functional parameter called the time symbol of Eq. (7) (t is replaced by s). In applications to mathematical physics equations, a function $\sigma(s)$ consists of all time-dependent terms of the equation under consideration: external forces, parameters of mediums, interaction functions, control functions, etc.; Chepyzhov and Vishik [7, 14, 15]; Sell [16]; Zgurovsky et al. [3] and references therein; see also Hale [17]; Ladyzhenskaya [18]; Mel'nik and Valero [19]; Kapustyan et al. [20]. In mentioned above papers and books it is assumed that the symbol σ of Eq. (7) belongs to a Hausdorff topological space Ξ_+ of functions defined on $\mathbb{R}_+$ with values in some complete metric space. Usually, in applications, the topology in the space Ξ_+ is a local convergence topology on any segment $[t_1, t_2] \subset \mathbb{R}_+$. Further, they consider the family of Eq. (7) with various symbols $\sigma(s)$ belonging to a set $\Sigma \subseteq \Xi_+$. The set Σ is called the symbol space of the family of Eq. (7). It is assumed that the set Σ, together with any symbol $\sigma(s) \in \Sigma$, contains all positive translations of $\sigma(s)$: $\sigma(t+s) = T(t)\sigma(s) \in \Sigma$ for any $t, s \geq 0$. The symbol space Σ is invariant with respect to the translation semigroup $\{T(t)\}_{t\geq 0}$: $T(t)\Sigma \subseteq \Sigma$ for any $t \geq 0$. To prove the existence of uniform trajectory attractor they suppose that the symbol space Σ with the topology induced from Ξ_+ is a compact metric space. Mostly in applications, as a symbol space Σ it is naturally to consider the hull of translation-compact function $\sigma_0(s)$ in an appropriate Hausdorff topological space Ξ_+. The direct realization of this approach for Problem (1) is problematic without any additional assumptions for parameters of Problem (1) and requires the translation-compactness of the symbol $\sigma(s) = A(s, \cdot)$ in some compact Hausdorff topological space of measurable multi-valued mappings acts from $\mathbb{R}_+$ to some metric space of pseudomonotone operators from $(V \to 2^{V^*})$ satisfying grows and signed assumptions. To avoid this technical difficulties we present the alternative approach for the existence and construction of the uniform trajectory attractor for all weak solutions for Problem (1). Note that Assumptions (I)–(V) are natural and guaranty, in the general case, only existence of weak solution for Cauchy problem on any finite time interval $[\tau, T] \subset \mathbb{R}_+$ and for any initial data form H; see, for example, Denkowski et al. [9]; Gasinski and Papageorgiou [21] etc.

The main result of this paper has the following form.

Theorem 3.1. *Let Assumptions* (I)–(V) *hold. Then there exists a uniform trajectory attractor $\mathcal{U} \subset \mathcal{K}^+$ of the translation semigroup $\{T(t)\}_{t\geq 0}$ on $\mathcal{K}^+$ in the induced topology from $C^{\mathrm{loc}}(\mathbb{R}_+; H)$. Moreover, there exists a compact in $C^{\mathrm{loc}}(\mathbb{R}_+; H)$ uniformly attracting set $\mathcal{P} \subset C^{\mathrm{loc}}(\mathbb{R}_+; H) \cap L_\infty(\mathbb{R}_+; H)$ for the united trajectory space $\mathcal{K}^+$ of Problem (1) in the topology of $C^{\mathrm{loc}}(\mathbb{R}_+; H)$ such that $\mathcal{U}$ coincides with ω-limit set of $\mathcal{P}$:*

$$\mathcal{U} = \bigcap_{t\geq 0} \mathrm{cl}_{C^{\mathrm{loc}}(\mathbb{R}_+;H)} \left[\bigcup_{h\geq t} T(h)\mathcal{P} \right]. \tag{8}$$

4 Proof of Theorem 3.1

Before the proof of Theorem 3.1 we provide some auxiliary constructions.

Assumptions (II) and (III) yield that there exist a positive constant $\alpha' > 0$ and a tr.-c. function c' in $L_1^{\mathrm{loc}}(\mathbb{R}_+)$ such that

$$A(t, u) \subseteq \mathbb{A}_{c'(t)}(u(t)) := \left\{ p \in V^* \,:\, \langle p, u\rangle_V \geq \alpha' \max\left\{ \|u\|_V^p; \|p\|_{V^*}^q \right\} - c'(t) \right\}$$

for any $u \in V$ and a.e. $t > 0$. Let $\mathcal{H}(c')$ be the hull of tr.-c. function c' in $L_1^{\mathrm{loc}}(\mathbb{R}_+)$, i.e. $\mathcal{H}(c') = \mathrm{cl}_{L_1^{\mathrm{loc}}(\mathbb{R}_+)}\{c'(\cdot + h) : h \geq 0\}$. This is a compact set in $L_1^{\mathrm{loc}}(\mathbb{R}_+)$; Chepyzhov and Vishik [7].

Let us consider the family of problems

$$y' = \mathbb{A}_\sigma(y), \quad \sigma \in \Sigma := \mathcal{H}(c'). \tag{9}$$

To each $\sigma \in \Sigma$ there corresponds a space of all globally defined on $[0, +\infty)$ weak solutions $\mathcal{K}_\sigma^+ \subset C^{\mathrm{loc}}(\mathbb{R}_+; H)$ of Problem (9). We set $\mathcal{K}_\Sigma^+ = \cup_{\sigma\in\Sigma}\mathcal{K}_\sigma^+$.

We remark that any element from $\mathcal{K}_\Sigma^+$ satisfies prior estimates.

Lemma 1. *There exist positive constants c_3 and c_4 such that for any $\sigma \in \Sigma$ and $y \in \mathcal{K}_\sigma^+$ the inequalities hold:*

$$\|y(t)\|_H^2 - 2\int_0^t \sigma(\xi)d\xi + 2\alpha'\int_s^t \|y(\xi)\|_V^p d\xi \leq \|y(s)\|_H^2 - 2\int_0^s \sigma(\xi)d\xi, \tag{10}$$

$$\|y(t)\|_H^2 \leq \|y(s)\|_H^2 e^{-c_3(t-s)} + c_4\int_s^t \sigma(\xi)e^{-c_3(t-\xi)}d\xi, \tag{11}$$

for any $t \geq s \geq 0$.

Proof. The proof naturally follows from conditions for the parameters of Problem (9) and Gronwall lemma.

Let us provide the result characterizing the compactness properties of solutions for the family of Problems (9).

Theorem 4.2. *Let $\{y_n\}_{n\geq 1} \subset \mathcal{K}_\Sigma^+$ be an arbitrary sequence, that is bounded in $L_\infty(\mathbb{R}_+; H)$. Then there exist a subsequence $\{y_{n_k}\}_{k\geq 1} \subset \{y_n\}_{n\geq 1}$ and an element $y \in \mathcal{K}_\Sigma^+$ such that*

$$\max_{t\in[\tau,T]} \|y_{n_k}(t) - y(t)\|_H \to 0, \quad k \to +\infty, \tag{12}$$

for any finite time interval $[\tau, T] \subset (0, +\infty)$.

Proof. For any $n \geq 1$ there exists $\sigma_n \in \Sigma$ such that $y_n \in \mathcal{K}_\sigma^+$. Furthermore, the definition of weak solution of evolution inclusion yields that for any $n \geq 1$ there exists $d_n \in L_q^{\mathrm{loc}}(\mathbb{R}_+; V^*)$ such that $y_n'(t) + d_n(t) = \bar{0}$ for a.e. $t > 0$. The definition of $\mathbb{A}_\sigma$ and estimates (10) and (11) provide that the sequence $\{y_n, y_n', d_n\}_{n\geq 1}$ is bounded in $L_p^{\mathrm{loc}}(\mathbb{R}_+; V)\times L_q^{\mathrm{loc}}(\mathbb{R}_+; V^*)\times L_q^{\mathrm{loc}}(\mathbb{R}_+; V^*)$. Since Σ is a compact set in $L_1^{\mathrm{loc}}(\mathbb{R}_+)$, Banach–Alaoglu theorem (cf. Zgurovsky et al. [10, Chap. 1]; Kasyanov [11]) yields that there exist a subsequence $\{y_{n_k}, d_{n_k}\}_{k\geq 1} \subset \{y_n, d_n\}_{n\geq 1}$ and elements $d \in L_q^{\mathrm{loc}}(\mathbb{R}_+; V^*)$, $y \in L_p^{\mathrm{loc}}(\mathbb{R}_+; V)$, such that $y' \in L_q^{\mathrm{loc}}(\mathbb{R}_+; V^*)$ and

$$\begin{aligned}
y_{n_k} &\to y && \text{weakly in } L_p^{\text{loc}}(\mathbb{R}_+; V),\\
y'_{n_k} &\to y' && \text{weakly in } L_q^{\text{loc}}(\mathbb{R}_+; V^*),\\
d_{n_k} &\to d && \text{weakly in } L_q^{\text{loc}}(\mathbb{R}_+; V^*),\\
y_{n_k} &\to y && \text{weakly in } C^{\text{loc}}(\mathbb{R}_+; H),\\
y_{n_k} &\to y && \text{in } L_2^{\text{loc}}(\mathbb{R}_+; H),\\
y_{n_k}(t) &\to y(t) && \text{in } H \text{ for a.e. } t > 0,\\
\sigma_{n_k} &\to \sigma && \text{in } L_1^{\text{loc}}(\mathbb{R}_+), \quad k \to +\infty.
\end{aligned} \tag{13}$$

Formula (12) follows from Zgurovsky et al. [3, Steps 1 and 5, p. 58]. We remark that in the proof we need to consider continuous and nonincreasing (by Lemma 1) functions on $\mathbb{R}_+$:

$$J_k(t) = \|y_{n_k}(t)\|_H^2 - 2\int_0^t \sigma_{n_k}(\xi)d\xi, \ J(t) = \|y(t)\|_H^2 - 2\int_0^t \sigma(\xi)d\xi, \ k \geq 1. \tag{14}$$

The two last statements in (13) imply $J_k(t) \to J(t)$, as $k \to +\infty$, for a.e. $t > 0$.

The definition of a weak solution of evolution inclusion (cf. Zgurovsky et al. [3, p. 58]) and (13) yield $y'(t) = -d(t)$ for a.e. $t > 0$. To finish the proof it is necessary to provide that

$$d(t) \in \mathbb{A}_{\sigma(t)}(y(t)) \text{ for a.e. } t > 0. \tag{15}$$

Let $\varphi \in C_0^\infty((0, +\infty))$, $\varphi \geq 0$. Then

$$\int_{\mathbb{R}_+} \varphi(t)\big(\alpha' \max\{\|y(t)\|_V^p; \|d(t)\|_{V^*}^q\} - \sigma(t)\big)dt \leq$$

$$\liminf_{k\to+\infty} \int_{\mathbb{R}_+} \varphi(t)\big(\alpha' \max\{\|y_{n_k}(t)\|_V^p; \|d_{n_k}(t)\|_{V^*}^q\} - \sigma_{n_k}(t)\big)dt \leq$$

$$\lim_{k\to+\infty} \int_{\mathbb{R}_+} \varphi(t)\langle d_{n_k}(t), y_{n_k}(t)\rangle_V dt = \lim_{k\to+\infty} \frac{1}{2}\int_{\mathbb{R}_+} \|y_{n_k}(t)\|_H^2 \frac{d}{dt}\varphi(t)dt =$$

$$\frac{1}{2}\int_{\mathbb{R}_+} \|y(t)\|_H^2 \frac{d}{dt}\varphi(t)dt = \int_{\mathbb{R}_+} \varphi(t)\langle d(t), y(t)\rangle_V dt,$$

where the first inequality holds, because the convex functional

$$(y, d) \to \int_{\mathbb{R}_+} \varphi(t)\big(\alpha' \max\{\|y(t)\|_V^p; \|d(t)\|_{V^*}^q\}\big)dt$$

is weakly lower semi-continuous on $L_p^{\text{loc}}(\mathbb{R}_+; V) \times L_q^{\text{loc}}(\mathbb{R}_+; V^*)$; the second one follows from the definition of $\mathbb{A}_\sigma$; the first and the third equalities follow from formula (3), because $y'_{n_k}(t) + d_{n_k}(t) = y'(t) + d(t) = \bar{0}$ for any $k \geq 1$ and a.e.

$t > 0$; the second equality holds, because $y_{n_k} \to y$ in $L_2^{\text{loc}}(\mathbb{R}_+; H)$, as $k \to +\infty$. As a nonnegative function $\varphi \in C_0^\infty((0, +\infty))$ is an arbitrary, then, by definition of $\mathbb{A}_\sigma$, formula (15) holds.

The theorem is proved.

Proof of Theorem 3.1. First, let us show that there exists a uniform trajectory attractor $\mathcal{U} \subset \mathcal{K}^+$ of the translation semigroup $\{T(t)\}_{t\geq 0}$ on $\mathcal{K}^+$ in the induced topology from $C^{\text{loc}}(\mathbb{R}_+; H)$. Lemma 1 and Theorem 4.2 yield that the translation semigroup $\{T(t)\}_{t\geq 0}$ has a compact absorbing (and, therefore, an uniformly attracting) set in the space of trajectories $\mathcal{K}_\Sigma^+$; Kasyanov [11, p. 215]. This set can be constructed as follows: (1) consider $\mathcal{P}$, the intersection of $\mathcal{K}_\Sigma^+$ with a ball in the space of bounded continuous functions on $\mathbb{R}_+$ with values in H, $C_b(\mathbb{R}_+; H)$, of sufficiently large radius; (2) shift the resulting set by any fixed distance $h > 0$. Thus, we obtain $T(h)\mathcal{P}$, a set with the required properties. Recall that the semigroup $\{T(t)\}_{t\geq 0}$ is continuous. Therefore, the set $\mathcal{P}_1 := \mathcal{P} \cap \mathcal{K}^+$ is a compact absorbing (and, therefore, an uniformly attracting) in the space $\mathcal{K}^+$ with the induced topology of $C^{\text{loc}}(\mathbb{R}_+; H)$. Then we can apply, for example, Theorem 3.1 from Melnik and Valero [22, p. 197]. In this case, the spaces E and E_0 coincide with H. In fact, here one can apply the classical theorem on the global attractor of a (unique) continuous semigroup in a complete metric space, the semigroup in question having a compact attracting (and, in particular, absorbing) set (see, for example, Babin and Vishik [23]; Temam [24]). In particular, formula (8) holds; cf. Babin and Vishik [23]; Melnik and Valero [22], Temam [24] etc.

5 Conclusions

For the class of nonautonomous differential-operator inclusions with pointwise pseudomonotone dependence between the defining parameters of the problem, the dynamics as $t \to +\infty$ of all global weak solutions defined on $[0, +\infty)$ is studied. The existence of a compact uniform trajectory attractor is proved. The results obtained allow one to study the dynamics of solutions for new classes of evolution inclusions related to nonlinear mathematical models of geophysical and socioeconomic processes and for fields with interaction functions of pseudomonotone type satisfying the power growth and sign conditions. For applications, one can consider new classes of problems with degeneracy, feedback control problems, problems on manifolds, problems with delay, stochastic partial differential equations, etc. (see Balibrea et al. [6]; Hu and Papageorgiou [25]; Gasinski and Papageorgiou [21]; Kasyanov [11]; Kasyanov et al. [26]; Mel'nik and Valero [19]; Denkowski et al. [9]; Gasinski and Papageorgiou [21]; Zgurovsky et al. [3]; etc.) involving differential operators of pseudomonotone type and the corresponding choice of the phase spaces.

Acknowledgements We thank Professor David Y. Gao for many years of cooperation.

References

1. Gajewski, H., Gröger, K., Zacharias, K.: Nichtlineare operatorgleichungen und operatordifferentialgleichungen. Akademie, Berlin (1978)
2. Migórski, S., Ochal, A.: Optimal control of parabolic hemivariational inequalities. J. Glob. Optim. **17**, 285–300 (2000)
3. Zgurovsky, M.Z., Kasyanov, P.O., Kapustyan, O.V., Valero, J., Zadoianchuk, N.V.: Evolution inclusions and variation Inequalities for Earth data processing III. Springer, Berlin (2012)
4. Clarke, F.H.: Optimization and Nonsmooth Analysis. Wiley, New York (1983)
5. Panagiotopoulos, P.D.: Inequality Problems in Mechanics and Applications. Convex and Nonconvex Energy Functions. Birkhauser, Basel (1985)
6. Balibrea, F., Caraballo, T., Kloeden, P.E., Valero, J.: Recent developments in dynamical systems: three perspectives. Int. J. Bifurcat. Chaos (2010). doi:10.1142/S0218127410027246
7. Chepyzhov, V.V., Vishik, M.I.: Evolution equations and their trajectory attractors. J. Math. Pures Appl. **76**, 913–964 (1997)
8. Migórski, S.: Boundary hemivariational inequalities of hyperbolic type and applications. J. Glob. Optim. **31**(3), 505–533 (2005)
9. Denkowski, Z., Migórski, S., Papageorgiou, N.S.: An Introduction to Nonlinear Analysis: Applications. Kluwer Academic/Plenum, Boston (2003)
10. Zgurovsky, M.Z., Mel'nik, V.S., Kasyanov, P.O.: Evolution Inclusions and Variation Inequalities for Earth Data Processing II. Springer, Berlin (2011)
11. Kasyanov, P.O.: Multivalued dynamics of solutions of autonomous operator differential equations with pseudomonotone nonlinearity. Math. Notes **92**, 205–218 (2012)
12. Kasyanov, P.O.: Multivalued dynamics of solutions of an autonomous differential-operator inclusion with pseudomonotone nonlinearity. Cybern. Syst. Anal. **47**, 800–811 (2011)
13. Chepyzhov, V.V., Vishik, M.I.: Trajectory and global attractors for 3D Navier-Stokes system. Mat. Zametki (2002). doi:10.1023/A:1014190629738
14. Chepyzhov, V.V., Vishik, M.I.: Trajectory attractors for evolution equations. C. R. Acad. Sci. Paris. Ser. I **321**, 1309–1314 (1995)
15. Chepyzhov, V.V., Vishik, M.I.: Trajectory attractor for reaction-diffusion system with diffusion coefficient vanishing in time. Discrete Contin. Dyn. Syst. **27**(4), 1498–1509 (2010)
16. Sell, G.R.: Global attractors for the three-dimensional Navier-Stokes equations. J. Dyn. Differ. Equ. **8**(12), 1–33 (1996)
17. Hale, J.K.: Asymptotic Behavior of Dissipative Systems. AMS, Providence (1988)
18. Ladyzhenskaya, O.A.: Attractors for Semigroups and Evolution Equations. Cambridge University Press, Cambridge (1991)
19. Mel'nik, V.S., Valero, J.: On global attractors of multivalued semiprocesses and nonautonomous evolution inclusions. Set-Valued Anal. (2000). doi:10.1023/A:1026514727329
20. Kapustyan, O.V., Kasyanov, P.O., Valero, J.: Pullback attractors for a class of extremal solutions of the 3D Navier-Stokes equations. J. Math. Anal. Appl. (2011). doi:10.1016/j.jmaa.2010.07.040
21. Gasinski, L., Papageorgiou, N.S.: Nonlinear Analysis. Series in Mathematical Analysis and Applications, vol. 9. Chapman & Hall/CRC, Boca Raton (2005)
22. Melnik, V.S., Valero, J.: On attractors of multivalued semi-flows and generalized differential equations. Set Valued Anal. **6**(1), 83–111 (1998)
23. Babin, A.V., Vishik, M.I.: Attractors of Evolution Equations. Nauka, Moscow (1989) (in Russian)
24. Temam, R.: Infinite-Dimensional Dynamical Systems in Mechanics and Physics. Applied Mathematical Sciences, vol. 68. Springer, New York (1988)
25. Hu, S., Papageorgiou, N.S.: Handbook of Multivalued Analysis. Volume II: Applications. Kluwer, Dordrecht (2000)

26. Kasyanov, P.O., Toscano, L., Zadoianchuk, N.V.: Regularity of weak solutions and their attractors for a parabolic feedback control problem. Set Valued Var. Anal. (2013). doi:10.1007/s11228-013-0233-8
27. Zgurovsky, M.Z., Kasyanov, P.O., Zadoianchuk (Zadoyanchuk), N.V.: Long-time behavior of solutions for quasilinear hyperbolic hemivariational inequalities with application to piezoelectricity problem. Appl. Math. Lett. **25**(10), 1569–1574 (2012)

A Contact Problem with Normal Compliance, Finite Penetration and Nonmonotone Slip Dependent Friction

Ahmad Ramadan, Mikäel Barboteu, Krzysztof Bartosz, and Piotr Kalita

Abstract In this work, we consider a static frictional contact problem between a linearly elastic body and an obstacle, the so-called foundation. This contact is described by a normal compliance condition of such a type that the penetration is restricted with unilateral constraint. The friction is modeled with a nonmonotone law. In order to approximate the contact conditions, we consider a regularized problem wherein the contact is modeled by a standard normal compliance condition without finite penetration. Next, we present a convergence result between the solution of the regularized problem and the original problem. Finally, we provide a numerical validation of this convergence result. To this end we introduce a discrete scheme for the numerical approximation of the frictional contact problems.

1 Introduction

The aim of this paper is to study frictional contact problems in which the contact is modeled with normal compliance of such a type that the penetration is restricted with unilateral constraint. In a physical point of view, this penetration can be assimilated to the flattening of the asperities on the contact interface. Furthermore, the friction is modeled with a nonmonotone law in which the friction bound depends on the tangential displacement, the penetration, and the size of the asperities. The behavior of the material is modeled with a linear elastic constitutive law. In the present paper we consider two frictional contact problems. The first problem is characterized by normal compliance in which the penetration is restricted by unilateral constraint and the second problem represents a regularization of the first problem by considering penetrations without restriction.

A. Ramadan (✉) • M. Barboteu
Université de Perpignan Via Domitia, 52 Avenue Paul Alduy, 66860 Perpignan, France
e-mail: ahmad.ramadan@univ-perp.fr; barboteu@univ-perp.fr

P. Kalita • K. Bartosz
Jagiellonian University, ul.Lojasiewiccza 6, 30348 Krakov, Poland
e-mail: piotr.kalita@ii.uj.edu.pl; krzysztof.bartosz@ii.uj.edu.pl

D. Gao et al. (eds.), *Advances in Global Optimization*, Springer Proceedings in Mathematics & Statistics 95, DOI 10.1007/978-3-319-08377-3_29

Our interest in this paper is to present the convergence of the solution of the regularized problem with nonmonotone friction to the solution of the original problem with normal compliance, finite penetration, and nonmonotone friction. After, we provide numerical simulations which illustrate the mechanical behavior of the contact model and the numerical validation of the convergence result.

The rest of the paper is structured as follows. In Sect. 2 we present the classical formulation of the contact problems, the variational formulation of the problems, the existence of the weak solution of the regularized problems, and the convergence result. Finally, in Sect. 3 we present the numerical solution of the problems and we provide some numerical simulations on an academic two-dimensional example including a numerical validation of the convergence result.

2 Mechanical Problems and Variational Formulations

In this section we describe the model for the nonmonotone frictional contact with normal compliance and finite penetration as well as a family of auxiliary models used for its approximation. The physical setting is as follows. A linearly elastic body occupies an open bounded connected set $\Omega \subset \mathbb{R}^d$ ($d \leq 3$ in applications) with a Lipschitz boundary Γ that is partitioned into three disjoint parts $\overline{\Gamma}_1$, $\overline{\Gamma}_2$, and $\overline{\Gamma}_3$ with Γ_1, Γ_2, and Γ_3 being relatively open, and meas $(\Gamma_1) > 0$. The body is clamped on Γ_1 and thus the displacement field vanishes there. A volume force of density f_0 acts in Ω and a surface traction of density f_2 acts on Γ_2. The body is in frictional contact with an obstacle on Γ_3. We consider $H = L^2(\Omega)^d = \{ u = (u_i) \mid u_i \in L^2(\Omega) \}$, $Q = \{ \sigma = (\sigma_{ij}) \mid \sigma_{ij} = \sigma_{ji} \in L^2(\Omega) \}$, $H_1 = \{ u = (u_i) \mid \varepsilon(u) \in Q \}$ et $Q_1 = \{ \sigma \in Q \mid \operatorname{Div} \sigma \in H \}$.

The classical formulation of the frictional contact problem considered is the following.

Problem. $\mathscr{P}_M$. Find a displacement field $u : \Omega \to \mathbb{R}^d$ and a stress field $\sigma : \Omega \to \mathbb{S}^d$ such that

$$\sigma = \mathscr{E}\varepsilon(u) \quad \text{in } \Omega, \tag{1}$$

$$\operatorname{Div} \sigma + f_0 = 0 \quad \text{in } \Omega, \tag{2}$$

$$u = 0 \quad \text{on } \Gamma_1, \tag{3}$$

$$\sigma \nu = f_2 \quad \text{on } \Gamma_2, \tag{4}$$

$$\sigma_\nu + p(u_\nu) \leq 0,\ u_\nu - g \leq 0,\ (\sigma_\nu + p(u_\nu))(u_\nu - g) = 0 \quad \text{on } \Gamma_3, \tag{5}$$

$$\left.\begin{array}{ll} |\sigma_\tau| \leq N(u_\nu)\mu(|u_\tau|) & \text{if } \ u_\tau = 0, \\ -\sigma_\tau = N(u_\nu)\mu(|u_\tau|)\frac{u_\tau}{|u_\tau|} & \text{if } \ u_\tau \neq 0, \end{array}\right\} \quad \text{on } \Gamma_3. \tag{6}$$

Condition (5) was first introduced in [1] and it was used in various papers, see [2] and the references therein where p is the compliance function. Condition (6) was introduced in [3] and $N(u_\nu)\mu(|u_\tau|)$ represents the magnitude of the limiting friction traction at which slip begins. In this case, the friction coefficient μ depends on the tangential displacement $|u_\tau|$ and the magnitude of the friction bound depends also on the penetrations and the size of the asperities via the function N defined by

$$N(x,\eta) = \begin{cases} 0 \ \text{ for } \ \eta \le 0, \\ S\frac{\eta}{g(x)} \ \text{ for } \ \eta \in (0, g(x)), \\ S \ \text{ for } \ \eta \ge g(x). \end{cases} \tag{7}$$

In the above formula the value $S \ \ge \ 0$ is a given value. Next we define the approximate problems corresponding to Problem $\mathscr{P}_M$. Let $n \in \mathbb{N}$.

Problem. $\mathscr{P}_M^n$. Find a displacement field $u^n : \Omega \to \mathbb{R}^d$ and a stress field $\sigma^n : \Omega \to \mathbb{S}^d$ such that (1)–(4) and (6) hold for $u = u^n$ and $\sigma = \sigma^n$, and

$$-\sigma_\nu^n \in \begin{cases} \{p(u_\nu^n)\} & \text{if } \ u_\nu^n < g, \\ [p(g), p(g) + nc_2] & \text{if } \ u_\nu^n = g, \\ \{p(g) + nc_2 + nc_3(u_\nu^n - g)\} & \text{if } \ u_\nu^n > g, \end{cases} \qquad \text{on } \Gamma_3. \tag{8}$$

In (8) c_2 and c_3 are arbitrary nonnegative constants such that $c_2 + c_3 > 0$.

Proceeding in a standard way, we obtain the following variational formulations of Problems $\mathscr{P}_M$ and $\mathscr{P}_M^n$. We consider $V = \{v \in H_1 \mid v = 0 \text{ on } \Gamma_1\}$, $K = \{v \in V,\ v_\nu \le g(x) \text{ on } \Gamma_3\}$, $B : V \to V^*$ tel que $\langle Bu, v\rangle = (\mathscr{E}\varepsilon(u), \varepsilon(v))_Q$ and f the element of V' such that $\langle f, v\rangle = \int_\Omega f_0 \cdot v\, dx + \int_{\Gamma_2} f_2 \cdot v\, d\,\Gamma$.

Problem. $\mathscr{P}_V$. Find the displacement field $u \in K$ and the friction density $\sigma_\tau \in L^2(\Gamma_3)^d$ such that for all $v \in K$ we have

$$\langle Bu - f, v - u\rangle + \int_{\Gamma_3} p(u_\nu)(v_\nu - u_\nu)\, d\,\Gamma \ge \int_{\Gamma_3} \sigma_\tau \cdot (v_\tau - u_\tau)\, d\,\Gamma, \tag{9}$$

$$\textit{with} \quad -\sigma_\tau \in N(u_\nu)\partial j_\tau(u_\tau) \ \textit{a.e. on } \Gamma_3 \tag{10}$$

where function $j_\tau : \mathbb{R}^d \to \mathbb{R}$ is defined by

$$j_\tau(\xi) = \int_0^{|\xi|} \mu(t)\, dt, \tag{11}$$

then under some assumptions, see [3] we can prove that the conditions (6) are equivalent to the subdifferential inclusion (10).

Problem. $\mathscr{P}_V^n$. Find the displacement field $u^n \in V$, friction density $\sigma_\tau^n \in L^2(\Gamma_3)^d$ and normal stress $\sigma_\nu^n \in L^2(\Gamma_3)$ such that for all $v \in V$ we have

$$\langle Bu^n - f, v\rangle = \int_{\Gamma_3} \sigma_\tau^n \cdot v_\tau + \sigma_\nu^n v_\nu \, d\,\Gamma, \tag{12}$$
$$\text{with} \quad -\sigma_\tau^n \in N(u_\nu^n)\partial j_\tau(u_\tau^n) \;\; a.e. \text{ on } \Gamma_3,$$
$$\text{and} \quad -\sigma_\nu^n \in \bar{p}(u_\nu^n) + \partial j_\nu^n(u_\nu^n) \;\; a.e. \text{ on } \Gamma_3.$$

Where $j_\nu^n(x,\eta) = \begin{cases} 0 & \text{if } \eta \le g(x), \\ nc_2\eta + \frac{nc_3}{2}(\eta - g(x))^2 & \text{if } \eta > g(x), \end{cases}$

and $\bar{p} : \Gamma_3 \times \mathbb{R} \to \mathbb{R}$ such that $\bar{p}(x,s) = \begin{cases} p(s) & \text{for } s \le g(x) \\ p(g(x)) & \text{for } s > g(x). \end{cases}$

Theorem 2.1. *Under some assumptions, see [3], Problem $\mathscr{P}_V^n$ has a solution for every $n \in \mathbb{N}$.*

Theorem 2.2. *Let $(u^n, \sigma_\tau^n, \sigma_\nu^n)$ be a solution of Problem $\mathscr{P}_V^n$, then under some assumptions, see [3], for a subsequence, we have $u^n \to u$ weakly in V, $\sigma_\tau^n \to \sigma_\tau$ weakly in $L^2(\Gamma_3;\mathbb{R}^d)$, where (u,σ_τ) is a solution of Problem $\mathscr{P}_V$.*

3 Numerical Solution

The numerical strategy presented in this section is based on a sequence of convex programming problems; more details can be found in [4]. We consider some materials for the discretization step. Let Ω a polyhedral domain, $\{\mathscr{T}^h\}$ a regular family of triangular finite element partitions of $\overline{\Omega}$. The space V is approximated by the finite dimensional space $V^h \subset V$ of continuous and piecewise affine functions, that is,

$$V^h = \{ v^h \in [C(\overline{\Omega})]^d \; : \; v^h|_T \in [P_1(T)]^d \quad \forall\, T \in \mathscr{T}^h,$$
$$v^h = 0 \text{ at the nodes on } \Gamma_1 \},$$

where $P_1(T)$ represents the space of polynomials of degree less or equal to one in T. For the discretization of the normal contact terms, we consider the spaces $X_\nu^h = \{ v_{\nu|\Gamma_3}^h \; : \; v^h \in V^h \}$; $X_\tau^h = \{ v_{\tau|\Gamma_3}^h \; : \; v^h \in V^h \}$ equipped with their usual norm. Let us consider the discrete spaces of piecewise constants $Y_\nu^h \subset L^2(\Gamma_3)$ and $Y_\tau^h \subset L^2(\Gamma_3)$ related, respectively, to the discretization of the normal stress σ_ν and the friction density σ_τ. We also introduce the function $\varphi : X_\nu^h \to (-\infty, +\infty]$ and the operator $L : X_\nu^h \to Y_\nu^h$ defined by

$$\varphi(u_\nu^h) = \int_{\Gamma_3} I_{\mathbb{R}_-}(u_\nu^h - g)\, d\Gamma, \quad \forall\, u_\nu^h \in X_\nu^h,$$

$$L : X_\nu^h \to Y_\nu^h \quad \langle L u_\nu^h, v_\nu^h \rangle_{Y_\nu^h, X_\nu^h} = \int_{\Gamma_3} p(u_\nu^h) v_\nu^h \, d\Gamma \quad \forall\, u_\nu^h, v_\nu^h \in X_\nu^h$$

where $I_{\mathbb{R}_-}$ represents the indicator function of the set $\mathbb{R}_- = (-\infty, 0]$.

The normal compliance condition with finite penetration (5) leads to the following discrete subdifferential inclusion

$$-\sigma_\nu^h \in \partial\varphi(\Pi_\nu^h u_\nu^h) + L\Pi_\nu^h u_\nu^h \quad \text{in} \quad Y_\nu^h.$$

The friction condition (6) leads to the following discrete subdifferential inclusion

$$-\sigma_\tau^h \in N(|\Pi_\nu^h u_\nu^h|)\mu(|\Pi_\tau^h u_\tau^h|)\partial|\Pi_\tau^h u_\tau^h| \quad \text{in} \quad Y_\tau^h,$$

where $\Pi_\nu^h : X_\tau^h \to Y_\nu^h$ and $\Pi_\tau^h : X_\tau^h \to Y_\tau^h$ represent, respectively, the boundary interpolation operators from X_ν^h to Y_ν^h and from X_τ^h to Y_τ^h (see [5]).

The numerical solution of the nonsmooth nonconvex variational problem $\mathscr{P}_V$ is based on the following iterative algorithm.

Let $\epsilon > 0$ and $u^{(0)}$ be given.

Then, for $k = 0, 1 \ldots,$

Problem $\mathscr{P}_{V_C^h}$. *Find a displacement field* $u^{h,(k+1)} \in V^h$,

a contact stress $\sigma_\nu^{h,(k+1)} \in Y_\nu^h$ *and a friction stress field* $\sigma_\tau^{h,(k+1)} \in Y_\tau^h$

such that, for $\forall\, v^h \in V^h$

$$\langle B u^{h,(k+1)} - f, v^h \rangle = \int_{\Gamma_3} \sigma_\nu^{h,(k+1)} v_\nu^h \, d\,\Gamma + \int_{\Gamma_3} \sigma_\tau^{h,(k+1)} \cdot v_\tau^h \, d\,\Gamma$$

$$\textit{with } -\sigma_\nu^{h,(k+1)} \in \partial\varphi(\Pi_\nu^h u_\nu^{h,(k+1)}) + L\Pi_\nu^h u^{h,(k+1)} \textit{ on } \Gamma_3$$

$$\textit{and } -\sigma_\tau^{h,(k+1)} \in N(|\Pi_\nu^h u_\nu^{h,(k)}|)\mu(|\Pi_h u_\tau^{h,(k)}|)\partial|\Pi_\tau^h u_\tau^{h,(k+1)}| \textit{ on } \Gamma_3$$

until $\|u^{h,(k+1)} - u^{h,(k)}\| \le \epsilon \|u^{h,(k)}\|$

and $\|\sigma^{h,(k+1)} - \sigma^{h,(k)}\|_{L^2(\Gamma_3)^d} \le \epsilon \|\sigma^{h,(k)}\|_{L^2(\Gamma_3)^d}$

Here, k represents the index of the iterative procedure. In Problem $\mathscr{P}_{V_C^h}$ the discrete stress σ^h on the contact boundary Γ_3 can be viewed as a Lagrange stress multiplier. This numerical strategy leads to the solution of a nonsmooth convex problem $\mathscr{P}_{V_C^h}$ at each iteration k. For the numerical treatment of the nonsmooth convex Problem $\mathscr{P}_{V_C^h}$ we use the penalized method for the normal compliance contact term combined with the augmented Lagrangean approach for the unilateral condition and Coulomb friction law. For details concerning this numerical treatment see [3].

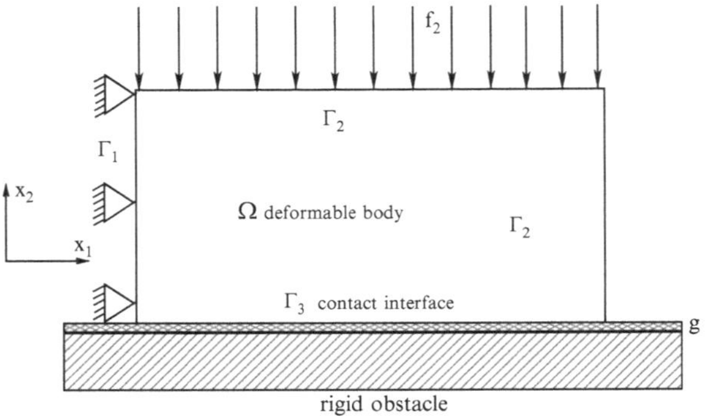

Fig. 1 Initial configuration of the two-dimensional example

Numerical Example. We consider the physical setting depicted in Fig. 1. There, $\Omega = (0, L_1) \times (0, L_2) \subset \mathbb{R}^2$ with $L_1,\ L_2 > 0$ and

$$\Gamma_1 = \{0\} \times [0, L_2],\ \Gamma_2 = (\{L_1\} \times [0, L_2]) \cup ([0, L_1] \times \{L_2\}),\ \Gamma_3 = [0, L_1] \times \{0\}.$$

The domain Ω represents the cross section of a three-dimensional deformable body subjected to the action of tractions in such a way that a plane stress hypothesis is assumed. On the part $\Gamma_1 = \{0\} \times [0, L_2]$ the body is clamped and, therefore, the displacement field vanishes there. Vertical tractions act on the part $[0, L_1] \times \{L_2\}$ of the boundary and the part $\{L_1\} \times [0, L_2]$ is traction free. No body forces are assumed to act on the body during the process. The body is in frictional contact with an obstacle on the part $\Gamma_3 = [0, L_1] \times \{0\}$ of the boundary.

We model the material's behavior with a constitutive law of the form (1) in which elasticity tensor $\mathscr{E}$ satisfies

$$(\mathscr{E}\tau)_{\alpha\beta} = \frac{E\kappa}{1-\kappa^2}(\tau_{11} + \tau_{22})\delta_{\alpha\beta} + \frac{E}{1+\kappa}\tau_{\alpha\beta}, \qquad 1 \leq \alpha, \beta \leq 2,$$

where E is the Young modulus, κ the Poisson ratio of the material, and $\delta_{\alpha\beta}$ denotes the Kronecker symbol. The friction is modeled by a nonmonotone law (6) in which the friction bound $N(u_\nu)\mu(|u_\tau|)$ depends on the depth of the penetration u_ν and on the tangential displacement $|u_\tau|$. For the simulations, the function $N : \mathbb{R} \to \mathbb{R}^+$ given in (7) is taken. Let us also consider the following friction coefficient $\mu : \mathbb{R}^d \to \mathbb{R}$:

$$\mu(|u_\tau|) = (a - b) \cdot e^{-\alpha |u_\tau|} + b, \tag{13}$$

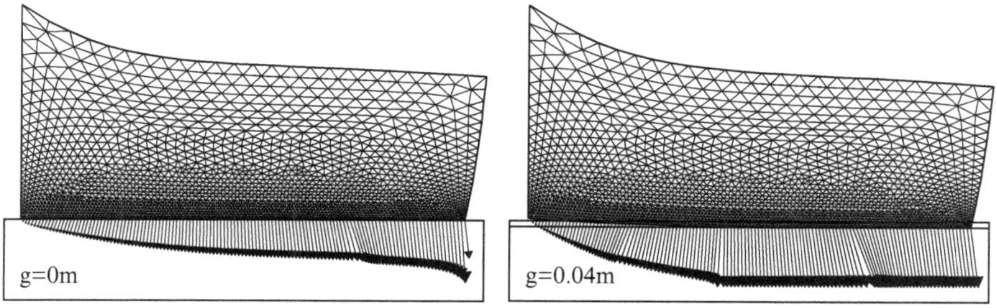

Fig. 2 Deformed meshes and frictional contact forces for $g = 0\,\text{m}$ and $g = 0.04\,\text{m}$

with $a, b, \alpha > 0$, $a \geq b$. For the computation below we use the following data:

$$
\begin{aligned}
&L_1 = 2m, \quad L_2 = 1m,\\
&E = 1000N/m^2, \quad \kappa = 0.3,\\
&f_0 = (0,0)N/m^2, \quad f_2 = \begin{cases} (0,0)\,N/m & \text{on}\{2\} \times [0,1],\\ (0,-300\,t)\,N/m & \text{on}[0,2] \times \{1\}, \end{cases}\\
&a = 1.5, \quad b = 0.5, \quad \alpha = 100, \quad S = 1N, p(u) = c_1 u_+, \quad c_1 = 100,\\
&\text{stopping criterion} : \epsilon = 10^{-6}.
\end{aligned}
$$

Our results are presented in Figs. 2, 3, and 4 and are described in what follows. First, in Fig. 2, the deformed configuration as well as the frictional contact forces is plotted both in the case $g = 0\,\text{m}$ and $g = 0.04\,\text{m}$, which represent, respectively, the case with a classical signorini unilateral contact and the case with normal compliance, finite penetration, and unilateral constraint.

In Fig. 3 we present the convergence of solution of problem $\mathscr{P}^n_{Vh}$ to the solution of problem $\mathscr{P}_{Vh}$. More precisely, we plot four deformed meshes and the associated frictional contact forces at four steps of convergence, for $n = 10,\ 100,\ 10^3,\ 10^4$. One can see that for $n = 10$ all the contact nodes are in strong penetration contact, whereas at $n = 10^4$ the contact nodes are into an admissible finite penetration, since the complete flattening of the asperities of size $g = 0.04\,m$ was reached.

For the numerical convergence we denote by (u^h_n, σ^h_n) and (u^h, σ^h) the discrete solution of the contact problems $\mathscr{P}^n_{Vh}$ and $\mathscr{P}_{Vh}$, respectively. The numerical estimations of the difference

$$\|u^h_n - u^h\|_V + \|\sigma^h_n - \sigma^h\|_Q,$$

for various values of the parameter n, are presented in Fig. 4. It results from here that this difference converges to zero when n tends toward infinity, which represents a numerical validation of the theoretical convergence result obtained in Theorem 2.2.

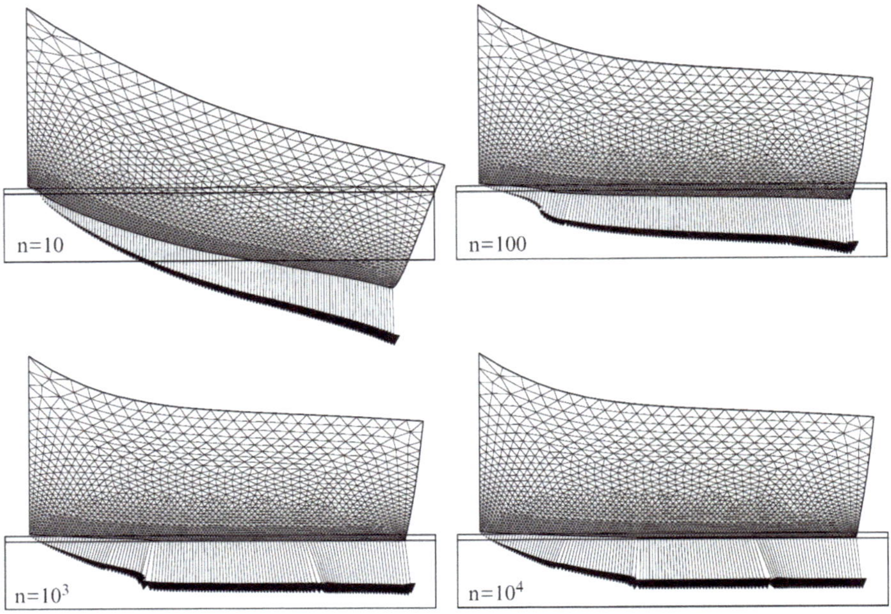

Fig. 3 Deformed meshes and frictional contact forces for $n = 10$, $n = 100$, $n = 10^3$, and $n = 10^4$

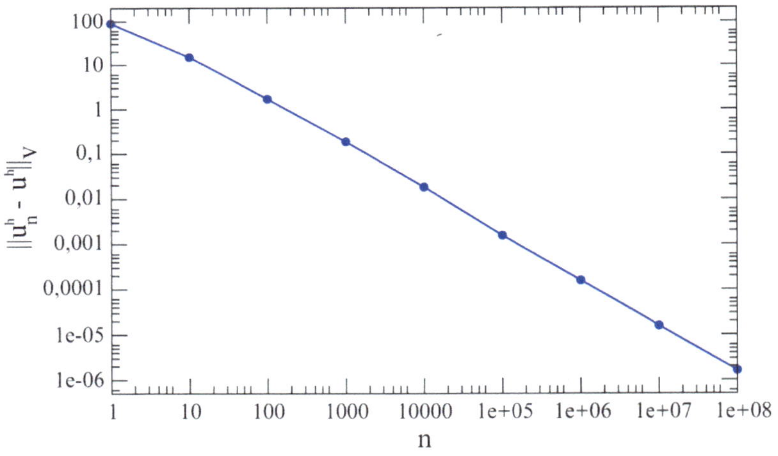

Fig. 4 Numerical validation of the convergence result in Theorem 2.2

Acknowledgements This research was supported by a Marie Curie International Research Staff Exchange Scheme Fellowship within the seventh European Community Framework Programme under Grant Agreement no. 2011-295118.

References

1. Jarušek, J., Sofonea, M.: On the solvability of dynamic elastic-visco-plastic contact problems. Zeitschrift für Angewandte Matematik und Mechanik, **88**, 3–22 (2008)
2. Sofonea, M., Matei, A.: History-dependent quasivariational inequalities arising in Contact Mechanics. Eur. J. Appl. Math. **22**, 471–491 (2011)
3. Barboteu, M., Bartosz, K., Kalita, P., Ramadan, A.: Analysis of a contact problem with normal compliance, finite penetration and non monotone slip dependent friction. Commun. Contemp. Math. **16**, 1350016 (2014)
4. Mistakidis, E.S., Panagiotopulos, P.D.: Numerical treatment of problems involving nonmonotone boundary or stress-strain laws. Comput. Struct. **64**, 553–565 (1997)
5. Khenous, H.B., Pommier, J., Renard, Y.: Hybrid discretization of the Signorini problem with Coulomb friction. Theoretical aspects and comparison of some numerical solvers. Appl. Numer. Math. **56**, 163–192 (2006)

A Class of Mixed Variational Problems with Applications in Contact Mechanics

Mircea Sofonea

Abstract We provide an existence result in the study of a new class of mixed variational problems. The problems are formulated on unbounded interval of time and involve history-dependent operators. The proof is based on generalized saddle point theory and various estimates, combined with fixed point arguments. Then, we consider a new mathematical model which describes the frictionless contact between a viscoelastic body and an obstacle. The process is quasistatic and the contact is modelled with a version of the normal compliance condition with unilateral constraint, which describes both the hardness and the softness of the foundation. We list the assumption on the data, derive a variational formulation of the problem, then we use our abstract result to prove its weak solvability.

1 Introduction

Mixed variational problems provide an useful framework in which a large number of problems involving unilateral constraints can be casted, analyzed, and solved numerically. For this reason, they are used both in Numerical Analysis, Optimization, Solid Mechanics and Fluid Mechanics, as well. The literature in the field was growing rapidly in the last decades. Existence and uniqueness results in the study of stationary mixed variational problems with Lagrange multipliers, together with various applications in Solid Mechanics, can be found in [3–5, 8, 10] and the references therein. Reference concerning the analysis of mixed variational problems associated with contact problems include [6, 7, 9], for instance.

The aim of this paper is twofold. The first one is to study the solvability of a new mixed variational problem involving Lagrange multipliers. The second one is to show how our abstract result can be used in the analysis of a mathematical model arising in Contact Mechanics. The paper is structured as follows. In Sect. 2 we introduce the mixed variational problem then we state and prove our main existence

M. Sofonea (✉)
Université de Perpignan Via Domitia, 52 Avenue Paul Alduy, 66860 Perpignan, France
e-mail: sofonea@univ-perp.fr

D. Gao et al. (eds.), *Advances in Global Optimization*, Springer Proceedings in Mathematics & Statistics 95, DOI 10.1007/978-3-319-08377-3_30

result, Theorem 2.1. In Sect. 3, we describe our mathematical model of contact, list the assumption on the data and derive its variational formulation. Then we use Theorem 2.1 to prove the weak solvability of the model.

We end this short introductory section with some notation. Everywhere in this paper we use r^+ for the positive part of r, $\mathbb{N}^*$ for the set of positive integers and $\mathbb{R}_+$ will represent the set of non negative real numbers, i.e. $\mathbb{R}_+ = [0, \infty)$. Notation (x, y) will represent an element of the product of the sets X and Y, denoted $X \times Y$. Given a normed space $(X, \|\cdot\|_X)$ we use the notation $C(\mathbb{R}_+; X)$ for the space of continuous functions defined on $\mathbb{R}_+$ with values on X. Also, for a subset $K \subset X$ we use the symbol $C(\mathbb{R}_+; K)$ for the set of continuous functions defined on $\mathbb{R}_+$ with values in K. Finally, if Y is a normed space and $\mathscr{R} : C(\mathbb{R}_+; X) \to C(\mathbb{R}_+; Y)$, then $\mathscr{R}\eta(t)$ represents the value of the function $\mathscr{R}\eta$ at the point t, i.e. $\mathscr{R}\eta(t) = (\mathscr{R}\eta)(t)$.

2 An Abstract Existence Result

Let $(X, (\cdot, \cdot)_X, \|\cdot\|_X)$, $(Y, (\cdot, \cdot)_Y, \|\cdot\|_Y)$ be two real Hilbert spaces and let $(Z, \|\cdot\|_Z)$ be a real normed space. We consider two operators $A : X \to X$, $\mathscr{R} : C(\mathbb{R}_+; X) \to C(\mathbb{R}_+; Z)$, a functional $\varphi : Z \times X \to \mathbb{R}$, a bilinear form $b : X \times Y \to \mathbb{R}$, a function $f : \mathbb{R}_+ \to X$, an element h of X and a set $\Lambda \subset Y$. We are interested in the problem of finding two functions $u : \mathbb{R}_+ \to X$ and $\lambda : \mathbb{R}_+ \to \Lambda$ such that, for each $t \in \mathbb{R}_+$, the following inequalities hold:

$$
\begin{aligned}
&(Au(t), v - u(t))_X + \varphi(\mathscr{R}u(t), v)_X - \varphi(\mathscr{R}u(t), u(t)) \\
&\qquad + b(v - u(t), \lambda(t)) \geq (f(t), v - u(t))_X \quad \forall\, v \in X,
\end{aligned} \tag{1}
$$

$$
b(u(t), \mu - \lambda(t)) \leq b(h, \mu - \lambda(t)) \qquad \forall\, \mu \in \Lambda. \tag{2}
$$

In the study of this problem we consider the following assumptions.

$$
\left\{
\begin{array}{l}
\text{(a) There exists } m_A > 0 \text{ such that} \\
\quad (Au - Av,\, u - v)_X \geq m_A \|u - v\|_X^2 \quad \forall\, u, v \in X. \\
\text{(b) There exists } L_A > 0 \text{ such that} \\
\quad \|Au - Av\|_X \leq L_A \, \|u - v\|_X \quad \forall\, u, v \in X.
\end{array}
\right. \tag{3}
$$

$$
\left\{
\begin{array}{l}
\text{For each } n \in \mathbb{N}^* \text{ there exists } r_n \geq 0 \text{ such that} \\
\quad \|\mathscr{R}u_1(t) - \mathscr{R}u_2(t)\|_Z \leq r_n \displaystyle\int_0^t \|u_1(s) - u_2(s)\|_X \, ds \\
\quad \forall\, u_1, u_2 \in C(\mathbb{R}_+; X), \ \forall\, t \in [0, n].
\end{array}
\right. \tag{4}
$$

$$\begin{cases} \text{(a) The function } \varphi(\eta,\cdot) : X \to \mathbb{R} \text{ is convex} \\ \qquad \text{and Lipschitz continuous, for any } \eta \in Z. \\ \text{(b) There exists } \alpha \geq 0 \text{ such that} \\ \qquad \varphi(\eta_1, v_2) - \varphi(\eta_1, v_1) + \varphi(\eta_2, v_1) - \varphi(\eta_2, v_2) \\ \qquad\quad \leq \alpha \, \|\eta_1 - \eta_2\|_X \|v_1 - v_2\|_X \quad \forall\, \eta_1,\, \eta_2 \in Z,\ v_1,\, v_2 \in X. \end{cases} \tag{5}$$

$$\begin{cases} \text{(a) There exists } M_b > 0 \text{ such that} \\ \qquad |b(v,\mu)| \leq M_b \|v\|_X \|\mu\|_Y \quad \forall\, v \in X,\ \mu \in Y. \\ \text{(b) There exists } m_b > 0 \text{ such that} \quad \inf\limits_{\mu \in Y, \mu \neq 0_Y} \sup\limits_{v \in X, v \neq 0_X} \dfrac{b(v,\mu)}{\|v\|_X \|\mu\|_Y} \geq m_b. \end{cases} \tag{6}$$

$$f \in C(\mathbb{R}_+; X), \qquad h \in X. \tag{7}$$

$$\Lambda \text{ is a closed convex subset of } Y \text{ that contains } 0_Y. \tag{8}$$

On these assumptions we have the following comments. First, (3) shows that A is a strongly monotone Lipschitz continuous operator. Next, following the terminology introduced in [12], (4) shows that $\mathscr{R}$ is a *history-dependent operator*. Finally, condition (6)(b) is the so-called inf-sup condition, used in the saddle point theory, see, for instance, [3–5, 8] and the references therein.

The solvability of problem (1)–(2) is given by the following result.

Theorem 2.1. *Assume* (3)–(8). *Then, there exists a couple of functions* $(u,\lambda) : \mathbb{R}_+ \to X \times Y$, *unique in* u, *such that* (1)–(2) *hold for all* $t \in \mathbb{R}_+$. *Moreover,* $u \in C(\mathbb{R}_+; X)$.

Proof. The proof of Theorem 2.1 is carried out in several steps, that we shortly describe in what follows.

(i) In the first step we consider $g \in X$, $z \in Z$ and, using arguments similar to those used in [2], we prove that there exist a couple $(u,\lambda) \in X \times \Lambda$, unique in u, such that

$$(Au, v-u)_X + \varphi(z,v) - \varphi(z,u) + b(v-u,\lambda) \geq (g, v-u)_X \quad \forall\, v \in X, \tag{9}$$

$$b(u, \mu - \lambda) \leq b(k, \mu - \lambda) \qquad \forall \mu \in \Lambda. \tag{10}$$

In addition, if (u_1, λ_1) and (u_2, λ_2) are two solutions of the problem (9)–(10) corresponding to the data $(g_1, z_1) \in X \times Z$ and $(g_2, z_2) \in X \times Z$, respectively, then there exists $c > 0$ which depends only on A and φ such that

$$\|u_1 - u_2\|_X \leq c\,(\|g_1 - g_2\|_X + \|z_1 - z_2\|_Z).$$

(ii) In the second step we consider an element $\eta \in C(\mathbb{R}_+; X)$ and introduce the notation $y_\eta = \mathscr{R}\eta \in C(\mathbb{R}_+; Z)$. Then we use the results in step i) to prove that there exists a couple of functions $(u_\eta, \lambda_\eta) : \mathbb{R}_+ \to X \times \Lambda$, unique in the first component, such that, for each $t \in \mathbb{R}_+$, the following inequalities hold:

$$(Au_\eta(t), v - u_\eta(t))_X + \varphi(y_\eta(t), v) - \varphi(y_\eta(t), u_\eta(t)) \tag{11}$$
$$+b(v - u_\eta(t), \lambda_\eta(t)) \geq (f(t), v - u_\eta(t))_X \qquad \forall\, v \in X,$$

$$b(u(t), \mu - \lambda_\eta(t)) \leq b(h, \mu - \lambda_\eta(t)) \qquad \forall\, \mu \in \Lambda. \tag{12}$$

Moreover, $u_\eta \in C(\mathbb{R}_+; X)$. In addition, if (u_1, λ_1) and (u_2, λ_2) are two solutions of problem (11)–(12) corresponding to the data $\eta_1, \eta_2 \in C(\mathbb{R}_+; X)$ then, for each positive integer n, we have

$$\|u_1(t) - u_2(t)\|_X \leq \frac{\alpha r_n}{m_A} \int_0^t \|\eta_1(s) - \eta_2(s)\|_X\, ds \qquad \forall\, t \in [0, n]. \tag{13}$$

(iii) In the next step we define the operator $\Theta : C(\mathbb{R}_+; X) \to C(\mathbb{R}_+; X)$ by equality $\Theta\eta = u_\eta$ for all $\eta \in C(\mathbb{R}_+; X)$. We use estimate (13) and a fixed point result obtained in [14] to prove that the operator Θ has a unique fixed point $\eta^* \in C(\mathbb{R}_+; X)$.

(iv) Let η^* be the unique fixed point of the operator Θ. Then, writing (11)–(12) for $\eta = \eta^*$ and using the equalities $u_{\eta^*} = \eta^*$, $y_{\eta^*} = \mathscr{R}\eta^*$, it follows that the couple $(u_{\eta^*}, \lambda_{\eta^*})$ is a solution of problem (1)–(2). Moreover, $u_{\eta^*} \in C(\mathbb{R}_+; X)$. The uniqueness of the solution in the first component follows from the uniqueness of the fixed point of the operator Θ, guaranteed by the step (iii).

□

3 A Viscoelastic Contact Model

In this section we introduce a model of frictionless contact which can be studied by using the abstract result presented in Sect. 2. The physical setting is as follows. A viscoelastic body occupies a bounded domain $\Omega \subset \mathbb{R}^d$ $(d = 2, 3)$, with the boundary Γ partitioned into three disjoint measurable parts Γ_1, Γ_2, Γ_3, such that $meas\, \Gamma_1 > 0$. We assume that Γ is Lipschitz continuous and we denote by $\boldsymbol{\nu}$ its unit outward normal, defined almost everywhere. The body is clamped on Γ_1 and, therefore, the displacement field vanishes there. A volume force of density $\boldsymbol{f}_0$ acts in Ω, surface tractions of density $\boldsymbol{f}_2$ act on Γ_2 and, finally, we assume that the body is in contact with a deformable foundation on Γ_3. The contact is frictionless and we model it with a version of the multivalued normal compliance condition with unilateral constraint. The process is quasistatic and we study it in the time interval $\mathbb{R}_+ = [0, \infty)$. The classical formulation of the problem is the following.

Problem 1. Find a displacement field $\boldsymbol{u} : \Omega \times \mathbb{R}_+ \to \mathbb{R}^d$ and a stress field $\boldsymbol{\sigma} : \Omega \times \mathbb{R}_+ \to \mathbb{S}^d$ such that

$$\boldsymbol{\sigma}(t) = \mathscr{A}\boldsymbol{\varepsilon}(\boldsymbol{u}(t)) + \int_0^t \mathscr{B}(t-s)\boldsymbol{\varepsilon}(\boldsymbol{u}(s))\,ds \quad \text{in } \Omega, \tag{14}$$

$$\operatorname{Div}\boldsymbol{\sigma}(t) + \boldsymbol{f}_0(t) = \boldsymbol{0} \quad \text{in } \Omega, \tag{15}$$

$$\boldsymbol{u}(t) = \boldsymbol{0} \quad \text{on } \Gamma_1, \tag{16}$$

$$\boldsymbol{\sigma}(t)\boldsymbol{\nu} = \boldsymbol{f}_2(t) \quad \text{on } \Gamma_2, \tag{17}$$

$$\boldsymbol{\sigma}_\tau(t) = \boldsymbol{0} \quad \text{on } \Gamma_3, \tag{18}$$

for all $t \in \mathbb{R}_+$, and there exists $\xi : \Gamma_3 \times \mathbb{R}_+ \to \mathbb{R}$ which satisfies

$$\left.\begin{array}{l} u_\nu(t) \le g,\ \sigma_\nu(t) + p(u_\nu(t)) + \xi(t) \le 0, \\ (u_\nu(t) - g)\Big(\sigma_\nu(t) + p(u_\nu(t)) + \xi(t)\Big) = 0, \\ 0 \le \xi(t) \le F\Big(\displaystyle\int_0^t u_\nu^+(s)\,ds\Big), \\ \xi(t) = 0 \text{ if } u_\nu(t) < 0,\ \xi(t) = F\Big(\displaystyle\int_0^t u_\nu^+(s)\,ds\Big) \text{ if } u_\nu(t) > 0 \end{array}\right\} \text{ on } \Gamma_3, \tag{19}$$

for all $t \in \mathbb{R}_+$.

Here and below $\mathbb{S}^d$ represents the space of second order symmetric tensors on $\mathbb{R}^d$ and, in order to simplify the notation, we do not indicate explicitly the dependence of various functions on the spatial variable $\boldsymbol{x} = (x_i)$; the indices i, j, k, l run between 1 and d and the summation convention over repeated indices is used; an index that follows a comma represents the partial derivative with respect to the corresponding component of the spatial variable, e.g. $u_{i,j} = \partial u_i/\partial x_j$; $\boldsymbol{\varepsilon}$ represents the deformation operator given by $\boldsymbol{\varepsilon}(\boldsymbol{v}) = (\varepsilon_{ij}(\boldsymbol{v}))$, $\varepsilon_{ij}(\boldsymbol{v}) = \frac{1}{2}(v_{i,j} + v_{j,i})$ and Div is the divergence operator, i.e. $\operatorname{Div}\boldsymbol{\sigma} = (\sigma_{ij,j})$.

Equation (14) represents the viscoelastic constitutive law of the material, already used in a large number of works, see, for instance, the books [11, 13] and the references therein. Equation (15) is the equilibrium equation and we use it here since the process is assumed to be quasistatic. Conditions (16) and (17) are the displacement and traction boundary conditions, respectively, and condition (18) represents the frictionless condition.

We now provide some comments on condition (19) in which $g \ge 0$ is a given bound for the penetration, p represents a positive function which vanishes for a negative argument and u_ν, σ_ν represent the normal displacement and the normal stress, respectively. This condition was used in [1] in the case when F vanishes and in [15] in the case when F is given. There, various mechanical interpretations related to this condition were provided. Here we restrict ourselves to recall that condition (19) describes the following features of the contact: when there is separation between the body's surface and the foundation then the normal stress vanishes; the penetration arises only if the absolute value of the normal stress reaches the critical value F; when there is penetration the contact follows a normal compliance-type condition but only up to the bound g and then, when this limit is reached, the contact follows a Signorini-type unilateral condition with the gap g. Note that, in contrast with [15], in this paper we assume that the yield value F depends on the history of the penetration, represented by the integral term in (19); this dependence describes the hardening and the softening properties of the foundation, makes the contact problem more general, and leads to a new and interesting mathematical model.

We turn now to the variational formulation of Problem 1. To this end we use the notation "$\cdot$" and $\|\cdot\|$ for the inner product and the Euclidean norm on $\mathbb{R}^d$ and $\mathbb{S}^d$, respectively, as well as the standard notation for the Lebesgue and Sobolev spaces associated with Ω and Γ. Moreover, we consider the spaces

$$V = \{ \boldsymbol{v} = (v_i) \in H^1(\Omega)^d : \boldsymbol{v} = \boldsymbol{0} \text{ on } \Gamma_1 \},$$
$$Q = \{ \boldsymbol{\tau} = (\tau_{ij}) \in L^2(\Omega)^{d\times d} : \tau_{ij} = \tau_{ji} \}$$

which are real Hilbert spaces endowed with their canonical inner products and the associated norms $\|\cdot\|_V$ and $\|\cdot\|_Q$, respectively.

For an element $\boldsymbol{v} \in V$ we still write $\boldsymbol{v}$ for the trace of $\boldsymbol{v}$ on the boundary and we denote by v_ν and $\boldsymbol{v}_\tau$ the normal and tangential components of $\boldsymbol{v}$ on Γ, given by $v_\nu = \boldsymbol{v}\cdot\boldsymbol{\nu}$, $\boldsymbol{v}_\tau = \boldsymbol{v} - v_\nu\boldsymbol{\nu}$. We also consider the space $S = \{\boldsymbol{w} = \boldsymbol{v}|_{\Gamma_3} : \boldsymbol{v} \in V\}$, where $\boldsymbol{v}|_{\Gamma_3}$ denotes the restriction of the trace of the element $\boldsymbol{v} \in V$ to Γ_3. Thus, $S \subset H^{1/2}(\Gamma_3;\mathbb{R}^d)$ where $H^{1/2}(\Gamma_3;\mathbb{R}^d)$ is the space of the restrictions on Γ_3 of traces on Γ of functions of $H^1(\Omega)^d$. It is known that S can be organized as a Hilbert space, in a canonical way. The dual of the space S will be denoted by D and the duality paring between D and S will be denoted by $\langle\cdot,\cdot\rangle_{\Gamma_3}$. For simplicity, we shall write $\langle\boldsymbol{\mu},\boldsymbol{v}\rangle_{\Gamma_3}$ instead of $\langle\boldsymbol{\mu},\boldsymbol{v}|_{\Gamma_3}\rangle_{\Gamma_3}$ when $\boldsymbol{\mu} \in D$ and $\boldsymbol{v} \in V$.

For a regular function $\boldsymbol{\sigma} \in Q$ we use the notation σ_ν and $\boldsymbol{\sigma}_\tau$ for the normal and the tangential traces, i.e. $\sigma_\nu = (\boldsymbol{\sigma}\boldsymbol{\nu})\cdot\boldsymbol{\nu}$ and $\boldsymbol{\sigma}_\tau = \boldsymbol{\sigma}\boldsymbol{\nu} - \sigma_\nu\boldsymbol{\nu}$. Finally, we denote by $\mathbf{Q}_\infty$ the space of fourth order tensor fields given by

$$\mathbf{Q}_\infty = \{ \mathscr{E} = (\mathscr{E}_{ijkl}) : \mathscr{E}_{ijkl} = \mathscr{E}_{jikl} = \mathscr{E}_{klij} \in L^\infty(\Omega), \quad 1 \le i,j,k,l \le d \},$$

and we recall that $\mathbf{Q}_\infty$ is a real Banach space with its usual norm.

In the study of the mechanical problem (14)–(19) we assume that the viscosity operator $\mathscr{A}$ and the relaxation tensor $\mathscr{B}$ satisfy the following conditions.

$$\left\{\begin{array}{l}\text{(a) } \mathscr{A} : \Omega \times \mathbb{S}^d \to \mathbb{S}^d.\\ \text{(b) There exists } L_{\mathscr{A}} > 0 \text{ such that}\\ \qquad \|\mathscr{A}(\boldsymbol{x}, \boldsymbol{\varepsilon}_1) - \mathscr{A}(\boldsymbol{x}, \boldsymbol{\varepsilon}_2)\| \le L_{\mathscr{A}} \|\boldsymbol{\varepsilon}_1 - \boldsymbol{\varepsilon}_2\| \quad \forall\, \boldsymbol{\varepsilon}_1, \boldsymbol{\varepsilon}_2 \in \mathbb{S}^d, \text{ a.e. } \boldsymbol{x} \in \Omega.\\ \text{(c) There exists } m_{\mathscr{A}} > 0 \text{ such that}\\ \qquad (\mathscr{A}(\boldsymbol{x}, \boldsymbol{\varepsilon}_1) - \mathscr{A}(\boldsymbol{x}, \boldsymbol{\varepsilon}_2)) \cdot (\boldsymbol{\varepsilon}_1 - \boldsymbol{\varepsilon}_2) \ge m_{\mathscr{A}} \|\boldsymbol{\varepsilon}_1 - \boldsymbol{\varepsilon}_2\|^2\\ \qquad \quad \forall\, \boldsymbol{\varepsilon}_1, \boldsymbol{\varepsilon}_2 \in \mathbb{S}^d, \text{ a.e. } \boldsymbol{x} \in \Omega.\\ \text{(d) The mapping } \boldsymbol{x} \mapsto \mathscr{A}(\boldsymbol{x}, \boldsymbol{\varepsilon}) \text{ is measurable on } \Omega, \text{ for any } \boldsymbol{\varepsilon} \in \mathbb{S}^d.\\ \text{(e) The mapping } \boldsymbol{x} \mapsto \mathscr{A}(\boldsymbol{x}, \boldsymbol{0}) \text{ belongs to } Q.\end{array}\right. \tag{20}$$

$$\mathscr{B} \in C(\mathbb{R}_+; \mathbf{Q}_\infty). \tag{21}$$

The densities of body forces and surface tractions are such that

$$\boldsymbol{f}_0 \in C(\mathbb{R}_+; L^2(\Omega)^d), \quad \boldsymbol{f}_2 \in C(\mathbb{R}_+; L^2(\Gamma_2)^d). \tag{22}$$

The normal compliance function p and the surface yield function F satisfy

$$\left\{\begin{array}{l}\text{(a) } p : \Gamma_3 \times \mathbb{R} \to \mathbb{R}_+.\\ \text{(b) There exists } L_p > 0 \text{ such that}\\ \qquad |p(\boldsymbol{x}, r_1) - p(\boldsymbol{x}, r_2)| \le L_p\, |r_1 - r_2| \quad \forall\, r_1,\, r_2 \in \mathbb{R}, \text{ a.e. } \boldsymbol{x} \in \Gamma_3.\\ \text{(c) The mapping } \boldsymbol{x} \mapsto p(\boldsymbol{x}, r) \text{ is measurable on } \Gamma_3, \text{ for any } r \in \mathbb{R}.\\ \text{(d) } p(\boldsymbol{x}, r) = 0 \text{ for all } r \le 0, \text{ a.e. } \boldsymbol{x} \in \Gamma_3.\end{array}\right. \tag{23}$$

$$\left\{\begin{array}{l}\text{(a) } F : \mathbb{R}_+ \to \mathbb{R}_+.\\ \text{(b) There exists } L_F > 0 \text{ such that}\\ \qquad |F(r_1) - F(r_2)| \le L_F |r_1 - r_2| \quad \text{for all } r_1, r_2 \in \mathbb{R}_+.\end{array}\right. \tag{24}$$

Finally, we assume that

$$\text{there exists } \boldsymbol{\theta} \in V \text{ such that } \theta_\nu = 1 \text{ a.e. on } \Gamma_3. \tag{25}$$

Next, we define the sets $K \subset V$ and $\Lambda \subset D$, the bilinear form $b : V \times D \to \mathbb{R}$, the function $\boldsymbol{f} : \mathbb{R}_+ \to V$ and the Lagrange multiplier $\boldsymbol{\lambda} : \mathbb{R}_+ \to \Lambda$ by equalities

$$K = \{\, \boldsymbol{v} \in V \;:\; v_\nu \le 0 \quad \text{a.e. on } \Gamma_3 \,\},$$

$$\Lambda = \{\, \boldsymbol{\mu} \in D \;:\; \langle \boldsymbol{\mu}, \boldsymbol{v} \rangle_{\Gamma_3} \le 0 \quad \forall\, \boldsymbol{v} \in K \,\},$$

$$b(\boldsymbol{v}, \boldsymbol{\mu}) = \langle \boldsymbol{\mu}, \boldsymbol{v} \rangle_{\Gamma_3} \quad \forall\, \boldsymbol{v} \in V,\ \boldsymbol{\mu} \in D,$$

$$(\boldsymbol{f}(t), \boldsymbol{v})_V = \int_\Omega \boldsymbol{f}_0(t) \cdot \boldsymbol{v}\, dx + \int_{\Gamma_2} \boldsymbol{f}_2(t) \cdot \boldsymbol{v}\, da \quad \forall\, \boldsymbol{v} \in V,\ t \in \mathbb{R}_+,$$

$$\langle \boldsymbol{\lambda}(t), \boldsymbol{w} \rangle_{\Gamma_3} = -\int_{\Gamma_3} (\sigma_\nu(t) + p(u_\nu(t)) + \xi(t)) w_\nu \, da \quad \forall\, \boldsymbol{w} \in S,\ t \in \mathbb{R}_+.$$

Then, using standard arguments based on integration by part combined with assumption (25), we obtain the following variational formulation of Problem 1.

Problem 2. Find a displacement field $\boldsymbol{u} : \mathbb{R}_+ \to V$ and a Lagrange multiplier $\boldsymbol{\lambda} : \mathbb{R}_+ \to \Lambda$ such that

$$\begin{aligned}(\mathscr{A}\boldsymbol{\varepsilon}(\boldsymbol{u}(t)), \boldsymbol{\varepsilon}(\boldsymbol{v}) - \boldsymbol{\varepsilon}(\boldsymbol{u}(t)))_Q + (\int_0^t \mathscr{B}(t-s)\boldsymbol{\varepsilon}(\boldsymbol{u}(s))\, ds, \boldsymbol{\varepsilon}(\boldsymbol{v}) - \boldsymbol{\varepsilon}(\boldsymbol{u}(t)))_Q \\ +(p(u_\nu(t)), v_\nu - u_\nu(t)))_{L^2(\Gamma_3)} + (F\Big(\int_0^t u_\nu^+(s)\, ds\Big), v_\nu^+ - u_\nu^+(t))_{L^2(\Gamma_3)} \\ +b(\boldsymbol{v} - \boldsymbol{u}(t), \boldsymbol{\lambda}(t)) \geq (\boldsymbol{f}(t), \boldsymbol{v} - \boldsymbol{u}(t))_V \quad \forall\, \boldsymbol{v} \in V,\end{aligned} \tag{26}$$

$$b(\boldsymbol{u}(t), \boldsymbol{\mu} - \boldsymbol{\lambda}(t)) \leq b(g\boldsymbol{\theta}, \boldsymbol{\mu} - \boldsymbol{\lambda}(t)) \quad \forall\, \boldsymbol{\mu} \in \Lambda, \tag{27}$$

for all $t \in \mathbb{R}_+$.

In the study of Problem 2 we have the following existence result.

Theorem 3.2. *Assume* (20)–(25). *Then, there exists a couple of functions* $(\boldsymbol{u}, \boldsymbol{\lambda}) : \mathbb{R}_+ \to V \times \Lambda$, *unique in* $\boldsymbol{u}$, *such that* (26)–(27) *hold for all* $t \in \mathbb{R}_+$. *Moreover,* $\boldsymbol{u} \in C(\mathbb{R}_+; V)$.

Proof. We define the operators $A : V \to V$, $\mathscr{R} : C(\mathbb{R}_+; V) \to C(\mathbb{R}_+; Q \times L^2(\Gamma_3))$ and the functional $\varphi : (Q \times L^2(\Gamma_3)) \times V \to \mathbb{R}$ by equalities

$$(A\boldsymbol{u}, \boldsymbol{v})_V = (\mathscr{A}\boldsymbol{\varepsilon}(\boldsymbol{u}), \boldsymbol{\varepsilon}(\boldsymbol{v}))_Q + (p(u_\nu), v_\nu)_{L^2(\Gamma_3)} \qquad \forall\, \boldsymbol{u},\, \boldsymbol{v} \in V,$$

$$\mathscr{R}\boldsymbol{u}(t) = (\int_0^t \mathscr{B}(t-s)\boldsymbol{\varepsilon}(\boldsymbol{u}(s))\, ds, F\Big(\int_0^t u_\nu^+(s)\, ds\Big)) \ \forall\, \boldsymbol{u} \in C(\mathbb{R}_+; V),\ t \in \mathbb{R}_+,$$

$$\varphi((\boldsymbol{\sigma}, \xi), \boldsymbol{v}) = (\boldsymbol{\sigma}, \boldsymbol{\varepsilon}(\boldsymbol{v}))_Q + (\xi^+, v_\nu^+)_{L^2(\Gamma_3)} \quad \forall\, (\boldsymbol{\sigma}, \xi) \in Q \times L^2(\Gamma_3),\ \boldsymbol{v} \in V.$$

Then it is easy to see that the couple $(\boldsymbol{u}, \boldsymbol{\lambda})$ is a solution of Problem 2 if and only if

$$\begin{aligned}(A\boldsymbol{u}(t), \boldsymbol{v} - \boldsymbol{u}(t))_V + \varphi(\mathscr{R}\boldsymbol{u}(t), \boldsymbol{v}) - \varphi(\mathscr{R}\boldsymbol{u}(t), \boldsymbol{u}(t)) \\ +b(\boldsymbol{v} - \boldsymbol{u}(t), \boldsymbol{\lambda}(t)) \geq (\boldsymbol{f}(t), \boldsymbol{v} - \boldsymbol{u}(t))_V \quad \forall\, \boldsymbol{v} \in V,\end{aligned} \tag{28}$$

$$b(\boldsymbol{u}(t), \boldsymbol{\mu} - \boldsymbol{\lambda}(t)) \leq b(g\boldsymbol{\theta}, \boldsymbol{\mu} - \boldsymbol{\lambda}(t)) \quad \forall\, \boldsymbol{\mu} \in \Lambda, \tag{29}$$

for all $t \in \mathbb{R}_+$.

We now apply Theorem 2.1 to the system (28)–(29) with $X = V$, $Y = D$, $Z = Q \times L^2(\Gamma_3)$ and $h = g\boldsymbol{\theta}$. To this end, we use assumptions (20) and (23) and the Sobolev trace theorem to see that the operator A verifies condition (3). Moreover, assumptions (21) and (24) show that the operator $\mathscr{R}$ satisfies condition (4) and, obviously, the functional φ verifies (5). Next, as showed, e.g., in [9], the bilinear form $b(\cdot, \cdot)$ is continuous and satisfies the "inf-sup" condition. We conclude from here that condition (6) holds. Also, taking into account assumption (22) it follows that $\boldsymbol{f} \in C(\mathbb{R}_+, V)$. Finally, (25) implies (7) and, obviously, condition (8) holds, too. Theorem 3.2 is now a direct consequence of Theorem 2.1. □

Let $(\boldsymbol{u}, \boldsymbol{\lambda})$ be a solution to Problem 2 and let $\boldsymbol{\sigma} : \mathbb{R}_+ \to Q$ be defined by (14). Then, the couple $(\boldsymbol{u}, \boldsymbol{\sigma})$ is called a weak solution to Problem 1. We conclude from Theorem 3.2 that, under assumptions (20)–(25), Problem 1 has a least unique weak solution $(\boldsymbol{u}, \boldsymbol{\sigma})$. Moreover the solution satisfies $\boldsymbol{u} \in C(\mathbb{R}_+; V)$, $\boldsymbol{\sigma} \in C(\mathbb{R}_+; Q)$.

Acknowledgements This research was supported by the Marie Curie International Research Staff Exchange Scheme Fellowship within the 7th European Community Framework Programme under Grant Agreement No. 295118.

References

1. Barboteu, M., Matei, A., Sofonea, M.: Analysis of quasistatic viscoplastic contact problems with normal compliance. Q. J. Mech. Appl. Math. **65**, 555–579 (2012)
2. Ciurcea, R., Matei, A.: Solvability of a mixed variational problem. Ann. Univ. Craiova **36**, 105–111 (2009)
3. Ekeland, I., Temam, R.: Convex Analysis and Variational Problems. Classics in Applied Mathematics, vol. 28. SIAM, Philadelphia (1999)
4. Haslinger, J., Hlaváček, I., Nečas, J.: Numerical methods for unilateral problems in solid mechanics. In: Ciarlet, P.G., Lions, J.-L. (eds.) Handbook of Numerical Analysis, vol. IV, pp. 313–485. North-Holland, Amsterdam (1996)
5. Hlaváček, I., Haslinger, J., Nečas, J., Lovíšek, J.: Solution of Variational Inequalities in Mechanics. Springer, New York (1988)
6. Hild, P., Renard, Y.: A stabilized Lagrange multiplier method for the finite element approximation of contact problems in elastostatics. Numer. Math. **115**, 101–129 (2010)
7. Hüeber, S., Wohlmuth, B.: An optimal a priori error estimate for nonlinear multibody contact problems. SIAM J. Numer. Anal. **43**, 156–173 (2005)
8. Lions, J.-L., Glowinski, R., Trémolières, R.: Numerical Analysis of Variational Inequalities. North-Holland, Amsterdam (1981)
9. Matei, A., Ciurcea, R.: Contact problems for nonlinearly elastic materials: weak solvability involving dual Lagrange multipliers. ANZIAM J. **52**, 160–178 (2010)
10. Reddy, B.D.: Mixed variational inequalities arising in elastoplasticity. Nonlinear Anal. Theory Methods Appl. **19**, 1071–1089 (1992)
11. Shillor, M., Sofonea, M., Telega, J.J.: Models and Analysis of Quasistatic Contact. Lecture Notes in Physics, vol. 655. Springer, Berlin (2004)
12. Sofonea, M., Matei, A., History-dependent quasivariational inequalities arising in contact mechanics. Eur. J. Appl. Math. **22**, 471–491 (2011)

13. Sofonea, M., Matei, A.: Mathematical Models in Contact Mechanics. London Mathematical Society Lecture Note Series, vol. 398. Cambridge University Press, Cambridge (2012)
14. Sofonea, M., Avramescu, C., Matei, A.: A Fixed point result with applications in the study of viscoplastic frictionless contact problems. Commun. Pure Appl. Anal. **7**, 645–658 (2008)
15. Sofonea, M., Han, W., Barboteu, M.: Analysis of a viscoelastic contact problem with normal compliance and unilateral constraint. Comput. Methods Appl. Mech. Eng. **264**, 12–22 (2013)

A Canonical Duality Approach for the Solution of Affine Quasi-Variational Inequalities

Vittorio Latorre and Simone Sagratella

Abstract We apply a sequential dual canonical transformation on the global optimization problem resulting from the reformulation of the Karush–Kuhn–Tucker conditions of affine quasi-variational inequalities (QVIs) using the Fischer-Burmeister complementarity function. Canonical duality is generally able to provide conditions for a critical point of the dual formulation to be the corresponding point of a global optimum of the original problem. By studying the new dual formulation it is possible to obtain properties that are not evident from the original one and that can be useful to develop new methods for the solution of (not necessarily affine) QVIs. The resulting formulation is canonically dual to the original in the sense that there is no duality gap between critical points of the original problem and those of the dual one.

Keywords Canonical duality • Quasi-variational inequalities • Complementarity

1 Introduction

Quasi-variational inequalities (QVIs) are a powerful modeling tool capable of describing complex equilibrium situations that can appear in different fields as generalized Nash games, mechanics, economics, statistics, and so on (see, e.g., [1–5]). In spite of their modeling power, relatively few studies have been devoted to the numerical solution of finite-dimensional QVIs (see, e.g., [5–11]), in particular in the recent paper [12] a solution method for QVIs based on solving their Karush–Kuhn–Tucker (KKT) conditions is proposed.

In this first stage paper we consider only affine QVIs, however theory presented here can be generalized to broader classes of QVIs. We define a point-to-set mapping K as a parametric set of linear inequality constraints: $K(x) := \{y \in \mathbb{R}^n \mid Ay + Bx - c \leq \mathbf{0}_m\}$, where $A, B \in \mathbb{R}^{m \times n}$ and $c \in \mathbb{R}^m$. Let $D \in \mathbb{R}^{n \times n}$

V. Latorre (✉) • S. Sagratella
Department of Computer Control and Management Engineering, Sapienza University of Rome, via Ariosto 25, 00185 Roma, Italy
e-mail: latorre@dis.uniroma1.it; sagratella@dis.uniroma1.it

D. Gao et al. (eds.), *Advances in Global Optimization*, Springer Proceedings in Mathematics & Statistics 95, DOI 10.1007/978-3-319-08377-3_31

and $e \in \mathbb{R}^n$, the Affine Quasi-Variational Inequality (AQVI) (A, B, c, D, e) is the problem of finding a point $x^* \in K(x^*)$ such that

$$(Dx^* + e)^T (y - x^*) \geq 0, \qquad \forall y \in K(x^*), \tag{1}$$

holds. A particularly well-known and studied case occurs when $K(x)$ is actually independent of x, in this case, the AQVI becomes an Affine Variational Inequality. For this latter problem, which is more specific than that considered in this paper, an extensive theory exists, see, for example, [13, 14].

We say that a point $x \in \mathbb{R}^n$ satisfies the KKT conditions if multipliers $\lambda \in \mathbb{R}^m$ exist such that

$$Dx + e + A^T\lambda = \mathbf{0}_n, \quad \mathbf{0}_m \leq \lambda \perp Ax + Bx - c \leq \mathbf{0}_m. \tag{2}$$

These KKT conditions for AQVIs parallel the classical KKT conditions for AVIs, see [14], and it is quite easy to show the following result, whose proof we omit.

Theorem 1. *If a point x, together with a suitable vector $\lambda \in \mathbb{R}^m$ of multipliers, satisfies KKT system* (2)*, then x is a solution of AQVI (A, B, c, D, e). Vice versa, if x is a solution of AQVI (A, B, c, D, e), then multipliers $\lambda \in \mathbb{R}^m$ exist such that the pair (x, λ) satisfies KKT conditions* (2).

A *complementarity function* is a function $\phi : \mathbb{R}^2 \to \mathbb{R}$ such that $\phi(a, b) = 0$ if and only if $a \geq 0$, $b \geq 0$, and $ab = 0$. There exist many types of complementarity functions, but the two most prominent ones are the minimum-function $\phi_{\min}(a, b) := \min\{a, b\}$ and the Fischer–Burmeister function $\phi_{\text{FB}}(a, b) := \sqrt{a^2 + b^2} - (a + b)$. It is well known that ϕ_{FB} has many useful properties, in particular it is semismooth, see [14].

In this paper we use function ϕ_{FB} to reformulate KKT conditions (2). Therefore it is not difficult to see that we can find a solution of system (2) by computing a global solution, with value equal to $-\frac{1}{2}e^T e$, of the following optimization problem

$$(\mathscr{P}) : \quad \min_{x,\lambda} \left\{ P(x, \lambda) := W(x, \lambda) + \frac{1}{2}(x, \lambda)^T M(x, \lambda) - f^T(x, \lambda) \right\}, \tag{3}$$

where for all $i = 1, \ldots, m$: $W(x, \lambda) := \frac{1}{2}\sum_{i=1}^{n} \phi_{\text{FB}}(\lambda_i, -g_i(x))^2$, $g_i(x) := A_{i*}x + B_{i*}x - c_i$, $M := \begin{pmatrix} D^T \\ A \end{pmatrix} \begin{pmatrix} D & A^T \end{pmatrix}$ and $f := -\begin{pmatrix} D^T \\ A \end{pmatrix} e$.

It is easy to see that function W is nonconvex, furthermore, since we are interested only in global minima, problem (3) is very hard to solve. In fact it is well known that not all critical points of P are solutions of the AQVI, see, e.g., [14, 15].

Canonical duality theory, developed from nonconvex analysis and global optimization [16, 17], is a potentially powerful methodology, which has been used successfully for solving a large class of challenging problems in biology, engineering, sciences [18–20], and recently in network communications [21, 22] and radial basis neural networks [23].

In this paper we use canonical duality theory in order to define new equivalent formulations for problem ($\mathscr{P}$). In particular we define first and second level total complementarity functions that in a suitable subset have the properties that any of their stationary points is a solution of the AQVI.

We use the following notation: $(a, b) \in \mathbb{R}^{n_a+n_b}$ indicates the column vector comprised by vectors $a \in \mathbb{R}^{n_a}$ and $b \in \mathbb{R}^{n_b}$; $\mathbb{R}^n_+ \subset \mathbb{R}^n$ denotes the set of nonnegative numbers; $\mathbb{R}^n_{++} \subset \mathbb{R}^n$ is the set of positive numbers; $\mathrm{sta}\{f(x) : x \in \mathscr{X}\}$ denotes the set of stationary points of function f in $\mathscr{X}$; given a matrix $Q \in \mathbb{R}^{a\times b}$ we indicate with Q_{i*} its i-th row and with Q_{*i} its i-th column; $\mathrm{diag}(a)$ denotes the (square) diagonal matrix whose diagonal entries are the elements of the vector a; $\circ$ denotes the Hadamard (componentwise) product operator.

2 Dual Canonical Transformation

In this section we use a sequential dual canonical transformation in order to define total complementarity and dual functions.

First of all, we introduce a semismooth operator $\xi := \Lambda(x, \lambda) : \mathbb{R}^{n+m} \to \mathscr{E}_0 \equiv \mathbb{R}^m$, which is defined as

$$\xi_i = \Lambda_i(x, \lambda_i) := \sqrt{\lambda_i^2 + g_i(x)^2} - \lambda_i + g_i(x), \qquad i = 1, \dots, m, \tag{4}$$

Furthermore we introduce a convex function $V_0 : \mathscr{E}_0 \to \mathbb{R}$ (associated with ξ), that is defined as

$$V_0(\xi) := \frac{1}{2} \sum_{i=1}^{n} \xi_i^2. \tag{5}$$

It is easy to see that

$$W(x, \lambda) = V_0(\Lambda(x, \lambda)) = V_0(\xi). \tag{6}$$

Furthermore, we introduce a dual variable

$$\sigma := \nabla V_0(\xi) = \xi, \tag{7}$$

which is defined on the range $\mathscr{S}_0 \equiv \mathbb{R}^m$ of $\nabla V_0(\cdot)$. Since the (duality) mapping (7) is invertible, i.e. ξ can be expressed as a function of σ, then the function $V_0(\xi)$ is said to be a canonical function on $\mathscr{E}_0$, see [16].

In order to define the total complementarity function in both primal and dual variables (x, λ, σ) we use a Legendre transformation [16]. Specifically the Legendre conjugate $V_0^*(\sigma) : \mathscr{S}_0 \to \mathbb{R}$ is defined as $V_0^*(\sigma) := \text{sta}\{\xi^T\sigma - V_0(\xi) : \xi \in \mathscr{E}_0\}$, which is equal to the function $\xi^T\sigma - V_0(\xi)$ in which ξ is fixed to a stationary point. Since $\xi^T\sigma - V_0(\xi)$ is a quadratic strictly concave function in ξ, then it is easy to see that its (unique) stationary point is $\bar{\xi} = \sigma$, and then

$$V_0^*(\sigma) = \bar{\xi}^T\sigma - V_0(\bar{\xi}) = \sigma^T\sigma - V_0(\sigma) \overset{(5)}{=} \frac{1}{2}\sum_{i=1}^{n}\sigma_i^2, \tag{8}$$

moreover we obtain that

$$V_0(\xi) = \xi^T\sigma - V_0^*(\sigma). \tag{9}$$

Since $W(x,\lambda) \overset{(6)}{=} V_0(\xi) \overset{(9)}{=} \xi^T\sigma - V_0^*(\sigma) \overset{(4),(8)}{=} \sum_{i=1}^{m}\sigma_i\left[\sqrt{\lambda_i^2 + g_i(x)^2} - \lambda_i + g_i(x)\right]$ $-\frac{1}{2}\sum_{i=1}^{m}\sigma_i^2$, we obtain the first level total complementarity function:

$$\begin{aligned}\Xi_0(x,\lambda,\sigma) := &\sum_{i=1}^{m}\left[\sigma_i\left(\sqrt{\lambda_i^2 + g_i(x)^2} - c_i\right) - \frac{1}{2}\sigma_i^2\right]\\ &+\frac{1}{2}(x,\lambda)^T M(x,\lambda) - \bar{f}(\sigma)^T(x,\lambda),\end{aligned} \tag{10}$$

where $\bar{f}(\sigma) := f + (-(A^T + B^T)\sigma, \sigma)$. It is easy to see that the total complementarity function Ξ_0 is strictly concave in σ for all (x,λ). Moreover Ξ_0 is convex in (x,λ) (although nonsmooth but only semi-smooth) for all $\sigma \in \mathbb{R}_+^m$, since $M \succeq 0$ and each function $\sqrt{\lambda_i^2 + g_i(x)^2}$ is convex in (x,λ).

According to canonical duality theory, in order to define a dual formulation it is convenient to operate another dual transformation to obtain a total complementarity function which is quadratic in the primal variables (x,λ). Then we have to get rid of the square root terms in the first level complementarity function (10), for this reason we define m second level nonlinear convex operators $\epsilon_i := \Gamma(x,\lambda_i) : \mathbb{R}^{n+1} \to \mathbb{R}_+$, which are defined as $\epsilon_i := \lambda_i^2 + g_i(x)^2$, for all $i = 1,\dots,m$, and m concave second level dual functions $V_{1,i} : \mathbb{R}_+ \to \mathbb{R}_+$ (associated with ϵ_i), that are defined as $V_{1,i}(\epsilon_i) := \sqrt{\epsilon_i}$, for all $i = 1,\dots,m$. Now we introduce m second level dual variables $\tau_i := \frac{\partial V_{1,i}(\epsilon_i)}{\partial \epsilon_i} = \frac{1}{2\sqrt{\epsilon_i}}$, for all $i = 1,\dots,m$, that are defined on $\mathbb{R}_+$. We denote by $\mathscr{E}_1 \equiv \mathbb{R}_+^m$ and $\mathscr{S}_1 \equiv \mathbb{R}_+^m$ the space of definition of $\epsilon := (\epsilon_i)_{i=1}^m$ and $\tau := (\tau_i)_{i=1}^m$, respectively. The Legendre conjugate of each second level dual function is $V_{1,i}^*(\tau_i) := \text{sta}\{\epsilon_i\tau_i - V_{1,i}(\epsilon_i) : \epsilon_i \in \mathbb{R}_+\} = -\frac{1}{4\tau_i}$, for all $i = 1,\dots,m$, and finally by using the same procedure used in the first level, we obtain the second level total complementarity function:

$$\Xi_1(x,\lambda,\sigma,\tau) := \frac{1}{2}(x,\lambda)^T G(\sigma,\tau)(x,\lambda) - \tilde{f}(\sigma,\tau)^T(x,\lambda) + \sum_{i=1}^{m}\left(-\frac{1}{2}\sigma_i^2 - c_i\sigma_i + \frac{\sigma_i}{4\tau_i} + c_i^2\sigma_i\tau_i\right),$$

where $G(\sigma,\tau) := M + 2\begin{pmatrix}(A^T + B^T)\operatorname{diag}(\sigma\circ\tau)(A+B) & \mathbf{0}_{n,m}\\ \mathbf{0}_{m,n} & \operatorname{diag}(\sigma\circ\tau)\end{pmatrix}$ and $\tilde{f}(\sigma,\tau) := f + (2(A^T + B^T)(c\circ\sigma\circ\tau) - (A^T + B^T)\sigma, \sigma)$. Note that we have obtained a (second level) total complementarity function Ξ_1 which is quadratic in the primal variables (x,λ). Furthermore Ξ_1 is strictly concave in σ for all (x,λ,τ), it is convex in τ for all $(x,\lambda)\in\mathbb{R}^{n+m}$, $\sigma\in\mathbb{R}^m_+$ and, assuming $G(\sigma,\tau)\succeq 0$, it is convex in (x,λ).

The dual function can be formulated in the following way, see [16]: $P^d(\sigma,\tau) := \operatorname{sta}\left\{\Xi_1(x,\lambda,\sigma,\tau) : (x,\lambda)\in\mathbb{R}^{n+m}\right\}$. Supposing that G is nonsingular and imposing to satisfy the first order conditions of Ξ_1 with respect to (x,λ), we can express the primal variables in function of the dual ones: $(x,\lambda) = G(\sigma,\tau)^{-1}\tilde{f}(\sigma,\tau)$. And then finally, we can define the dual function

$$P^d(\sigma,\tau) = -\frac{1}{2}\tilde{f}(\sigma,\tau)^T G(\sigma,\tau)^{-1}\tilde{f}(\sigma,\tau) + \sum_{i=1}^{m}\left(-\frac{1}{2}\sigma_i^2 - c_i\sigma_i + \frac{\sigma_i}{4\tau_i} + c_i^2\sigma_i\tau_i\right).$$

3 Canonical Complementarity, Existence and Uniqueness

In this section we present some properties of the dual and the total complementarity functions defined in the previous section, along with their relations with the primal function.

The following theorem describes relations between critical points of the primal and the dual functions, the proof is omitted.

Theorem 2 (Primal Analytic Solution). *Let $(\bar{\sigma},\bar{\tau})$ be a critical point for P^d and suppose that $G(\bar{\sigma},\bar{\tau})$ is nonsingular and that $\bar{\sigma}_i \neq 0$ for all $i = 1,\ldots,m$, then*

$$(\bar{x},\bar{\lambda}) = G(\bar{\sigma},\bar{\tau})^{-1}\tilde{f}(\bar{\sigma},\bar{\tau}) \tag{11}$$

is a critical point for $P(x,\lambda)$.

Conversely, let $(\bar{x},\bar{\lambda})$ be a critical point for P and suppose that $\bar{\lambda}_i^2 + g_i(\bar{x})^2 \neq 0$ for all $i = 1,\ldots,m$, then the point $(\bar{\sigma},\bar{\tau})\in\mathbb{R}^{2m}$, where for all $i = 1,\ldots,m$

$$\bar{\sigma}_i = \sqrt{\bar{\lambda}_i^2 + g_i(\bar{x})^2} - \bar{\lambda}_i + g_i(\bar{x}), \qquad \bar{\tau}_i = \frac{1}{2\sqrt{\bar{\lambda}_i^2 + g_i(\bar{x})^2}}, \tag{12}$$

is a critical point for $P^d(\sigma,\tau)$ if $G(\bar{\sigma},\bar{\tau})$ is nonsingular.

In Theorem 2 the total complementarity functions Ξ_0 and Ξ_1 are not considered. However in the proof of Theorem 2 the first order optimality conditions of the first and the second level total complementarity functions are implicitly used in order to link primal and dual ones. The following theorem links first order conditions of Ξ_0 and Ξ_1 with those of P and P^d, the proof is omitted.

Theorem 3. *The following statements hold:*

1. *let $(\bar{x}, \bar{\lambda}, \bar{\sigma}, \bar{\tau})$ be a critical point for Ξ_1 then:*
   a. *if $\bar{\sigma}_i \neq 0$ for all $i = 1, \ldots, m$ then $(\bar{x}, \bar{\lambda}, \bar{\sigma})$ is a critical point for Ξ_0 and (12) holds;*
   b. *if $G(\bar{\sigma}, \bar{\tau})$ is nonsingular, then $(\bar{\sigma}, \bar{\tau})$ is a critical point for P^d and (11) holds;*
2. *let $(\bar{x}, \bar{\lambda}, \bar{\sigma})$ be a critical point for Ξ_0 then $(\bar{x}, \bar{\lambda})$ is a critical point for P.*

The following theorem defines the values of the functions in their critical points, proof omitted.

Theorem 4 (Complementarity Dual Principle). *Let $(\bar{x}, \bar{\lambda}, \bar{\sigma}, \bar{\tau})$ be a critical point for Ξ_1 such that $\bar{\sigma}_i \neq 0$ for all $i = 1, \ldots, m$ and $G(\bar{\sigma}, \bar{\tau})$ is nonsingular then*

$$P(\bar{x}, \bar{\lambda}) = \Xi_0(\bar{x}, \bar{\lambda}, \bar{\sigma}) = \Xi_1(\bar{x}, \bar{\lambda}, \bar{\sigma}, \bar{\tau}) = P^d(\bar{\sigma}, \bar{\tau}).$$

Theorem 4 shows that function P^d is canonically dual to P in the sense that the duality gap between their critical points is zero. It also shows that there is no gap between the critical points of Ξ_0 and Ξ_1 and those of the primal and the dual functions.

The following theorems are useful for realizing a solution method for QVIs. In fact in Theorem 5 a variable subset is defined in which any stationary point of Ξ_0 is a solution of the QVI. While Theorem 6 is a (weaker) extension for Ξ_1. For simplicity in Theorem 5 we suppose the strict complementarity of the primal solution, however it is not difficult to prove the theorem also without this assumption. Proofs of both theorems are omitted.

Theorem 5. *Suppose that (x^*, λ^*) exists such that $P(x^*, \lambda^*) = -\frac{1}{2}e^T e$ and $(\lambda_i^*)^2 + g_i(x^*)^2 \neq 0$ for all $i = 1, \ldots, m$ then*

1. *$(x^*, \lambda^*, \mathbf{0}_m)$ is a critical point for Ξ_0 and $\Xi_0(x^*, \lambda^*, \mathbf{0}_m) = -\frac{1}{2}e^T e$;*
2. *all points $(x, \lambda, \sigma) \in \mathbb{R}^n \times \mathbb{R}^m \times \{\mathbb{R}^m_+ \setminus \{\mathbf{0}_m\}\}$ are not critical for Ξ_0.*

Theorem 6. *Suppose that (x^*, λ^*) exists such that $P(x^*, \lambda^*) = -\frac{1}{2}e^T e$ and $(\lambda_i^*)^2 + g_i(x^*)^2 \neq 0$ for all $i = 1, \ldots, m$ then*

1. $(x^*, \lambda^*, \mathbf{0}_m, \tau^*)$ *is a critical point for* Ξ_1 *and* $\Xi_1(x^*, \lambda^*, \mathbf{0}_m, \tau^*) = -\frac{1}{2}e^T e$, *where* $\tau_i^* = \frac{1}{2\sqrt{(\lambda_i^*)^2 + g_i(x^*)^2}}$ *for all* $i = 1, \dots, m$;
2. *all points* $(x, \lambda, \sigma, \tau) \in \mathbb{R}^n \times \mathbb{R}^m \times \mathbb{R}^m_{++} \times \mathbb{R}^m_{++}$ *are not critical for* Ξ_1.

Theorems 5 and 6 suggest that an interior point method convergent to a stationary point of total complementarity functions Ξ_0 and Ξ_1 which moves only in the positive space of σ and τ can be an effective strategy for solving QVIs. This will be a subject of future research.

We want to remark also that Theorem 5 implies that all stationary points of P in the subset $(x, \lambda) \in \mathbb{R}^n \times \mathbb{R}^m$ where $\phi_{\text{FB}}(\lambda_i, -g_i(x)) \geq 0$, for all $i = 1, \dots, m$, are solutions of the AQVI. To the best of our knowledge this is a new result that can be useful for developing new solution methods for QVIs.

4 Example

Let us consider the AQVI (A, B, c, D, e) where $A = -1$, $B = -1$, $c = -2$, $d = 1$ and $e = 1$. It is easy to see that it has a unique solution $x^* = 1$. Let us write its KKT conditions:

$$x + 1 - \lambda = 0$$
$$0 \leq \lambda \perp -2x + 2 \leq 0,$$

whose unique solution is $(x^*, \lambda^*) = (1, 2)$. Then we can formulate the primal function and the first level total complementarity function:

$$\begin{aligned} P(x, \lambda) &= \frac{1}{2}\left(\sqrt{\lambda^2 + (-2x + 2)^2} - \lambda - 2x + 2\right)^2 \\ &\quad + \frac{1}{2}(x^2 - 2x\lambda + \lambda^2) + x - \lambda, \\ \Xi_0(x, \lambda, \sigma) &= \sigma\left(\sqrt{\lambda^2 + (-2x + 2)^2} - \lambda - 2x + 2\right) - \frac{1}{2}\sigma^2 \\ &\quad + \frac{1}{2}(x^2 - 2x\lambda + \lambda^2) + x - \lambda. \end{aligned}$$

Figure 1 shows that $(1, 2)$ is a stationary point for P and it also gives a picture of how point $(1, 2, 0)$ is a saddle point for Ξ_0. Moreover it holds that $P(1, 2) = \Xi_0(1, 2, 0) = -\frac{1}{2}$.

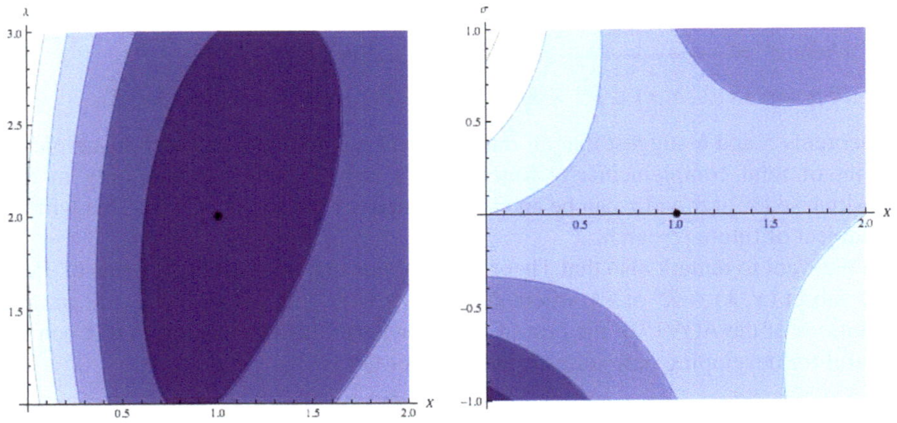

Fig. 1 Primal function plot (*left side figure*) and first level total complementarity function plot with λ fixed to 2 (*right side figure*)

References

1. Baiocchi, C., Capelo, A.: Variational and Quasivariational Inequalities: Applications to Free Boundary Problems. Wiley, New York (1984)
2. Mosco, U.: Implicit variational problems and quasi variational inequalities. In: Gossez, J., Lami Dozo, E., Mawhin, J., Waelbroeck, L. (eds.) Nonlinear Operators and the Calculus of Variations. Lecture Notes in Mathematics, vol. 543, pp. 83–156. Springer, Berlin (1976)
3. Outrata, J., Kocvara, M.: On a class of quasi-variational inequalities. Optim. Methods Softw. **5**, 275–295 (1995)
4. Outrata, J., Kocvara, M., Zowe, J.: Nonsmooth Approach to Optimization Problems with Equilibrium Constraints. Kluwer Academic Publishers, Dordrecht/Boston (1998)
5. Pang, J.-S., Fukushima, M.: Quasi-variational inequalities, generalized Nash equilibria, and multi-leader-follower games. Comput. Manag. Sci. **2**, 21–56 (2005) (Erratum: ibid 6, 373–375 (2009))
6. Chan, D., Pang, J.-S.: The generalized quasi-variational inequality problem. Math. Oper. Res. **7**, 211–222 (1982)
7. Fukushima, M.: A class of gap functions for quasi-variational inequality problems. J. Ind. Manag. Optim. **3**, 165–171 (2007)
8. Harms, N., Kanzow, C., Stein, O.: Smoothness Properties of a Regularized Gap Function for Quasi-Variational Inequalities. Preprint 313, Institute of Mathematics, University of Würzburg, Würzburg (March 2013)
9. Nesterov, Y., Scrimali, L.: Solving strongly monotone variational and quasi-variational inequalities. CORE Discussion Paper 2006/107, Catholic University of Louvain, Center for Operations Research and Econometrics (2006)
10. Noor, M.A.: On general quasi-variational inequalities. J. King Saud Univ. **24**, 81–88 (2012)
11. Ryazantseva, I.P.: First-order methods for certain quasi-variational inequalities in Hilbert space. Comput. Math. Math. Phys **47**, 183–190 (2007)
12. Facchinei, F., Kanzow, C., Sagratella, S.: Solving quasi-variational inequalities via their KKT conditions. Math. Prog. Ser. A (2013). doi:10.1007/s10107-013-0637-0
13. Cottle, R.W., Pang, J.-S., Stone, R.E.: The Linear Complementarity Problem. Academic, New York (1992)

14. Facchinei, F., Pang, J.-S.: Finite-Dimensional Variational Inequalities and Complementarity Problems, vols. I and II. Springer, New York (2003)
15. Dreves, A., Facchinei, F., Kanzow, C., Sagratella, S.: On the solution of the KKT conditions of generalized Nash equilibrium problems. SIAM J. Optim. **21**, 1082–1108 (2011)
16. Gao, D.Y.: Duality Principles in Nonconvex Systems: Theory, Methods and Applications. Kluwer Academic Publishers, Dordrecht (2000)
17. Gao, D.Y.: Canonical dual transformation method and generalized triality theory in nonsmooth global optimization. J. Glob. Optim. **17**(1/4), 127–160 (2000)
18. Gao, D.Y.: Canonical duality theory: theory, method, and applications in global optimization. Comput. Chem. **33**, 1964–1972 (2009)
19. Wang, Z.B., Fang, S.C., Gao, D.Y., Xing, W.X.: Canonical dual approach to solving the maximum cut problem. J. Glob. Optim. **54**, 341–352 (2012)
20. Zhang, J., Gao, D.Y., Yearwood, J.: A novel canonical dual computational approach for prion AGAAAAGA amyloid fibril molecular modeling. J. Theor. Biol. **284**, 149–157 (2011)
21. Gao, D.Y., Ruan, N., Pardalos, P.M.: Canonical dual solutions to sum of fourth-order polynomials minimization problems with applications to sensor network localization. In: Pardalos, P.M., Ye, Y.Y., Boginski, V., Commander, C. (eds.) Sensors: Theory, Algorithms and Applications. Springer, New York (2010)
22. Ruan, N., Gao, D.Y.: Global optimal solutions to a general sensor network localization problem. Perform. Eval. (2013, to appear). Published online at http://arxiv.org/submit/654731
23. Latorre, V., Gao, D.Y.: Canonical dual solution to nonconvex radial basis neural network optimization problem. Neurocomputing **134**, 189–197 (2013)

Numerical Analysis for a Class of Non Clamped Contact Problems

Oanh Chau

Abstract We study a class of dynamic thermal sub-differential contact problems with friction, for long memory viscoelastic materials, without the clamped condition, which can be put into a general model of system defined by a second order evolution inequality, coupled with a first order evolution equation. After statement of an existence and uniqueness result, we present a fully discrete scheme for numerical approximations and analysis of error order estimate.

Keywords Long memory thermo-visco-elasticity • Sub-differential contact condition • Non clamped condition • Dynamic process • Evolution inequality • Numerical analysis

1 Introduction

In the years 1970 until 1988, Duvaut and Lions, followed by Nečas and Hlaváček, Martins, Oden and Kikuchi, Panagiotopoulos, Ciarlet were the first to study contact problems for elastic or viscoelastic materials within the variational formulation framework, see [1–5]. Since multiple problems are still open in contact mechanics, which remains today a challenge and presents an active domain of research: for example the uniqueness of static Signorini Coulomb frictional problems and the solvability of dynamic problems for purely elastic bodies.

By taking into account the parameter of the temperature field, Figueiredo and Trabucho investigated thermoelastic and thermo-viscoelastic models, using Galerkin approximation method combined with a regularization and compactness technique, see [6]. Later quasi-static thermal contact problems were analyzed in [7], where the friction is described by a general normal damped response condition, there the existence and uniqueness of weak solution has been established. Further extensions to non convex contact conditions with non-monotone and possible multi-valued constitutive laws led to the domain of non-smooth thermo-viscoelastic frictional contact, within the framework of the so-called hemivariational inequalities, see, e.g., [8].

O. Chau (✉)
University of La Réunion, 97715 Saint-Denis Messag cedex 9, La Réunion, France
e-mail: oanh.chau@univ-reunion.fr

D. Gao et al. (eds.), *Advances in Global Optimization*, Springer Proceedings in Mathematics & Statistics 95, DOI 10.1007/978-3-319-08377-3_32

Here we are dealing with infinitesimal models with thermal effects in the framework of thermo-viscoelasticity. Infinitesimal frictional models are widely used in contact literature, see, e.g., the recent book [9]. These models are good approximations in the framework of linearized deformations, with some limitations on the impenetrability of mass condition, and on the appropriated conservation laws of thermodynamics, by neglecting some quadratic deformation term generated by heat.

This work is a companion paper of the results obtained in [10]. In [10] we studied a class of dynamic long memory viscoelastic thermal problems, without the usual clamped condition, where the contact is governed by a general sub-differential condition, putting then the problem into a coupled system, defined by a second order evolution inequality and a differential equation. The main difficulty is that Korn's inequality cannot be applied any more, which a-priori is leading to non coercive difficulties. For this proposal, following the technic already developed by [2] for Coulomb's friction models, we use the inertial term of the dynamic process to compensate the loss of coerciveness in the a priori estimates. Then using monotonicity, convexity, and fixed point methods, we prove an existence and uniqueness result, followed by some numerical simulations. However no numerical analysis in [10] was considered. Here, in order to complete the studies in the later reference, following the approximation methods developed in [11], we present a fully discrete scheme for the approximation of the solution fields, and we elaborate a general analysis of error estimates. In particular we prove the convergence of the numerical scheme with estimation of the speed of convergence, under additional regularity assumptions on the solution fields. The paper is organized as follows. In Sect. 2 we describe the mechanical problem and derive the variational formulation, then we state our main existence and uniqueness result. In Sect. 3, we introduce a fully discrete approximation scheme, and show an optimal order error estimate.

2 Main Existence and Uniqueness Result

The physical setting is as follows. A viscoelastic body occupies the domain Ω with surface Γ that is partitioned into two disjoint measurable parts, Γ_F and Γ_c. Let $[0, T]$ be the time interval of interest, where $T > 0$. We assume that a volume force of density $\mathbf{f}_0$ acts in $\Omega \times (0, T)$ and that surface tractions of density $\mathbf{f}_F$ act on $\Gamma_F \times (0, T)$. The body may come in contact with an obstacle, the foundation, over the potential contact surface Γ_c.

We use here the usual Lebesgue spaces $L^p(\Omega)$, $1 \le p \le +\infty$ and the Sobolev space $H^1(\Omega) = W^{1,2}(\Omega)$.

Let us introduce now some specific functional spaces, see details in [9, 11]:

$$H = \left(L^2(\Omega)\right)^d, \qquad \mathcal{H} = \{\boldsymbol{\sigma} = (\sigma_{ij}) \mid \sigma_{ij} = \sigma_{ji} \in L^2(\Omega),\ 1 \le i, j \le d\},$$

$$H_1 = \{\mathbf{u} \in H \mid \boldsymbol{\varepsilon}(\mathbf{u}) \in \mathcal{H}\}, \qquad \mathcal{H}_1 = \{\boldsymbol{\sigma} \in \mathcal{H} \mid \mathrm{Div}\, \boldsymbol{\sigma} \in H\}.$$

Here $\boldsymbol{\varepsilon} : H_1 \longrightarrow \mathscr{H}$ and Div : $\mathscr{H}_1 \longrightarrow H$ are the deformation and the divergence operators, respectively, defined by:

$$\boldsymbol{\varepsilon}(\mathbf{u}) = (\varepsilon_{ij}(\mathbf{u})), \quad \varepsilon_{ij}(\mathbf{u}) = \frac{1}{2}(u_{i,j} + u_{j,i}), \quad \mathrm{Div}\, \boldsymbol{\sigma} = (\sigma_{ij,j}).$$

The spaces H, $\mathscr{H}$, H_1 and $\mathscr{H}_1$ are real Hilbert spaces endowed with the canonical inner products.

Then the long memory thermo-viscoelastic problem with sub-differential contact condition can be formulated as follows.

Problem Q**:** Find a displacement field $\mathbf{u} : \Omega \times [0, T] \longrightarrow \mathbb{R}^d$ and a stress field $\boldsymbol{\sigma} : \Omega \times [0, T] \longrightarrow S_d$ and a temperature field $\theta : \Omega \times [0, T] \longrightarrow \mathbb{R}_+$ such that for a.e. $t \in (0, T)$:

$$\boldsymbol{\sigma}(t) = \mathscr{A}\boldsymbol{\varepsilon}(\dot{\mathbf{u}}(t)) + \mathscr{G}\boldsymbol{\varepsilon}(\mathbf{u}(t)) + \int_0^t \mathscr{B}(t-s)\,\boldsymbol{\varepsilon}(\mathbf{u}(s))\,ds - \theta(t)\,C_e \quad \text{in} \quad \Omega. \tag{1}$$

$$\ddot{\mathbf{u}}(t) = \mathrm{Div}\,\boldsymbol{\sigma}(t) + \mathbf{f}_0(t) \quad \text{in} \quad \Omega. \tag{2}$$

$$\boldsymbol{\sigma}(t)\nu = \mathbf{f}_F(t) \quad \text{on} \quad \Gamma_F. \tag{3}$$

$$\mathbf{u}(t) \in U, \quad \varphi(\mathbf{w}) - \varphi(\dot{\mathbf{u}}(t)) \geq -\boldsymbol{\sigma}(t)\boldsymbol{\nu} \cdot (\mathbf{w} - \dot{\mathbf{u}}(t)) \quad \forall \mathbf{w} \in U \quad \text{on } \Gamma_c. \tag{4}$$

$$\dot{\theta}(t) - \mathrm{div}(K_c\, \nabla\theta(t)) = -c_{ij}\,\frac{\partial \dot{u}_i}{\partial x_j}(t) + q(t) \quad \text{on} \quad \Omega. \tag{5}$$

$$-k_{ij}\,\frac{\partial \theta}{\partial x_j}(t)\,n_i = k_e\,(\theta(t) - \theta_R) \quad \text{on} \quad \Gamma_c. \tag{6}$$

$$\theta(t) = 0 \quad \text{on} \quad \Gamma_F. \tag{7}$$

$$\theta(0) = \theta_0 \quad \text{in} \quad \Omega. \tag{8}$$

$$\mathbf{u}(0) = \mathbf{u}_0, \quad \dot{\mathbf{u}}(0) = \mathbf{v}_0 \quad \text{in} \quad \Omega. \tag{9}$$

Here the general relation (4) is a sub-differential boundary condition, where $\mathscr{D}(\Omega)^d \subset U$ represents the set of contact admissible test functions, see examples in [10]. To derive the variational formulation of the mechanical problems (1)–(9) we need additional notations. Thus, let V denote the closed subspace of H_1 defined by

$$\mathscr{D}(\Omega)^d \subset V = H_1 \cap U.$$

$$E = \{\eta \in H^1(\Omega),\ \eta = 0 \quad \text{on} \quad \Gamma_F\}, \quad F = L^2(\Omega).$$

On V we consider the inner product given by

$$(\mathbf{u}, \mathbf{v})_V = (\boldsymbol{\varepsilon}(\mathbf{u}), \boldsymbol{\varepsilon}(\mathbf{v}))_{\mathscr{H}} + (\mathbf{u}, \mathbf{v})_H \qquad \forall\, \mathbf{u},\, \mathbf{v} \in V,$$

and let $\|\cdot\|_V$ be the associated norm, i.e.

$$\|\mathbf{v}\|_V^2 = \|\boldsymbol{\varepsilon}(\mathbf{v})\|_{\mathscr{H}}^2 + \|\mathbf{v}\|_H^2 \qquad \forall\, \mathbf{v} \in V.$$

The space E is endowed with the canonical inner product of $H^1(\Omega)$. We denote by E' the dual space of E.

In the study of the mechanical problem (1)–(9), we assume that the viscosity operator $\mathscr{A} : \Omega \times S_d \longrightarrow S_d$, $(\mathbf{x}, \boldsymbol{\tau}) \longmapsto (a_{ijkh}(\mathbf{x})\, \tau_{kh})$ is linear on the second variable and satisfies the usual properties of ellipticity and symmetry, i.e.

$$\left\{\begin{array}{l} \text{(i) } a_{ijkh} \in W^{1,\infty}(\Omega); \\ \text{(ii) } \mathscr{A}\boldsymbol{\sigma}\cdot\boldsymbol{\tau} = \boldsymbol{\sigma}\cdot\mathscr{A}\boldsymbol{\tau} \;\; \forall \boldsymbol{\sigma}, \boldsymbol{\tau} \in S_d, \text{ a.e. in } \Omega; \\ \text{(iii) there exists } m_{\mathscr{A}} > 0 \text{ such that} \\ \qquad \mathscr{A}\boldsymbol{\tau}\cdot\boldsymbol{\tau} \geq m_{\mathscr{A}}\, |\boldsymbol{\tau}|^2 \quad \forall \boldsymbol{\tau} \in S_d, \text{ a.e. in } \Omega. \end{array}\right. \tag{10}$$

The elasticity operator $\mathscr{G} : \Omega \times S_d \longrightarrow S_d$ satisfies:

$$\left\{\begin{array}{ll} \text{(i)} & \text{there exists } L_{\mathscr{G}} > 0 \text{ such that} \\ & |\mathscr{G}(\mathbf{x}, \boldsymbol{\varepsilon}_1) - \mathscr{G}(\mathbf{x}, \boldsymbol{\varepsilon}_2)| \leq L_{\mathscr{G}} |\boldsymbol{\varepsilon}_1 - \boldsymbol{\varepsilon}_2| \\ & \forall \boldsymbol{\varepsilon}_1, \boldsymbol{\varepsilon}_2 \in S_d,\ a.e.\ \mathbf{x} \in \Omega\,; \\ \text{(ii)} & \mathbf{x} \longmapsto \mathscr{G}(\mathbf{x}, \boldsymbol{\varepsilon}) \text{ is Lebesgue measurable on } \Omega, \forall \boldsymbol{\varepsilon} \in S_d\,; \\ \text{(iii)} & \text{the mapping } \mathbf{x} \longmapsto \mathscr{G}(\mathbf{x}, \mathbf{0}) \in \mathscr{H}. \end{array}\right. \tag{11}$$

The relaxation tensor $\mathscr{B} : [0,T]\times\Omega\times S_d \longrightarrow S_d, (t, \mathbf{x}, \boldsymbol{\tau}) \longmapsto (B_{ijkh}(t,\mathbf{x})\, \tau_{kh})$ satisfies

$$\left\{\begin{array}{l} \text{(i) } B_{ijkh} \in W^{1,\infty}(0,T; L^\infty(\Omega)); \\ \text{(ii) } \mathscr{B}(t)\boldsymbol{\sigma}\cdot\boldsymbol{\tau} = \boldsymbol{\sigma}\cdot\mathscr{B}(t)\boldsymbol{\tau} \\ \qquad \forall \boldsymbol{\sigma}, \boldsymbol{\tau} \in S_d, \text{ a.e. } t \in (0,T), \text{ a.e. in } \Omega. \end{array}\right. \tag{12}$$

We suppose the body forces and surface tractions satisfy

$$\mathbf{f}_0 \in W^{1,2}(0,T;H), \qquad \mathbf{f}_F \in W^{1,2}(0,T;L^2(\Gamma_F)^d). \tag{13}$$

For the thermal tensors and the heat sources density, we suppose that

$$C_e = (c_{ij}), \quad c_{ij} = c_{ji} \in L^\infty(\Omega), \quad q \in W^{1,2}(0,T;L^2(\Omega)). \tag{14}$$

The boundary thermal data satisfy

$$k_e \in L^\infty(\Omega;\mathbb{R}^+), \quad \theta_R \in W^{1,2}(0,T;L^2(\Gamma_c)). \tag{15}$$

The thermal conductivity tensor verifies the usual symmetry and ellipticity: for some $c_k > 0$ and for all $(\xi_i) \in \mathbb{R}^d$,

$$K_c = (k_{ij}), \quad k_{ij} = k_{ji} \in L^\infty(\Omega), \quad k_{ij}\, \xi_i \xi_j \geq c_k\, \xi_i \xi_i. \tag{16}$$

We assume that the initial data satisfy the conditions

$$\mathbf{u}_0 \in V, \quad \mathbf{v}_0 \in V \cap H_0^2(\Omega)^d, \quad \theta_0 \in E \cap H_0^2(\Omega). \tag{17}$$

On the contact surface, the following frictional contact function

$$\psi(\mathbf{w}) := \int_{\Gamma_c} \varphi(\mathbf{w})\, da$$

verifies

$$\begin{cases} \text{(i) } \psi : V \longrightarrow \mathbb{R} \quad \text{is well defined, continuous, and convex;} \\ \text{(ii) there exists a sequence of differentiable convex functions} \\ \qquad (\psi_n) : V \longrightarrow \mathbb{R} \quad \text{such that} \quad \forall \mathbf{w} \in L^2(0,T;V), \\ \qquad \int_0^T \psi_n(\mathbf{w}(t))\, dt \longrightarrow \int_0^T \psi(\mathbf{w}(t))\, dt,\ n \longrightarrow +\infty; \\ \text{(iii) for all sequence } (\mathbf{w}_n) \text{ and } \mathbf{w} \text{ in } W^{1,2}(0,T;V) \text{ such that} \\ \qquad \mathbf{w}_n \rightharpoonup \mathbf{w},\ \mathbf{w}'_n \rightharpoonup \mathbf{w}' \text{ weakly in } L^2(0,T;V); \\ \qquad \text{then } \liminf_{n\longrightarrow+\infty} \int_0^T \psi_n(\mathbf{w}_n(t))\, dt \geq \int_0^T \psi(\mathbf{w}(t))\, dt; \\ \text{(iv) } \forall \mathbf{w} \in V, \quad (\mathbf{w} = 0 \text{ on } \Gamma_c \Longrightarrow \forall n \in \mathbb{N},\ \psi'_n(\mathbf{w}) = 0_{V'}). \end{cases} \tag{18}$$

Here $\psi'_n(\mathbf{v})$ denotes the Fréchet derivative of ψ_n at $\mathbf{v}$.

To continue, using Green's formula and under some set of assumptions (see [10]), we obtain the variational formulation of the mechanical problem Q in abstract form as follows.

Problem QV**:** Find $\mathbf{u} : [0,T] \to V$, $\theta : [0,T] \to E$ satisfying a.e. $t \in (0,T)$:

$$\begin{cases} \langle \ddot{\mathbf{u}}(t) + A\,\dot{\mathbf{u}}(t) + B\,\mathbf{u}(t) + C\,\theta(t),\ \mathbf{w} - \dot{\mathbf{u}}(t)\rangle_{V'\times V} \\ +(\int_0^t B(t-s)\,\boldsymbol{\varepsilon}(\mathbf{u}(s))\, ds, \boldsymbol{\varepsilon}(\mathbf{w}) - \boldsymbol{\varepsilon}(\dot{\mathbf{u}}(t)))_{\mathcal{H}} + \psi(\mathbf{w}) - \psi(\dot{\mathbf{u}}(t)) \\ \qquad \geq \langle \mathbf{f}(t),\ \mathbf{w} - \dot{\mathbf{u}}(t)\rangle_{V'\times V} \quad \forall \mathbf{w} \in V; \\ \dot{\theta}(t) + K\,\theta(t) = R\dot{\mathbf{u}}(t) + Q(t) \qquad \text{in} \quad E'; \\ \mathbf{u}(0) = \mathbf{u}_0, \quad \dot{\mathbf{u}}(0) = \mathbf{v}_0, \quad \theta(0) = \theta_0. \end{cases}$$

Here, the operators and functions A, $B : V \longrightarrow V'$, $C : E \longrightarrow V'$, $\psi : V \longrightarrow \mathbb{R}$, $K : E \longrightarrow E'$, $R : V \longrightarrow E'$, $\mathbf{f} : [0,T] \longrightarrow V'$, and $Q : [0,T] \longrightarrow E'$ are defined by $\forall \mathbf{v} \in V$, $\forall \mathbf{w} \in V$, $\forall \tau \in E$, $\forall \eta \in E$:

$$\begin{cases} \langle A\,\mathbf{v},\mathbf{w}\rangle_{V'\times V} = (\mathscr{A}(\boldsymbol{\varepsilon}\mathbf{v}),\boldsymbol{\varepsilon}\mathbf{w})_{\mathscr{H}}; \\ \langle B\,\mathbf{v},\mathbf{w}\rangle_{V'\times V} = (\mathscr{G}(\boldsymbol{\varepsilon}\mathbf{v}),\boldsymbol{\varepsilon}\mathbf{w})_{\mathscr{H}}; \\ \langle C\tau,\mathbf{w}\rangle_{V'\times V} = -(\tau\, C_e,\,\boldsymbol{\varepsilon}\mathbf{w})_{\mathscr{H}}; \\ \psi(\mathbf{w}) := \int_{\Gamma_c}\varphi(\mathbf{w})\,da; \\ \langle \mathbf{f}(t),\mathbf{w}\rangle_{V'\times V} = (\mathbf{f}_0(t),\mathbf{w})_H + (\mathbf{f}_F(t),\mathbf{w})_{(L^2(\Gamma_F))^d}; \\ \langle Q(t),\eta\rangle_{E'\times E} = \int_{\Gamma_c} k_e\,\theta_R(t)\,\eta\,dx + \int_\Omega q(t)\,\eta\,dx; \\ \langle K\,\tau,\eta\rangle_{E'\times E} = \sum_{i,j=1}^d \int_\Omega k_{ij}\,\frac{\partial\tau}{\partial x_j}\,\frac{\partial\eta}{\partial x_i}\,dx + \int_{\Gamma_c} k_e\,\tau\cdot\eta\,da; \\ \langle R\,\mathbf{v},\eta\rangle_{E'\times E} = -\int_\Omega c_{ij}\,\frac{\partial v_i}{\partial x_j}\,\eta\,dx. \end{cases}$$

Then we obtain the main existence and uniqueness result stated as follows.

Theorem 2.1. *Assume that (10)–(18) hold, then there exists a unique solution* $\{\mathbf{u},\theta\}$ *to problem* QV *with the regularity:*

$$\begin{cases} \mathbf{u} \in W^{2,2}(0,T;V) \cap W^{2,\infty}(0,T;H); \\ \theta \in W^{1,2}(0,T;E) \cap W^{1,\infty}(0,T;F). \end{cases} \tag{19}$$

Proof. See details in [10]. □

3 Analysis of a Numerical Scheme

In this section, we study a fully discrete numerical approximation scheme of the variational problem QV. For this purpose, we suppose in the following that the conditions on the Result 1 are satisfied. In particular, we have

$$\mathbf{f} \in C([0,T];V'), \quad Q \in C([0,T];E').$$

Let $\{\mathbf{u},\theta\}$ be the unique solution of the problem QV, and introduce the velocity variable

$$\mathbf{v}(t) = \dot{\mathbf{u}}(t), \quad \forall t \in [0,T].$$

Then

$$\mathbf{u}(t) = \mathbf{u}_0 + \int_0^t \mathbf{v}(s)\,ds, \quad \forall t \in [0,T].$$

From Theorem 2.1 we see that $\{\mathbf{v},\theta\}$ verify for all $t \in [0,T]$:

$$\begin{aligned} &\langle \dot{\mathbf{v}}(t) + A\,\mathbf{v}(t) + B\,\mathbf{u}(t) + C\,\theta(t), \mathbf{w} - \mathbf{v}(t)\rangle_{V'\times V} \\ &+ (\textstyle\int_0^t B(t-s)\,\boldsymbol{\varepsilon}(\mathbf{u}(s))\,ds, \boldsymbol{\varepsilon}(\mathbf{w}) - \boldsymbol{\varepsilon}(\dot{\mathbf{u}}(t)))_{\mathscr{H}} + \psi(\mathbf{w}) - \psi(\mathbf{v}(t)) \\ &\quad \geq \langle \mathbf{f}(t), \mathbf{w} - \mathbf{v}(t)\rangle_{V'\times V}, \quad \forall\,\mathbf{w} \in V. \end{aligned} \tag{20}$$

$$\langle \dot{\theta}(t), \eta \rangle_F + \langle K\,\theta(t), \eta \rangle_{E' \times E} = \langle R\mathbf{v}(t), \eta \rangle_{E' \times E} + \langle Q(t), \eta \rangle_{E' \times E}, \quad \forall\, \eta \in E. \tag{21}$$

$$\mathbf{u}(0) = \mathbf{u}_0, \quad \mathbf{v}(0) = \mathbf{v}_0, \quad \theta(0) = \theta_0, \tag{22}$$

with the regularity:

$$\begin{cases} \mathbf{v} \in W^{1,2}(0, T; V) \cap W^{1,\infty}(0, T; H); \\ \theta \in W^{1,2}(0, T; E) \cap W^{1,\infty}(0, T; F). \end{cases} \tag{23}$$

To continue, we make the following additional assumptions on the solution and contact function:

$$\begin{cases} \mathbf{v} \in W^{2,2}(0, T; H); \\ \theta \in W^{2,2}(0, T; F); \\ \psi \text{ is Lipschitz continuous on } V. \end{cases} \tag{24}$$

Now let $V^h \subset V$ and $E^h \subset E$ be a family of finite dimensional subspaces, with $h > 0$ a discretization parameter. We divide the time interval $[0, T]$ into N equal parts: $t_n = n\,k$, $n = 0, 1, \ldots, N$, with the time step $k = T/N$. For a continuous function $\mathbf{w} \in C([0, T]; X)$ with values in a space X, we use the notation $\mathbf{w}_n = \mathbf{w}(t_n) \in X$.

Then from (20) to (22) we introduce the following fully discrete scheme.

Problem P^{hk}**.** Find $\mathbf{v}^{hk} = \{\mathbf{v}_n^{hk}\}_{n=0}^N \subset V^h$, $\theta^{hk} = \{\theta_n^{hk}\}_{n=0}^N \subset E^h$ such that

$$\mathbf{v}_0^{hk} = \mathbf{v}_0^h, \quad \theta_0^{hk} = \theta_0^h \tag{25}$$

and for $n = 1, \cdots, N$,

$$\begin{aligned} &\left(\tfrac{\mathbf{v}_n^{hk} - \mathbf{v}_{n-1}^{hk}}{k}, \mathbf{w}^h - \mathbf{v}_n^{hk}\right)_H + \langle A\,\mathbf{v}_n^{hk}, \mathbf{w}^h - \mathbf{v}_n^{hk}\rangle_{V' \times V} + \langle B\,\mathbf{u}_{n-1}^{hk}, \mathbf{w}^h - \mathbf{v}_n^{hk}\rangle_{V' \times V} \\ &+ \langle C\,\theta_{n-1}^{hk}, \mathbf{w}^h - \mathbf{v}_n^{hk}\rangle_{V' \times V} + \psi(\mathbf{w}^h) - \psi(\mathbf{v}_n^{hk}) \\ &+ (k \textstyle\sum_{m=0}^{n-1} \mathscr{B}(t_n - t_m)\, \boldsymbol{\varepsilon}(\mathbf{u}_m^{hk}), \boldsymbol{\varepsilon}(\mathbf{w}^h) - \boldsymbol{\varepsilon}(\mathbf{v}_n^{hk}))_{\mathscr{H}} \\ &\quad \geq \langle \mathbf{f}_n, \mathbf{w}^h - \mathbf{v}_n^{hk}\rangle_{V' \times V}, \quad \forall\, \mathbf{w}^h \in V^h. \end{aligned} \tag{26}$$

$$\begin{aligned} &\left(\tfrac{\theta_n^{hk} - \theta_{n-1}^{hk}}{k}, \eta^h\right)_F + \langle K\,\theta_n^{hk}, \eta^h\rangle_{E' \times E} \\ &\quad = \langle R\,\mathbf{v}_n^{hk}, \eta^h\rangle_{E' \times E} + \langle Q_n, \eta^h\rangle_{E' \times E}, \quad \forall\, \eta^h \in E^h. \end{aligned} \tag{27}$$

where

$$\mathbf{u}_n^{hk} = \mathbf{u}_{n-1}^{hk} + k\,\mathbf{v}_n^{hk}, \quad \mathbf{u}_0^{hk} = \mathbf{u}_0^h. \tag{28}$$

Here $\mathbf{u}_0^h \in V^h$, $\mathbf{v}_0^h \in V^h$, $\theta_0^h \in E^h$ are suitable approximations of the initial values $\mathbf{u}_0, \mathbf{v}_0, \theta_0$.

For $n = 1, \ldots, N$, suppose that $\mathbf{u}_{n-1}^{hk}$, $\mathbf{v}_{n-1}^{hk}$, θ_{n-1}^{hk} are known, then we calculate $\mathbf{v}_n^{hk}$ by (26), θ_n^{hk} by (27) and $\mathbf{u}_n^{hk}$ by (28). Hence the discrete solution $\mathbf{v}^{hk} \subset V^h$, $\theta^{hk} \subset E^h$ exists and is unique.

As a typical example, let us consider $\Omega \subset \mathbb{R}^d$, $d \in \mathbb{N}^*$, a polygonal domain. Let $\mathscr{T}^h$ be a regular finite element partition of Ω. Let $V^h \subset V$ and $E^h \subset E$ be the finite element space consisting of piecewise polynomials of degree $\leq \alpha$, with $\alpha \geq 1$, according to the partition $\mathscr{T}^h$.

Finally we assume now the following additional data and solution regularities:

$$\begin{cases} \mathbf{u}_0 \in H^{\alpha+1}(\Omega)^d; \\ \mathbf{v} \in C([0,T]; H^{2\alpha+1}(\Omega)^d), \quad \dot{\mathbf{v}} \in L^1(0,T; H^{\alpha}(\Omega)^d); \\ \theta \in C([0,T]; H^{\alpha+1}(\Omega)), \quad \dot{\theta} \in L^1(0,T; H^{\alpha}(\Omega)). \end{cases} \tag{29}$$

We now turn to an error analysis of the numerical solution. The main result of this section is the following.

Theorem 3.2. *We keep the assumptions of Theorem 2.1. Under the additional assumptions (24) and (29), we obtain the error estimate for the corresponding discrete solution:*

$$\begin{aligned} &\max\nolimits_{0\leq n\leq N} \|\mathbf{v}_n - \mathbf{v}_n^{hk}\|_H + \Big(k \textstyle\sum_{n=0}^N \|\mathbf{v}_n - \mathbf{v}_n^{hk}\|_V^2\Big)^{1/2} \\ &+ \max\nolimits_{0\leq n\leq N} \|\theta_n - \theta_n^{hk}\|_F + \Big(k \textstyle\sum_{n=0}^N \|\theta_n - \theta_n^{hk}\|_E^2\Big)^{1/2} \leq c\,(h^\alpha + k). \end{aligned}$$

In particular, for $\alpha = 1$, we have

$$\begin{aligned} &\max\nolimits_{0\leq n\leq N} \|\mathbf{v}_n - \mathbf{v}_n^{hk}\|_H + \Big(k \textstyle\sum_{n=0}^N \|\mathbf{v}_n - \mathbf{v}_n^{hk}\|_V^2\Big)^{1/2} \\ &+ \max\nolimits_{0\leq n\leq N} \|\theta_n - \theta_n^{hk}\|_F + \Big(k \textstyle\sum_{n=0}^N \|\theta_n - \theta_n^{hk}\|_E^2\Big)^{1/2} \leq c\,(h + k). \end{aligned}$$

Proof. Here we use methods based on technics and results developed in [11, 12], where we refer for details. □

These last results guaranty then the convergence of the numerical scheme under regularity assumptions.

References

1. Ciarlet, P.G.: Mathematical Elasticity. Three-Dimensional Elasticity, vol. 1. North-Holland, Amsterdam (1988)
2. Duvaut, G., Lions, J.L.: Les Inéquations en Mécanique et en Physique. Dunod, Paris (1972)
3. Kikuchi, N., Oden, J.T.: Contact Problems in Elasticity. SIAM, Philadelphia (1988)
4. Martins, J.A.C., Oden, J.T.: Existence and uniqueness results for dynamic contact problems with nonlinear normal and friction interface laws. Nonlinear Anal. **11**, 407–428 (1987)

5. Panagiotopoulos, P.D.: Inequality Problems in Mechanics and Applications. Birkhäuser, Basel (1985)
6. Figueiredo, I., Trabucho, L.: A class of contact and friction dynamic problems in thermoelasticity and in thermoviscoelasticity. Int. J. Eng. Sci. **33**, 45–66 (1995)
7. Awbi, B., Chau, O.: Quasistatic thermovisoelastic frictional contact problem with damped response. Appl. Anal. **83**(6), 635–648 (2004)
8. Denkowski, Z., Migórski, S.: A system of evolution hemivariational inequalities modeling thermoviscoelastic frictional contact. J. Nonlinear Anal. **60**, 1415–1441 (2005)
9. Sofonea, M., Matei, A.: Variational Inequalities with Applications: A Study of Antiplane Frictional Contact Problems. Advances in Mechanics and Mathematics, vol. 18. Springer, New York (2009)
10. Adly, S., Chau, O.: On some dynamical thermal non clamped contact problems. Math. Program. Ser. B (2013). doi:10.1007/s1010701306579
11. Han, W., Sofonea, M.: Quasistatic Contact Problems in Viscoelasticity and Viscoplasticity. American Mathematical Society and International Press, Providence (2002)
12. Ciarlet, P.G.: The Finite Element Method for Elliptic Problems. North-Holland, Amsterdam (1978)

Part VI
Numerical Optimization

A Newton-CG Augmented Lagrangian Method for Convex Quadratically Constrained Quadratic Semidefinite Programs

Xin-Yuan Zhao, Tao Cai, and Dachuan Xu

Abstract This paper presents a Newton-CG augmented Lagrangian method for solving convex quadratically constrained quadratic semidefinite programming (QCQSDP) problems. Based on the Robinson's CQ, the strong second order sufficient condition, and the constraint nondegeneracy conditions, we analyze the global convergence of the proposed method. For the inner problems, we prove the equivalence between the positive definiteness of the generalized Hessian of the objective functions in those inner problems and the constraint nondegeneracy of the corresponding dual problems, which guarantees the superlinear convergence of the inexact semismooth Newton-CG method to solve the inner problem. Numerical experiments show that the proposed method is very efficient to solve the large-scale convex QCQSDP problems.

Keywords Quadratically constrained quadratic semidefinite programs • Augmented Lagrangian • Semismoothness • Newton-CG method • Iterative solver

1 Introduction

Let $\mathbf{S}^n$ be the space of all $n \times n$ real symmetric matrices equipped with a scalar product $\langle \cdot, \cdot \rangle$ which is the usual Frobenius inner product in $\mathbf{S}^n$ and its induced norm $\| \cdot \|$ and $\mathbf{S}^n_+$ be the cone of all $n \times n$ symmetric positive semidefinite matrices. The notation $X \succeq 0$ means that X is a symmetric positive semidefinite matrix. The convex quadratically constrained quadratic semidefinite programming (QCQSDP) problem takes the form

$$(P) \quad \begin{array}{ll} \min & q_0(X) := \frac{1}{2}\langle X, \mathcal{A}_0(X)\rangle + \langle B_0, X\rangle \\ \text{s.t.} & q_i(X) := \frac{1}{2}\langle X, \mathcal{A}_i(X)\rangle + \langle B_i, X\rangle + c_i \geq 0, \ i = 1, \cdots, m \\ & X \succeq 0, \end{array} \tag{1}$$

X.-Y. Zhao • T. Cai • D. Xu (✉)
Department of Applied Mathematics, Beijing University of Technology,
100 Pingleyuan, Chaoyang District, Beijing 100124, People's Republic of China
e-mail: xyzhao@bjut.edu.cn; ct@emails.bjut.edu.cn; xudc@bjut.edu.cn

D. Gao et al. (eds.), *Advances in Global Optimization*, Springer Proceedings in Mathematics & Statistics 95, DOI 10.1007/978-3-319-08377-3_33

where $\mathcal{A}_0 : \mathbf{S}^n \longrightarrow \mathbf{S}^n$ is a self-adjoint positive semidefinite linear operator in $\mathbf{S}^n$, $\mathcal{A}_i : \mathbf{S}^n \rightarrow \mathbf{S}^n, i = 1, \cdots, m$, is a self-adjoint negative semidefinite linear operator in $\mathbf{S}^n$; $X, B_i \in \mathbf{S}^n, c_i \in \mathbf{R}$ is a scalar, let $q(X) := (q_1(X), \cdots, q_m(X))$, let the matrix representation of operator $\mathcal{A}_i$ under this transform be A_i, for any X, we have $\mathcal{A}_i(X) = A_i X A_i$ or $\mathcal{A}_i(X) = X$. $\mathcal{A}_i$ is self-adjoint positive (negative) semidefinite means that A_i is symmetric positive (negative) semidefinite matrix.

The Lagrangian dual form of the (P) problem is

$$(D) \quad \begin{array}{ll} \max & g_0(y, Z) := \inf\limits_{X \in \mathbf{S}^n} L_0(X, y, Z) \\ \text{s.t.} & y \geq 0, Z \succeq 0, \end{array} \tag{2}$$

where the Lagrangian function $L_0 : \mathbf{S}^n \times \mathbf{R}^m \times \mathbf{S}^n \rightarrow \mathbf{R}$ of (P) is defined as

$$L_0(X, y, Z) = q_0(X) - \langle y, q(X) \rangle - \langle Z, X \rangle.$$

Given a penalty parameter $\sigma > 0$, the augmented Lagrangian function for the convex QCQSDP problem (P) is defined as

$$\begin{aligned} Ł_\sigma(X, y, Z) = &\tfrac{1}{2}\langle X, \mathcal{A}_0(X) \rangle + \langle B_0, X \rangle \\ &+ \tfrac{1}{2\sigma}[\| \Pi_{\mathbf{R}^m_+ \times \mathbf{S}^n_+}[(y, Z) - \sigma(q(X), X)] \|^2 - \|(y, Z)\|^2]. \end{aligned} \tag{3}$$

where $(X, y, Z) \in \mathbf{S}^n \times \mathbf{R}^m \times \mathbf{S}^n$ and for any $(y, Z) \in \mathbf{S}^n \times \mathbf{R}^m$, $\Pi_{\mathbf{R}^m_+ \times \mathbf{S}^n_+}$ is the metric projection onto $\mathbf{R}^m_+ \times \mathbf{S}^n_+$ at (y, Z). For any $\sigma \geq 0$, $L_\sigma(X, y, Z)$ is convex in X and concave in $(y, Z) \in \mathbf{R}^m \times \mathbf{S}^n$.

For a given nondecreasing sequence of numbers $\{\sigma_k\}$,

$$0 < \sigma_1 \leq \sigma_2 \leq \cdots \leq \sigma_k \leq \cdots \leq \sigma_\infty \leq +\infty$$

and an initial multiplier $(y^0, Z^0) \in \mathbf{R}^m \times \mathbf{S}^n$, the augmented Lagrangian method for solving problem (P) and its dual (D) generates sequences $\{X^k\} \subset \mathbf{S}^n, \{y^k\} \subset \mathbf{R}^n$, and $\{Z^k\} \subset \mathbf{S}^n$ as follows:

$$\begin{cases} X^{k+1} \approx \arg\min\limits_{X \in \mathbf{S}^n} L_{\sigma_k}(X, y^k, Z^k), \\ y^{k+1} = \Pi_{\mathbf{R}^m_+}[y^k - \sigma_k q(X^k)], \\ Z^{k+1} = \Pi_{\mathbf{S}^n_+}[Z^k - \sigma_k X^k], \\ \sigma_{k+1} = \rho\sigma_k \text{ or } \sigma_{k+1} = \sigma_k. \end{cases} \tag{4}$$

For solving the problem (1), Sun and Zhang [1] proposed the modified alternate direction method with less computational effort. The numerical experiments in Zhao [2] showed that the semismooth Newton-CG augmented Lagrangian method was very efficient for large-scale linear and convex quadratic SDPs with the linear constraints, the numerical experiments showed that the proposed method is very efficient. It inspires us to extend the proposed method to solve the convex QCQSDP.

In order to achieve higher accuracy, we extend the Newton-CG augmented Lagrangian method to solve the convex QCQSDP problem (1) with superlinear convergence.

The organization of this paper is as follows. In Sect. 2.1, we show the strong second order sufficient condition and constraint nondegeneracy for the convex QCQSDP problems. In Sects. 2.2–2.3, a semismooth Newton-CG augmented lagrangian method was designed to solve the convex QCQSDP problems. Finally, we report numerical results of some QCQSDP problems solved by the proposed algorithm in Sect. 3.

2 A Newton-CG Augmented Lagrangian Method

2.1 Preliminaries

From [3, 4], the augmented Lagrangian method can be expressed in terms of the method of multipliers for (D), for the sake of subsequent developments, we introduce related conditions.

Assumption 3. *For the problem (P), there exists an $a \in R$ such that the level set $\{X \in \mathbf{S}^n \mid q_0(X) \leq a, X \in F(P)\}$ is nonempty and bounded.*

The first order optimality condition, namely the Karush–Kuhn–Tucker (KKT) condition of the problem (P) is

$$\begin{cases} \mathcal{A}(X) + B_0 - \sum\limits_{i=1}^{m} y_i(\mathcal{A}_i(X) + B_i) - Z = 0, \\ y \geq 0, q(X) \geq 0, \langle y, q(X)\rangle = 0, \\ X \succeq 0, Z \succeq 0, \langle X, Z\rangle = 0. \end{cases} \tag{5}$$

For any KKT triple $(X, y, Z) \in \mathbf{S}^n \times \mathbf{R}^m \times \mathbf{S}^n$ satisfying (5), we call $X \in \mathbf{S}^n$ a stationary point and (y, Z) a Lagrangian multiplier with respect to X. Let $\mathcal{M}(X)$ be the set of all Lagrangian multipliers at X. Assume that $\overline{X}$ is an optimal solution to (P), $\mathcal{M}(\overline{X})$ is nonempty and bounded since the following Robinson's constraint qualification holds at $\overline{X}$.

Assumption 4. *Let $\overline{X}$ be a feasible solution to the convex QCQSDP problem (P). Robinson's constraint qualification (CQ) [5] is said to hold at $\overline{X}$ if*

$$\begin{pmatrix} \mathcal{J}q(\overline{X}) \\ \mathcal{I} \end{pmatrix} \mathbf{S}^n + \begin{pmatrix} \mathbf{T}_{\mathbf{R}^m_+}(q(\overline{X})) \\ \mathbf{T}_{\mathbf{S}^n_+}(\overline{X}) \end{pmatrix} = \begin{pmatrix} \mathbf{R}^m \\ \mathbf{S}^n \end{pmatrix}, \tag{6}$$

or, equivalent to slater condition,

$$\exists \overline{X} \succ 0 \; s.t \; q(\overline{X}) \geq 0 \tag{7}$$

where $\mathbf{T}_K(s)$ is the tangent cone of K at s.

By the introduction of the constrain nondegeneracy for sensitivity and stability in optimization and variational inequalities in [5], we have the following formula for the problem (P).

Assumption 5. *Let $\overline{X}$ be a feasible solution to the convex CQCQSDP problem (P) and $(\overline{y}, \overline{Z}) \in \mathcal{M}(\overline{X})$. We say that the primal constrain nondegeneracy holds at $\overline{X}$ to the problem (P) if*

$$\begin{pmatrix} \mathcal{J}q(\overline{X}) \\ \mathcal{I} \end{pmatrix} \mathbf{S}^n + \begin{pmatrix} lin[\mathbf{T}_{\mathbf{R}^m_+}(q(\overline{X}))] \\ lin[\mathbf{T}_{\mathbf{S}^n_+}(\overline{X})] \end{pmatrix} = \begin{pmatrix} \mathbf{R}^m \\ \mathbf{S}^n \end{pmatrix}, \tag{8}$$

or, equivalently,

$$\mathcal{J}q(\overline{X})[lin[\mathbf{T}_{\mathbf{S}^n_+}(\overline{X})]] + lin[\mathbf{T}_{\mathbf{R}^m_+}(q(\overline{X}))] = \mathbf{R}^m \tag{9}$$

To discuss the rate of convergence, we introduce a strong form of the strong second order sufficient condition for nonlinear programming, which is an extension from nonlinear semidefinite programming introduced by Sun [6].

Assumption 6. *Let $\overline{X}$ be a feasible solution to the convex CQCQSDP problem (P) and $(\overline{y}, \overline{Z}) \in \mathcal{M}(\overline{X})$, if the primal constrain nondegeneracy (9) holds at $\overline{X}$, we say that the strong second order sufficient condition holds at $\overline{X}$ if*

$$\langle B, (\mathcal{A}_0 - \sum_{i=1}^{n} y_i \mathcal{A}_i)(B)\rangle - \Upsilon_{\overline{X}}(\overline{Z}, B) > 0, \forall B \in aff(\mathcal{C}(\overline{X}))\backslash\{0\}, \tag{10}$$

where $aff(\mathcal{C}(\overline{X}))$ denotes the affine hull of $\mathcal{C}(\overline{X})$, the critical cone $\mathcal{C}(\overline{X})$ of the problem (P) at $\overline{X}$ given by

$$\mathcal{C}(\overline{X}) = \{B \in \mathcal{S}^n | B \in \mathcal{T}_{\mathcal{S}^n_+}(\overline{X}), \mathcal{J}q(\overline{X})(B) \in \mathcal{T}_{\mathcal{R}^m_+}(q(\overline{X})), \langle \mathcal{J}q_0(\overline{X}), B\rangle = 0\}, \tag{11}$$

and the linear-quadratic function $\Upsilon_{\overline{X}} := X \times X \to \mathbf{R}$ is defined by

$$\Upsilon_{\overline{X}}(\overline{Z}, \overline{B}) := 2\langle \overline{Z}, BX^{\dagger}B\rangle, \tag{12}$$

where $X^{\dagger}$is the Moore–Penrose pseudo-inverse of X.

2.2 A Semismooth Newton-CG Method for Inner Problem

To apply the augmented Lagrangian method (4) to solve the problem (P) and problem (D), for some $(y, Z) \in \mathbf{R}^m \times \mathbf{S}^n$ and $\sigma > 0$, we need to consider the following form of the convex inner problem

$$\min\{\varphi(X) := L_\sigma(X, y, Z) \mid X \in \mathbf{S}^n\}. \tag{13}$$

For the existence of the optimal solution to inner problem (13), we need the following condition:

Assumption 7. *For the inner problem (13), there exists an* $\alpha^0 \in \mathcal{R}$ *such that the level sets* $\{X \in \mathcal{S}^n | \varphi(X) \leq \alpha^0\}$ *is nonempty and bounded.*

Under Assumption 7, from the introduction, we know that $\varphi(\cdot)$ is a continuously differentiable convex function, and for any $(X, y, Z) \in \mathbf{S}^n \times \mathcal{R}^m \times \mathbf{S}^n$,

$$\nabla\varphi(X) = \mathcal{A}_0(X) + B_0 - \mathcal{J}q(X)^*[\Pi_{\mathcal{R}^m_+}(y - \sigma q(X))] - \Pi_{\mathbf{S}^n_+}(Z - \sigma X). \quad (14)$$

But $\nabla\varphi(X)$ is not continuously differentiable because the $\Pi_{\mathcal{K}}(\cdot)$ is strongly semismooth [7], we can use the semismooth Newton-CG method [2] to solve the following nonlinear equation

$$\nabla\varphi(X) = 0, \quad \text{for any} \quad (y, Z) \in \mathcal{M}(X). \quad (15)$$

Choose $X^0 \in \mathcal{S}^n$. Then the algorithm can be stated as follows.

Algorithm 6. *[* $SNCG(X^0, y, Z, \sigma)$ *]*

Step 0. *Given* $\mu \in (1, 1/2)$, $\bar{\eta} \in (0, 1)$, $\tau \in (0, 1]$, $\tau_1, \tau_2 \in (0, 1)$ *and* $\delta \in (0, 1)$
Step 1. *For* $j = 0, 1, 2, \cdots$

Step 1.1. *Given a maximum number of CG iterations* $n_j > 0$, *compute*

$$\eta_j := min(\bar{\eta}, \|\nabla\varphi(X^j)\|^{1+\tau}).$$

Apply the practical CG Algorithm to find an approximation solution d^j *to*

$$\hat{V}(X^j + \varepsilon_j \mathcal{I})d = -\nabla\varphi(X^j), \quad (16)$$

where $\hat{V}(X^j) \in \hat{\partial}^2\varphi(X^j)$ *defined as the generalized Hessian of* φ *at* X, *and* $\varepsilon_j := \tau_1 min\{\tau_2, \|\nabla\varphi(X^j)\|\}$

Step 1.2. *Set* $\alpha_j = \delta^{m_j}$, *where* m_j *is the first nonnegative integer* m *for which*

$$(\hat{\varphi}(X^j + \alpha_j d^j) \leq \varphi(X^j) + \mu\alpha_j \langle\nabla\varphi(X^j), d^j\rangle. \quad (17)$$

Step 1.3. *Set* $X^{j+1} = X^j + \alpha_j d^j$.

Its convergence analysis can be conducted in a similar way to that in Qi and Sun [8]. We state these results in the next theorem, whose proof is omitted and deferred to the journal version.

Theorem 7. *Suppose that Assumption (7) holds for problem (13), then the SNCG algorithm is well defined and the generated iteration* $\{X^j\}$ *convergence to the unique optimal solution* X^* *of the problem (13), the rate of convergence for the SNCG algorithm is of order* $(1 + \tau)$.

2.3 A NAL Method for the Convex QCQSDPs

In this section, for any $k \geq 0$, let $\varphi_{\sigma_k}(\cdot) \equiv L_\sigma(\cdot, y^k, Z^k)$. Since the inner problems can be solved inexactly, we use the stopping criteria considered by Rockafellar [3,4] for terminating the SNCG algorithm. Consequently, the semismooth Newton-CG augmented Lagrangian algorithm will be introduced to solve the convex CQCQSDP problem (P) and (D).

Algorithm 8. *[NAL Algorithm]*

Step 0. *Given* $(X^0, y^0, Z^0) \in \mathbf{S}^n \times \mathbf{R}^m \times \mathbf{S}^n, \sigma_0 > 0$, *a threshold* $\hat{\sigma} \geq \sigma_0 > 0$ *and* $\rho > 1$.
Step 1. *For* $k = 0, 1, 2 \cdots$

Step 1.1. *Starting with* X^k *as the initial point, apply the SNCG algorithm to* $\varphi_{\sigma_k}(\cdot)$ *to find* $X^{k+1} = SNCG(X^k, y^k, Z^k, \sigma_k)$.
Step 1.2. *Updating* $y^{k+1} = \Pi_{\mathbf{R}^m_+}(y^k - \sigma_k q(X^{k+1}))$, $Z^{k+1} = \Pi_{\mathbf{S}^n_+}[Z^k - \sigma_k X^k]$,
Step 1.3. *If* $\sigma_k \leq \hat{\sigma}$, $\sigma_{k+1} = \rho\sigma_k$ *or* $\sigma_{k+1} = \sigma_k$.

Similarly to Rockafellar [3, 4], we can get the global convergence of the NAL algorithm.

3 Numerical Experiment

Applying the Newton-CG augmented Lagrangian method for solving the convex QCQSDP problems, we consider the following feasibility conditions of the primal and the dual problems:

$$R_P = \frac{\|q(X) - S\|}{\max\{1, \|c\|\}}, \quad R_D = \frac{\|\mathcal{A}_0(X) + B_0 - \sum_{i=1}^{m} \xi_i(\mathcal{A}_i(X) + B_i) - \zeta\|}{\max\{1, \|B_0\|\}},$$

where $S = (\Pi_{\mathcal{R}^m_+}(W) - W)/\sigma$, $W = \xi - \sigma q(X)$. The NAL algorithm termination criterion is

$$\max\{R_P, R_D\} \leq tol = 10^{-6}.$$

For the inner problem, the iterative steps up to 50; the practical CG algorithm for solving the Newton equation has the maximum number of iterations up to 500.

3.1 The Random Convex QCQSDPs

Consider the following random convex quadratically constraint quadratic programming problem

$$\begin{aligned} \min\ & q_0(X) := \frac{1}{2}\langle X, \mathcal{A}_0(X)\rangle + \langle B_0, X\rangle \\ \text{s.t.}\ & q_i(X) := \frac{1}{2}\langle X, \mathcal{A}_i(X)\rangle + \langle B_i, X\rangle + c_i \geq 0,\ i = 1, \cdots, m, \\ & X \succeq 0. \end{aligned} \tag{18}$$

Applying the NAL algorithm for the problem (18), we obtain the numerical result in Table 1.

3.2 Robust Estimation of Covariance Matrix Problem

The robust estimation of covariance matrix problem:

$$\begin{aligned} \min\ & q_0(X) := \frac{1}{2}\|X - C\|^2 \\ \text{s.t.}\ & \frac{1}{2}\|X - C\|^2 \leq \varepsilon, \\ & \langle A_i, X\rangle = b_i, i = 1, \cdots, p, \\ & \langle A_i, X\rangle \geq b_i, i = p+1, \cdots, q, \\ & X \succeq 0. \end{aligned} \tag{19}$$

We apply the NAL algorithm to solve the above robust estimation of covariance matrix problem and show the numerical result in Table 2.

Acknowledgements The research of the first author is supported by NSF of China (No. 11101016). The third author's research is supported by Scientific Research Common Program of Beijing Municipal Commission of Education (No. KM201210005033) NSF of China (No. 11371001), and China Scholarship Council.

Table 1 Numerical results for the random convex QCQSDPs

N	M	Iter	Itersub	Pcg	Pobj	Dobj	Rp	Rd	Gap	Time (s)
10	500	12	7.3	18.7	−8.86e+5	−8.86e+5	2.41e−12	5.13e−10	4.6611e−3	219.71
10	1,000	17	4.2	14.6	8.52e−12	8.52e−12	8.31e−12	1.68e−11	1.04e−5	407.45
100	50	17	12.5	18.4	−1.44e+6	−1.44e+6	2.39e−11	1.68e−10	4.40e−5	101.15
100	100	18	14.1	15.9	−0.04	−0.04	1.61e−10	5.90e−10	0.04	634.05
500	10	16	6.4	15.9	−2.84e+4	−2.84e+4	8.52e−12	5.81e−14	9.98e−5	128.96
500	50	15	8.9	17.6	−1.76e+4	−1.76e+4	4.69e−13	5.69e−13	1.75e−4	256.19

Table 2 Numerical results for robust estimation of covariance matrix

n	q	Iter	Itersub	Pcg	Pobj	Dobj	Rp	Rd	Gap	Time (s)
10	500	16	10.4	12.8	1.06e−4	1.06e−4	3.78e−8	3.78e−8	1.51e−4	89.71
100	100	18	17.5	14.6	9.19e−5	9.19e−5	3.01e−8	4.09e−8	9.19e−5	534.05
500	10	17	9.5	16.2	1.35e−4	1.35e−4	5.36e−8	3.39e−8	1.35e−4	89.82
500	100	16	8.5	16.8	1.51e−4	1.51e−4	6.36e−8	4.61e−8	1.51e−4	734.64

References

1. Sun, J., Zhang, S.: A modified alternating direction method for convex quadratically constrained quadratic semidefinite programs. Eur. J. Oper. Res. **207**, 1210–1220 (2010)
2. Zhao, X.Y.: A semismooth Newton-CG augmented Lagrangian method for large scale linear and convex quadratic semidefinite programming. Ph.D. thesis, National University of Singapore (2009)
3. Rockafellar, R.T.: Monotone operators and the proximal point algorithm. SIAM J. Control Optim. **14**, 877–898 (1976)
4. Rockafellar, R.T.: Augmented Lagrangians and applications of the proximal point algorithm in convex programming. Math. Oper. Res. **1**, 97–116 (1976)
5. Bonnans, J.F., Shapriro, A.: Perturbation Analysis of Optimization Problems. Springer, New York (2000)
6. Sun, D.F.: The strong second-order sufficient condition and constraint nondegeneracy in nonlinear semidefinite programming and their implications. Math. Oper. Res. **31**, 761–776 (2006)
7. Sun, D.F., Sun, J.: Semismooth matrix-valued functions. Math. Oper. Res. **27**, 150–169 (2002)
8. Qi, L.Q., Sun, J.: A nonsmooth version of Newton's method. Math. Program. **58**, 353–367 (1993)

A Novel Hybrid SP-QPSO Algorithm Using CVT for High Dimensional Problems

Ghazaleh Taherzadeh, Chu Kiong Loo, and Ling Teck Chaw

Abstract In this work, a novel hybrid population-based algorithm, named SP-QPSO has been introduced by combining Shuffled Complex Evolution with PCS (SP-UCI) and Quantum Particle Swarm Optimization (QPSO). The main purpose of this algorithm is to improve the efficiency of optimization task in both low and high dimensional problems. SP-QPSO is using the main strategy of SP-UCI by constructing complexes and monitoring their dimensionality, then evolving each complex based on QPSO. In this algorithm the initialization of point is done using Centroidal Voronoi Tessellations (CVT) to ensure that points visit the entire search space. Twelve popular benchmark functions are employed to evaluate the SP-QPSO performance in 2, 10, 50, 100, and 200 Dimensions. The results show that the proposed algorithm performed better in most functions.

Keywords Shuffled complex evolution with PCA (SP-UCI) • Quantum particle swarm optimization (QPSO) • Centroidal voronoi tessellations (CVT)

1 Introduction

Many population-based algorithms have been introduced and proposed during last few years. Shuffled Complex Evolution with PCA (SP-UCI) [1] known as the version of Shuffled Complex algorithm is applicable for high dimensional which is one of the population-based algorithms. In this algorithm populations are divided into several complexes. Modified competitive complex evolution (MCCE) search strategy is applied in each complex to find the optimum solution and the sort the complex based on the produced results. In each iteration, problem's dimensionality is checked and the lost dimensions are restored. Although MCCE method can handle the large number of simplex, but incrementing the number of simplex leads to having the high probability of convergence in local minima.

G. Taherzadeh (✉) • C.K. Loo • L.T. Chaw
Faculty of Computer Science and Information Technology, University of Malaya, Kuala Lumpur, Malaysia
e-mail: ghzl.thr@gmail.com; ckloo.um@um.edu.my

D. Gao et al. (eds.), *Advances in Global Optimization*, Springer Proceedings in Mathematics & Statistics 95, DOI 10.1007/978-3-319-08377-3_34

On the other hand, one of the famous population-based algorithms named PSO is used. PSO is inspired from the birds or fish social life. PSO was introduced by Eberhrt and Kennedy in 1995 [2]. However it also has premature convergence especially in complex search spaces. Throughout the years, some modifications have been done on standard PSO to avoid this problem and increase its proficiency such as [3–5]. QPSO is one of the variations of standard PSO which achieved better performance than the other versions [6]. QPSO is working with a set of random particles or agents as well as PSO. However QPSO is using a wave function instead of position and velocity in the search space.

Hybrid algorithms are offered to improve the efficiency of existing optimization algorithms. Large numbers of hybrid optimization algorithms are presented by researchers [7–9]. In this work, a new hybrid optimization algorithm named SP-QPSO for high dimensional problem has been introduced. Twelve benchmark functions in 2, 10, 50, 100, and 200 dimensions are used and the results are analyzed and compared with the results of SP-UCI and QPSO.

2 Hybrid SP-QPSO Algorithm

In this section, overview of QPSO and SP-UCI is discussed briefly and the proposed algorithm is explained in detail.

2.1 QPSO Overview

QPSO is introduced by Yang, Wang, and Jiao in 2004 to enhance the main disadvantage of standard PSO. In QPSO, selecting value of constriction factor and acceleration confidents leads the particles to have cyclic trajectory or global convergence. As QPSO follows standard PSO strategy, the agents are investigating around the feasible search space to find the optimum solution in each iteration. In each stage, particles keep following the best solution it has found so far. The best value of each particle is called *pbest* (Personal Best) and the best value among all the particles is known as global best or *gbest*. However in QPSO, the location of particle is changed using wave function W(x,t) in place of position and velocity. In QPSO, particles are initialized first and move toward the global best using below equations. The potential wells are constructed between two points by Eqs. (1), (2), and (3).

$$P_d = \varphi .\ p_{id} + (1 - \varphi)\ .\ p_{gd}, \quad 0 < \varphi < 1 \tag{1}$$

$$Q\left(||p - x||\right) = \frac{1}{L}\ e^{-2||p-x||/L} \tag{2}$$

$$D\left(\|p-x\|\right)=e^{-2\|p-x\|/L} \tag{3}$$

The evolution equation is based on Mont Carlo method using Eq. (4) and the mean of pbest position using Eq. (5).

$$x=P\pm\frac{L}{2}\ln\left(\frac{1}{u}\right)\qquad u=r\ and\quad(0,1) \tag{4}$$

$$mbest=\frac{1}{M}\sum_{i=1}^{M}p_{i,n}^{j}\quad(1\leq j\leq N) \tag{5}$$

Using Quantum manner helps PSO to fly away from the local optimum/ minimum.

2.2 *SP-UCI Overview*

In 2011, Chu [1] introduced a new optimization algorithm for high dimensional problems. The algorithm implemented based on shuffled complex evolution-University of Arizona (SCE-UA). According to the results, SCE-UA [10] performs well in state the fitness landscape with countless regional minimum, unidentified roughness, and discontinuities. Thus, SP-UCI has been proposed to check the dimension of variables in each iteration and the "population degeneration." SP-UCI guarantees that the population is able to explore the whole feasible space in each loop and entire evolution progress. SP-UCI method implies four concepts: (1) The complex shuffling scheme, (2) Population dimensionality monitoring and restoration, (3) MCCE strategy, (4) Multinomial resampling.

SP-UCI begins the optimization task with a population of points spread randomly in the search space and stored in an array $D=\{x_i, f_i,\quad i=1,\ \ldots, s\}$, which $i=1$ signifies the point with the smallest function value. After initialization of the points, population (array D) split into p complexes $A^1, \ldots, A^p$, each containing m points and are called simplex Eq. (6). Each simplex is required to undertake the process independently.

$$A^k=\left\{x_j^k, f_j^k \middle| x_j^k=x_{k+p(j-1)},\ f_j^k=f_{k+p(j-1)},\ j=1,\ldots,m\right\} \tag{6}$$

1	2	3					.....	N
78	63	91	13	49	66	54	.....	26

In each iteration, complexes are shuffled together and new complexes construct in a way that the information achieved by every individual complex is shared. In every stage, the results stored in D are sorted and replaced. The shuffling and constructing new complexes are replicated until meeting the end criteria.

2.3 Hybrid SP-QPSO Structure

According to [11], a number of hybridization techniques are offered to build a new hybrid algorithm. In this work, high-level hybridization is used as there is no similarity in both algorithm's functionality and their internal work. This hybrid algorithm of SP-UCI and QPSO proposed to improve the performance of optimization task in high dimensional problems and named SP-QPSO. In addition, the initialization of agents in the search space is accomplished by Centroidal Voronoi Tessellations (CVT) to ensure that agents are distributed over the feasible space. Refer to Qiang Du [12], this method is recognized as a way to partition the search space into sections. A generator is employed to produce a number of points in the space and after that, partitioning the space is taking place based on the individual point's adjacency to the generator [13]. In this work, CVT is used to generate the initial point for the first time which is exploited in SP-UCI part.

2.3.1 Hybrid SP-QPSO Description

Finding best fitness value is an essential task in each optimization algorithm. In this paper the strategy of dividing population into complexes and monitoring dimensionality of each agent is obtained from SP-UCI algorithm. In this hybrid algorithm each complex is consider as a population separately. In each complex, according to QPSO strategy, particles are looking for the best solution found so far and the global best for the whole population. Proposed algorithm is presented below in detail.

(1) Initializing p as number of complex, m as number of points in each complex and sample size of $s = pm$ and considered as $X_1, \ldots, X_s$ in the search space.
(2) Partitioning and initializing sample points over the feasible space using Halton step method of CVT and calculating the function value for each point.
(3) Sorting the points based on function value and store them in an array $D = \{X_i, f_i, i = 1, \ldots, s\}$.
(4) Dividing D into p complexes $A^1, \ldots, A^p$, consisting m points. $A^k = \{X_j^k, f_j^k | X_j^k = X_{k+p(j-1)}, f_j^k = f_{k+p(j-1)}, j = 1, \ldots, m\}$.
(5) Evolve each complex A^k separately using QPSO.
    (a) Initializing x as population size and itr maximum number of iteration.
    (b) Build a sub-swarm containing x points from the $Y_1{}^k, \ldots, Y_q{}^k$ in A^k based on their function value and store in $E^k = \{Y_i{}^k, v_i{}^k, i = 1, \ldots, q\}$ where Y_i^k is the particle and $v_i{}^k$ is representing function value. Like the strategy of QPSO, each particle has its own best visited position so far.
    (c) Finding the best value by comparing all the personal bests and assigning it to the global best.

(d) Calculating *mbest* and the movements of particle toward the global best for the next *itr.*
(e) Repeating mentioned process until meeting the stopping criteria.

(6) Applying Multinormal resampling to help search over the rough fitness landscapes.
(7) Shuffling the complex and replace the $A^1, \ldots, A^p$ into D and sort it.
(8) Reaping the process until stopping criteria satisfied.

This algorithm improved the results of high dimensional problems and also helps the particles to converge and search the whole feasible space. On the other hand, the CVT initialization assists to partition the search and distributes the points to make sure that search space is visited by the points.

3 Experimental Results and Analysis

In this section, to evaluate the performance of proposed hybrid algorithm, 12 benchmark functions are employed. The algorithm is tested in 2, 10, 50, 100, and 200 dimensions. Comparison is taking place with the results of QPSO and SP-UCI. Bench mark functions are named Ackley, Griewank, Quarticnoise, Rastrigin, Rosenbrock, Sphere, Step, Schwefel1_2, Schwefel2_21, Schwefel2_22, Penalized1, and Penalized2 [14, 15].

This evaluation is done based on minimization and is compared according to the function's mean and minimum value. Parameters such as QPSO population size, Number of Iteration in QPSO, Number of complex in SP-QPSO are initialized as: 20, 300, and 2 respectively. The other parameters are initialized as it was by default. The achieved results from SP-QPSO, SP-UCI, and QPSO are presented and compared in Table 1.

The results are averaged over 20 runs and the minimum results in each function are specified in bold type.

From the statistical point of view, SP-QPSO achieved minimum results as the number of dimensions increased. According to the result table, SP-QPSO obtained minimum result in Ackley, Griewank, Quarticnoise, Rastrigin, Sphere, and Penalized1 in both low and high dimensions. Whereas, in some functions, like Rosen brock, Schwefel1_2, Schwefel2_21, and Schwefel2_22, SP-QPSO reached to the minimum results as the number of dimensions increased. Among these 12 functions SP-QPSO could not perform well on the remaining functions.

To show the performance of SP-QPSO, for an instance, Quarticnoise function Comparison graph is illustrated. However, to have a clear view of graph, amount of data deduced from the achieved values by QPSO (Fig. 1).

The graph shows the performance comparison between three algorithms and demonstrates that the SP-QPSO performs better than the other two algorithms.

Table 1 The minimum and average function values of the QPSO, SP-UCI, and SP-QPSO (CVT)

			Ackley	Griewank	Quarticnoise	Rastrigin	Rosenbrock	Sphere	Step	Schwefel1_2	Schwefel2_21	Schwefel2_22	Penalized1	Penalized2
Dim = 2	QPSO	Mean	0.9892	0.0207	62.6029	0.0534	0.0018	41.7804	81	5.28E−27	7.68E−24	4.71E−31	2.36E−31	1.35E−32
		Min	**8.88E−16**	1.75E−07	58.532	0	8.77E−10	26.2144	81	**1.39E−58**	**3.77E−36**	**1.24E−37**	**2.36E−31**	**1.35E−32**
	SP-UCI	Mean	2.43E−05	8.20E−03	0.0497	2.89E−01	2.78E−12	2.27E−12	4.11E−12	6.68E−10	2.35E−05	2.72E−05	2.44E−05	2.45E−05
		Min	5.77E−06	1.14E−11	0.0179	2.63E−11	**4.81E−14**	1.02E−13	1.59E−13	3.84E−12	5.54E−06	4.27E−06	1.09E−05	5.53E−06
	SP-QPSO with CVT	Mean	2.68E−05	8.50E−03	0.0377	1.49E−01	0.1204	1.76E−11	3.91E−07	3.33E−05	5.50E−03	2.61E−07	3.59E−14	2.10E−08
		Min	**8.88E−16**	**0.00E+00**	**0.0165**	**0**	1.99E−09	**4.81E−33**	**0**	5.35E−21	9.16E−19	1.11E−24	**2.36E−31**	**1.35E−32**
Dim = 10	QPSO	Mean	0.0015	0.1327	2.3645	4.4441	18.4442	8.60E−08	5.27E−08	70.12105	0.244515	12.57062	1.47E−04	1.49E−03
		Min	1.27E−06	4.68E−02	1.6476	2.0042	4.8187	1.26E−11	3.53E−13	11.4227	0.0049	0.000193	4.65E−11	7.92E−11
	SP-UCI	Mean	1.16E−01	3.61E−02	2.0958	1.48E+01	0.399	1.53E−10	1.62E−09	7.14E−08	1.83E−02	1.22E−02	1.44E−09	4.99E−09
		Min	4.14E−05	3.78E−08	1.5457	2.12E−08	**3.78E−09**	2.10E−11	3.30E−10	8.60E−09	1.08E−03	4.43E−04	2.11E−10	6.81E−09
	SP-QPSO with CVT	Mean	4.44E−15	1.59E−02	1.4428	0	3.334	2.46E−70	0	1.07E−14	2.08E−15	1.13E−37	4.71E−32	1.35E−32
		Min	**4.44E−15**	**0.00E+00**	**1.2947**	**0**	2.76E+00	**3.35E−79**	**0**	**1.03E−16**	**1.54E−18**	**4.15E−43**	**4.71E−32**	**1.35E−32**
Dim = 50	QPSO	Mean	12.5423	25.6276	34.3123	73.5271	2.28E+04	8.7341	7.8183	6.51E+04	51.18791	1.14E+23	3.04E+03	2.74E+08
		Min	7.97E+00	1.43E+01	21.8024	52.0523	2080	2.6492	2.5011	5.00E+04	39.2352	8.67E+02	1170	3.08E+07
	SP-UCI	Mean	6.02E−01	2.30E−03	17.5997	0.6965	3.5945	8.40E−10	3.89E−08	6.18E−06	5.38E−04	4.27E−03	1.24E−02	4.03E−08
		Min	7.55E−05	2.27E−07	16.4872	7.90E−08	**7.33E−07**	6.26E−10	**2.45E−08**	**9.99E−07**	6.36E−04	3.87E−03	7.82E−10	3.73E−08
	SP-QPSO with CVT	Mean	4.44E−15	4.10E−03	15.2781	0	47.1277	5.02E−100	3.60E+02	4.32E−05	3.81E−06	2.34E−59	9.42E−33	1.35E−32
		Min	**4.44E−15**	**0.00E+00**	**14.2332**	**0**	4.69E+01	**7.03E−108**	348	1.58E−05	**9.36E−07**	**5.00E−64**	**9.42E−33**	**1.35E−32**
Dim = 100	QPSO	Mean	21.1535	2.7816	206.2966	495.679	402800	2.61E+03	202500	237350	96.73044	5.73E+89	5.45E+09	8.32E+09
		Min	2.11E+01	2.05E+00	116.6865	366.029	247000	2600	202500	186000	93.7467	4.19E+57	2.72E+09	3.69E+09
	SP-UCI	Mean	6.34E−01	2.60E−03	39.031	9.95E−02	50.2501	1.73E−09	1.64E−07	3.17E−02	1.45E−03	2.49E−02	7.80E−03	9.79E−03
		Min	8.75E−05	2.00E−07	36.2352	2.20E−07	**7.23E−06**	1.28E−09	**1.01E−07**	**2.78E−05**	1.24E−03	2.36E−02	1.09E−09	5.36E−08
	SP-QPSO with CVT	Mean	4.44E−15	7.40E−04	35.3738	0	97.0983	3.01E−115	8.37E+03	6.50E−03	1.60E−03	4.25E−74	4.71E−33	1.35E−32
		Min	**4.44E−15**	**0.00E+00**	**34.3087**	**0**	9.70E+01	**1.41E−121**	8.37E+03	3.30E−03	**4.78E−04**	**3.29E−76**	**4.71E−33**	**1.35E−32**
Dim = 200	QPSO	Mean	21.237	9.9229	2.19E+03	1.53E+03	2710000	5236	810000	969100	98.37613	2.93E+299	5.01E+10	5.97E+10
		Min	2.12E+01	9.11E+00	1.53E+03	1007	1650000	5220	810000	758000	97.2069	2.28E+190	3.74E+10	4.22E+10
	SP-UCI	Mean	9.00E−01	2.30E−03	83.1895	9.00E−07	344	4.35E−09	7.90E−07	2.72E+00	2.35E−01	2.88E−01	7.00E−03	5.49E−03
		Min	1.30E−04	3.72E−07	79.0395	7.17E−07	2.14E+02	3.00E−09	**8.40E−07**	2.72E+00	**2.25E−01**	3.45E−01	1.42E−09	**1.81E−07**
	SP-QPSO with CVT	Mean	4.44E−15	1.10E−03	77.395	0	196.9244	1.38E−134	8.37E+03	2.74E−02	4.55E+00	1.47E−92	2.36E−33	1.95E+01
		Min	**4.44E−15**	**0.00E+00**	**76.7474**	**0**	**1.97E+02**	**6.33E−137**	8.37E+03	**2.74E−02**	4.55E+00	**1.47E−92**	**2.36E−33**	1.95E+01

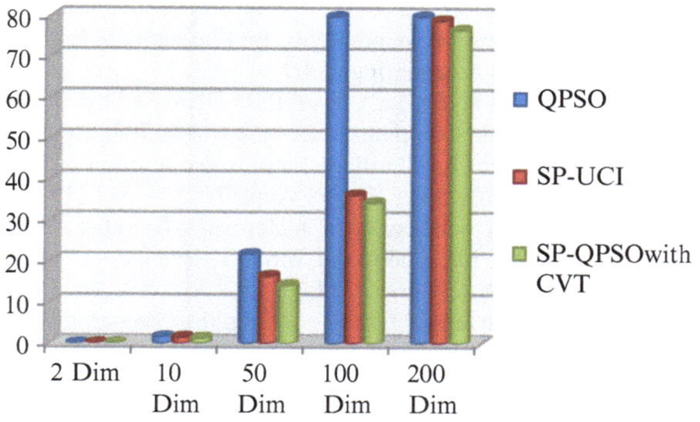

Fig. 1 Comparison of QPSO, SP-UCI, SP-QPSO on the Quarticnoise function in 2, 10, 50, 100, and 200 dimension

4 Conclusion

In this paper, a new hybrid optimization algorithm named SP-QPSO proposed and evaluated in both low dimension and high dimension problems. The main purpose of this algorithm is to apply the strategy of QPSO for each complex in SP-UCI algorithm. Evaluation is done using twelve benchmark functions in 2, 10, 50, 100, and 200. Results illustrate that the hybrid algorithm performed better in most functions specially for solving high dimensional problems.

Acknowledgments This study was funded by University of Malaya Research Grant (UMRG) project RG115-12ICT project title of Creative Learning for Emotional Expression of Robot Partners Using Interactive Particle Swarm Optimization.

References

1. Chu, W., Gao, X., Sorooshian, S.: A new evolutionary search strategy for global optimization of high-dimensional problems. Inf. Sci. **181**, 4909–4927 (2011)
2. Kennedy, J., Eberhart, R.: Particle swarm optimization. In: Proceedings of IEEE International Conference on Neural Networks, Piscataway, pp. 1942–1948. 1995
3. Kendall, G., Su, Y.: A particle swarm optimization approach in the construction of optimal risky portfolios. In: Proceedings of the 23rd IASTED International Multi-Conference Artificial Intelligence and Applications, Innsbruck, Austria, pp. 324–344. 2005
4. Zielinski, K., Laur, R.: Constrained single-objective optimization using particle swarm optimization. In: IEEE Congress on Evolutionary Computation, pp. 23–39. 2006
5. Kennedy, J.F., Eberhart, R.C., Shi, Y.: Swarm Intelligence. Morgan Kaufmann Publication, San Francisco (2001)
6. Chen, J., Yang, D., Feng, Z.: A novel quantum particle swarm optimizer with dynamic adaptation. J. Comput. Inf. Syst. **8**, 5203–5210 (2012)
7. Kaoa, Y., Zahara, E.: A hybrid genetic algorithm and particle swarm optimization for multimodal functions. Appl. Soft Comput. **8**, 849–857 (2008)

8. Xiao, Y., Song, X., Yao, Z.: Improved ant colony optimization with particle swarm optimization operator solving continuous optimization problems. In: International Conference on Computational Intelligence and Software Engineering, 2009
9. Li, L., Xue, B., Niu, B., Tan, L., Wang, J.: A novel PSO-DE-based hybrid algorithm for global optimization. In: Lecture Notes in Computer Science, pp. 785–793. Springer, Berlin, 2008
10. Duan, Q., Gupta, V., Sorooshian, S.: Shuffled complex evolution approach for effective and efficient global minimization. J. Optimiz. Theory Appl. **76**, 501–521 (1993)
11. Talbi, E.: A taxonomy of hybrid metaheuristics. J. Heuristics **8**, 541–546 (2002)
12. Du, Q., Faber, V., Gunzburger, M.: Centroidal voronoi tessellations: applications and algorithms. Soc. Ind. Appl. Math. **41**, 637–676 (1999)
13. Richards, M., Ventura, D.: Choosing a starting configuration for particle swarm optimization. In: IEEE International Joint Conference on Neural Networks, vol. 3, pp. 2309–2312. 2004
14. Dieterich, J.M., Hartke, B.: Empirical review of standard benchmark functions using evolutionary global optimization, (2012) arXiv preprint arXiv:1207.4318
15. http://www.rforge.net/doc/packages/hydroPSO/test_functions.html. Cited 20 Feb 2013

A Filter-Genetic Algorithm for Constrained Optimization Problems

Junjie Tang and Wei Wang

Abstract A filter-genetic method for constrained optimization problems is presented. It uses the filter technique instead of a fitness function to determine the merits of individuals. The method not only ensures the optimization of the offspring, but also avoids selecting the penalty parameter of a penalty function, which often leads to computational instability. And the numerical results are listed in the end.

Keywords Global optimization • Constrained problems • Genetic algorithms • Filter technique

1 Introduction

Genetic algorithm was first proposed by Professor Holland of University of Michigan in 1975. The main idea of the method is that it searches randomly and globally and makes an assessment of the fitness function for each individual. By using selection, crossover, and mutation, it evolves the solution of the problem until an optimal solution is found or the generation size is reached [1].

In genetic method, the fitness function is used to distinguish the pros and cons of individuals in the groups which directly affect the effectiveness of the algorithm. The traditional genetic algorithm is suitable for solving the constraint programming problem whose variable belongs to a box. It is difficult for it to deal with constrained optimization problems. Now the genetic algorithm for constrained nonlinear programming commonly uses the penalty function method as the fitness function. The main point of the penalty function method is merging the constraint function to the objective function in some form so that the original problem becomes an unconstrained problem. But due to the issue of penalty factor selection, it is often easy to cause premature convergence or random walk. The algorithms often need a lot of calculation but still may not get the correct results. So nowadays, a lot of

J. Tang • W. Wang (✉)
Department of Mathematics, East China university of science and technology, Shanghai, 200237, China
e-mail: tjjannnal@163.com; wangwei@ecust.edu.cn

D. Gao et al. (eds.), *Advances in Global Optimization*, Springer Proceedings in Mathematics & Statistics 95, DOI 10.1007/978-3-319-08377-3_35

researchers mainly consider combining the genetic algorithms with other algorithms like particle swarm optimization, tabu search algorithms, and so on [2–4].

Next let's make an overview of the filter method which is another practical technology for constrained programmings.

The filter method was first proposed by Fletcher in a plenary talk at the SIAM Optimization Conference in Victoria in May 1996 [5]. Fletcher and Leyffer introduced the concept and algorithms of filter in the literature [6]. This method considers two functions $f(x)$ and $g(x)$ as two competing objectives and denotes the filter sub-elements in a number of pairs (f, g).

Definition 1.1 we say the pair $(f(x_k),\ g(x_k))$ dominates the pair $(f(x_j), g(x_j))$ if and only if $f(x_k) \leq f(x_j)$ and $g(x_k) \leq g(x_j)$.

Definition 1.2 Filter is a set of pairs (f, g) such that no pair dominates another pair.

Sometimes we say that the point x_k is in the Filter, which means the pair $(f(x_k),$ $g(x_k))$ is in the Filter.

To prevent a point accepted by the filter being too close to each other and thus reducing the search speed, Fletcher added an envelope line around the current filter. Our algorithm also sets a "fully down" conditions. A new iteration is acceptable if and only if it satisfies:

$$f(x_k) \leq f(x_j) - \gamma g(x_j) \quad \text{or} \quad g(x_k) \leq \beta g(x_j)$$

for all $\forall (f_j, g_j) \in \mathcal{F}$, where $0 < \gamma < \beta < 1$ are constants.

Assuming that the current filter is $\mathcal{F}_\mathrm{k}$ and the newly generated point is x_k, if the pair $(f(x_k), g(x_k))$ cannot be dominated by any point in the filter, it is said that x_k can be accepted by the filter. Add the pair into the filter and remove the pairs dominated by x_k. The process of Filter update can be expressed as [7]: The set of pairs dominated by $(f(x_k), g(x_k))$ can be denoted as

$$D_k = \left\{(f_d, g_d) \,\middle|\, f_d \geq f_k, g_d \geq g_k, (f_d, g_d) \in \mathcal{F}_k\right\}$$

Then the updated Filter is

$$\mathcal{F}_{k+1} = \mathcal{F}_k \cup \{(f_k, g_k)\} \backslash D_k$$

2 Filter-Genetic Algorithm

Many optimization problems in engineering, management and some other fields of science can be classified as problems of extreme value with constraints.

Consider nonlinear constrained programming

$$\begin{cases} \min\ f(x) \\ \text{s.t. } c_j(x) \leq 0, \quad j \in E \end{cases} \tag{1}$$

Where $x \in R^n$ is the independent variable, $f: R^n \to R$ is the objective function, $c_j(x): R^n \to R$ is the constraint function. $E = \{1, 2, \ldots, p\}$ is the constraint indicators set. For problem (1), if there is a $x^* \in X$, where X is the feasible domain of problem (1) such that $f(x^*) \leq f(x)$ for any $x \in X$, then x^* is the global optimal solution of f(x) in the feasible domain of R^n, and $f(x^*)$ is the global optimum.

Many genetic algorithms for solving problem (1) are based on the penalty function used as a fitness function. We all know that it is easy to have computational instability because of the penalty function parameters.

In this paper, we combine the genetic algorithms and filter methods, i.e. using Filter methods' filterability to determine the merits of individuals. We consider the value of the objective function and the violation constrained function in the constrained optimization problems as two competing objectives. The Filter element (f, g) is defined as

$$f = f(x), \quad g = \sum_{i \in I} max\left(0, c_j(x)\right)$$

When $g(x_k) = 0$, x_k is in the feasible domain.

Before each generation, evaluate each individual in the parent population with filters. And add some individuals into filter which cannot be dominated by any others and remove the individuals dominated by the others. In order to avoid premature convergence and improve search efficiency, make all the individuals in the filter and some in the parent population form the new population (to ensure the population size is always the same). Use crossover and mutation on the new population to get the offspring. Considering that as long as $f(x_k)$ or $g(x_k)$ has a decline, x_k may be distinguished as a better point and be added into the filter. But only when $g(x_k) = 0$, x_k is in the feasible domain. So when the size of the filter set reaches a certain value, we need to perform feasibility restoration. It removes the individuals far away from the feasible domain boundaries to ensure the filter's determining reliability.

Based on the ideas above, we propose the following algorithm:

Step0 Generate the initial population $P(0) = \{x_1^0, x_2^0, \ldots, x_N^0\}$ randomly, population size N, Genetic algebra record variables $t = 0$, the largest genetic algebra G, the crossover operator p_c, the mutation operator p_m. Initialize the filter set as $\mathcal{F}$.

Step1 Extend the filter set. Check all the points in P(t). If the pair (f, g) to which the current point x_j^t (the jth individual of the tth generation) correspond is accepted by the filter, add x_j^t to filter set $\mathcal{F}$ and remove any individuals dominated by x_j^t;

Step2 Let m be the number of individuals in filter $\mathcal{F}$. If $m \leq N$, select randomly $(N - m)$ of individuals from $P(t)$ to form $P'(t)$, set $P(t) = \mathcal{F} + P'(t)$ and update $P(t)$, go to **Step3**; If $m > N$, go to **Step2.1** for the feasibility restoration;

Step2.1 Set a small enough $\varepsilon > 0$. Remove the individuals in the filter which $g > \varepsilon$ and update the filter set. Let n be the number of the rest of the individuals. If $n \leq N$, elect randomly $(N - n)$ of individuals from $P(t)$ to form $P'(t)$, set $P(t) = \mathcal{F} + P'(t)$ and update $P(t)$, go to **Step3**; If $n > N$, make an ascending sort of $\mathcal{F}$ by f which corresponds to the residual individual. Take the first N individuals and update the filer, set $P(t) = \mathcal{F}$, go to **Step3**;

Step3 Set all the individuals x_j^t in the $P(t)$ as the parents. Use the crossover operator p_c acting on each x_j^t to generate a new offspring. Then use the mutation operator p_m acting on the offspring to generate a new set of individuals O^t, set $P(t+1) = O^t$, go to **Step4**;

Step4 $t = t + 1$, if $t < G$, go to **Step1**; otherwise, go to **Step5**;

Step5 Check all the points in $\mathcal{F}$ to find point x^* which makes $g = 0$ and f minimum. Output x^* as the global optimal solution and $f(x^*)$ as the global optimum.

Because of the filterability and good stability of the filter, in actual process, the crossover probability should be larger; the mutation probability should be smaller.

3 Properties of the Algorithm

Assumption 3.1 The probability individual x^t in each generation population $P(t)$ mutates into any other individual y is not less than $\varepsilon(t)$, where $\varepsilon(t)$ is a constant greater than 0 and possibly related to t.

Definition 3.1 For any individual x, we say individual y is accessed from x by crossover and mutation [8], if the probability of x evolving to y by crossover and mutation operator is greater than 0.

Definition 3.2 For any individual x, we say individual y is ε accuracy accessed from x by crossover and mutation [8, 9], if the probability of x evolving to x' by crossover and mutation operator satisfying $\|y - x'\|_\infty \leq \varepsilon$ is greater than 0, where ε is any small positive number.

Definition 3.3 Filter set $\mathcal{F}_\mathrm{n}$ is called a constraint violations orderly filter set, if $\mathcal{F}_\mathrm{n}$ is sorted ascending by constraint violations g, where the filter set $\mathcal{F}_\mathrm{n}$ denotes the filter of *nth* generation population P(n).

Definition 3.4 For two constraint violations orderly filter sets $\mathcal{F}_n$, $\mathcal{F}_m$, we say $\mathcal{F}_n$ is better than $\mathcal{F}_m$, if the first individual $\mathcal{F}_n(1)$ in $\mathcal{F}_n$ can dominate the first individual $\mathcal{F}_m(1)$ in $\mathcal{F}_m$.

Theorem 3.1 *For two constraint violations orderly filter sets $\mathcal{F}_n$, $\mathcal{F}_m$, if $\mathcal{F}_n$ is better than $\mathcal{F}_m$, then $n \geq m$.*

Proof. Because $\mathcal{F}_n$ is better than $\mathcal{F}_m$, then $\mathcal{F}_n(1)$ can dominate $\mathcal{F}_m(1)$. Suppose $\mathcal{F}_n(1)$ is the pair (f_{n1}, g_{n1}), $\mathcal{F}_m(1)$ is the pair (f_{m1}, g_{m1}). According to the definition of the filter, the following formulas can be established:

$$\begin{cases} f_{n1} \leq f_{m1} \\ g_{n1} \leq g_{m1} \end{cases}.$$

□

Proof by Contradiction. If $n \geq m$ is not established, that is $n < m$. Consider the $(n+1)th$ generation. If some new filter pairs can dominate (f_{n1}, g_{n1}) when the filter set is updated, then there must be one pair that can dominate (f_{n1}, g_{n1}) and cannot be dominated by the others. Remove (f_{n1}, g_{n1}) and make the new pair located in the first place of $\mathcal{F}_{n+1}$, denoted as $(f_{(n+1)1}, g_{(n+1)1})$; If there are no pairs to dominate (f_{n1}, g_{n1}), then keep (f_{n1}, g_{n1}) and it will continue to be located in the first place of $\mathcal{F}_{n+1}$, denoted as $(f_{(n+1)1}, g_{(n+1)1}) = (f_{n1}, g_{n1})$; And so on, when generating to the $(m-1)th$ generation, suppose $(f_{(m-1)1}, g_{(m-1)1})$ is the retained filter pair after multiple updates. Because of the transitivity of filter domination, $(f_{(m-1)1}, g_{(m-1)1})$ can dominate (f_{n1}, g_{n1}) and also (f_{m1}, g_{m1}). Therefore for the mth generation, (f_{m1}, g_{m1}) cannot be accepted by the filter. This contradicts our assumption. Theorem is proved. □

Using the Theorem 3.1 above, we can easily get the following theorem:

Theorem 3.2 *Constraint violations orderly filter sets sequence* $\mathcal{F}_1, \mathcal{F}_2, \ldots, \mathcal{F}_n, \ldots$ *is monotonous, that is* $\forall\, n$, $\mathcal{F}_{n+1}$ *is not inferior to* $\mathcal{F}_n$.

Theorem 3.3 *For a constraint violations orderly filter sets sequence* $\{\mathcal{F}_n\}$ *and constants* $0 < \gamma < \beta < 1$, *we have* $\lim_{n\to\infty} g_{n1} = 0$.

Proof. Because the sequence $\{\mathcal{F}_n\}$ is monotonous, for $\forall\, n, m$, if $n < m$, then (f_{m1}, g_{m1}) can dominate or equal to (f_{n1}, g_{n1}). Because of the dominating condition of filter, if $g_{m1} \neq g_{n1}$, then $g_{m1} \leq \beta g_{n1}$.

Taking a sub-sequence $\{\mathcal{F}_{nk}\}$ from $\{\mathcal{F}_n\}$ where the first individual is different, and then we get $\lim_{k\to\infty} g_{n_k 1} \leq \lim_{k\to\infty} \beta^k g_{n_1 1}$. So we get $\lim_{n\to\infty} g_{n1} = 0$ because of $\beta < 1$. Theorem is proved.

Theorem 3.3 shows our algorithm can always find a feasible solution. Let's prove that our algorithm can always find the optimal feasible solution.

Definition 3.5 We call a population sequence P(1), P(2), ..., P(n) is monotonous for $\forall$ n, if the current optimal solution of P(n) is non-inferior to the current optimal solution of P(n − 1), or not worse than the current optimal solution of P(n − 1) at least. The so-called current optimal solution is the individuals in the current population which constraint violations closest to the boundary and has a smaller function value.

We now introduce the following lemma which has been proved by a lot of people [8, 10, 11].

Lemma 3.1 *The evolutionary algorithm is said to converge with probability 1 to the optimal solution set with ε accuracy, if the following conditions are satisfied:*

(1) *For any two points* x and x' *in the feasible domain,* x' *is ε accuracy up from* x *by crossover and mutation;*
(2) *Population sequence* P(1), P(2), ..., P(n) ... *is monotonous.*

Theorem 3.4 For a given $\varepsilon > 0$, our algorithm converges with probability 1 to the optimal solution set with ε accuracy.

Proof. According to Assumption 3.1, we know our algorithm satisfies the condition (1) of Lemma 3.1. In fact, our algorithm takes the non-uniform mutation operator, and condition (1) is clearly satisfied. Besides, according to Theorem 3.2, the constraint violations orderly filter sets sequence $\{\mathcal{F}_n\}$ is monotonous. It always guarantees $\lim_{k\to\infty} g_{n1} = 0$ due to Theorem 3.3. Then $\exists\,\mathbb{N}$, we have $g_{k1} = 0$ for any $\forall\, k > N$. Moreover, the first individual of the constraint violations orderly filter sets sequence from the *k-th* generation must be the feasible solution. Due to the Monotonous of $\mathrm{f_{k1}}$, (f_{k1}, g_{k1}) in each generation is the current optimal solution of the corresponding $P(k)$. At this time the corresponding $\{P(k)\}$ of the constraint violations orderly filter sets sequence $\{\mathcal{F}_k\}$ satisfies condition (2) of Lemma 3.1. So our algorithm converges with probability 1 to the optimal solution set with ε accuracy. □

4 Numerical Calculations

In this paper we selected several test functions, using the linear crossover operator and non-uniform mutation operator. The following results are obtained by MATLAB programming:

Example 4.1 [9, 12]

$$\min f(x) = \left(x_1^2 + x_2 - 11\right)^2 + \left(x_1 + x_2^2 - 7\right)^2$$
$$\text{s.t.} \begin{cases} 4.84 - x_1^2 - (x_2 - 2.5)^2 \le 0 \\ (x_1 - 0.05)^2 + (x_2 - 2.5)^2 - 4.84 \le 0, \\ 0 \le x_1, x_2 \le 6 \end{cases}$$

Take the population size of 20 and the generation size of 300, mutation operator $\mathrm{p_m} = 0.05$ and boundary parameters $\beta = 0.7$, $\gamma = 0.2$, the global minimum is at $x^* = (2.24683, 2.38191)$ where $f(x^*) = 13.59084$.

Example 4.2 [9, 13]

$$\min f(x) = (x_1 - 10)^3 + (x_2 - 20)^3$$
$$\text{s.t.} \begin{cases} 100 - (x_1 - 5)^2 - (x_2 - 5)^2 \le 0 \\ (x_1 - 6)^2 + (x_2 - 5)^2 - 82.81 \le 0, \\ 13 \le x_1 \le 100, 0 \le x_2 \le 100 \end{cases}$$

Take the population size of 20 and the generation size of 300, mutation operator $p_m = 0.01$ and boundary parameters $\beta = 0.7$, $\gamma = 0.2$, the global minimum is at $x^* = (14.09500, 0.84296)$ where $f(x^*) = -6961.81397$.

Example 4.3 [12, 13]

$$\min\ f(x) = (x_1-10)^2 + 5(x_2-12)^2 + x_3^4 + 3(x_4-11)^2 + 10x_5^6 + 7x_6^2 + x_7^4 - 4x_6x_7 - 10x_6 - 8x_7$$

$$\text{s.t.} \begin{cases} -127 + 2x_1^2 + 3x_2^4 + x_3 + 4x_4^2 + 5x_5 \le 0 \\ -282 + 7x_1 + 3x_2 + 10x_3^2 + x_4 - x_5 \le 0 \\ -196 + 23x_1 + x_2^2 + 6x_6^2 - 8x_7 \le 0 \\ 4x_1^2 + x_2^2 - 3x_1x_2 + 2x_3^2 + 5x_6 - 11x_7 \le 0 \\ -10 \le x_i \le 10, i = 1, \cdots, 7 \end{cases}$$

Take the population size of 70 and the generation size of 3,000, mutation operator $p_m = 0.01$ and boundary parameters $\beta = 0.7$, $\gamma = 0.2$, the global minimum is at $x^* = (2.32445, 1.95024, -0.46924, 4.36911, -0.61834, 1.05586, 1.594152)$ where $f(x^*) = 680.63361$。

Example 4.4 [12, 13]

$$\min\ f(x) = 5\sum_{i=1}^{4} x_i - 5\sum_{i=1}^{4} x_i^2 - \sum_{i=5}^{13} x_i$$

$$\text{s.t.} \begin{cases} 2x_1 + 2x_2 + x_{10} + x_{11} \le 10 \\ 2x_1 + 2x_3 + x_{10} + x_{12} \le 10 \\ 2x_2 + 2x_3 + x_{11} + x_{12} \le 10 \\ -8x_1 + x_{10} \le 0 \\ -8x_2 + x_{11} \le 0 \\ -8x_3 + x_{12} \le 0 \\ -2x_4 - x_5 + x_{10} \le 0 \\ -2x_6 - x_7 + x_{11} \le 0 \\ -2x_8 - x_9 + x_{12} \le 0 \\ 0 \le x_i \le 1, i = 1, \cdots, 9 \\ 0 \le x_j \le 100, j = 10, 11, 12 \\ 10 \le x_{13} \le 1 \end{cases}$$

Take the population size of 130 and the generation size of 5,000, mutation operator $p_m = 0.01$ and boundary parameters $\beta = 0.7$, $\gamma = 0.2$, the global minimum is at $x^* = (1.00000, 1.00000, 1.00000, 1.00000, 1.00000, 1.00000, 1.00000, 1.00000, 1.00000, 3.00000, 3.00000, 3.00000, 1.00000)$ where $f(x^*) = -15.00000$

Acknowledgment This research was supported by the National Natural Science Foundation of China (No: 11271128).

References

1. Holland, J.H.: Adaptation in Natural and Artificial Systems: An Introductory Analysis with Applications to Biology, Control, and Artificial Intelligence. University of Michigan Press, Ann Arbor (1975)
2. Abd-EI-Wahed, W.F., Mousa, A.A., El-Shorbagy, M.A.: Integrating particle swarm optimization with genetic algorithms for solving nonlinear optimization problems. J. Comput. Appl. Math. **235**(5), 1446–1453 (2011)
3. Fevrier Valdez, Patricia Melin, Oscar Castillo: An improved evolutionary method with fuzzy logic for combining particle swarm optimization and genetic algorithms. Appl. Soft Comput. **11**(2), 2625–2632 (2011)
4. Sung-Kwun, O., Han-Jong, J., Witold, P.: A comparative experimental study of type-1/type-2 fuzzy cascade controller based on genetic algorithms and particle swarm optimization. Expert Syst. Appl. **38**(9), 11217–11229 (2011)
5. Fletcher, R., Leyffer, S., Toint P.L.: A brief history of filter methods. J. SIAG/OPT Views and News. **18**(1), 2–12 (2006)
6. Fletcher, R., Leyffer, S.: Nonlinear programming without a penalty function. Math. Program. **91**(2), 239–269 (2002)
7. SU, K.: Trust-region filter method with NCP function. J. Syst. Sci. Math. Sci. **28**(12), 1525–1534 (2008)
8. Rudolph, G., Agapie, A.: Convergence properties of some multi-objective evolutionary algorithm[C]. In: Ali Zalzala (ed.) Proceedings of the Congress on Evolutionary Computation. pp. 1010–1016. IEEE Press, Piscataway (2000)
9. Chun-an, L.I.U.: Multi-objective genetic algorithm for nonlinear programming problem and its convergence. Comput. Eng. Appl. **27**(5), 27–29, 79 (2006)
10. Hanne: On the convergence of multiobjective evolutionary algorithms. Eur. J. Oper. Res. **117**(3), 553–564 (1999)
11. Rudolph: Convergence properties of canonical genetic algorithms. IEEE Trans. Neural Netw. **5**(1), 96–101 (1994)
12. Kalyanmoy Deb: An efficient constraint handing method for genetic algorithms. Comput. Methods Appl. Mech. Eng. **186**, 311–338 (2000)
13. Runarsson, T.P., Yao, X.: Stochastic ranking for constrained evolutionary optimization. J. Evol. Comput. IEEE Trans. **4**(3), 284–294 (2000)

A Modified Neural Network for Solving General Singular Convex Optimization with Bounded Variables

Rendong Ge, Lijun Liu, and Jinzhi Wang

Abstract Singular nonlinear optimization problem has been the difficulty for optimization, which is frequently encountered in practical applications. People have been using numerical iteration methods to deal with the singular problem previously, but the numerical instability and large calculation amount have not been able to be resolved. Based on neural network, this paper puts forward a continuity solution with any rank defect, by using the Augmented Lagrangian method and Projection to form a stable model. By using LaSalle's invariance principle, it is shown that the solution of the proposed network model is convergent. The numerical simulation also further confirmed the effectiveness of the method.

1 Introduction

The nonlinear optimization model with any rank defects is a class of important singular problem, many numerical methods studying singular problems involve the special case. Schnabel and Dan Feng [1–3] get some numerical accomplishments in applying Tensor methods to solve unconstrained optimization under rank one defect. The author and Pro. Xia have done some work on numerical solution [4,5]. For large-scale computational problems, the computation of the classical numerical method is still far from satisfactory.

In recent years, neural network approaches were proposed to deal with classical nonlinear optimization problems. Xia and Wang [6] presented neural networks for solving nonlinear convex optimization with bounded constraints and box constraints. Recently, projection neural networks for solving constrained Optimization Problems are developed [7] and recurrent neural networks for solving nonconvex optimization problem have been also studied [8]. It is regrettable that the study of singular nonlinear optimization problems in the neural network method has not been involved.

R. Ge (✉) • L. Liu • J. Wang
School of Science, Dalian Nationalities University, Dalian 116600, People's Republic of China
e-mail: bgrbgg@163.com; liulijun@dlnu.edu.cn; wjz@dlnu.edu.cn

D. Gao et al. (eds.), *Advances in Global Optimization*, Springer Proceedings in Mathematics & Statistics 95, DOI 10.1007/978-3-319-08377-3_36

This paper is organized as follows. In Sect. 2, the singular nonlinear convex optimization problem and its equivalent formulations are described. In Sect. 3, a recurrent neural network model is proposed to solve such singular nonlinear optimization problems. Global convergence of the proposed neural network is analyzed. Finally, in Sect. 4, several illustrative examples are presented to evaluate the effectiveness of the proposed neural network method.

2 Problem Formulation and Neural Design

Let $\Omega = \{x \in R^n | l \leq x \leq h\}$. Assume $f(x) : \Omega \to R$ is a continuous differentiable convex function. Consider the following unconstrained convex programming problem

$$\begin{aligned} &\min \ f(x) \\ &\text{s.t.} \ \ l \leq x \leq h \end{aligned} \tag{1}$$

which can be easily transformed to equivalent non-negative bounded convex programming problem by using the such transformation as $u = x - l$,

$$\begin{aligned} &\min \ f(x) \\ &\text{s.t.} \ \ 0 \leq x \leq c. \end{aligned} \tag{2}$$

Let x^* be the unique optimal solution to (2). We will discuss the solution of (2) under the following assumptions.

Assumption A1. *$f(x)$ is both strictly convex and four times continuous differentiable. For optimum point x^*, $rank(\nabla^2 f(x^*)) = n - s, (0 < s << n)$ and*

$$Null(\nabla^2 f(x^*)) = \{\mu_1, \mu_2, \cdots, \mu_s\}, U = (\mu_1, \mu_2, \cdots, \mu_s).$$

Assumption A2. *For $x \neq x^*$, there exists $u^T \nabla^2 f(x) u > 0$ for any nonzero $u \in R^n$. Moreover, $\| \nabla^2 f(x) \|$ and $\| \nabla^3 f(x) \|$ are all uniformly bounded.*

Lemma 1. *For any matrix P with full column rank such that $P^T U$ is nonsingular, $\nabla^2 f(x^*) + PP^T$ is nonsingular.*

Proof. It is easy to verify this result, thus its proof is omitted here for the sake of saving space.

Assumption A3. *There exists a $q \neq 0$, as setting $\mu = U(P^T U)^{-1} q$, such that for any $v \in Null(\nabla^2 f(x^*))$, $f^{(4)}(x^*) \cdot \mu^2 v^2 > 0$. (The reason for this assumption can be found, for example, in [4].)*

Define function $F(x)$ as follows:

$$F(x) = f(x) + \lambda h(x),$$

where $h(x) = \mu(x)\nabla^2 f(x)\mu(x)$ and $\mu(x) = (\nabla^2 f(x) + PP^T)^{-1}Pq$, for any $q \neq 0$ and $P^T U$ is nonsingular. It can be proved that $\mu = U(P^T U)^{-1}q \in Null(\nabla^2 f(x^*))$ is a solution for the following equation

$$(\nabla^2 f(x) + PP^T)\mu(x) = Pq.$$

According to the definition of $F(x)$, we have the following important result.

Lemma 2. *For any $\lambda > 0$, the Hessian matrix $\nabla^2 F(x^*)$ is positive definite. Moreover, if λ is small enough, then $\nabla^2 F(x)$ is positive definite for any $x \in R^n$.*

Proof. This conclusion can be easily proved according to the results in [4] under Assumptions A2 and A3 Thus the proof is omitted here.

Because the Hessian matrix of $f(x)$ is singular at x^* for (2), it is generally difficult to obtain ideal convergence results by conventional optimization algorithm (see [2, 4, 5]). In order to overcome this difficulty, we first establish equivalent unconstrained convex optimization problem as follows:

$$\begin{aligned} &\min\ F(x) \\ &\text{s.t.}\ \ 0 \le x \le c, \end{aligned} \tag{3}$$

for which we can establish the equivalent lemma as follows:

Lemma 3. *x^* is a solution of (2) if and only if x^* is a solution of (3).*

On the difficulty caused by computing the matrix inverse for further consideration, we turn optimization problem (3) into the following equivalent constrained optimization problem.

$$\begin{cases} \min g(x, y) = f(x) + \lambda y^T \nabla^2 f(x) y \\ \quad s.t.(\nabla^2 f(x) + PP^T)y = Pq. \end{cases} \tag{4}$$

Define Lagrange function of (4) as follows:

$$\tilde{L}(x, y, z) = f(x) + \lambda y^T \nabla^2 f(x) y + z^T[(\nabla^2 f(x) + PP^T)y - Pq]$$

By Assumptions A2 and Lemma 2, it is easy to know that the function $g(x, y)$ is strictly convex. Based on the Karush–Kuhn–Tucker sufficient conditions, the KKT point $(\hat{x}, \hat{y})$ of the formula (4) is a unique optimal solution of the optimization problem (4) and there exits $\hat{z} \in R^n$ satisfies the following condition:

$$\begin{cases} (\nabla_x \tilde{L}(x,y,z))_i \begin{cases} \geq 0, \text{ if } x_i = 0, \\ \leq 0, \text{ if } x_i = c_i, \\ = 0, \text{ if } 0 < x_i < c_i, \end{cases} \quad (i = 1, 2, \cdots n) \\ \nabla_y \tilde{L}(x,y,z) = 0, \\ \nabla_z \tilde{L}(x,y,z) = 0, \\ x \in \Omega = \{x \in R^n | 0 \leq x \leq c\}. \end{cases}$$

Equivalently, the point $(\hat{x}, \hat{y}, \hat{z})$ satisfies the following condition,

$$\begin{cases} (x - \hat{x})^T (\nabla_x \tilde{L}(\hat{x}, \hat{y}, \hat{z})) \geq 0, \ x \in \Omega \\ \nabla_y \tilde{L}(\hat{x}, \hat{y}, \hat{z}) = 0, \nabla_z \tilde{L}(\hat{x}, \hat{y}, \hat{z}) = 0 \end{cases} \tag{5}$$

In order to discuss the constrained programming problem (4), first, we define a augmented Lagrangian function of (4) as follows:

$$L(x, y, z, k) = f(x) + \lambda y^T \nabla^2 f(x) y + \\ z^T [(\nabla^2 f(x) + PP^T) y - Pq] + \tfrac{k}{2} ||(\nabla^2 f(x) + PP^T) y - Pq||^2, x \in \Omega,$$

where $k > 0$ is a penalty parameter and z is an approximation of the Lagrange multiplier vector. Hence, the problem (4) can be solved by the stationary point of the following problem,

$$\min_{x \in \Omega, y, z \in R^n} L(x, y, z, k) \tag{6}$$

Then, the condition (5) can be written as

$$\begin{cases} (x - \hat{x})^T \nabla_x L(\hat{x}, \hat{y}, \hat{z}, k) \geq 0, \quad x \in \Omega, \\ \nabla_y L(\hat{x}, \hat{y}, \hat{z}, k) = 0, \\ \nabla_z L(\hat{x}, \hat{y}, \hat{z}, k) = 0. \end{cases} \tag{7}$$

Now, we introduce the projection function as follows:

$$P_\Omega : R^n \to \Omega, \ P_\Omega(u) = (P_1(u), P_2(u), \cdots, P_n(u))$$

where

$$P_i(u) = \begin{cases} 0, \ u_i < 0, \\ u_i, \ u_i \in [0, c_i], \\ c_i, \ u_i > c_i. \end{cases} \tag{8}$$

From the projection conclusion as shown in [10], the first inequality of (7) can be equivalently represented as

$$P_{\Omega}(\hat{x} - \alpha\nabla_x L(\hat{x}, \hat{y}, \hat{z}, k)) - \hat{x} = 0, \forall\alpha > 0.$$

So the optimal solution of (4) and the stationary point of (6) meet with the conditions

$$\begin{cases} x = P_{\Omega}(x - \alpha\nabla_x L(x, y, z, k)), \forall\alpha > 0, \\ \nabla_y L(x, y, z, k) = 0, \\ (\nabla_x^2 f(x) + PP^T)y = q. \end{cases} \tag{9}$$

3 Stability Analysis of the Neural Network Model

By Theorems 8 and 9 in [9], there exists a constant $k > 0$, such that if $c^* = (x^*, y^*, z^*)$ is an optimal solution of the problem (6), then (x^*, y^*) is an optimal solution of the problem (4) and

$$\min_{x\in\Omega, y,z\in R^n} L(x, y, z, k) = f(x^*).$$

Notice that $L(x, y, z) \equiv L(x, y, z, k)$. By the Lagrange function defined above, we can describe the neural network model by the following nonlinear dynamic system for solving (10). The logical graph is shown in Fig. 1.

$$\begin{cases} \dfrac{dx}{dt} = P_{\Omega}(x - \alpha\nabla x L(x, y, z)) - x \\ \quad = P_{\Omega}(x - \alpha(\nabla f(x) + \lambda\nabla^3 f(x)yy + \nabla^3 f(x)yz + k\nabla^3 f(x)y((\nabla^2 f(x) + \\ \quad\quad PP^T)y - Pq))) - x \\ \dfrac{dv}{dt} = -\beta\nabla y L(x, y, z) \\ \quad = -\beta(2\lambda\nabla^2 f(x)y + (\nabla^2 f(x) + PP^T)z + k(\nabla^2 f(x) + PP^T)((\nabla^2 f(x) + \\ \quad\quad PP^T)y - Pq)) \\ \dfrac{dw}{dt} = -\beta\nabla z L(x, y, z) = -\beta((\nabla^2 f(x) + PP^T)y - Pq) \\ y_j = s(v_j), \quad j = 1, 2, \ldots, n \\ z_k = s(w_k), \quad k = 1, 2, \ldots, n \end{cases} \tag{10}$$

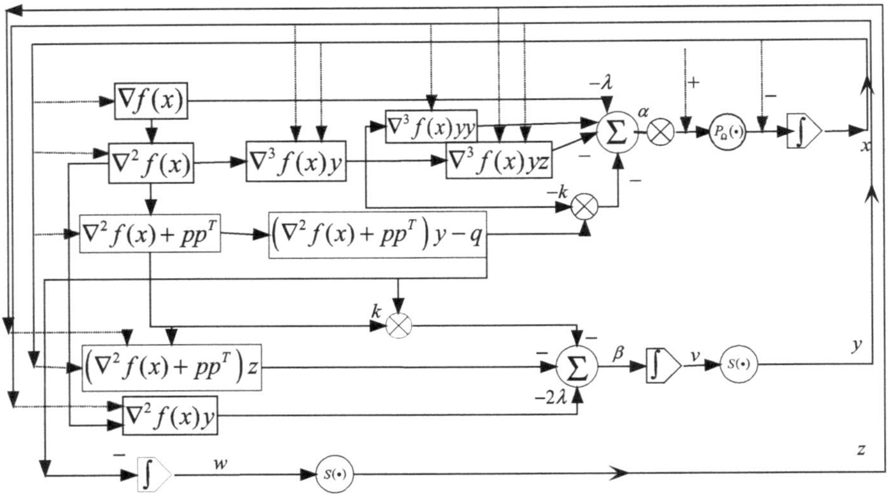

Fig. 1 Logical graph of the proposed neural network model

where

$$\nabla f(x) = (f_1(x), f_2(x), \cdots, f_n(x))^T, \alpha > 0, \beta > 0$$
$$\nabla^3 f(x)y = (\nabla^2 f_1(x)y, \nabla^2 f_2(x)y, \cdots, \nabla^2 f_n(x)y)^T,$$

and the activation function $s(\cdot)$ is continuously differentiable and satisfies that $s'(\cdot) > 0$.

It is easy to see that if $c^* = (x^*, y^*, z^*)$ is an optimal solution of the problem (6), then it is an equilibrium point of network (10). Conversely, if (x^*, y^*, z^*) is a equilibrium point of network (10), it must be KKT point of original problem (4). To analyze the convergence of the neural network (10), the following lemmas are first introduced (see [10]).

Lemma 4. *Assume that the set $\Omega \subset R^n$ is a closed convex set, then the following two inequalities hold,*

$$(P_\Omega(x') - y')^T(x - P_\Omega(x')) \geq 0, \ \forall x' \in R^n, \ y' \in \Omega$$
$$\|P_\Omega(x') - P_\Omega(y')\| \leq \|x' - y'\|, \ \forall x', \ y' \in R^n$$

where $P_\Omega : R^n \rightarrow \Omega$ is a projection operator defined as $P_\Omega(\gamma) = \min_{\varsigma \in \Omega} \|\gamma - \varsigma\|$.

Lemma 5. *For any initial point $(x(t_0), v(t_0), w(t_0)) \in R^{3n}$, there exists a unique continuous solution $(x(t), v(t), w(t)) \in R^{3n}$ for (10). Moreover, $x(t) \in \Omega$ provided that $x(t_0) \in \Omega$. The equilibrium point of (10) solves (5).*

Proof. By Assumption A1, $P_\Omega(x - \alpha\nabla_x L(x, y, z, k)) - x, \nabla_y L(x, y, z, k)$ and $\nabla^2 f(x) + PP^T)y - Pq$ are locally Lipschitz continuous. According to local existence theorem of ordinary differential equation, there exists a unique continuous solution $(x(t), v(t), w(t))$ of (10) for (t_0, T).

Next, let initial point $x(t_0) \in \Omega$. Since $\frac{dx}{dt} + x = P_\Omega(x - \alpha\nabla x L(x, y, z))$, we have

$$\int_{t_0}^{t} (\frac{dx}{dt} + x)e^s ds = \int_{t_0}^{t} e^s P_\Omega(x - \alpha\nabla x L(x, y, z))ds$$

Or equivalently, $x(t) = e^{-(t-t_0)}x(t_0) + e^{-t}\int_{t_0}^{t} e^s P_\Omega(x - \alpha\nabla x L(x, y, z))ds$.

So, $x(t) \geq 0$ provided that $x(t_0) \geq 0$, $P_\Omega(x - \alpha\nabla_x L(x, y, z, k)) \geq 0$, and since

$$\begin{aligned} x(t) &= e^{-(t-t_0)}x(t_0) + e^{-t}\textstyle\int_{t_0}^{t} e^s P_\Omega(x - \alpha\nabla x L(x, y, z))ds \\ &\leq e^{-(t-t_0)}x(t_0) + e^{-t}(e^t - e^{t_0})c = c - (c - x(t_0))e^{-(t-t_0)} \leq c \end{aligned}.$$

Thus, $x(t) \in \Omega$ provided that $x(t_0) \in \Omega$.

Before establishing the convergence theorem, we need the property of the following augmented Lagrangian function.

Assumption A4. *$L(x, y, z)$ satisfies the local monotone property of the following definition about x.*

$$(x - x^*)^T(\nabla L_x(x, y, z) - \nabla_x L(x^*, y^*, z^*)) \geq 0.$$

Now we are ready to establish the stability and the convergence results of network (10).

Theorem 1. *Assume that $f(x) : R^n \rightarrow R$ is strictly convex and the fourth differentiable, and $c^* = (x^*, y^*, z^*)$ is a global optimal solution of the problem (6), if the initial point $(x(t_0), y(t_0), z(t_0))$ with $x(t_0) \in \Omega$ is chosen in a small neighborhood of the equilibrium point, then the proposed neural network of (10) is stable in the sense of Lyapunov and globally convergent to the stationary point (x^*, y^*, z^*), where x^* is the optimal solution of (2).*

Proof. The proof is omitted.

4 Numerical Examples

In order to verify the effectiveness of the presented algorithm in this paper, three examples were selected from the literature [11].

For the first example, it is easy to verify that the Hessian matrix of the object function $f_p(x) = \frac{1}{2}\|F(x)\|^2$ is rank two deficiency at the minimizer x^*. For the last two examples, the corresponding Hessian matrix is nonsingular at the minimizer. In order to adapt them to the rank two deficiency's case, we have adopt the same procedure as proposed in [1] by introducing function transformation as follows:

$$F(x) := F(x) - \nabla F(x^*)A(A^T A)^{-1} A^T (x - x^*) \tag{11}$$

where x^* is the root of $F(x) = 0$ and

$$F(x) : R^n \to R^m, A \in R^{n\times 1}, A^T = \begin{pmatrix} 1 & 1 & 1 & 1 & \cdots & 1 \\ 1 & -1 & 1 & -1 & \cdots & (-1)^n \end{pmatrix}.$$

Now we can construct the relevant objection function $f_p(x)$

$$f_p(x) = \frac{1}{2}\|F(x)\|^2$$

for which its Hessian matrix being rank two deficiency and it can be checked that the root of the original $F(x)$ is the minimizer of the constructed $f_p(x)$.

Ex.1: Powell Singular Function:

a. $n = 4, m = 4$
b. $f_1(x) = x_1 + 10x_2$
$f_2(x) = 5^{1/2}(x_3 - x_4)$
$f_3(x) = (x_2 - 2x_3)^2$
$f_4(x) = 10^{1/2}(x_1 - x_4)^2$
c. $f = 0$ at the minimizer $x^* = (0, 0, 0, 0)$.

Ex.2: Beale Function:

a. $n = 2, m = 3$
b. $f_1(x) = y_1 - x_1(1 - x_2)$
$f_2(x) = y_2 - x_1(1 - x_2^2)$
$f_3(x) == y_3 - x_1(1 - x_2^3)$
$y1 = 1.5; y2 = 2.25; y3 = 2.625$
c. $f = 0$ at the minimizer $x^* = (3, 0.5)$.

Ex.3: Modified Broyden Tridiagonal Function:

a. $n = 4, m = 4$
b. $f_1(x) = (3 - 2x_1)x_1 - 2x_2 + 1$
$f_2(x) = (3 - 2x_2)x_2 - x_1 - 2x_3 + 2$
$f_3(x) == (3 - 2x_3)x_3 - x_2 - 2x_4 + 2$
$f_4(x) = (3 - 2x4)x4 - x3$
c. $f = 0$ at the minimizer $x^* = (1, 1, 1, 1)$.

Meanwhile, we compare the dynamic behavior of the proposed model with the classical projection gradient system [7] as follows:

$$\frac{dx}{dt} = P_{\Omega}(x - \alpha \nabla f_p(x)) - x, \alpha > 0 \tag{12}$$

for which the Hessian matrix at the minimizer is generally assumed to be nonsingular.

We use Matlab 7.0 to simulate the dynamics of the corresponding systems. The integral curves are obtained by using the ODE function ode15s for the numerical integration. For the proposed system (10) (**PR**) and the classical projection gradient system (12) (**PG**), we have chosen the same initial point to numerically solve the ODEs.

For Ex.1, we choose $x_0 = (01, 0.9, rand(1,2) * 12), \alpha = 1, \beta = 10$ for both **PR** and **PG** and let y_0 and z_0 be some random values between 0 and 12. The other parameters **PR** are chosen as $l = 0, h = 20, \lambda = 0.0000001, k = 3{,}000$. The results are shown in Figs. 2 and 3. It can be seen in Fig. 2, that the integral curves

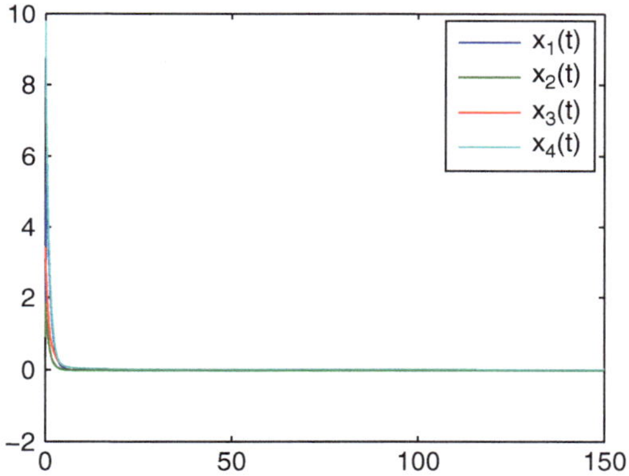

Fig. 2 Trajectory of $x(t)$ for **PR** (Ex.1)

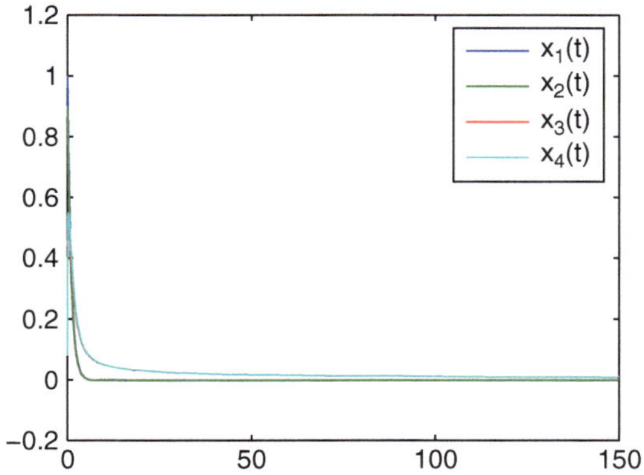

Fig. 3 Trajectory of $x(t)$ for **PG** (Ex.1)

response of **PR** converge to f_p's minimizer $x^* = (0, 0, 0, 0)$. On the contrary, as shown in Fig. 3, the curves of **PG** failed to converge to f_p's minimizer with the same initial point.

For Ex.2, we choose $l = 0, h = 3, \lambda = 0.0000001; x_0 = rand(1, 6); k = 5{,}000, \alpha = \beta = 1$. Similar results are obtained, i.e., the proposed system **PR** successfully found the minimizer $x^* = (3, 0.5)$ while the classical system **PG** failed. The results are shown in Figs. 4 and 5, respectively.

Figures 6 and 7 show the corresponding results for Ex.3 with initial conditions chosen as $x_0 = (-1, -0.5, -0.5, -1), l = 0, h = 1, \lambda = 0.0000001, k = 5{,}000, \alpha = 10, \beta = 100$. The proposed system **PR** finally got the minimizer $x^* = (1, 1, 1, 1)$, while the system **PR** got stuck all the time.

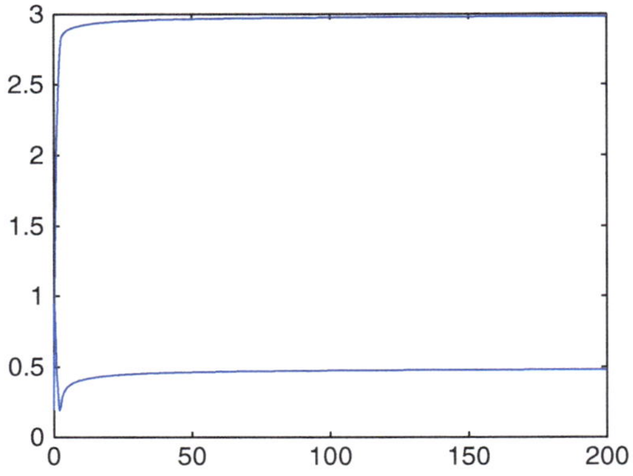

Fig. 4 Trajectory of $x(t)$ for **PR** (Ex.2)

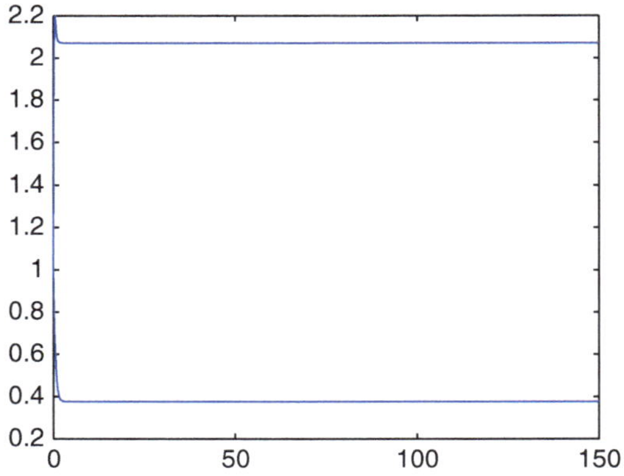

Fig. 5 Trajectory of $x(t)$ for **PG** (Ex.2)

5 Concluding Remarks

Singular nonlinear convex optimization problems have been traditionally studied by classical numerical methods. In this paper, a novel neural network model was established to solve such a difficult problem. Under some mild assumptions, the unconstrained nonlinear optimization problem is turned into a constrained optimization problem. By establishing the relationship between KKT points and the augmented Lagrange function, a neural network model is successfully obtained. Global analysis with illustrative examples supports the presented results.

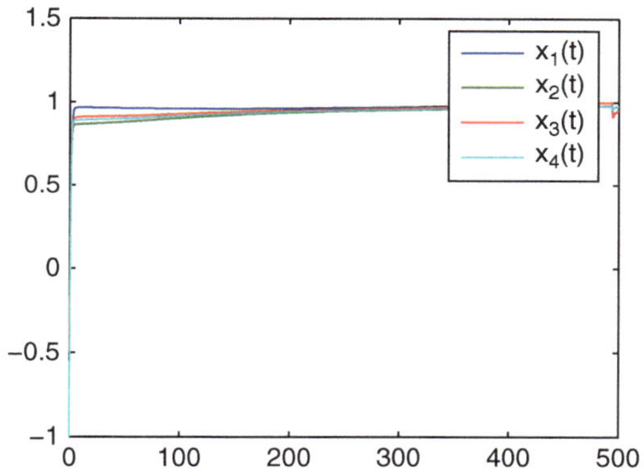

Fig. 6 Trajectory of $x(t)$ for **PR** (Ex.3)

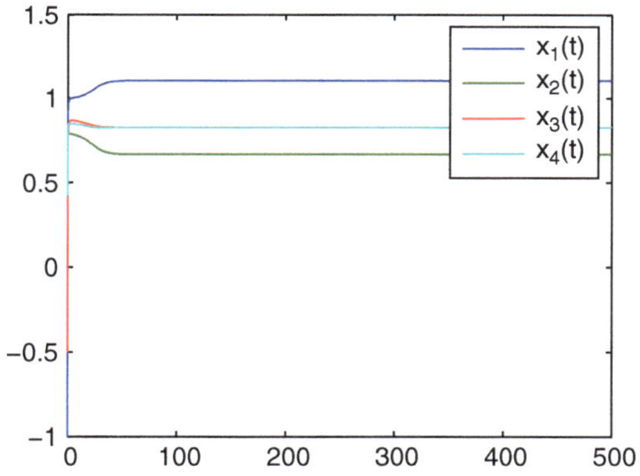

Fig. 7 Trajectory of $x(t)$ for **PG** (Ex.3)

References

1. Schnabel, R.B., Chow, T.-T.: Tensor methods for unconstrained optimization using second derivatives. SIAM J. Optim. **1**(3), 293–315 (1991)
2. Feng, D., Schnabel, R.B.: Tensor methods for equality constrained optimization. SIAM J. Optim. **6**(3), 653–673 (1996)
3. Bouaricha, A.: Tensor methods for large sparse unconstrained optimization. SIAM J. Optim. **7**(3), 732–756 (1997)
4. Ge, R., Xia, Z.: Solving a type of modified BFGS algorithm with any rank defects and the local Q-superlinear convergence properties. J. Comput. Appl. Math. **22**, 1–2 (2006)
5. Ge, R., Xia, Z.: A type of modified BFGS algorithm with rank defects and its global convergence in convex minimization. J Pure Appl. Math. Adv. Appl. **3**, 17–35 (2010)
6. Xia, Y.S., Wang, J.: On the stability of globally projected dynamical systems. J. Optim. Theory Appl. **106**, 129–150 (2000)
7. Xia, Y.S., Leung, H., Wang, J.: A projection neural network and its application to constrained optimization problems. IEEE Trans. Circuits Syst. I **49**, 447–458 (2002)
8. Sun, C.Y., Feng, C.B.: Neural Networks for Nonconvex Nonlinear Programming Problems: A Switching Control Approach. Lecture Notes in Computer Science, Springer, vol. 3496, pp. 694–699 (2005)
9. Du, X., Yang, Y., Li, M.: Further studies on the Hestenes-Powell augmented lagrangian function for equality constraints in nonlinear programming problems. OR Trans. **10**, 38–46 (2006)
10. Kinderlehrer, D., Stampacchia, G.: An Introduction to Variational Inequalities and Their Applications. Academic, New York (1980)
11. More, J.J., Garbow, B.S., Hillstrom, K.E.: Testing unconstrained optimization software. ACM Trans. Math. Softw. **7**(1), 19–31 (1981)

A Teaching–Learning-Based Cuckoo Search for Constrained Engineering Design Problems

Jida Huang, Liang Gao, and Xinyu Li

Abstract A new hybrid algorithm named teaching–learning-based Cuckoo Search (TLCS) is proposed for constrained optimization problems. The TLCS modifies the Cuckoo Search (CS) based on the teaching–learning-based Optimization (TLBO) and then is applied for constrained engineering design problems. Experimental results on several well-known constrained engineering design problems demonstrate the effectiveness, efficiency, and robustness of the proposed TLCS. Moreover, the TLCS obtains some solutions better than those previously reported in the literature.

Keywords Cuckoo search • TLBO • TLCS • Constrained optimization • Engineering design

1 Introduction

Many real-world engineering design problems involve a number of constraints, and these problems can be classified as constrained optimization problems [1, 2]. The general constrained optimization problem is stated as follows:

$$\min f\left(\vec{x}\right)$$
$$s.t. \begin{cases} g_p\left(\vec{x}\right) \le 0, & p = 1, 2, \ldots, n_g \\ h_q\left(\vec{x}\right) = 0, & q = 1, 2, \ldots, n_h \\ L_i \le x_i \le U_i, & i = 1, 2, \ldots, n \end{cases} \tag{1}$$

where $\vec{x} = [x_1, x_2, \ldots, x_n]^{\mathrm{T}}$ denotes the decision solution vector, n_g is the number of inequality constraints and n_h is the number of equality constraints, L_i and U_i are the lower bound and the upper bound of the variables.

J. Huang • L. Gao • X. Li (✉)
State Key Laboratory of Digital Manufacturing Equipment and Technology, Huazhong University of Science and Technology, Wuhan, China
e-mail: jideny@163.com; gaoliang@mail.hust.edu.cn; lixinyu@mail.hust.edu.cn

D. Gao et al. (eds.), *Advances in Global Optimization*, Springer Proceedings in Mathematics & Statistics 95, DOI 10.1007/978-3-319-08377-3_37

It is difficult to solve such constrained optimization problems. In the past two decades, many methods have been proposed to solve constrained optimization problems and the constraint-handling techniques is one of the major concerns when solving such problems. Using the penalty functions [3] is one of the most common constraints handing techniques. The idea is to transform the constrained optimization problem into an unconstraint one by adding a certain value to objective function value based on the constraint violations presenting in a solution [4]. It has been widely used for its simpleness and effectiveness.

In order to tackle the fine-tuning parameters required by traditional penalty functions, some works have been dedicated to the self-adaptive mechanism to improve the performance of the algorithms [5]. Using a co-evolution model to adapt the penalty factors, Coello [4, 6] proposed a self-adaptive penalty approach based on a genetic algorithm (GA). Hamida and Schoenauer [7] proposed an adaptive segregational constraint-handling technique with EA. Mezura et al. [5] proposed an approach where a self-adaptive parameter controlling for the differential evolution (DE) parameters and also for the parameters introduced by the constraint-handling mechanism. Brest et al. [5] proposed a self-adaptive differential evolution algorithm in constrained real-parameter optimization. However, in these methods, there are many parameters need to be considered in order to further a better performance of the algorithms, the process of finding an optimum set of parameters is arduous, thus the adaption of these methods is limited.

We have proposed an effective hybrid algorithm named teaching–learning-based Cuckoo Search (TLCS) for continuous optimization problems lately [8]. In order to extend its applications range, it has been combined with the penalty function to optimize the engineering design problems in this paper. The rest of this paper is organized as follows: Sect. 2 provides a basic framework of the proposed TLCS. Experimental results based on some engineering design problems and comparisons with previously reported resulted are presented in Sect. 3, and the conclusions and future work in Sect. 4.

2 The Proposed TLCS

For the proposed TLCS, it has a strong global search ability along with a fast convergence rate, and the method could be suitable for a broad spectrum of problem domains. In order to adopt this effective algorithm in the constrained optimization problems, the penalty function is combined with TLCS for solving the engineering design problems. The framework of the proposed method is as in Fig. 1.

As in the framework, the penalty function is used to transform the constrained optimization problem into an unconstraint one firstly; then for solutions to be abandoned in the CS will perform lévy flight to generate new solutions; for other better solutions, we use the teaching–learning-based optimization (TLBO) to enhance the local search ability of CS. Thus, the algorithm becomes more practical

```
Utilize penalty function to transform the objective function into f(x), x = (x_1, ... , x_d)^T
Generate initial population of n solutions x_i (i = 1, 2, ..., n)
while (t < Max Generation) or (stop criterion)
    for all solutions to be abandoned do
        Perform Lévy flight from x_i to generate new solution x_new,i
        x_i ← x_new,i
        f_i ← f_new,i
    end for
    for all of the top solutions do
        X_new,i=X_old,i+r_i(X_teacher-(T_F)Mean)
        Student-Learning-Process
        x_i ← x_new,i
        f_i ← f_new,i
    end for
    A fraction (p_a) of worse solutions are abandoned and new ones are built
    Rank the solutions and find the current best
end while
Postprocess results and visualization
```

Fig. 1 Pseudo code of TLCS

for a wider range of applications but without losing attractive features of the original CS. The lévy flight and the teaching–learning process are presented as follows.

2.1 Lévy Flight

Cuckoo Search (CS) is a new meta-heuristic search algorithm, based on cuckoo bird's behavior [9]. The algorithm is inspired by the reproduction strategy of cuckoos. For a maximization problem, the quality or fitness of a solution can simply be proportional to the objective function. Other forms of fitness can be defined in a similar way to the fitness function in genetic algorithms.When generating new solutions x^{t+1} for cuckoo i, a Lévy flight is performed:

$$x_i^{t+1} = x_i^t + \alpha \oplus L\acute{e}vy\ (\lambda) \tag{2}$$

where $\alpha > 0$ is the step size which should be related to the scales of the problem. In most cases, we can use $\alpha = 1$. The product $\oplus$ means entry-wise multiplications. Lévy flights essentially provide a random walk while their random steps are drawn from a Lévy distribution for large steps

$$L\acute{e}vy\ (\lambda) \sim u = t^{-\lambda}, \quad (1 < \lambda \leq 3) \tag{3}$$

which has an infinite variance with an infinite mean. Here the consecutive jumps/steps of a cuckoo essentially form a random walk process which obeys a power-law step-length distribution with a heavy tail.

2.2 *Teaching–Learning Process*

TLBO is a population based method. The algorithm mimics the teaching-learning ability of teacher and learners in a class room [10]. The working of TLBO is divided into two parts, "Teacher phase" and "Learner phase." Working of this two phases are explained below.

(1) Teacher phase

It is the first part of the algorithm where learners learn from the teacher. During this phase a teacher tries to increase the mean result of the class room from any value M_i to his or her level (i.e., T_A). But practically it is not possible and a teacher can move the mean of the class room M_1 to any other value M_2 which is better than M_1 depending on his or her capability. Considered M_j be the mean and T_i be the teacher at any iteration i. Now T_i will try to improve existing mean M_j towards it so the new mean will be T_i designated as M_{new} and the difference between the existing mean and new mean is given by:

$$Difference\_Mean_i = r_i \left(M_{new} - T_F M_i \right) \quad (4)$$

where T_F is the teaching factor which decides the value of mean to be changed, and r_i is the random number in the range [0, 1]. Value of T_F can be either 1 or 2 which is a heuristic step and it is decided randomly with equal probability as:

$$T_F = round\left[1 + rand\,(0,1)\,\{2-1\}\right] \quad (5)$$

Based on this *Difference_Mean*, the existing solution is updated according to the following expression:

$$X_{new,i} = X_{old,i} + Difference\_Mean_i \quad (6)$$

(2) Learner phase

It is the second part of the algorithm where learners increase their knowledge by interaction among themselves. A learner interacts randomly with other learners for enhancing his or her knowledge. A learner learns new things if the other learner has more knowledge than him or her. Mathematically the learning phenomenon of this phase is expressed below.

At any iteration i, considering two different learners X_i and X_j where $i \neq j$,

$$X_{new,i} = X_{old,i} + r_i\left(X_i - X_j\right) \quad if\ f\left(X_i\right) < f\left(X_j\right) \tag{7}$$

$$X_{new,i} = X_{old,i} + r_i\left(X_j - X_i\right) \quad if\ f\left(X_j\right) < f\left(X_i\right) \tag{8}$$

Accept X_{new} if it gives better function value.

3 Experimental Results and Analysis

The performance of the proposed TLCS is investigated on several well-known constrained engineering design problems. These selected examples have linear and non-linear constraints, and have been well studied previously by a variety of techniques. The results are compared with some good reported results which solved by EA-based methods and other traditional mathematical programming methods.

For each testing problem, the parameters of the TLCS are set as follows: the population size $n = 20$, $p_a = 0.25$, the Max Generation = 1,000. As to the penalty function, we use Eq. (9) to transform the constrained optimization problem into an unconstraint one.

$$F(x) = f(x) + w_1 \times (sum\_viol)^2 + w_2 \times num\_viol \tag{9}$$

where $f(x)$ is the objective function, *sum_viol* denotes the sum of all the amounts by which the constraints are violated, *num_viol* denotes the number of constraints violation, $w_1 = 1e10$, $w_2 = 1e15$.

3.1 Welded Beam Design

The welded beam design problem is taken from Rao et al. [11], in which a welded beam is designed for minimum cost subject to constraints on shear stress (τ), bending stress in the beam (θ), buckling load on the bar (P_c), end deflection of the beam (δ), and side constraints. There are four design variables as shown in Fig. 2, i.e., $h(x_1)$, $l(x_2)$, $t(x_3)$, and $b(x_4)$.

The range of the design variables for this problem is: $0.1 \leq x_1 \leq 2$, $0.1 \leq x_2 \leq 10$, $0.1 \leq x_3 \leq 10$, $0.1 \leq x_4 \leq 2$. And the TLCS is run 30 times independently. The approaches applied to this problem include geometric programming [12], genetic algorithm with binary representation and traditional penalty function [13], a GA-based co-evolution model [6], a feasibility-based tournament selection scheme

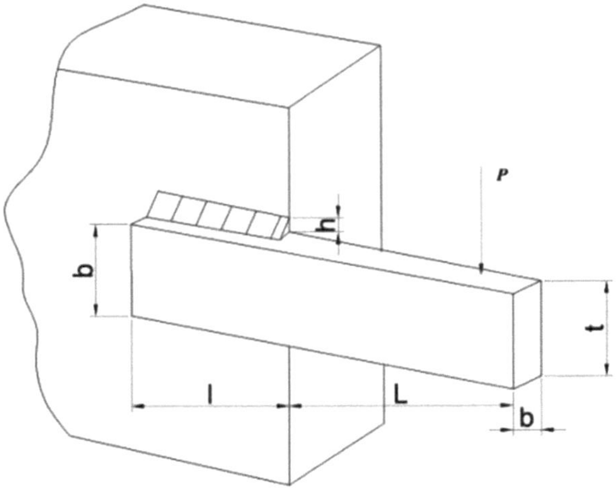

Fig. 2 Welded beam design problem

Table 1 Comparison of the best solution found by different methods

Design variables	TLCS	Ragsdell and Phillips [12]	Deb [13]	Coello [6]	Coello [4]	He and Wang [14]
$x_1(h)$	0.205730	0.245500	0.248900	0.208800	0.205986	0.202369
$x_2(l)$	3.470489	6.196000	6.173000	3.420500	3.471328	3.544214
$x_3(t)$	9.036624	8.273000	8.178900	8.997500	9.020224	9.048210
$x_4(b)$	0.205730	0.245500	0.253300	0.210000	0.206480	0.205723
$g_1(x)$	−7.5463e−06	−5743.82652	−5758.60378	−0.337812	−0.074092	−12.83980
$g_2(x)$	−0.000120	−4.715097	−255.576901	−353.902604	−0.266227	−1.247467
$g_3(x)$	−1.0334e−08	0.000000	−0.004400	−0.001200	−0.000495	−0.001498
$g_4(x)$	−3.432984	−3.020289	−2.982866	−3.411865	−3.430043	−3.429347
$g_5(x)$	−0.080730	−0.120500	−0.123900	−0.083800	−0.080986	−0.079381
$g_6(x)$	−0.235540	−0.234208	−0.234160	−0.235649	−0.235514	−0.235536
$g_7(x)$	−9.4039e−07	−3604.275002	−4465.27093	−363.23238	−58.66644	−11.68136
$f(x)$	**1.724852**	2.385937	2.433116	1.748309	1.728226	1.728024

inspired by the multi-objective optimization techniques [4], and a co-evolutionary particle swarm optimization [14]. The best solutions obtained by these methods and TLCS are listed in Table 1, and their statistical results are shown in Table 2.

From Table 1, it can be seen that the best feasible solution found by TLCS is better than the best solutions found by other techniques. From Table 2, it can be seen that the average searching quality of TLCS is also better than those of other methods, and even the worst solution found by TLCS is better than the best solution found by other methods. In addition, the standard deviation of the results by TLCS in 30 independent runs is very small.

Table 2 Statistical results of different methods for welded beam design problem

Method	Best	Mean	Worst	Std Dev
TLCS	**1.724852**	**1.724884**	**1.725761**	**0.000166**
Ragsdell and Phillips [12]	2.385937	N/A	N/A	N/A
Deb [13]	2.433116	N/A	N/A	N/A
Coello [6]	1.748309	1.771973	1.785835	0.011220
Coello [4]	1.728226	1.792654	1.993408	0.074713
He and Wang [14]	1.728024	1.748831	1.782143	0.012926

Fig. 3 Tension/compression string design problem

3.2 A Tension/Compression String Design

This problem is from Belegundu et al. [2], which needs to minimize the weight (i.e., $f(x)$) of a tension/compression spring (as shown in Fig. 3) subject to constraints on minimum deflection, shear stress, surge frequency, limits on outside diameter and on design variables. The design variables are the mean coil diameter $D(x_2)$, the wire diameter $d(x_1)$, and the number of active coils $P(x_3)$.

The range of the design variables for this problem is: $0.05 \leq x_1 \leq 2$, $0.25 \leq x_2 \leq 1.3$, $2 \leq x_3 \leq 15$. And the TLCS is run 30 times independently. The approaches applied to this problem include eight different numerical optimization techniques [2], a numerical optimization technique called constraint correction at constant cost [1], a GA-based co-evolution model [6], a feasibility-based tournament selection scheme [4], and a co-evolutionary particle swarm optimization [14]. The best solutions obtained by these methods and TLCS are listed in Table 3, and their statistical results are shown in Table 4.

From Table 3, it can be seen that the best feasible solution found by TLCS is better than the best solutions found by other techniques. From Table 4, it can be seen that the average searching quality of TLCS is also better than those of other methods. In addition, the standard deviation of the results by TLCS in 30 independent runs is also very small.

Table 3 Comparison of the best solution found by different methods

Design variables	TLCS	Belegundu [2]	Arora [1]	Coello [6]	Coello [4]	He and Wang [14]
x_1 (d)	0.05159855	0.050000	0.053396	0.051480	0.051989	0.051728
x_2 (D)	0.3545443	0.315900	0.399180	0.351661	0.363965	0.357644
x_3 (P)	11.41753	14.250000	9.185400	11.632201	10.890522	11.244543
$g_1(x)$	−4.951595e−14	−0.000014	−0.000019	−0.002080	−0.000013	−0.000845
$g_2(x)$	−1.952749e−11	−0.003782	−0.000018	−0.000110	−0.000021	−1.2600e−05
$g_3(x)$	−4.049472	−3.938302	−4.123832	−4.026318	−4.061338	−4.051300
$g_4(x)$	−0.7292381	−0.756067	−0.698283	−4.026318	−0.722698	−0.727090
$f(x)$	**0.0126654**	0.0128334	0.0127303	0.0127048	0.0126810	0.0126747

Table 4 Statistical results of different methods for tension/compression string design problem

Method	Best	Mean	Worst	Std Dev
TLCS	**0.0126654**	**0.0126822**	**0.012734**	**1.749364e-005**
Belegundu [2]	0.0128334	N/A	N/A	N/A
Arora [1]	0.0127303	N/A	N/A	N/A
Coello [6]	0.0127048	0.0127690	0.012822	3.939000e−005
Coello [4]	0.0126810	0.0127420	0.012973	5.900000e−005
He and Wang [14]	0.0126747	0.012730	0.012924	5.198500e−005

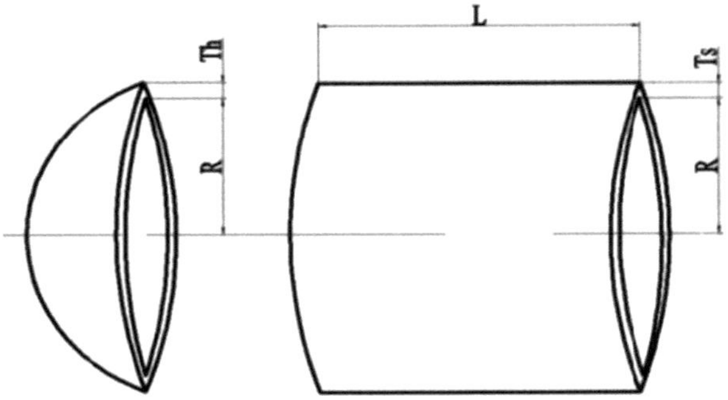

Fig. 4 *Center* and *end* section of pressure vessel design problem

3.3 A Pressure Vessel Design

In this problem, the objective is to minimize the total cost $f(x)$, including the cost of the material, forming, and welding. A cylindrical vessel is capped at both ends by hemispherical heads as shown in Fig. 4. There are four design variables: $T_s(x_1$, thickness of the shell), T_h (x_2, thickness of the head), $R(x_3$, inner radius), and $L(x_4$, length of the cylindrical section of the vessel, not including the head). Among the four variables, T_s and T_h are integer multiples of 0.0625 in that are the available thicknesses of rolled steel plates, and R and L are continuous variables.

The range of the design variables for this problem is: $1 \leq x_1 \leq 99$, $1 \leq x_2 \leq 99$, $10 \leq x_3 \leq 200$, $10 \leq x_4 \leq 200$. And the TLCS is run 30 times independently. The approaches applied to this problem include genetic adaptive search [15], an augmented Lagrangian multiplier approach [16], a branch and bound technique [17], a GA-based co-evolution model [6], a feasibility-based tournament selection scheme [4], and a co-evolutionary particle swarm optimization [14]. The best solutions obtained by these methods and TLCS are listed in Table 5, and their statistical results are shown in Table 6.

From Table 5, it can be seen that the best feasible solution found by TLCS is better than the best solutions found by other techniques. From Table 6, it can be seen that the average searching quality of TLCS is also better than those of other methods and even the worst solution found by CPSO is better than the best solutions found by Sandgren [17], Kannan and Kramer [16], and Deb [15].

Based on the above simulation results and comparisons, it can be concluded that TLCS is of superior searching quality and robustness for constrained engineering design problems. Moreover, our proposed TLCS is more effective than GA-based co-evolution by Coello [6]. So it can be pointed out that TLCS is a better alternative for constrained optimization.

4 Conclusions

In this paper, a new hybrid algorithm is applied to solve the constrained engineering design problems. The results are compared with other previously reported results and it shows that the proposed method outperforms technics on its effectiveness, efficiency, and robustness. Future work is adopting this new method on wider spectrum problems such as manufacturing parameters optimization problems. In addition, the parallel implementation of TLCS and its application on constrained combinatorial optimization problems will be studied.

Acknowledgment This research work is supported by the National Basic Research Program of China (973 Program) under grant no. 2011CB706804.

Table 5 Comparison of the best solution found by different methods

Design variables	TLCS	Sandgren [17]	Kannan and Kramer [16]	Deb [15]	Coello [6]	Coello [4]	He and Wang [14]
$x_1(T_s)$	0.812500	1.125000	1.125000	0.937500	0.812500	0.812500	0.812500
$x_2(T_h)$	0.437500	0.625000	0.625000	0.500000	0.437500	0.437500	0.437500
$x_3(R)$	42.09845	47.700000	58.291000	48.329000	40.323900	42.097398	42.091266
$x_3(L)$	176.6366	117.701000	43.690000	112.679000	200.000000	176.654050	176.7465
$g_1(x)$	−1.1102e−16	−0.204390	−0.000016	−0.004750	−0.034324	−0.000020	−0.000139
$g_2(x)$	−0.03588083	−0.169942	−0.068904	−0.038941	−0.052847	−0.035891	−0.035949
$g_3(x)$	−2.9104e−10	−54.226012	−21.220104	−3652.8768	−27.105845	−27.886075	−116.382700
$g_4(x)$	−63.3634	−122.2990	−196.3100	−127.3210	−40.0000	−63.345953	−63.253500
$f(x)$	**6059.7143**	8129.1036	7198.0428	6410.3811	6288.7445	6059.9463	6061.0777

Table 6 Statistical results of different methods for pressure vessel design problem

Method	Best	Mean	Worst	Std Dev
TLCS	**6059.7143**	**6088.5879**	6381.8220	79.1187
Sandgren [17]	8129.1036	N/A	N/A	N/A
Kannan and Kramer [16]	7198.0428	N/A	N/A	N/A
Deb [15]	6410.3811	N/A	N/A	N/A
Coello [6]	6288.7445	6293.8432	**6308.1497**	**7.4133**
Coello [4]	6059.9463	6177.2533	6469.3220	130.9297
He and Wang [14]	6061.0777	6147.1332	6363.8041	86.4545

References

1. Arora, J.S.: Introduction to Optimum Design. McGraw-Hill, New York (1989)
2. Belegundu, A.D.: A Study of Mathematical Programming Methods for Structural Optimization. Department of Civil and Environmental Engineering, University of Iowa, Iowa City (1982)
3. Smith, A.E., Coit, D.W.: Constraint handling techniques—penalty functions. In: Bäck, T., Fogel, D.B., Michalewicz, Z. (eds.) Handbook of Evolutionary Computation, pp. C 5.2:1–C 5.2:6. Oxford University Press, Institute of Physics Publishing, Oxford (1997)
4. Coello, C.A.C.: Theoretical and numerical constraint-handling techniques used with evolutionary algorithms: a survey of the state of the art. Comput. Methods Appl. Mech. Eng. **191**, 1245–87 (2002)
5. Brest, J., Zumer, V., Maucec, M.S.: Self-adaptive differential evolution algorithm in constrained real-parameter optimization. In: 2006 IEEE Congress on Evolutionary Computation (CEC'2006), IEE, Vancouver, pp. 919–926 (2006)
6. Coello, C.A.C.: Use of a self-adaptive penalty approach for engineering optimization problems. Comput. Ind. **41**, 113–127 (2000)
7. Hamida, S.B., Schoenauer, M.: ASCHEA: new results using adaptive segregational constraint handling. In: Fogel, D.B. et al. (Eds.), Proceedings of the 2002 Congress on Evolutionary Computation IEEE Service Center, Piscataway, pp. 884–889 (2002)
8. Huang, J., Li, X., Gao, L.: A new hybrid algorithm for unconstrained optimisation problems. Int. J. Comput. Appl.Technol. **46**(3), 187–194 (2013)
9. Yang, X.S., Deb, S.: Cuckoo search via Lévy flights. In: Proceedings of World Congress on Nature & Biologically Inspired Computing (NaBIC 2009, India), IEEE Publications, USA, pp. 210–214 (2009)
10. Rao, R.V., Savsani, V.J., Vakharia, D.P.: Teaching–learning-based optimization: a novel method for constrained mechanical design optimization problems. Comput. Aided Des. **43**(3), 303–315 (2011)
11. Rao, S.S.: Engineering Optimization. Wiley, New York (1996)
12. Ragsdell, K.M., Phillips, D.T.: Optimal design of a class of welded structures using geometric programming. ASME J. Eng. Ind. **98**(3), 1021–1025 (1976)
13. Deb, K.: Optimal design of a welded beam via genetic algorithms. AIAA J. **29**(11), 2013–2015 (1991)
14. He, Q., Wang, L.: An effective co-evolutionary particle swarm optimization for constrained engineering design problems. Eng. Appl. Artif. Intel. **20**, 89–99 (2007)
15. Deb, K.: GeneAS: a robust optimal design technique for mechanical component design. In: Dasgupta, D., Michalewicz, Z. (eds.) Evolutionary Algorithms in Engineering Applications, pp. 497–514. Springer, Berlin (1997)

16. Kannan, B.K., Kramer, S.N.: An augmented Lagrange multiplier based method for mixed integer discrete continuous optimization and its applications to mechanical design. J. Mech. Des. **116**, 318–320 (1994) (Transactions of the ASME)
17. Sandgren, E.: Nonlinear integer and discrete programming in mechanical design. In: Proceedings of the ASME Design Technology Conference, Kissimine, pp. 95–105 (1988)

Leader-Following Consensus of Second-Order Multi-Agent Systems with Switching Topologies

Li Xiao, Xi Shi, and Huaqing Li

Abstract A leader-following consensus problem of second-order multi-agent systems with switching topologies is considered in this thesis. The consensus problem of the multi-agent systems is converted to the stability problem of the error dynamical system here. By studying the stability properties of error switched systems consisting of both Hurwitz stable and unstable via the average dwell time approach, the necessary condition for the agents reaching leading-following consensus is obtained. It is found that if the average dwell time is chosen sufficiently large and the total activation time of unstable subsystems is relatively small compared with that of Hurwitz stable subsystems, the multi-agent systems can reach leader-following consensus. Finally, the effectiveness of the theoretical findings is demonstrated through some numerical examples.

Keywords Consensus • Multi-agent • Leader-following • Switching topologies

1 Introduction

Recent literatures have witnessed steadily increasing recognition and attention of coordinated motion of mobile agents across a broad range of disciplines [1, 2]. Research on multi-agent coordinated control problems benefits many practical applications of networked cyber-physical systems [3–11]. A fundamental approach to achieve cooperative control is consensus [3]. Olfati-Saber and Murray [4] presented a systematic framework to analyze the first-order consensus algorithms. Ren and Beard [5] generalized the results of [4] and presented more relaxed condition for the topology of directed networks, that is, interaction graph has a

L. Xiao (✉) • X. Shi
College of Computer Science, Chongqing University, Chongqing 400044, PR, China

Department of Mathematics and Information Engineering, Chongqing University of Education, Chongqing 400067, PR, China
e-mail: xsxiaoli@163.com

H. Li
College of Computer Science, Chongqing University, Chongqing 400044, PR, China

D. Gao et al. (eds.), *Advances in Global Optimization*, Springer Proceedings in Mathematics & Statistics 95, DOI 10.1007/978-3-319-08377-3_38

directed spanning tree. Sun et al. [6] discussed the first-order average consensus problem of dynamic agents with multiple time-varying communication delays. Lu et al. [7] studied the first-order consensus problem over directed networks with arbitrary finite communication delays and nonlinear couplings. Recently, the second-order consensus problem of multi-agent systems has received increasing attention [8–13] and the references therein. The extension of consensus algorithms from first-order to second-order is non-trivial [8], and the second-order consensus problem is more complicated and challenging than the first-order case.

For most of the existing consensus protocols, the final common value to be achieved is a function of initial states of all the agents and is inherently a prior unknown constant [14, 15]. However in real applications, it is often required that all the agents converge to a desired common value. To solve this problem, leader-following consensus protocols of multi-agent systems have been proposed, where the leader is a particular agent, whose motion is independent of the other agents and followed by the other ones. The leader-following consensus has been an active area of research [10, 16–20]. Jadbabaie et al. [18] considered such a leader-following consensus problem and proved that if all the agents were jointly connected with their leader, their states would converge to that of the leader as time goes on. Peng and Yang [19] discussed the leader-following consensus problem with a varying-velocity leader and time-varying delays. Distributed observer design for leader-following control of multi-agent networks was considered in [10, 20] provided a rigorous proof for the consensus using an extension of LaSalle's invariance principle.

In this paper, we investigate the leader-following consensus problem of second-order multi-agent systems with switching topologies. The consensus problem of the multi-agent systems is converted to the stability problem of the error dynamical system here. By studying the stability properties of error switched systems consisting of both Hurwitz stable and unstable via the average dwell time approach, the necessary condition for the agents reaching leading-following consensus is obtained. It is found that if the average dwell time is chosen sufficiently large and the total activation time of unstable subsystems is relatively small compared with that of Hurwitz stable subsystems, the multi-agent systems can reach leader-following consensus.

The rest of this paper is organized as follows. In Sect. 2, some preliminaries on the graph theory and the model formulation are given. The main results are established in Sect. 3. In Sect. 4, a numerical example is simulated to verify the theoretical analysis. Conclusions are finally drawn in Sect. 5.

2 Preliminaries and Model

2.1 Graph Theory

We consider a system consisting of N agents labeled by agents 1 to N and a leader labeled by 0.

Let $g=(V,\ \varepsilon,\ A)$ be a weighted digraph of order N with the set of nodes $V=\{1,\ 2,\ \ldots\ ,N\}$, set of edges $\varepsilon \subseteq V\times V$, and a weighted adjacency matrix $A=(a_{ij})_{N\times N}$ with non-negative elements, where a_{ij} is a non-negative element if $\varepsilon_{ij}\in\varepsilon$, while $a_{ij}=0$ if $\varepsilon_{ij}\notin\varepsilon$. $(i,\ j)\in\varepsilon$ means that node i can directly receive information from node j. The set of neighbors is denoted by $N_i=\{j\in V:\ (i,j)\in\varepsilon,\ i\neq j\}$. Let (generally nonsymmetrical) Laplacian matrix $L=(l_{ij})_{N\times N}$ associated with directed network g be defined by

$$l_{ij}=\begin{cases}\sum_{k=1,\ k\neq i}^{N}a_{ik} & i=j,\\ -a_{ij} & i\neq j,\end{cases}$$

which ensure the diffusion property $\sum_{j=1}^{N}l_{ij}=0$.

We use $\overline{g}=\left(\overline{V},\ \overline{\varepsilon},\ \overline{A}\right)$ to model the network topology in this case, where $\overline{V}=\{0,1,\ 2,\ \ldots\ ,N\}$, $\varepsilon\subseteq\overline{\varepsilon}$, $\overline{A}$ is the adjacency matrix for agents $0,1,\ 2,\ \ldots\ ,N$. To depict whether agents are connected to the leader in digraph $\overline{g}$, we define the leader adjacency matrix $B=diag\{b_1,\ b_2,\ \ldots\ ,b_N\}$ associated with $\overline{g}$, where $b_i=a_{i0}>0$ if the leader is the neighbor of node i and $b_i=0$ otherwise. From the above, we can see that the relationships between matrices $\overline{A},\ A,\ B_1$ can be described as $\overline{A}=\begin{bmatrix}0 & \mathbf{0}_N^{\mathrm{T}}\\ B_1 & A\end{bmatrix}$, where $B_1=[b_1,\ b_2,\ \ldots\ ,b_N]^T$. Throughout this paper, we always assume that the weights of all edges are 1.

2.2 Model Description

Consider the following second-order agents networks of N agents:

$$\begin{cases}\dot{x}_i(t)=v_i(t),\\ \dot{v}_i(t)=u_i(t), & i=1,\ 2,\ldots,\ N.\end{cases}\tag{1}$$

where $x_i(t)\in R^n$ and $\upsilon_i(t)\in R^n$ are the position and velocity states of the ith agent, respectively.

The dynamics of the leader is expressed as follows:

$$\begin{cases}\dot{x}_0(t)=v_0(t)\\ \dot{v}_0(t)=0\end{cases}\tag{2}$$

where $v_o(t)$ is a constant indicating the desired constant velocity.

Our control goal is to let all the agents follow the leader asymptotically with the velocity of all agents converging to $v_o(t)$, namely, $x_i(t)\to x_0(t), v_i(t)\to v_0(t)$ as $t\to\infty$. In this paper, the following neighbor-based protocol is used:

$$u_i = \gamma(\sum_{j \in N_i} a_{ij}(x_j(t) - x_i(t)) + b_i(x_0(t) - x_i(t)))$$
$$+ \eta(\sum_{j \in N_i} a_{ij}(v_j(t) - v_i(t)) + b_i(x_0(t) - x_i(t))) \quad (3)$$

Let $p_i = x_i - x_0, q_i = v_i - v_0, i = 1, 2, \ldots N$. Then (3) can be rewritten as

$$u_i = \gamma \sum_{j \in N_i} a_{ij}\left(p_j(t) - p_i(t)\right) + \eta \sum_{j \in N_i} a_{ij}\left(q_j(t) - q_i(t)\right) - \gamma b_i p_i(t) - \eta b_i q_i(t) \quad (4)$$

The system (1), (2) can be expressed as follows:

$$\begin{cases} \dot{p}(t) = q(t) \\ \dot{q}(t) = -\gamma L_{\sigma(t)} p(t) - \eta L_{\sigma(t)} q(t) - \gamma B_{\sigma(t)} p(t) - \eta B_{\sigma(t)} q(t) \end{cases} \quad (5)$$

where, $p(t) = (p_1(t), p_2(t), \ldots, p_N(t))^T$, $q(t) = (q_1(t), q_2(t), \ldots, q_N(t))^T$. Let $\varepsilon = (p(t)^T, q(t)^T)^T$,then

$$\dot{\varepsilon}(t) = F_{\sigma(t)}\begin{pmatrix} 0_{N \times N} & I_N \\ -\gamma L_{\sigma(t)} - \gamma B_{\sigma(t)} & -\eta L_{\sigma(t)} - \eta B_{\sigma(t)} \end{pmatrix}\varepsilon(t) \quad (6)$$

$$\dot{\varepsilon}(t) = F_{\sigma(t)}\varepsilon(t) \quad (7)$$

3 Main Results

To describe the subsystems, a piecewise constant function of time is defined as

$$\sigma(t) : [0, \infty) \to \Lambda = \{1, 2, \ldots M\}$$

where $M > 1$ denotes the number of topologies. The $F_{\sigma(t)}, \sigma(t) \in \{1, 2, \ldots, M\}$ is the error subsystems corresponding the M topologies.

For any switching signal $\sigma(t)$ and any $t \geq \tau \geq 0$, let $N_\sigma(\tau, t)$ denote the number of switchings over the interval (τ, t). Let $\Psi[\tau_a, N_0]$ denotes the set of all switching signals satisfying

$$N_\sigma(\tau, t) \leq N_o + \frac{t - \tau}{\tau_a}$$

where N_o denotes the chatter bound, and the constant τ_a is called the average dwell time. The idea there is that there may exist some consecutive switching separated by less than τ_a, but the average interval between consecutive switching is no less than τ_a. The method used in [21] is introduced for the stability analysis of system (7) so that a sufficient condition is given to ensure that the second-order consensus of system (1) can be reached.

Definition 1 [21]. For certain switching signal $\sigma(t)$, the switched system (7) is said to be globally exponentially stable, if there exists the constants a and λ so that $\|\varepsilon(t)\| \leq e^{a-\lambda(t-t_0)t} \|\varepsilon(0)\|$ holds for all $t \geq t_o$. Since both Hurwitz stable and unstable subsystems $F_{\sigma(t)}, \sigma(t) \in \{1, 2, \ldots, M\}$ exist in (7), without loss of generality, we assume that $F_1, F_2, \ldots F_r$ are unstable and remaining matrices $F_{r+1}, F_{r+2}, \ldots F_M$ are Hurwitz stable. Then there always exist a set of scalars $\beta_i > 0$ and α_i such that

$$\begin{cases} \left\| e^{F_i t} \right\| \leq e^{\alpha_i + \beta_i t}, & 1 \leq i \leq r \\ \left\| e^{F_i t} \right\| \leq e^{\alpha_i - \beta_i t}. & r+1 \leq i \leq M \end{cases} \tag{8}$$

Let $\beta^+ = \min_{r+1 \leq q \leq M} \beta^q$, $\beta^- = \max_{1 \leq q \leq r} \beta^q$.

For the switching signal $\sigma(t)$, we define $T^+(\tau, t)$ and $T^-(\tau, t)$ denote the total activation time of the stable subsystems and the unstable subsystems during the interval $[\tau, t), t > \tau \geq 0$.

Assumption 1. Choosing a scalar $\beta^* \in (\beta, \beta^+)$ arbitrarily for any given $\beta \in (0, \beta^+)$, determine the switching signal $\sigma(t)$, so that $\frac{T^+(t_0,t)}{T^-(t_0,t)} \geq \frac{\beta^- + \beta^*}{\beta^+ - \beta^*}$ holds for any $t > t_0$.

Theorem 1. *Suppose Assumption* **1** *holds, then there is a finite positive constant* τ_a^* *such that the switched system* (7) *is globally exponentially stable with stability degree* β *over* $\Psi[\tau_a, N_0]$ *for any average dwell time* $\tau_a \geq \tau_a^*$ *and the chatter bound* $N_0 > 0$.

Proof. Let $t_1, t_2, \ldots$ denote the time points at which switching occurs, $m_j \in \{1, 2, \ldots, M\}$ denote the value of $\sigma(t)$ on $[t_{j-1}, t_j)$. Then, for t satisfying $t_0 < \cdots < t_i \leq t \leq_{i+1}$, we have:

$$\varepsilon(t) = e^{F_{m_i+1}(t-t_i)} e^{F_{m_i}(t_i-t_{i-1})} \cdots e^{F_{m_1}(t_1-t_0)} \varepsilon_0 \tag{9}$$

where $\varepsilon_0 = \varepsilon(t_0)$.

From the (8), we collect the terms of Hurwitz stable and unstable subsystems respectively. Thus,

$$\|\varepsilon(t)\| \leq \left(\prod_{q=1}^{i+1} e^{\alpha_{m_q}}\right) e^{\beta^- T^-(t_0,t) - \beta^+ T^+(t_0,t)} \|\varepsilon_0\|. \tag{10}$$

Let $\alpha = \max_{q\in\Lambda}\alpha_q, \omega = e^{\alpha}$, one has

$$\begin{aligned}\|\varepsilon(t)\| &\leq e^{(i+1)\alpha+\beta^{-}T^{-}(t_0,t)-\beta^{+}T^{+}(t_0,t)}\|\varepsilon_0\| \\ &= \omega e^{\alpha N_{\sigma}(t_0,t)+\beta^{-}T^{-}(t_0,t)-\beta^{+}T^{+}(t_0,t)}\|\varepsilon_0\|\end{aligned} \tag{11}$$

From the Assumption **1**, we obtain

$$\beta^{-}T^{-}(t_0,t) - \beta^{+}T^{+}(t_0,t) \leq -\beta^{*}\left(T^{-}(t_0,t) + T^{+}(t_0,t)\right) = -\beta^{*}(t-t_0) \tag{12}$$

Combining (11) and (12), we have:

$$\|\varepsilon(t)\| \leq \omega e^{\alpha N_{\sigma}(t_0,t)-\beta^{*}(t-t_0)}\|\varepsilon_0\| \tag{13}$$

When $\alpha \leq 0$, since $\omega = e^{\alpha} > 0$, we have

$$\omega e^{\alpha N_{\sigma}(t_0,t)-\beta^{*}(t-t_0)}\|\varepsilon_0\| \leq e^{-\beta^{*}(t-t_0)}\|\varepsilon_0\| \leq e^{-\beta(t-t_0)}\|\varepsilon_0\|$$

By (13), we can get

$$\|\varepsilon(t)\| \leq e^{-\beta(t-t_0)}\|\varepsilon_0\| \tag{14}$$

When $\alpha > 0$, based on $N_{\sigma}(\tau,t) \leq N_o + \frac{t-\tau}{\tau_a}$, the following inequation holds

$$N_{\sigma}(t_0,t) \leq N_0 + \frac{(t-t_0)}{\tau_{\alpha}^{*}} \tag{15}$$

on the interval $[t_0, t)$ as $\tau_a \geq \tau_a^{*}$.

Let $\tau_{\alpha}^{*} = \frac{\alpha}{\beta^{*}-\beta}$, $N_0 = \frac{\theta}{\alpha}$, it gives $N_{\sigma}(t_0,t) \leq \frac{\theta}{\alpha} + \frac{(t-t_0)}{\left(\frac{\alpha}{\beta^{*}-\beta}\right)}$, which is equivalent to $\alpha N_{\sigma}(t_0,t) - \beta^{*}(t-t_0) \leq \theta - \beta(t-t_0)$. Thus,

$$e^{\alpha N_{\sigma}(t_0,t)-\beta^{*}(t-t_0)} \leq e^{\theta-\beta(t-t_0)} \tag{16}$$

Based on the above, from (13), one obtains

$$\|\varepsilon(t)\| \leq e^{\theta-\beta(t-t_0)}\|\varepsilon_0\| \tag{17}$$

Since θ is arbitrary, N_0 can also be specified arbitrarily; we conclude that the system (7) is globally exponentially stable with stability degree β over $\Psi[\tau_a, N_0]$ for any average dwell time $\tau_a \geq \tau_a^{*}$ and chatter bound N_0. Proof is completed. □

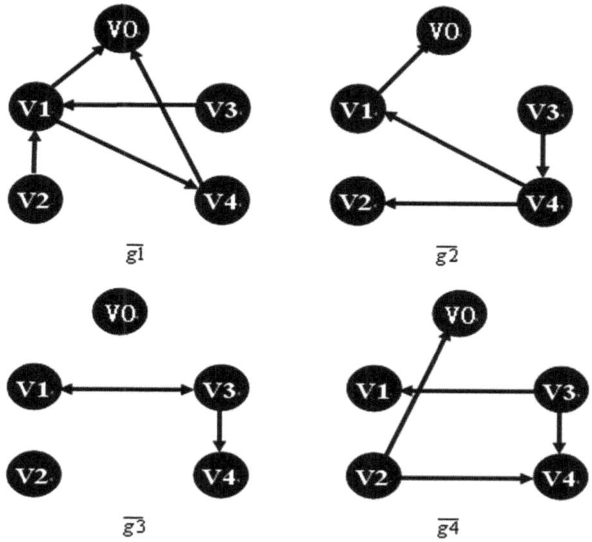

Fig. 1 Switching topologies: $\overline{gi}, i = 1, 2, 3, 4$

4 Numerical Simulations

Consider a leader-following consensus with four switching topologies; see Fig. 1.

Let $\gamma = 2, \eta = 1$. Similar to example 1, F_1, F_2, F_3, F_4 can be easily computed, and it has that $F1$ and $F2$ is Hurwitz stable while $F3$ and $F4$ is unstable. From (8), $\alpha_1 = 1.8, \beta_1 = 0.4$, $\alpha_2 = 1.9, \beta_2 = 0.4$, $\alpha_3 = 1.3, \beta_1 = 0.2$, and $\alpha_4 = 1.0, \beta_4 = 0.2$. Since $\beta^+ = 0.4, \beta^- = 0.2$, we choose $\beta = 0.03, \beta* = 0.2$, then the condition in Assumption **1** will require $\frac{T^+(t_0,t)}{T^-(t_0,t)} \geq 2$, $\tau_a^* = 10.5$. Then under the switching sequence showed in Fig. 2, the multi-agent system (1) can reach the consensus, where the evolutions of positions and velocities of all agents are shown in Figs. 3 and 4.

5 Conclusions

In this paper, we studied some stability properties of the error system of second-order multi-agent systems with switching topologies where both Hurwitz stable and unstable subsystems exit. A sufficient condition for the agents reaching leading-following consensus is obtained. It is found that if the average dwell time is chosen

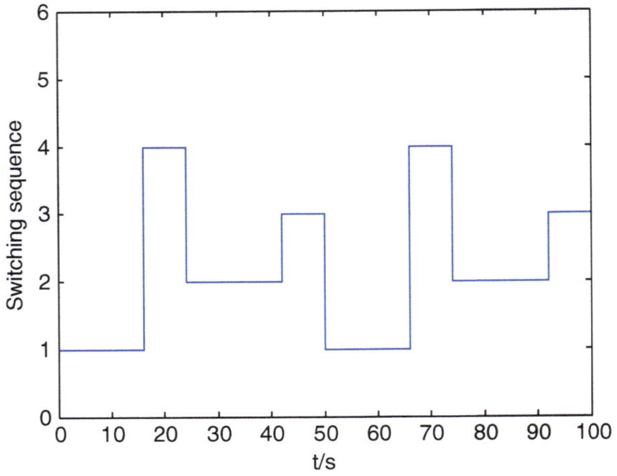

Fig. 2 Switching sequence with four topologies

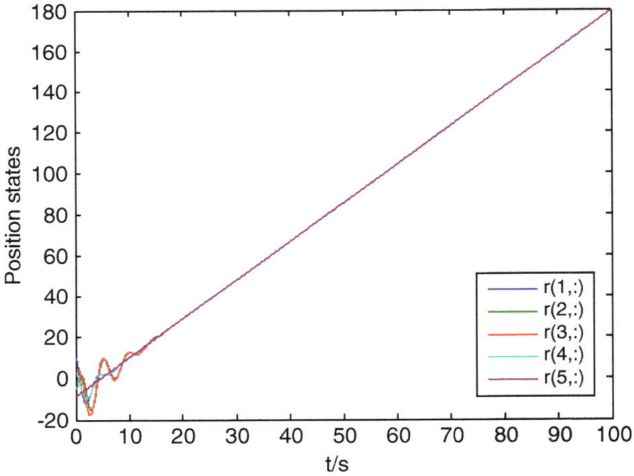

Fig. 3 Position states of the agents in the multi-agent system (1) with four switching topologies

sufficiently large and the total activation time of unstable subsystems is relatively small compared with that of Hurwitz stable subsystems, the multi-agent systems can reach leader-following consensus.

Acknowledgments The work is supported by the Fundamental Research Funds for the Central Universities (No. CDJXS11182239), the National Natural Science Foundation of China (No. 60973114), the Foundation of Chongqing University of Education (No. KY201318B), and the Foundation project of CQCSTC (No. cstc2014jcyjA40041).

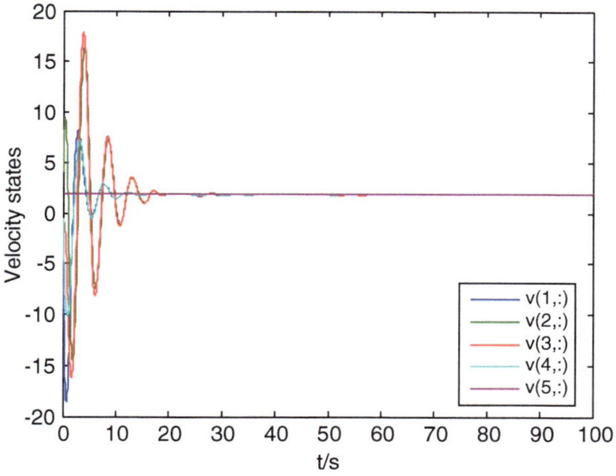

Fig. 4 Velocity states of the agents in the multi-agent system (1) with four switching topologies

References

1. Hu, J.P., Hong, Y.G.: Leadering-following coordination of multi-agent systems with coupling time delays. Physica A **374**, 853–863 (2007)
2. Ren, W., Beard, R., Atkins, E.: Information consensus in multivehicle cooperative control: collective group behavior through local interaction. IEEE Contr. Syst. Mag. **27**, 71–82 (2007)
3. Lin, P., Jia, Y.M.: Consensus of second-order discrete-time multi-agent systems with nonuniform time-delays and dynamically changing topologies. Automatica **45**, 2145–2158 (2009)
4. Olfati-Saber, R., Murray, R.M.: Consensus problems in networks of agents with switching topology and time-delays. IEEE Trans. Automat. Contr. **49**, 1520–1533 (2004)
5. Ren, W., Beard, R.W.: Consensus seeking in multi-agent systems under dynamically changing interaction topologies. IEEE Trans. Automat. Contr. **50**, 655–661 (2005)
6. Sun, Y., Wang, L., Xie, G.: Average consensus in networks of dynamic agents with switching topologies and multiple time-varying delays. Syst. Contr. Lett. **57**, 175–183 (2008)
7. Lu, J., Ho, D.W.C., Kurths, J.: Consensus over directed static networks with arbitrary finite communication delays. Phys. Rev. E. **80**, 066121-1–066121-7 (2009)
8. Ren, W., Atkins, E.: Distributed multi-vehicle coordinated control via local information exchange. Int. J. Robust Nonlinear Contr. **17**, 1002–1033 (2007)
9. Xie, G., Wang, L.: Consensus control for a class of networks of dynamic agents. Int. J. Robust Nonlinear Contr. **17**, 941–959 (2007)
10. Hong, Y., Chen, G., Bushnell, L.: Distributed observers design for leader-following control of multi-agent networks. Automatica **44**, 846–850 (2008)
11. Sun, Y., Wang, L.: Consensus problems in networks of agents with double integrator dynamics and time-varying delays. Int. J. Contr. **82**, 1937–1945 (2009)
12. Tian, Y., Liu, C.: Robust consensus of multi-agent systems with diverse input delays and asymmetric interconnection perturbations. Automatica **45**, 1347–1353 (2009)
13. Lin, P., Jia, Y.: Consensus of second-order discrete-time multi-agent systems with nonuniform time-delays and dynamically changing topologies. Automatica **45**, 2154–2158 (2009)
14. Cortés, J.: Distributed algorithms for reaching consensus on general functions. Automatica **44**, 726–737 (2008)

15. Jiang, F.C., Xie, G.M., Wang, L., Chen, X.J.: The χ-consensus problem of high- order multi-agent systems with fixed and switching topologies. Asian J. Contr. **10**, 246–253 (2008)
16. Hammel, D.: Formation flight as an energy saving mechanism. Israel J. Zool. **41**, 261–278 (1995)
17. Andersson, M., Wallander, J.: Kin selection and reciprocity in flight formation. Behav. Ecol. **15**, 158–162 (2004)
18. Jadbabaie, A., Lin, J., Morse, A.S.: Coordination of groups of mobile autonomous agents using nearest neighbor rules. IEEE Trans. Automat. Contr. **48**, 943–948 (2007)
19. Peng, K., Yang, Y.P.: Leader-following consensus problem with a varying-velocity leader and time- varying delays. Physica A **388**, 193–208 (2008)
20. Cheng, D., Wang, J., Hu, X.: An extension of Lasall's invariance principle and its application to multi-agent consensus. IEEE Trans. Automat. Contr. **53**, 1765–1770 (2008)
21. Zhai, G.S., Hu, B., Yasuda, K., et al.: Stability analysis of switched systems with stable and unstable subsystems: an average dwell time approach. In: Proceedings of American Control Conference, American Automatic Control Council, Chicago, lllinois, pp. 200–204 (2002)
22. Yu, W.W., Chen, G.R., Ming, C.: Consensus in directed networks of agents with nonlinear dynamics. IEEE Trans. Automat. Contr. **56**, 1436–1441 (2011)

A Semismooth Newton Multigrid Method for Constrained Elliptic Optimal Control Problems

Jun Liu, Tingwen Huang, and Mingqing Xiao

Abstract A multigrid scheme is proposed for solving the Schur complement linear systems arising in each Newton iteration when the semi-smooth Newton method is applied to solve control-constrained elliptic optimal control problems. Numerical experiments are performed to illustrate the high efficiency of our proposed method. Computation simulation shows that the convergence rate is quite robust as the regularization parameter approaches to zero.

Keywords Semismooth Newton method • Multigrid method • Elliptic optimal control • Schur complement

1 Introduction

The optimization and control of the models of complex processes in natural sciences, engineering, and economics often leads to optimal control problems governed by partial differential equations with control constraints [1–3]. During the last decade, increasing varieties of applications necessitate an extensive study for developing efficient and robust numerical methods to approximate the optimal solutions [4–6].

As a prototype application, we consider the following linear-quadratic elliptic optimal control problem with an objective functional of tracking type

$$\begin{aligned} &\inf_{u\in\mathcal{U}_{ad}} \quad J(u) = \tfrac{1}{2}\|z(u) - d\|^2_{L^2(\Omega)},\\ &s.t. \quad -\Delta z = u + g \text{ in } \Omega, \quad z = 0 \text{ on } \partial\Omega, \end{aligned}$$

J. Liu • M. Xiao (✉)
Department of Mathematics, Southern Illinois University, Carbondale, IL 62901, USA
e-mail: junliu2010@siu.edu; mxiao@siu.edu

T. Huang
Texas A&M University at Qatar, PO Box 23874, Doha, Qatar
e-mail: tingwen.huang@qatar.tamu.edu

D. Gao et al. (eds.), *Advances in Global Optimization*, Springer Proceedings in Mathematics & Statistics 95, DOI 10.1007/978-3-319-08377-3_39

where $\Omega = (0, 1)^2$, $g, d \in L^2(\Omega)$, and the set of admissible controls

$$\mathscr{U}_{ad} = \{u \in L^2(\Omega) |\ \underline{u} \leq u \leq \bar{u},\ a.e. \text{ in } \quad \Omega\}$$

with $\underline{u}, \bar{u} \in L^\infty(\Omega)$. It is well known [2] that the minimizer to the above problem may not be unique. The uniqueness usually requires the objective functional to be coercive in terms of the control function. A useful strategy of approximating an optimal control u^* is to introduce a Tikhonov regularization term as following:

$$\begin{aligned} &\inf_{u \in \mathscr{U}_{ad}} \quad J_\alpha(u) = \tfrac{1}{2}\|z - d\|^2_{L^2(\Omega)} + \tfrac{\alpha}{2}\|u\|^2_{L^2(\Omega)}, \\ &s.t. \quad -\Delta z = u + g \text{ in } \Omega, \quad z = 0 \text{ on } \partial\Omega, \end{aligned} \tag{1}$$

where $\alpha > 0$ is the regularization parameter. With this setting, $J_\alpha(u)$ is coercive and thus its global minimizer u_α is unique. Furthermore, we have $u_\alpha \to u^*$ as $\alpha \to 0$.

The optimal solution of (1) is characterized by the optimality system [2]:

$$\begin{aligned} -z + \Delta p \quad &= -d \quad \text{in } \Omega, \quad p = 0 \quad \text{on } \partial\Omega. \\ \Delta z + \Phi(p/\alpha) &= -g \quad \text{in } \Omega, \quad z = 0 \quad \text{on } \partial\Omega; \end{aligned} \tag{2}$$

where $\Phi(\cdot)$ is element-wise defined by $\Phi(f) = \min\{\underline{u}, \max\{\bar{u}, f\}\}$.

In [4, 7], the authors presented a full approximation storage (FAS) multigrid scheme with a projected Gauss–Seidel smoother. Its convergence rate with a fixed step size $h = 2^{-7}$ appeared to be deteriorated when $\alpha \leq 10^{-6}$. The authors argued that sufficiently fine mesh size h should be used to resolve the switching structure in order to obtain better convergence rate. However, such a remedy would be very costly. In [8], the authors theoretically proved, for an unconstrained model, their proposed W-cycle multigrid method has an α-independent convergence rate which is given by $\alpha \geq ch^4$ for some constant c. A similar condition was given in [9], where the authors developed a multigrid method with a block preconditioned Richardson iteration as smoother for solving the KKT system arising in each iteration of a primal-dual active-set method [10]. It was shown [11] that the primal-dual active-set method for solving the optimality (2) is theoretically equivalent to the semi-smooth Newton method. However, the primal-dual active-set approach is required to introduce an additional parameter that limits the computational efficiency when α is sufficiently small. It is known that the semi-smooth Newton method, under suitable assumptions, has a solid provable mesh-independent convergence [12]. This motivates us to improve the semi-smooth Newton iteration by using multigrid method, which differs from [9] by discarding the Lagrange multipliers. This simple but critical difference allows us to obtain a more robust Gauss–Seidel smoother for solving the Schur complement part. Our proposed approach appears to be able to handle those cases when $\alpha \to 0$ in an efficient way.

Our paper is organized as follows. In Sect. 2 we present a new approach by combining the semismooth Newton iteration and the multigrid method to approximate the solution of (2). The discretization is based on finite difference.

Numerical examples are given in Sect. 3 to demonstrate the effectiveness of the proposed approach. Concluding remarks are presented in Sect. 4.

2 Semismooth Newton Multigrid Method

In this section, we focus on the numerical scheme for solving the optimality system (2). This is a typical optimize-then-discretize approach. Let us discretize Ω uniformly as

$$\Omega_h = \{(x_i, y_j) = (ih, jh)| \quad i = 0, 1, 2, \cdots, n, j = 0, 1, 2, \cdots, n.\}$$

with step size $h = 1/n$. Denote the approximations to $f(x, y)$ over all grid points (x_i, y_j) by a grid function $\tilde{f}_h$ defined on Ω_h. We define the discrete L_h^2 norm [7] of a grid function $\tilde{f}_h$ on Ω_h by $\|\tilde{f}_h\|_2 = \left(h^2 \sum_{(x_i,y_j)\in\Omega_h} \tilde{f}_h(x_i, y_j)^2\right)^{1/2}$. A standard second order five-point finite difference approximation to (2) gives

$$\begin{aligned} -z_h + \Delta_h p_h \quad &= -d_h, \\ \Delta_h z_h + \Phi(\tfrac{1}{\alpha} p_h) &= -g_h \end{aligned} \tag{3}$$

where $\Delta_h = (I \otimes A_h) + (A_h \otimes I)$ with $A_h = \text{tridiag}(1, -2, 1)/h^2$ and f_h is the vectorization [13] of the matrix $\mathbb{F}(f_h)$ defined by $[\mathbb{F}(f_h)]_{i,j} = \tilde{f}_h(x_i, y_j)$. Here we also define $\|f_h\|_2 := \|\tilde{f}_h\|_2$ as above. For any vector $v \in \mathbb{R}^m$, the diagonal matrix $\mathscr{D}(v) \in \mathbb{R}^{m\times m}$ is defined as $[\mathscr{D}(v)]_{ii} = v_i$ for $1 \le i \le m$.

Let X, Y be Banach spaces and S be an open subset of X. The mapping $F : S \mapsto Y$ is called slantly differentiable [11, 14] in the open subset $U \subset S$ if there exists a family of mappings $G : U \mapsto \mathscr{L}(X, Y)$ such that

$$\lim_{\delta x\to 0} \tfrac{1}{\|\delta x\|} \|F(x + \delta x) - F(x) - G(x + \delta x)\delta x\| = 0.$$

for every $x \in U$. We call such G a slanting function of F.

Theorem 2.1 ([11]). *Suppose that x^* is a solution of $F(x) = 0$ and F is slantly differentiable in an open neighborhood U containing x^* with slanting function $G(x)$. If $G(x)$ is nonsingular and $\|G(x)^{-1}\|$ is bounded for all $x \in U$, then the Newton iteration*

$$x^{k+1} = x^k - G(x^k)^{-1}F(x^k), \quad k = 0, 1, 2, \cdots \tag{4}$$

converges super-linearly to x^*, provided that $x^0 - x^*$ is sufficiently small.

Rewrite (3) as

$$F(z_h, p_h) := \begin{bmatrix} -z_h + \Delta_h p_h + d_h \\ \Delta_h z_h + \Phi(\tfrac{1}{\alpha} p_h) + g_h \end{bmatrix} = 0.$$

Then it's easy to verify that F has a slanting function

$$G(z_h, p_h) = \begin{bmatrix} -I & \Delta_h \\ \Delta_h & \frac{1}{\alpha}\mathscr{D}\left(\chi(\frac{1}{\alpha}p_h)\right) \end{bmatrix}$$

with the element-wise defined characterization function

$$\chi(f(x)) = \begin{cases} 1, & \text{if } \underline{u}(x) < f(x) < \bar{u}(x); \\ 0, & \text{otherwise.} \end{cases}$$

Then the semi-smooth Newton method [5, 6] to (3) iterates according to

$$\begin{bmatrix} z_h^{k+1} \\ p_h^{k+1} \end{bmatrix} = \begin{bmatrix} z_h^k \\ p_h^k \end{bmatrix} - G(z_h^k, p_h^k)^{-1} F(z_h^k, p_h^k), \quad k = 0, 1, 2, \cdots$$

where the initials (z_h^0, p_h^0) is chosen as the unconstrained solution. For a 2-by-2 partitioned nonsingular matrix with a nonsingular $(1, 1)$ block we have [13]

$$\begin{bmatrix} A_{11} & A_{12} \\ A_{21} & A_{22} \end{bmatrix}^{-1} = \begin{bmatrix} I & -A_{11}^{-1}A_{12} \\ 0 & I \end{bmatrix} \begin{bmatrix} A_{11}^{-1} & 0 \\ 0 & (A_{22} - A_{21}A_{11}^{-1}A_{12})^{-1} \end{bmatrix} \begin{bmatrix} I & 0 \\ -A_{21}A_{11}^{-1} & I \end{bmatrix}.$$

Using this formula, it is straightforward to get the uniformly bound

$$\left\| G(z_h^k, p_h^k)^{-1} \right\|_2 \leq (2 + \|\Delta_h^{-2}\|_2) < 2 + \tfrac{1}{16^2},$$

which verifies the non-singularity and boundedness condition in Theorem 2.1. Denote $\mathrm{D}_h^k := \mathscr{D}\left(\chi(\frac{1}{\alpha}p_h^k)\right)$. In each iteration, we need to solve the symmetric system

$$\begin{bmatrix} -I & \Delta_h \\ \Delta_h & \frac{1}{\alpha}\mathrm{D}_h^k \end{bmatrix} \begin{bmatrix} \delta z^k \\ \delta p^k \end{bmatrix} = F(z_h^k, p_h^k). \tag{5}$$

When (5) is only approximately solved, it gives the inexact Newton method [15]. The review paper [16] summarized modern numerical methods for solving (5). However, the numerical computation of (5) becomes more challenging as $\alpha \to 0$ since the system tends to be more ill-conditioned. The fixed-point iterative methods usually deteriorates or fails to converge when α becomes small (about $O(10^{-3})$).

In the following, we illustrate how to solve (5) numerically by making use of multigrid method. Applying the partitioned inverse formula to (5) gives

$$\begin{bmatrix} \delta z^k \\ \delta p^k \end{bmatrix} = \begin{bmatrix} I & \Delta_h \\ 0 & I \end{bmatrix} \begin{bmatrix} -I & 0 \\ 0 & (\frac{1}{\alpha}\mathrm{D}_h^k + \Delta_h^2)^{-1} \end{bmatrix} \begin{bmatrix} I & 0 \\ \Delta_h & I \end{bmatrix} F(z_h^k, p_h^k).$$

Therefore, we only need to solve the following symmetric positive definite linear system comes from the Schur complement part

$$(\tfrac{1}{\alpha}\mathrm{D}_h^k + \Delta_h^2) w_h = f_h. \tag{6}$$

The systems (6) also become very ill-conditioned as $\alpha \to 0$. The multigrid method has been successfully employed to solve ill-conditioned Toeplitz systems [17]. Unfortunately, here the underlying system (6) is not Toeplitz. Therefore, to develop some suitable approximation scheme becomes necessary.

For a given linear system discretized with finest mesh-size h

$$S_h w_h = f_h,$$

the μ-cycle multigrid scheme can be delineated as Algorithm 1 [19]. The case $\mu = 1$ and $\mu = 2$ is known as V-cycle and W-cycle, respectively. It is customary to accelerate above μ-cycle multigrid scheme by combining with a nested iteration. The corresponding full multigrid (FMG) method is described in Algorithm 2 [18, 19].

We choose the red-black Gauss–Seidel iteration as the smoother smooth. Note that Jacobi and Gauss–Seidel methods for solving biharmonic equations are discussed in [20]. Moreover, we define the restriction operator I_h^H from the full-weighting averaging with stencil $\frac{1}{16}\begin{bmatrix} 1 & 2 & 1 \\ 2 & 4 & 2 \\ 1 & 2 & 1 \end{bmatrix}$ and the prolongation operator I_H^h from linear interpolation with stencil $\frac{1}{4}\begin{bmatrix} 1 & 2 & 1 \\ 2 & 4 & 2 \\ 1 & 2 & 1 \end{bmatrix}$. The remaining tricky part is

Algorithm 1	The general μ-cycle multigrid	
Steps	$w_h := \mathsf{MG}(h, S_h, w_0, f_h)$	
-	IF ($h == h_0$)	
(1)	Solve:	$S_h w_h = f_h$
-	ELSE	
(2)	Pre-smooth ν_1 times:	$w_h := \mathsf{smooth}^{\nu_1}(S_h, w_0, f_h)$
(3)	Restriction:	$r_H := I_h^H(f_h - S_h w_h)$
(4)	Initialize correction:	$\delta_H := 0$
(5)	Recursion μ times:	$\delta_H := \mathsf{MG}^{\mu}(H, S_H, \delta_H, r_H)$
(6)	Prolongation:	$\delta_h := I_H^h \delta_H$
(7)	Correction:	$w_h := w_h + \delta_h$
(8)	Post-smooth ν_2 times:	$w_h := \mathsf{smooth}^{\nu_2}(S_h, w_h, f_h)$
-	ENDIF	
(9)	RETURN w_h.	

Algorithm 2	The recursive μ-cycle full multigrid	
Steps	$w_h := \mathsf{FMG}(h, S_h, f_h)$	
–	IF $(h == h_0)$	
(1)	Solve:	$S_h w_h = f_h$
–	ELSE	
(2)	Restriction:	$f_H := I_h^H f_h$
(3)	Recursion:	$w_H := \mathsf{FMG}(H, S_H, f_H)$
(4)	Prolongation:	$w_h := I_H^h w_H$
(5)	One μ-cycle iteration:	$w_h := \mathsf{MG}(h, S_h, w_h, f_h)$
–	ENDIF	
(6)	RETURN w_h.	

to construct the coarse grid operator S_H. Unlike the algebraic way, we perform coarsening in a geometric way by performing the finite difference discretization with a coarse step-size H. There are two possible approaches to coarse the non-smoothness operator D_h^k. We first apply the restriction operator I_h^H to the adjoint variable p_h^k to obtain

$$\mathrm{D}_H^k := \mathscr{D}\left(\chi(\tfrac{1}{\alpha} I_h^H p_h^k)\right)$$

which fails to achieve a satisfactory convergence rate in our simulations. This is mainly due to losing too much information during coarsening. We thus place the restriction operator I_h^H after the characterization of p_h^k by χ, that is,

$$\mathrm{D}_H^k := \mathscr{D}\left(I_h^H \chi(\tfrac{1}{\alpha} p_h^k)\right), \tag{7}$$

which gives the corresponding coarse operator as $S_H := \mathrm{D}_H^k + \Delta_H$. In summary, we have defined the smoothing operator smooth, the restriction operator I_h^H, the prolongation operator I_H^h, and the coarse grid operator S_H.

3 Numerical Results

All numerical experiments are implemented using MATLAB on a laptop with Intel(R) Pentium(R) CPU P6100@2.00 GHz and 6 GB RAM. We set $H = 2h$, $h_0 = 2^{-2}$, $w_0 = 0$ and $(\mu, \nu_1, \nu_2) = (2, 4, 4)$. Numerical experiments show that the W-cycle performs robuster than the V-cycle. Usually, one FMG iteration suffices to reduce the error of the approximation to the level of discretization. Thus, we only perform one FMG iteration plus one additional W-cycle iteration to approximately solve (6) at each Newton iteration. The stopping criterion for the outer Newton iteration is

Table 1 The results of semi-smooth Newton multigrid method with $\alpha = 10^{-8}$

n	$\|z - z_h\|_2$	EOC	$\|p - p_h\|_2$	EOC	$\|u - u_h\|_2$	EOC	Iter	CPU
2^{-2}	3.60e−03		1.17e−01		5.59e−01		4	0.04
2^{-3}	8.53e−04	2.1	2.65e−02	2.1	2.08e−01	1.4	4	0.05
2^{-4}	1.91e−04	2.2	6.48e−03	2.0	7.40e−02	1.5	4	0.14
2^{-5}	4.54e−05	2.1	1.61e−03	2.0	2.62e−02	1.5	4	0.35
2^{-6}	1.12e−05	2.0	4.02e−04	2.0	9.27e−03	1.5	4	0.96
2^{-7}	2.78e−06	2.0	1.00e−04	2.0	3.28e−03	1.5	3	2.40
2^{-8}	6.95e−07	2.0	2.51e−05	2.0	1.16e−03	1.5	3	9.22
2^{-9}	1.74e−07	2.0	6.27e−06	2.0	4.07e−04	1.5	4	44.00

$$\left\| \begin{bmatrix} z_h^{k+1} \\ p_h^{k+1} \end{bmatrix} - \begin{bmatrix} z_h^{k} \\ p_h^{k} \end{bmatrix} \right\|_2 \leq 10^{-8}.$$

Notice that different stopping criterion may result in different iteration numbers. The experimental order of convergence (EOC) is estimated by EOC $= \log_2(e_{2h}/e_h)$, where e_h denotes the discrete L_h^2 error norms of z_h, p_h, and u_h, respectively.

Example 1 ([4]). Let $\underline{u} = -1$ and $\bar{u} = 1$. Choose $d(x, y)$ and $g(x, y)$ such that $z = \sin(\pi x)\sin(\pi y)$ and $p = \sin(2\pi x)\sin(2\pi y)$.

Table 1 shows the error norms, convergence orders, iteration numbers, and the computation times for our proposed method applied to Example 1 with $\alpha = 10^{-8}$. First, one can see that the discretize L_h^2 norm errors for state and adjoint approximations showing expected second-order convergence and the control exhibits $\frac{3}{2}$-order convergence. Second, the number of semismooth Newton iterations to fulfill the stop criterion is mesh-independent. Last, the computational time does present a roughly fourfold increasing since the number of unknowns quadruples from one level to the next, which numerically verifies the optimal $O(N)$ complexity of one FMG iteration.

Table 2 reports how the required numbers of Newton iterations are influenced by the regularization parameter α and discretization step-size h. The almost staggering iteration numbers under all cases illustrate that our proposed method is very robust with respect to the regularization parameter α. Numerically, Table 2 tells that violating the condition $\alpha \geq ch^4$ [8, 9] does not necessarily deteriorate the convergence rate of our proposed method. We believe this advantage comes from our critical step of applying multigrid to the reduced Schur complement system (6) instead the full Newton system (5), which permits an effective and robust Gauss–Seidel smoother.

Example 2 ([7]). Let $\underline{u} = -30$, $\bar{u} = 30$, $g = 0$, and $d(x, y) = \sin(4\pi x)\sin(2\pi y)$. Table 3 displays the convergence iteration numbers for solving Example 2. We do observe slightly increasing of the iteration numbers for fix h as $\alpha \to 0$. This

Table 2 The numbers of semi-smooth Newton iterations for varying h and α appending (Example 1)

$h\backslash\alpha$	10^{-2}	10^{-4}	10^{-6}	10^{-8}	10^{-10}	10^{-12}	10^{-14}
2^{-2}	2	2	4	4	4	4	4
2^{-3}	4	4	4	4	4	4	4
2^{-4}	4	4	4	4	4	4	4
2^{-5}	3	4	4	4	4	4	4
2^{-6}	3	4	4	4	4	4	4
2^{-7}	3	3	3	3	3	3	3
2^{-8}	3	3	3	3	3	3	3
2^{-9}	3	3	3	4	3	3	3

Table 3 The numbers of semi-smooth Newton iterations for varying h and α appending (Example 2)

$h\backslash\alpha$	10^{-2}	10^{-4}	10^{-6}	10^{-8}	10^{-10}	10^{-12}	10^{-14}
2^{-2}	1	1	1	1	1	1	1
2^{-3}	1	3	3	3	3	3	3
2^{-4}	1	3	4	4	4	4	4
2^{-5}	1	3	4	5	5	5	5
2^{-6}	1	3	4	5	6	6	6
2^{-7}	1	3	4	5	7	8	9
2^{-8}	1	3	4	5	6	10	13
2^{-9}	1	3	4	5	7	9	15

is reasonable since one FMG iteration may only provide solutions with limited accuracy to (6), especially for those ill-conditioned cases with $\alpha \leq 10^{-12}$.

4 Concluding Remarks

We have developed a full multigrid scheme for accelerating the semi-smooth Newton method which can be applied to the control-constrained elliptic optimal control problems. Compared to existing methods, numerical results show that our proposed method has a robust convergence rate with respect to the regularization parameter.

Acknowledgements This work was supported by National Priority Research Project NPRP 4-451-2-168 funded by Qatar National Research Foundation.

References

1. Hinze, M., Pinnau, R., Ulbrich, M., Ulbrich, S.: Optimization with PDE Constraints. Springer, New York (2009)
2. Lions, J.L.: Optimal Control of Systems Governed by Partial Differential Equations. Springer, New York (1971)
3. Tröltzsch, F.: Optimal Control of Partial Differential Equations. AMS, Providence (2010)
4. Borzì, A., Schulz, V.: Computational Optimization of Systems Governed by Partial Differential Equations. SIAM, Philadelphia (2012)
5. Ito, K., Kunisch, K.: Lagrange Multiplier Approach to Variational Problems and Applications. SIAM, Philadelphia (2008)
6. Ulbrich, M.: Semismooth Newton Methods for Variational Inequalities and Constrained Optimization Problems in Function Spaces. SIAM, Philadelphia (2011)
7. Borzì, A., Kunisch, K.: A multigrid scheme for elliptic constrained optimal control problems. Comput. Optim. Appl. **31**, 309–333 (2005)
8. Schöberl, J., Simon, R., Zulehner, W.: A robust multigrid method for elliptic optimal control problems. SIAM J. Numer. Anal. **49**(4), 1482–1503 (2011)
9. Engel, M., Griebel, M.: A multigrid method for constrained optimal control problems. J. Comput. Appl. Math. **235**(15), 4368–4388 (2011)
10. Bergounioux, M., Ito, K., Kunisch, K.: Primal-dual strategy for constrained optimal control problems. SIAM J. Control Optim. **37**(4), 1176–1194 (1999)
11. Hintermüller, M., Ito, K., Kunisch, K.: The primal-dual active set strategy as a semismooth Newton method. SIAM J. Optim. **13**(3), 865–888 (2002)
12. Hintermüller, M., Ulbrich, M.: A mesh-independence result for semismooth Newton methods. Math. Program. **101**(1, Ser. B), 151–184 (2004)
13. Horn, R.A., Johnson, C.R.: Matrix Analysis, 2nd edn. Cambridge University Press, Cambridge (2013)
14. Chen, X., Nashed, Z., Qi, L.: Smoothing methods and semismooth methods for nondifferentiable operator equations. SIAM J. Numer. Anal. **38**(4), 1200–1216 (2000)
15. Dembo, R., Eisenstat, S., Steihaug, T.: Inexact newton methods. SIAM J. Numer. Anal. **19**(2), 400–408 (1982)
16. Benzi, M., Golub, G.H., Liesen, J.: Numerical solution of saddle point problems. Acta Numer. **14**, 1–137 (2005)
17. Chan, R.H., Chang, Q.S., Sun, H.W.: Multigrid method for ill-conditioned symmetric Toeplitz systems. SIAM J. Sci. Comput. **19**(2), 516–529 (1998)
18. Briggs, W.L., Henson, V.E., McCormick, S.F.: A Multigrid Tutorial, 2nd edn. SIAM, Philadelphia (2000)
19. Saad, Y.: Iterative Methods for Sparse Linear Systems, 2nd edn. SIAM, Philadelphia (2003)
20. Rodrigue, G., Varga, R.: Convergence rate estimates for iterative solutions of the biharmonic equation. J. Comput. Appl. Math. **24**(1–2), 129–146 (1988)

Modified DIRECT Algorithm for Scaled Global Optimization Problems

Qunfeng Liu, Jianxiong Zhang, and Fen Chen

Abstract DIRECT is a popular deterministic algorithm for global optimization problems. It can find the basins of attraction for global or local optima efficiently, especially when dimension is small. Recently, we have proposed a class of modified DIRECT algorithms to eliminate the sensitivities of the original DIRECT to linear scaling of the objective function. In this paper, we devote to find a specific algorithm with best performance among this class. We compare the performance of the modified DIRECT algorithms on the GKLS test set. Numerical results show that DIRECT-median performs outstanding among this class. What is more, numerical results also show that DIRECT-median can find solutions with high accuracy much more efficiently than the original DIRECT.

Keywords Global optimization • DIRECT algorithm • Linear scaled objective function

1 Introduction

Global optimization problems come from a broad arrange of applications, such as engineering design, financial analysis, chemistry and biology computations, etc. [1–7]. However, due to the lacking of global optimal condition, it is hard to develop efficient algorithms for global optimization problems.

In this paper, we focus on the DIRECT algorithm for the following bound constrained global optimization problems

$$\min_{x \in \Omega} f(x), \tag{1}$$

where $\Omega = \{x \in \mathbb{R}^n | -\infty < l_i \leq x_i \leq u_i < \infty, l_i < u_i, i = 1, \ldots, n\}$, and the dimension n is often small [8–10].

DIRECT is the abbreviation of DIviding RECTangle. Through dividing the feasible region Ω into more and more small hyperrectangles, DIRECT is able to

Q. Liu (✉) • J. Zhang • F. Chen
Dongguan University of Technology, Dongguan, China
e-mail: liuqf@dgut.edu.cn

D. Gao et al. (eds.), *Advances in Global Optimization*, Springer Proceedings in Mathematics & Statistics 95, DOI 10.1007/978-3-319-08377-3_40

find the attraction basin of the global optimal solution efficiently [11–14]. Due to its convergence guaranteeing [8, 15], DIRECT is popularly used in applications. For example, DIRECT and its several variants have been adopted in some popular packages, e.g., the Tomlab package [16, 17].

In [11,18], it has been pointed out that DIRECT is sensitive to the additive scaling of the objective function. In [11], a parameter adaptive approach has been proposed to eliminate the sensitivity. Such approach is substituted with another approach in [18]. The new approach is easy to implement and adds no additive parameter.

Motivated by Finkel and Kelley [18], we have proved in [19] that the original DIRECT's sensitivities to the additive scaling of the objective function stem from the definition of the potential optimal hyperrectangle (POH). Based on the theoretical proof, we proposed a class of modified DIRECT algorithms which eliminate the whole sensitivities to any kind of linear scaling of the objective function. We denote this class of modified DIRECT algorithms as MDIRECT.

In this paper, we compare the performance of different MDIRECT algorithms on the popular GKLS test function set [20, 21]. Our main purpose is to select one MDIRECT algorithm with best performance.

We organize this paper as follows. In Sect. 2, we review the MDIRECT algorithm briefly. In Sect. 3, we introduce the GKLS test function set. In Sect. 4, we report the numerical results. In the final section, we give some conclusions.

2 Modified DIRECT Algorithms

MDIRECT adopts the same approach as DIRECT to divide the feasible region. The only difference between MDIRECT and DIRECT is that they adopt different ways to identify the POHs, which contains the optimal solution potentially, for further dividing.

Specifically, the original DIRECT adopts the following definition of POH.

Definition 7. *Given $\epsilon > 0$ and the index set of the hyperrectangles S. Let f_{min} be the known minimal function value, and c_i, σ_i be the center and the size of the ith hyperrectangle. If there is a constant $\gamma > 0$, such that the following inequalities hold*

$$f(c_j) - \gamma\sigma_j \leq f(c_i) - \gamma\sigma_i, \quad \forall i \in S, \tag{2a}$$

$$f(c_j) - \gamma\sigma_j \leq f_{min} - \epsilon|f_{min}|, \tag{2b}$$

then the jth hyperrectangle is a POH.

In [19], we have proved that Definition 7 makes the original DIRECT be sensitive to linear scaling to the objective functions. Therefore, MDIRECT adopts a new definition of POH, which is modified slightly from the above definition.

Definition 8. *Given $\epsilon > 0$ and the index set of the hyperrectangles S. Let f_{min} be the known minimal function value, and c_i, σ_i be the center and the size of the ith* *hyperrectangle. If there is a constant* $\gamma > 0$*, such that the following inequalities hold*

$$f(c_j) - \gamma\sigma_j \leq f(c_i) - \gamma\sigma_i, \quad \forall i \in S, \tag{3a}$$

$$f(c_j) - \gamma\sigma_j \leq f_{min} - \epsilon|f_{min} - f_{cc}|, \tag{3b}$$

then the j*th hyperrectangle is a POH. In (3b),* f_{cc} *is any convex combination of gathered function values.*

Based on Definition 8, MDIRECT can eliminate the whole sensitivities to any linear scaling of the objective function. See [19] for more details.

From Definition 8 we obtain a class of MDIRECT algorithms, each with different f_{cc}. Denote F as the set of the gathered function values, then

- let f_{cc} equals the median of F, we obtain the DIRECT-median algorithm [18];
- let f_{cc} be the average of $\{f | Q_1 \leq f \leq Q_3\} \subset F$, where Q_1 and Q_3 are two quartiles of F, then we obtain the DIRECT-a algorithm [19].

Besides DIRECT-median and DIRECT-a, there are still many other MDIRECT algorithms through defining different f_{cc}. For example, let f_{cc} equal the maximal value or the minimal value or the average value of F, then we obtain different MDIRECT algorithms. However, both the maximal value and the average value are sensitivity, while the minimal value is often too small. Therefore, these choices are not proper.

In this paper, we mainly compare the performance of DIRECT-a and DIRECT-median. In [19], preliminary numerical results on a small test function set show that DIRECT-a performs slightly better than DIRECT-median. In this paper, we will compare their performance on a much larger test function set, the GKLS test set.

3 The GKLS Test Set

The GKLS test set proposed in [20, 21] is designed specially for comparing of optimization algorithms for bound constrained global optimization problems.

GKLS contains three classes of test functions: non-differentiable (ND-type) functions, continuously differentiable (D-type) functions, and twice continuously differentiable (D2-type) functions. Each test class consists of 100 functions. All test functions in GKLS are generated by defining a convex quadratic function (paraboloid) and then systematically distorts randomly selected parts of this function by polynomials in order to introduce local minima.

In order to determine a class, the user defines the following parameters:

- problem dimension,
- number of local minima including the paraboloid min and the global min,
- global minimum value,

- distance from the paraboloid vertex to the global minimizer,
- radius of the attraction region of the global minimizer.

All other necessary parameters are generated randomly for all 100 functions of the class. The C code of GKLS can be downloaded from the following website: http://si.deis.unical.it/~yaro/GKLS.html.

4 Numerical Experiments

In this section, we compare the performance of the following algorithms on the GKLS test set.

- DIRECT: The original DIRECT algorithm proposed in [8].
- DIRECT-median: The first modified DIRECT algorithm for scaled global optimization problems, which was proposed in [18] firstly and then proved theoretically to be insensitive to linear scaling in [19].
- DIRECT-a: A modified DIRECT algorithm proposed in [19] for scaled global optimization problems.

The last two algorithms have different f_{cc}, see Sect. 2 or [18, 19] for more details.

4.1 Experiments Setup

In our experiments, we adopt default parameters of GKLS except the dimension of function. Because DIRECT is only suitable for small scale problems, we test 300 2D functions and 300 5D functions generated from GKLS. Specifically, for each type (ND, D and D2) and each dimension, we test 100 functions. Totally, we test 600 functions.

We use the codes of DIRECT written by D.E. Finkel in Matlab. The codes can be downloaded from the following website: http://www4.ncsu.edu/~ctk/Finkel_Direct/. See [22] for its user-guide. Because GKLS is written in C, we have written an interface program for DIRECT and GKLS.

We compare the number of function evaluations needed for each algorithm to achieve convergence. The convergence is defined in the following way

$$\frac{f_{min} - f_{global}}{|f_{global}|} < perror, \tag{4}$$

where f_{global} is the known global optima, f_{min} is the best function value obtained by each algorithm. The accuracy parameter $perror$ is a given small positive constant. In our experiments, for each function, we test six different $perror$ $(1e{-}k, k = 2, \ldots 7)$.

The budget of function evaluations is 15,000 in our experiments. If the convergence is not achieved within 15,000 function evaluations, then the algorithm stops and the actual number of function evaluations (slightly larger than 15,000) is accepted as the computational cost. This will bias to the original DIRECT algorithm.

For each type of functions, we select two functions randomly and illustrate these three algorithms' performance on them with a budget of 100,000 function evaluations.

4.2 *Numerical Results*

First, for each type of functions, we list how many functions can be solved by these three algorithms. Second, we report how the accuracy constant *perror* affects the performance. Third, we report how the modified DIRECT improves (or degenerates) the original DIRECT's performance. These numerical results are summarized in Table 1 (where DIRECT-median is denoted as DIRECT-m).

In Table 1, for each type of functions, the first three rows show how many functions have been solved by each algorithm; the second three rows show how the accuracy constant *perror* affects the performance of these three algorithms; and the last three rows show how the robustness affects the performance.

Specifically, the ratios

$$\sum_{i=1}^{200} \frac{N^i(perror)}{N^i(1e-2)}, \quad perror = 1e-2, \ldots, 1e-7$$

are summarized in the second three rows for each type of functions, where $N^i(perror)$ denotes the number of function evaluations needed for the i-th function to achieve convergence defined by (4).

The following ratios

$$\sum_{i=1}^{200} \frac{n^i(perror)}{n^i_D(perror)}, \quad perror = 1e-2, \ldots, 1e-7$$

are summarized in the last three rows for each type of functions, where $n^i(perror)$ denotes the number of function evaluations needed for the i-th function to achieve convergence for three algorithms, while $n^i_D(perror)$ denotes the number of function evaluations needed for the i-th function to achieve convergence for DIRECT.

From Table 1 we can see that, for each type of functions, all algorithms can solve more than 95 % of the functions when $perror \geq 1e-4$. However, when *perror* decreases, more and more functions cannot be solved. For example, when $perror = 1e-7$, both DIRECT-a and DIRECT-median can only solve about 50 % of the functions, while DIRECT-median can solve more than 70 %.

What is more, when $peero \geq 1e-5$, the smoothness of functions seems to bring no benefit for both DIRECT and MDIRECTs. However, when $perror < 1e-5$,

Table 1 Numerical results for three algorithms on three different types of functions

	perror =	1e−2	1e−3	1e−4	1e−5	1e−6	1e−7
ND-type	DIRECT	200	200	200	200	120	100
	DIRECT-a	200	200	200	191	133	102
	DIRECT-m	200	200	200	190	179	143
	DIRECT	1	1.3562	1.8753	3.4694	13.1155	40.8566
	DIRECT-a	1	1.3549	1.9007	3.6477	9.4539	15.2362
	DIRECT-m	1	1.3564	1.9054	2.9708	4.6672	9.2237
	DIRECT	1	1	1	1	1	1
	DIRECT-a	1.0092	1.0105	1.0265	1.1825	0.9284	0.6581
	DIRECT-m	1.0157	1.0171	1.0344	1.0250	0.6544	0.5270
D-type	DIRECT	195	193	192	189	101	100
	DIRECT-a	195	193	191	187	121	102
	DIRECT-m	194	193	191	186	173	151
	DIRECT	1	1.3418	1.6994	2.3825	12.5539	32.5093
	DIRECT-a	1	1.3415	1.7180	2.8970	9.9906	13.5826
	DIRECT-m	1	1.3421	1.7184	2.6873	3.9212	7.0313
	DIRECT	1	1	1	1	1	1
	DIRECT-a	1.0039	1.0042	1.0160	1.3038	0.8804	0.6694
	DIRECT-m	1.0079	1.0085	1.0202	1.2812	0.6064	0.4858
D2-type	DIRECT	193	192	190	190	146	101
	DIRECT-a	192	191	189	189	153	108
	DIRECT-m	192	191	189	189	180	160
	DIRECT	1	1.3047	1.5940	2.0881	7.6973	17.8755
	DIRECT-a	1	1.3024	1.6081	2.3846	6.4490	12.3556
	DIRECT-m	1	1.3020	1.6095	2.2823	3.6164	5.7218
	DIRECT	1	1	1	1	1	1
	DIRECT-a	1.0105	1.0105	1.0210	1.1831	1.1077	0.8143
	DIRECT-m	1.0165	1.0164	1.0275	1.1622	0.9120	0.5666

smoothness brings clear benefits for these algorithms, especially for DIRECT-median.

From the second three rows for each type of functions we can see that more and more computational costs are needed to achieve convergence as *perror* decreases. For example, for DIRECT, the computational cost when $perror = 1e - 7$ is dozens of times larger than that when $perror = 1e - 2$. For DIRECT-median, it is no more than ten times larger. It is clear that the computational costs increase more rapidly for the original DIRECT than modified DIRECTs.

It needs to note that if the convergence cannot be achieved within 15,000 function evaluations, then the algorithm stops and the number of function evaluations at that time is simply taken as the computational cost. Because DIRECT-median solved much more problems than DIRECT when $perror \leq 1e - 6$, thus the numerical results are biased to DIRECT. In other words, we can say that the modified DIRECT especially DIRECT-median performs much better than DIRECT.

From the last three rows for each type of functions we can see that the modified DIRECT algorithms sometimes degenerate the performance of DIRECT when the required accuracy is low ($perror \geq 1e-5$). However, when the required accuracy is high ($perror < 1e-5$), DIRECT-a and DIRECT-median perform much better than DIRECT. For example, when $perror = 1e-7$, the needed computational cost for DIRECT-median is almost half of that for DIRECT.

In order to compare the performance of these three algorithms deeply, we illustrate the residual histories of some selected functions. First, we generate two integer numbers no more than 100 randomly. As a result, 56 and 97 are generated. Then the No.56 2D function and the No.97 5D function are selected for ND-type functions, D-type functions, and D2-type functions, respectively. Because these algorithms have similar performance on ND-type functions, D-type functions, and D2-type functions, here we only provide the illustration for ND-type.

Figure 1 shows the residual histories of the No.56 2D ND-type function and the No.97 5D ND-type function, respectively. The budget of function evaluations is 100,000. In Fig. 1, vertical coordinate shows the logarithm of the residual, which is defined by the difference between the minimal value found by the algorithm and the global optimal value.

From the left subfigure in Fig. 1 we can see that all these three algorithms achieve convergence for the No.56 2D ND-type function. DIRECT-median and DIRECT-a perform very well, their computational costs are no more than 4,000, while DIRECT needs more than 75,000. From the right subfigure we can see that both DIRECT and DIRECT-a do not achieve convergence within 100,000 function evaluations, while DIRECT-median solves it within 30,000 function evaluations.

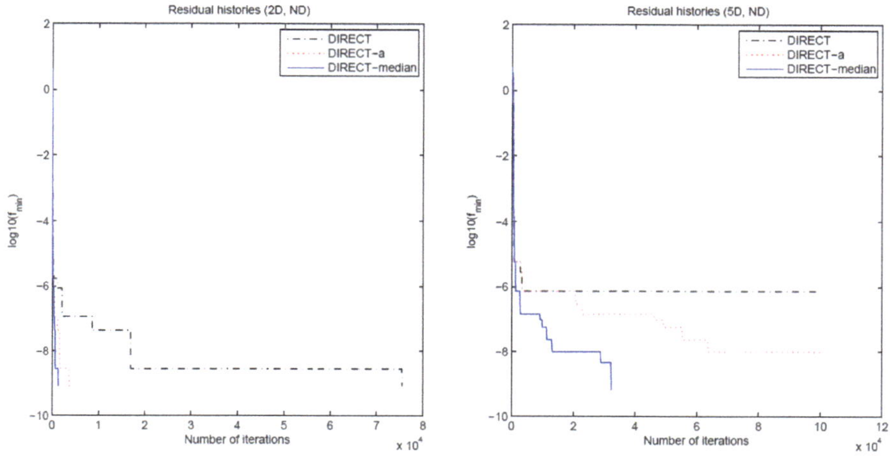

Fig. 1 Residual history of No.56 2D ND-type function and No.97 5D ND-type function

Both subfigures show that when the required accuracy is low, there is no clear advantage for modified DIRECT algorithms. However, as the required accuracy increases, their advantages are more and more clear.

5 Conclusions

In this paper, we compared the performance of a class of modified DIRECT algorithms on the GKLS test function set. Our main purpose was to find the best one among this class of modified DIRECT algorithms.

GKLS was designed specially for comparison of algorithms for bound constrained global optimization problems. It was very suitable for our comparison.

Totally 600 small-scale test functions were tested in our experiments. The numerical results showed that, although modified DIRECT may degenerate the performance of the original DIRECT when the required accuracy is low, they performed very well when the required accuracy is high. Moreover, our numerical results show that DIRECT-median seemed to the best modified DIRECT algorithm.

Acknowledgements This work was supported by NSF of China (No.11271069) and MOE (Ministry of Education in China) Project of Humanities and Social Sciences (Project No.13YJC630095).

References

1. Floudas, C.A., Gounaris, C.E.: A review of recent advances in global optimizations. J. Glob. Optim. **45**, 3–38 (2009)
2. Floudas, C.A.: Deterministic Global Optimization: Theory, Methods and Applications. Kluwer Academic, Dordrecht (2000)
3. Hendrix, E.M.T., G.-Tóth, B.: Introduction to Nonlinear and Global Optimization. Springer, New York (2010)
4. Horst, R., Tuy, H.: Global Optimization: Deterministic Approaches. Springer, Berlin (1996)
5. Liuzzi, G., Lucidi, S., Piccialli, V.: A DIRECT-based approach exploiting local minimizations for the solution of large-scale global optimization problems. Comput. Optim. Appl. **45**(2), 353–375 (2010)
6. Locatelli, M., Schoen, F.: Local search based heuristics for global optimization: atomic clusters and beyond. Eur. J. Oper. Res. **222**(1), 1–9 (2012)
7. Sun, W.T., Dong, Y.: Study of multiscale global optimization based on parameter space partition. J. Glob. Optim. **49**(1), 149–172 (2011)
8. Jones, D.R., Perttunen, C.D., Stuckman, B.E.: Lipschitzian optimization without the Lipschitz constant. J. Optim. Theory Appl. **79**(1), 157–181 (1993)
9. Jones, D.R.: DIRECT global optimization algorithm. In: The Encyclopedia of Optimization. Kluwer Academic, Dordrecht (1999)
10. Pošík, P.: BBOB-Benchmarking the DIRECT global optimization algorithm. In: Proceedings of the 11th Annual Conference Companion on Genetic and Evolutionary Computation Conference, pp. 2315–2320 (2009)

11. Finkel, D.E.: Global optimization with the DIRECT algorithm. Ph.D thesis, North Carolina State University (2005)
12. Gablonsky, J.M., Kelley, C.T.: A locally-biased form of the DIRECT algorithm. J. Glob. Optim. **21**, 27–37 (2001)
13. Ljungberg, K., Holmgren, S.: Simultaneous search for multiple QTL using the global optimization algorithm DIRECT. Bioinformatics **20**(12), 1887–1895 (2004)
14. Sasena, M., Papalambros, P., Goovaerts, P.: Global optimization of problems with disconnected feasible Regions via SurrogateModeling. In: 9th AIAA/ISSMO Symposium on Multidisciplinary Analysis and Optimization, Atlanta (2002)
15. Finkel, D.E., Kelley, C.T.: Convergence analysis of the DIRECT algorithm. Technical Report CRSC-TR04-28, North Carolina State University, Center for Research in Scientific Computation (2004)
16. Björkman, M., Holmström, K.: Global optimization using the DIRECT algorithm in Matlab. Adv. Medel. Optim. **1**(2), 17–37 (1999)
17. Holmström, K.: The TOMLAB optimization environment in Matlab. Adv. Medel. Optim. **1**(1), 47–69 (1999)
18. Finkel, D.E., Kelley, C.T.: Additive scaling and the DIRECT algorithm. J. Glob. Optim. **36**, 597–608 (2006)
19. Liu, Q.: Linear scaling and the DIRECT algorithm. J. Glob. Optim. **56**, 1233–1245 (2013)
20. Gaviano, M., Kvasov, D.E., Lera, D., Sergeyev, Ya.D.: Algorithm 829: software for generation of classes of test functions with known local and global minima for global optimization. ACM Trans. Math. Softw. **9**(4), 469–480 (2003)
21. Knuth, D.: The Art of Computer Programming. Seminumerical Algorithms, vol. 2, 3rd edn. Addison-Wesley, Reading (1997)
22. Finkel, D.E.: DIRECT Optimization User Guide. Center for Research and Scientific Coputation CRSC-TR03-11, North Carolina State University, Raleigh (2003)

A Fast Tabu Search Algorithm for the Reliable P-Median Problem

Qingwei Li and Alex Savachkin

Abstract Lean distribution networks have been facing an increased exposure to risk of unpredicted disruptions causing significant economic forfeitures. At the same time, the existing literature features very few studies which examine the impact of facility fortification for improving network reliability. In this paper, we present a reliable P-median problem (RPMP) for planning distribution networks against disruptions. We consider heterogeneous facility failure probabilities, one layer of supplier backup, and facility fortification within a finite budget. The RPMP is formulated as nonlinear integer programming model and we develop a Tabu search heuristic that is capable of solving large size problems efficiently.

1 Motivation

In the last few decades, lean manufacturing and just-in-time inventory philosophy have been widely implemented by global enterprises. According to [1], 36 % of U.S. companies and 70 % of UK companies are using lean as their principle to become more efficient. Inasmuch as such reductionism has boosted the operational efficiency of the companies, it has also elevated their risk exposure to unpredicted disruptions. Such disruptions, as triggered by forces of nature, process hazards, and human intervention can have a potential to entail staggering economic ramifications. This is evidenced by the following sample of recent multi-billion enterprise forfeitures lost to disrupted distribution networks.

The foot-and-mouth disease scare in the U.K. in 2001 caused the USA to ban meat imports from European Union [2]. The ban costed European exporters as much as $458 million a year in sales. In 2003, the outbreak of the deadly lung disease SARS disrupted the manufacturing and importing of furniture made in China, which accounted for about 15 % of all furniture sold in the USA [3]. After the terrorist

Q. Li (✉) • A. Savachkin
University of South Florida, 4202 E Fowler Ave., Tampa, FL, USA
e-mail: qli4@mail.usf.edu; alexs@usf.edu

D. Gao et al. (eds.), *Advances in Global Optimization*, Springer Proceedings in Mathematics & Statistics 95, DOI 10.1007/978-3-319-08377-3_41

attacks on September 11, 2001, all U.S. borders were closed and all flights were canceled. This lockdown forced Ford Motors to idle several assemble lines, as components supplied from outside the USA were delayed [4]. In that quarter, Ford Motor's output was down by 13 % compared to its production plan.

As seen from examples above, it is clear that disruptions can halt production, paralyze distribution of products/service, and cause multi-million dollar losses. There is an utmost need for an enterprise distribution network planning that can offer more robustness than what are available today.

2 Status of Current Literature

In literature, only a small fraction of the proliferate supply chain literature studies the impact of disruptions, most of which focuses on localized issues such as inventory management with supply/demand disruptions. Traditional research efforts on distribution network design are narrowed down at minimizing total operational costs and constructing cost by leveraging economics of scale. These works often yield results that overconcentrate resources and the resulting optimal solutions can be very unstable when disruptions occur.

Only a few works in the literature have been dedicated to the design of reliable distribution networks to hedge the impact of random disruptions. The work is pioneered by Eiselt et al. [5]. The main application area of the model discussed is computer networks. The authors present an algorithm to solve the problem and optimally locate nodes for an undirected simple network. The limitation of the work is the assumption that only one node fails at a time.

Snyder and Daskin [6] presented two reliability models for facility location: a reliable P-median and a reliable uncapacitated fixed-charge location model. Cui et al. [7] and Li et al. [8] relaxed the assumption of equal failure probabilities to location specific probabilities. Li and Ouyang [9] further expanded this direction by considering correlated and site specific probabilities. Li et al. [10] looked at a more integrated supply chain design problem and considered the case in which both the supplier and retailers are disrupted randomly. Peng et al. [11] introduced the P-robustness criterion so that the designed supply network performs well in both disrupted and normal conditions. A hybrid metaheuristic algorithm was proposed.

A few recent articles have taken the analysis one step further and examined the impact of facility fortification on the reliability of *existing* networks. Church et al. [12] examined two related network interdiction problems: the r-interdiction median and the r-interdiction covering model. Both models are based on the P-median problem. The only study of P-median problem design with fortification is by Li et al. [8]. This paper aims at developing an efficient solution algorithm that runs faster comparing to [8]. This could allow problems with much larger sizes being solved in reasonable amount of time.

3 Model Formulation

Define I to be the set of customers, J the set of potential facility locations, and P the number of facilities to open. Each customer $i \in I$ has a demand h_i. Let $d_{ij} \geq 0$ be the cost of transporting one unit of demand from facility location $j \in J$ to customer i (with the convention that $d_{ii} = 0\ \forall i$). Associated with each facility j is the failure probability $0 \leq q_j \leq 1$. The events of facility failures are assumed to be independent [6, 7, 13]. Once a facility fails, it becomes unavailable. Each customer is assigned a primary supplier and a different backup supplier (as in [13]). While [13] required each backup facility to be "totally reliable" (i.e., always available), we consider that for any customer, the probability of a simultaneous failure of its primary and backup supplier is negligible.

Our model incorporates facility fortification whereby reliability of facilities can be improved at some cost. We assume that if a facility is fortified, it becomes non-failable. The total cost of fortifying facility j includes the setup cost and the variable cost components. The setup cost S_j is a fixed cost required to implement facility fortification (examples include the costs of R&D, contract negotiation, overhead, personnel training, etc.). The variable fortification cost varies with the amount of reliability improvement of the facility. Examples include the cost of acquiring and installing the units of protective measures, the cost of procurement and storage of backup inventory, and the cost of hiring extra workforce. We define r_j as the cost associated with the unit reduction in the failure probability of facility j. Our model incorporates a total available fortification budget B. Finally, the facilities are assumed to have unlimited capacity (as in [6, 7, 13]).

$$X_j = \begin{cases} 1, \text{if a facility is opened at location } j; \\ 0, \text{otherwise.} \end{cases}$$

$$Y_{ij0} = \begin{cases} 1, \text{if customer } i \text{ has facility } j \text{ as the primary supplier;} \\ 0, \text{otherwise.} \end{cases}$$

$$Y_{ij1} = \begin{cases} 1, \text{if customer } i \text{ has facility } j \text{ as the backup supplier;} \\ 0, \text{otherwise.} \end{cases}$$

$$Z_j = \begin{cases} 1, \text{if facility } j \text{ is fortified;} \\ 0, \text{otherwise.} \end{cases}$$

We formulate the problem as follows:

$$\begin{aligned}\textbf{(RPMP)}\ \text{minimize} \sum_{i \in I} \sum_{j \in J} \Big[& h_i d_{ij} Y_{ij0} (1 - q_j (1 - Z_j)) \\ & + h_i d_{ij} Y_{ij1} \sum_{r \in J,\, r \neq j} q_r Y_{ir0} (1 - Z_r) \Big]\end{aligned}$$

subject to

$$\sum_{j \in J} Y_{ij0} = 1,\ \forall i \in I \tag{1a}$$

$$\sum_{j \in J} Y_{ij1} = 1,\ \forall i \in I \tag{1b}$$

$$Y_{ij0} + Y_{ij1} \leq X_j,\ \forall i \in I,\ j \in J \tag{1c}$$

$$\sum_{j \in J} X_j = P \tag{1d}$$

$$\sum_{j \in J} (S_j + r_j q_j) Z_j \leq B \tag{1e}$$

$$X_j, Z_j \in \{0, 1\},\ \forall j \in J \tag{1f}$$

$$Y_{ij0}, Y_{ij1} \in \{0, 1\},\ \forall i \in I,\ j \in J. \tag{1g}$$

The objective function of the reliable P-median problem (RPMP) is the expected total transportation cost associated with satisfying the demands of all customers. Constraint (1a) and (1b), respectively, assures that each customer is assigned only one primary and one backup supplier. Constraint (1c) serves two purposes. First, it guarantees that only open facility can serve as a supplier. It also assures that for each customer, the primary and backup suppliers are different facilities. Constraint (1d) demands P facilities to be opened. Constraint (1e) is the total fortification budget constraint. Finally, (1f) and (1g) are the integrality constraints.

4 A Tabu Search Heuristic

The RPMP is $\mathcal{NP}$-hard [8] and has a nonlinear objective function. The Lagrangian relaxation based algorithm developed in [8] runs relatively slow. This motivated us to develop a Tabu search solution heuristic that is capable of solving large problem fast. First, we have the following properties of this problem.

For simplicity, let $Q_u = q_u(1 - Z_u)$ and S be a location solution if $S \subset J, \| S \| = P$. We further assume that $q_j \leq 0.5$. Otherwise the facilities would be highly unreliable and out of the research scope of this paper.

Theorem 1. *If customer $i \in I$ is assigned to supplier $p \in S$ as its primary supplier and to supplier $b \in S$ as its backup supplier in an optimal solution S, then $d_{ip} < d_{ib}$.*

Proof. For customer $i \in I$, suppose $d_{ip} < d_{ib}$.
If assign p as primary supplier, b as backup supplier, the transportation cost is: $d_{ip}(1 - Q_p) + d_{ib} Q_p$. If assign b as primary supplier, p as backup supplier, the transportation cost is: $d_{ib}(1 - Q_b) + d_{ip} Q_b$.

The difference between the two is:
$[d_{ip}(1-Q_p)+d_{ib}Q_p]-[d_{ib}(1-Q_b)+d_{ip}Q_b] = d_{ip}(1-Q_p-Q_b)+d_{ib}(Q_p-1+Q_b) = d_{ip}(1-Q_p-Q_b)-d_{ib}(1-Q_p-Q_b) = (d_{ip}-d_{ib})(1-Q_p-Q_b) < 0$ □

Theorem 2. *Given a solution S, customer $i \in I$ is assigned to supplier $p \in S$ as its primary supplier if d_{ip} is the minimum of $d_{ir}, \forall r \in S$; customer $i \in I$ is assigned to supplier $b \in S$ as its backup supplier if d_{ib} is the second minimum of $d_{ir}, \forall r \in S$.*

Proof. Assume customer i select $u, v \in S$ as its primary and backup supplier. Let $p, b \in S$ be i's two closest suppliers such that $d_{ip} < d_{ib}$. From Theorem 1, we know that $d_{iu} < d_{iv}$. The distance mapping from node u, v, p, b to customer i can only fall into three cases below:

1. $d_{\cdot p} \leq d_{\cdot b} \leq d_{\cdot u} \leq d_{\cdot v}$. Due to the objective function is a convex combination of d_{iu} and d_{iv}, it's clear that any point between p and b is smaller than any point between u and v. Therefore (u, v) is not optimal in this case.
2. $d_{\cdot p} \leq d_{\cdot b} = d_{\cdot u} \leq d_{\cdot v}$. Similar to the proof for case 1.
3. $d_{\cdot p} = d_{\cdot u} \leq d_{\cdot b} \leq d_{\cdot v}$. $[d_{iu}(1-Q_u) + d_{iv}Q_u] - [d_{ip}(1-Q_p) + d_{ib}Q_p]$ $= [d_{ip}(1-Q_p) + d_{iv}Q_p] - [d_{ip}(1-Q_p) + d_{ib}Q_p] = d_{iv}Q_p - d_{ib}Q_p > 0$. Therefore (u, v) is not optimal in this case.
   So we conclude that if p, b are the two closest locations, then (p, b) results in the optimal solution.

□

Theorem 3. *If S is optimal solution without fortification, S is also optimal with fortification.*

Proof. From the proofs of Theorems 1 and 2, we conclude that customers are assigned according to their distance to opened suppliers. In particular, those assignments are independent of their radiabilities. Therefore, an optimal solution before fortification is also optimal after fortification. □

4.1 Heuristic Algorithm

Based on the above properties, once we know the supplier locations, we can have all customers assigned. Fortification then is made by solving the RPMP with **X** and **Y** fixed, which is a simple knapsack problem and can be handled by any of the off-the-shelf solvers. A heuristic algorithm has been developed to solve the model.

Initial solution. The initial solution was found using greedy dropping, i.e., starting with a solution containing all locations, and iteratively removing the locations yielding the least increase in the objective function value. The search stopped when P locations were remaining. This initial solution is passed to the Tabu search process.

Tabu list and dynamic time. A tabu list TL is maintained so that if a location is in the list, it cannot be added to the current solution. The initial length of the Tabu list is L_T. The number of iterations that a location staying in the Tabu list is defined as T_i. T_i is initialized to be L_T and each time a location i is added to the Tabu list, we let $T_i = T_i n$, where n is a constant multiplier and $n > 1$. Therefore i is forced to stay in the Tabu list for $\lfloor T_i \rfloor$ iterations. This forces a location to stay in the Tabu list longer if it appears in solutions frequently. Thus the length of the Tabu list is dynamic. To allow a location to be able to re-enter solutions, we let $T_i = L_T$ if $T_i \geq 20$.

Aspiration criteria. The aspiration criteria specifies whether a location can be added to the current solution even if it is tabued [14]. If adding a tabued location resulted in a solution better than the best known so far, the location is removed from the list.

Escape mechanism. A large number of the previous solutions are stored in a vector V. We let the length of this vector to be L_V. For each elements S in V, we record the number of occurrences O_S. If a solution has been repeated for a certain number of times O_{max}, a escape mechanism is activated to avoid cycling in the search [15]. In our case, the algorithm randomly replaces locations in the current solution with outside locations.

Terminating criteria. We terminate the algorithm if number of iterations or the number of iterations without improving the best solution exceeds its respective predetermined threshold.

Let S be the set of locations in a solution to the RPMP. Note that when fixing S in the RPMP, Y and Z can be determined by Theorem 2 and then solving a knapsack problem. Let $F(S)$ be the objective function value of RPMP associated with S. The best feasible solution found is denoted as S^*.

5 Computational Performance

We tested the performance of the RPMP heuristic on six datasets containing 30, 49, 100, and 150 nodes, respectively. Transportation cost d_{ij} was taken as the Euclidean distance between nodes i and j. For the purpose of this testbed, we let sets I and J be equal. The failure probabilities q_j were randomly generated from $U \sim [0, 0.05]$. The facility failure probabilities q_j were randomly generated from Uni$(0, 0.05)$. The variable fortification costs r_j were generated from Uni$(0, 3000)$. The upper bounds for reliability improvement u_j were generated from Uni$(0, q_j)$. The algorithm was tested for the values of fortification budget B ranging between 0 and 210. The algorithm was coded in C++ and run on a 64-bit Linux machine with a 2.8 GHz Duo core CPU and 4.0 GB of physical RAM. CPLEX10 was used to solve the knapsack problem. The parameters of the Tabu search heuristic are shown in Table 1.

Results for the developed heuristic using four datasets are listed in Table 2. The feasible solutions found by using the heuristics is benchmarked with the lower bound obtained by the Lagrangian relaxation based algorithm (LR) in [8].

Table 1 Tabu search heuristic parameters

Parameter	Value
Initial length of the Tabu list (L_T)	10
Multiplier n	1.06
Maximum number of iterations	1,000
Maximum number of iterations without improvement	200
Length of vector V (L_V)	40
Maximum number of repeated solutions (O_{max})	15

Table 2 Testbed performance results for the RPMP algorithm

		$P = 5$					$P = 8$				
					Time (s)					Time (s)	
Nodes	B	LB	Obj.	Gap	LR	Tabu	LB	Obj.	Gap	LR	Tabu
30	0	3,694.3	3,694	0.00	0.7	3.4	2,192.5	2,241	0.02	0.5	6.21
30	20	3,573.9	3,652	0.02	0.8	4.2	2,144.1	2,246	0.05	0.9	8.68
30	60	3,366.2	3,564	0.06	0.7	6.1	2,044.4	2,213	0.08	0.6	6.96
30	120	3,283.5	3,541	0.08	0.5	4.0	1,981.5	2,088	0.05	0.5	7.31
49	0	8,826.4	8,870	0.00	3.5	3.6	5,874.6	5,878	0.00	2.1	6.34
49	20	8,704.7	8,725	0.00	8.9	3.6	5,772.2	5,788	0.00	3.1	5.66
49	60	8,538.4	8,888	0.04	29.3	4.6	5,678.3	5,703	0.00	3.5	6.92
49	120	8,325.5	8,443	0.01	10.3	4.9	5,580.9	5,689	0.02	27.0	6.01
100	0	17,594.4	18,012	0.02	56.3	15.5	12,711.3	12,780	0.01	27.3	39.50
100	20	17,328.8	17,512	0.01	27.1	23.4	12,571.7	12,657	0.01	32.3	43.94
100	60	16,977.0	17,352	0.02	64.1	30.6	12,311.3	12,481	0.01	27.6	52.52
100	120	16,810.0	17,352	0.03	31.1	33.5	12,198.9	12,413	0.02	46.8	56.21
150	0	20,136.7	20,237	0.00	98.3	42.9	14,682.7	14,756	0.00	143.2	86.39
150	20	19,881.7	20,084	0.01	120.8	45.6	14,659.3	14,949	0.02	190.4	89.86
150	60	19,653.5	19,816	0.01	221.3	45.1	14,436.3	14,896	0.03	406.9	92.98
150	120	19,512.0	19,816	0.02	171.8	37.4	14,389.7	14,826	0.03	872.2	90.80

The abbreviation LB stands for the lower bound and Obj stands for the objective function value of the best feasible solution found by the heuristic. The gap is the difference (in %) between Obj and LB.

From the table above, we can see that our heuristic algorithm performs well compared to the lower bound available. The developed heuristic algorithm overall is faster than the LR algorithm in obtaining feasible solutions and the gap is reasonably small. We see the gap between LB and Obj becomes smaller as the size of the problem increases.

Comparing with the LR algorithm in [8], the computational time of our heuristics increases with the problem size but in a significantly slower trend. This enables us to solve large size problems in a reasonable amount of time. When applied to a randomly generated 500 node problem, the developed heuristic obtained a feasible solution within 600 s, while the LR algorithm failed in 1 h.

6 Conclusion

In this paper, we developed a nonlinear integer programming model for a RPMP. We assume heterogenous facility failure probabilities and one layer of supplier backing up. To achieve reliability of the network, a one level backing up mechanism and facility fortification with respect to a budget are incorporated in the model. We developed a Tabu search heuristic to overcome the computational difficulties for larger size problems. Our computational experiments show that the developed heuristic is efficient to solve large size problems and the gap between the feasible solution and the known lower bound is reasonably small.

References

1. Bragg, S.: ARC advisory group's strategy report. http:www.manh.com/library/MANH-Lean_WhitePaper.pdf (2004)
2. Marquis, C.: Meat from Europe is banned by U.S. as illness spreads. The New York Times, 14 March 2001
3. Koncius, J.: SARS hits the furniture industry; China is crucial to the U.S. market. The Washington Post, 29 May 2003
4. Sheffi, Y., Rice, J.B.: A supply chain view of the resilient enterprise. MIT Sloan Manag. Rev. **47**(1), 41–48 (2005)
5. Eiselt, H.A., Gendreau, M., Laporte, G.: Optimal location of facilities on a network with an unreliable node or link. Inf. Process. Lett. **58**, 71–74 (1996)
6. Snyder, L.V., Daskin, M.S.: Reliability models for facility location: the expected failure cost case. Transp. Sci. **39**, 400–416 (2005)
7. Cui, T., Ouyang, Y., Shen, Z.J.M.: Reliable facility location design under the risk of disruptions. Oper. Res. **58**, 998–1011 (2010)
8. Li, Q., Zeng, B., Savachkin, A.: Reliable facility location design under disruptions. Comput. Oper. Res. **40**, 901–909 (2013)
9. Li, X., Ouyang, Y.: A continuum approximation approach to reliable facility location design under correlated probabilistic disruptions. Transp. Res. Part B **44**, 535–548 (2010)
10. Li, Q., Shen, M., Snyder, L.V.: The effect of supply disruptions on supply chain design decisions. Transp. Sci. **44**, 274–289 (2010)
11. Peng, P., Snyder, L.V., Liu, Z., Lim, A.: Design of reliable logistics networks with facility disruptions. Transp. Res. Part B Methodol. **45**, 1190–1211 (2011)
12. Church, R.L., Scaparra, M.P., Middleton, R.S.: Identifying critical infrastructure: the median and covering facility interdiction problems. Ann. Assoc. Am. Geogr. **94**, 491–502 (2004)
13. Lim, M., Daskin, M.S., Bassamboo, A., Chopra, S.: A facility reliability problem: formulation, properties, and algorithm. Naval Res. Logist. **57**, 58–70 (2010)
14. Rolland, E., Schilling, D.A., Current, J.R.: An efficient tabu search procedure for the P-median problem. Eur. J. Oper. Res. **96**, 329–342 (1996)
15. Zhang, R., Yun, W.Y., Moon, I.: A reactive tabu search algorithm for the multi-depot container truck transportation problem. Transp. Res. Part E **45**, 904–914 (2009)

Part VII
Stochastic Models and Simulation

On the Implementation of a Class of Stochastic Search Algorithms

Jiaqiao Hu and Enlu Zhou

Abstract We propose a stochastic approximation approach for implementing a class of random search-based optimization algorithms called the model-based methods. The approach makes efficient use of the past sampling information as the search progresses and can significantly reduce the number of function evaluations needed to obtain high quality solutions. We illustrate our approach through a specific algorithm called Model-based Annealing Random Search with Stochastic Averaging (MARS-SA), which maintains the per-iteration sample size at a small constant value. We present the global convergence property of MARS-SA and report on numerical results.

Keywords Global optimization • Stochastic approximation • Model-based annealing random search

1 Introduction

We consider global optimization problems of the following general form:

$$x^* \in \arg\max_{x \in \mathbb{X}} H(x), \tag{1}$$

where $\mathbb{X}$ is a nonempty compact solution space in $\Re^n$ and $H : \mathbb{X} \to \Re$ is a deterministic objective function. Central to the context of our discussion is that the objection function H need not be available in any explicit form and there is little structural information that can be exploited in optimizing H. This setting has necessitated the development of various stochastic search techniques that rely only on the system performance measures in searching for improved solutions.

J. Hu (✉)
State University of New York at Stony Brook, Stony Brook, NY 11794, USA
e-mail: jqhu@ams.sunysb.edu

E. Zhou
Georgia Institute of Technology, Atlanta, GA 30332, USA
e-mail: enlu.zhou@isye.gatech.edu

D. Gao et al. (eds.), *Advances in Global Optimization*, Springer Proceedings in Mathematics & Statistics 95, DOI 10.1007/978-3-319-08377-3_42

The model-based methods [1, 2] are a class of stochastic search algorithms that have recently found many successful applications to hard optimization problems [3–6]. These algorithms, including ant colony optimization [3], estimation of distribution algorithms [4], the cross-entropy method [5], and model reference adaptive search (MRAS) [7], are based on repeatedly sampling from and updating an underlying probability distribution function over the feasible region. The fundamental idea is to construct a sequence of (idealized) distribution functions $\{g_k\}$ over $\mathbb{X}$ with the hope that the sequence will converge to a limiting distribution assigning most of its probability mass to the set of optimal solutions. In actual implementation, the distribution g_k is often approximated by a surrogate (theoretical) distribution based on sampled objective function values, because the construction of $\{g_k\}$ requires the explicit form of H, which may not be available a priori. The practical performance of model-based algorithms relies heavily on whether the $\{g_k\}$ sequence can be closely approximated by the surrogate distributions. Therefore, effective implementations of these algorithms typically require hundreds or even thousands of candidate solutions to be generated and evaluated at each iteration.

In this paper, we aim to improve the computational efficiency of model-based algorithms by reducing the number of candidate solutions generated per iteration. In particular, we propose a general stochastic approximation implementation of model-based algorithms, which allows the underlying algorithm to construct surrogate distributions by using the past sampling information in a recursive manner. We illustrate the approach through a specific model-based algorithm called Model-based Annealing Random Search with Stochastic Averaging (MARS-SA), which improves the MARS algorithm proposed in [8] by including an additional stochastic approximation component. We show that the MARS-SA algorithm converges globally even when the per iteration sample size is fixed at a small constant value. Our numerical results indicate that the new algorithm can be more efficient (in terms of the number of performance evaluations) than the original MARS algorithm.

2 A Stochastic Averaging Approach

In an attempt at solving (1), a model-based algorithm proceeds at each iteration k by constructing a probability distribution function g_k over $\mathbb{X}$ and then randomly sampling candidate solutions from g_k. There are several commonly used approaches to construct the $\{g_k\}$ sequence (cf. e.g., [9]), all of which can be expressed abstractly in terms of the following recursion:

$$g_k(x) = \frac{\ell(H(x))g_{k-1}(x)}{E_{g_{k-1}}[\ell(H(X))]} \quad \forall\, x \in \mathbb{X},\ k = 1, 2, \ldots, \tag{2}$$

where $\ell(\cdot)$ is a positive, increasing performance function that depends on the specific algorithm, $X \sim g_{k-1}$ is a random variable taking values from $\mathbb{X}$, and $E_g[\cdot]$ denotes the expectation with respect to g. Intuitively, by assigning more probability mass to solutions with better performance, this way of constructing $\{g_k\}$ ensures that high quality solutions will be sampled with large probabilities as k increases.

However, since g_k depends on H, directly constructing the sequence is intractable unless the entire solution space can be enumerated. This implementation difficulty has been addressed in a number of papers [5,7,10], and a common solution is to use a surrogate distribution to approximate g_k by projecting g_k onto a family of parameterized distributions $\{f_\theta : \theta \in \Theta \subseteq \Re^d\}$. This is carried out at each iteration by finding an optimal parameter θ_k that minimizes the Kullback-Leibler (KL) divergence between g_k and f_θ, i.e.,

$$\theta_k = \arg\min_{\theta \in \Theta} \left(\mathscr{D}(g_k | f_\theta) := E_{g_k}\left[\ln \frac{g_k(X)}{f_\theta(X)} \right] \right). \tag{3}$$

In practice, the parameterized distribution f_θ is often taken from the natural exponential family (NEF) of the form $f_\theta(x) = \exp\left(\theta^T \Gamma(x) - K(\theta)\right)$, $\forall\, \theta \in \Theta$, where Θ is the natural parameter space, $\Gamma(\cdot) : \Re^n \to \Re^d$ is a continuous mapping, and $K(\theta) = \ln \int_{\mathbb{X}} \exp\left(\theta^T \Gamma(x)\right) \nu(dx)$ is a normalization constant with ν being the Lebesgue/discrete measure on $\mathbb{X}$. This allows an analytical closed-form solution to (3) for an arbitrary g_k, making the implementation of the underlying algorithms tractable. Note that $K(\theta)$ is convex and analytic on Θ with gradient $\nabla_\theta K(\theta) = E_{f_\theta}[\Gamma(X)]$. In addition, the mean parameter function defined by $m(\theta) := \nabla_\theta K(\theta) = E_{f_\theta}[\Gamma(X)]$ is a one-to-one invertible mapping of θ.

By expanding (2), we can write g_k in the following equivalent form

$$g_k(x) = \frac{\ell_k(H(x)) g_0(x)}{E_{g_0}[\ell_k(H(X))]}, \tag{4}$$

where ℓ_k is an appropriate iteration-varying function that depends on ℓ. Substituting (4) into (3), it is easy to see that θ_k in (3) can be obtained by solving the following problem

$$\theta_k = \arg\max_{\theta \in \Theta} \left(Q_k(\theta) := \int_{x \in \mathbb{X}} \ell_k(H(x)) \ln f_\theta(x) \nu(dx) \right). \tag{5}$$

In model-based algorithms, the integral involved in the Q-function $Q_k(\theta)$ is estimated by generating N i.i.d. candidate solutions $X_{k-1}^1, \ldots, X_{k-1}^N$ from $f_{\theta_{k-1}}$ (i.e., the approximation of g_{k-1} at iteration $k-1$), and then replacing $Q_k(\theta)$ by its sample average approximation $\bar{Q}_k(\theta) := \frac{1}{N} \sum_{i=1}^N \frac{\ell_k(H(X_{k-1}^i))}{f_{\theta_{k-1}}(X_{k-1}^i)} \ln f_\theta(X_{k-1}^i)$. Although $\bar{Q}_k(\theta)$ is an unbiased estimator of $Q_k(\theta)$, the corresponding optimization step will lead to an estimator of θ_k that is biased for any finite sample size N, because the optimal solution to (5) involves a ratio of integrals/expectations. Consequently, common implementations of these algorithms either require hundreds or even thousands of candidate solutions to be generated per iteration [5], or require the use of a sample size N that increases at least polynomially with k in order to reduce the ratio bias effect [7–9].

In this paper, we propose an approach to address this bias issue. The basic idea is to replace the sample average $\bar{Q}_k(\theta)$ with the stochastic averaging procedure

$$\hat{Q}_k(\theta) = (1-\beta_{k-1})\hat{Q}_{k-1}(\theta) + \beta_{k-1}\frac{1}{N}\sum_{i=1}^{N}\frac{\ell_k(H(X_{k-1}^i))}{f_{\theta_{k-1}}(X_{k-1}^i)}\ln f_\theta(X_{k-1}^i) \quad (6)$$

with $\hat{Q}_1(\theta) := \frac{1}{N}\sum_{i=1}^{N}\left(\ell_1(H(X_0^i))/f_{\theta_0}(X_0^i)\right)\ln f_\theta(X_0^i)$, where β_k is a step size constant satisfying $\beta_k \in (0,1]\ \forall\, k$. Note that this procedure incrementally updates the current estimate of the Q-function as new sampling information becomes available at each iteration. In addition, due to the recursive nature of (6), all candidate solutions generated in the previous iterations contribute to the estimation of the Q-function $Q_k(\theta)$. Consequently, it is reasonable to expect that the number of samples per iteration N can be significantly reduced or even be held at a small constant value.

It is interesting to note that when f_θ belongs to NEFs, $\hat{Q}_k(\theta)$ can be expressed as a linear function in terms of the parameter vector θ and the function $K(\theta)$:

$$\hat{Q}_k(\theta) = \theta^T S_k - K(\theta)R_k,$$

where S_k and R_k can be computed via the following recursions: $S_k = S_{k-1} + \beta_{k-1}\left(\frac{1}{N}\sum_{i=1}^{N}\frac{\ell_k(H(X_{k-1}^i))}{f_{\theta_{k-1}}(X_{k-1}^i)}\Gamma(X_{k-1}^i) - S_{k-1}\right)$, $R_k = R_{k-1} + \beta_{k-1}\left(\frac{1}{N}\sum_{i=1}^{N}\frac{\ell_k(H(X_{k-1}^i))}{f_{\theta_{k-1}}(X_{k-1}^i)} - R_{k-1}\right)$, and we define $S_1 := \frac{1}{N}\sum_{i=1}^{N}\frac{\ell_1(H(X_0^i))}{f_{\theta_0}(X_0^i)}\Gamma(X_0^i)$ and $R_1 := \frac{1}{N}\sum_{i=1}^{N}\frac{\ell_1(H(X_0^i))}{f_{\theta_0}(X_0^i)}$. Thus, by substituting $\hat{Q}_k(\theta)$ for $Q_k(\theta)$ in (5), we have the following optimization problem:

$$\theta_k = \arg\max_{\theta\in\Theta}\left(\theta^T S_k - K(\theta)R_k\right),$$

whose unique closed-form solution (assuming that θ_k is an interior point of Θ) is given by $m(\theta_k) = \frac{S_k}{R_k}$ orequivalently $\theta_k = m^{-1}\left(\frac{S_k}{R_k}\right)$.

3 MARS with Stochastic Averaging

In this section, we incorporate the stochastic averaging idea introduced in Sect. 2 into the recent MARS algorithm [8] and propose a new algorithm we call MARS with stochastic averaging (MARS-SA). The MARS algorithm can be viewed as an implementable version of the Annealing Adaptive Search (AAS) algorithm proposed in [11]. The idea of AAS is to iteratively approximate the global optimal solution x^* of (1) by sampling candidate solutions at each iteration k from the Boltzmann distribution

$$g_k(x) = \frac{e^{H(x)/T_k}}{\int_{\mathbb{X}} e^{H(x)/T_k}\nu(dx)}, \tag{7}$$

where T_k is an iteration-dependent temperature parameter that approaches zero as $k \to \infty$. As T_k becomes smaller, the distribution g_k will be more concentrated on the set of optimal solutions, so that better solutions will be sampled with larger probabilities as the search proceeds. Note that the Boltzmann distribution (7) can be written in the form of (4) by letting $\ell_k(H(x)) := e^{H(x)/T_k}$.

The MARS algorithm avoids the explicit sampling from the Boltzmann distribution by using an NEF distribution family $\{f_\theta(\cdot), \theta \in \Theta\}$ to approximate the $\{g_k\}$ sequence. At each iteration k, new candidate solutions are generated from the sampling distribution that minimizes the KL-divergence $\mathscr{D}(\tilde{g}_{k+1}|f_\theta)$, where $\tilde{g}_{k+1}$ is the mixture of the Boltzmann distribution g_{k+1} [cf. (7)] with the sampling distribution f_{θ_k} obtained in the previous iteration, i.e., $\tilde{g}_{k+1}(x) = \alpha_k g_{k+1}(x) + (1-\alpha_k) f_{\theta_k}(x)$, $\alpha_k \in (0,1]\ \forall\, k$. Intuitively, this smoothing scheme ensures that the new sampling distribution $f_{\theta_{k+1}}$ obtained via minimizing $\mathscr{D}(\tilde{g}_{k+1}|f_\theta)$ will not significantly deviate from the current sampling distribution f_{θ_k}, making the performance of the algorithm less sensitive to the choices of the annealing schedule $\{T_k\}$. The optimization problem $\mathscr{D}(\tilde{g}_{k+1}|f_\theta)$ can be solved analytically and it can be shown (cf. [8,9]) that the mean parameter functions obtained at successive iterations satisfy the following recursion:

$$m(\theta_{k+1}) = \alpha_k \frac{\int_{\mathbb{X}} e^{H(x)/T_{k+1}}\Gamma(x)\nu(dx)}{\int_{\mathbb{X}} e^{H(x)/T_{k+1}}\nu(dx)} + (1-\alpha_k)m(\theta_k). \tag{8}$$

In the original implementation of MARS, the integrals in (8) are approximated by their corresponding sample averages based on sampled solutions. This results in a ratio bias that accumulates as the search proceeds. Therefore, existing theoretical analysis of MARS requires the use of a sample size N that increases polynomially with k for convergence.

In the MARS-SA algorithm, we eliminate the polynomially increasing computational requirement by embedding the stochastic averaging approach discussed in Sect. 2 into MARS. So a primary difference between MARS-SA and the original MARS algorithm lies in how the integrals in (8) are approximated.

Step 0: Select an annealing schedule $\{T_k\}$, gain sequences $\{\alpha_k\}$ and $\{\beta_k\}$, and a constant sample size $N \geq 1$. Set $S_0 = R_0 = 0$ and $\alpha_0 = \beta_0 = 1$. Set $k = 0$.
Step 1: Generate N i.i.d. solutions $\Lambda_k = \{X_1, \ldots, X_N\}$ from $f_{\theta_k}(x)$.
Step 2: Update S_{k+1} and R_{k+1} according to the recursions:

$$S_{k+1} = S_k + \beta_k \left(\frac{1}{N} \sum_{x \in \Lambda_k} \frac{e^{H(x)/T_{k+1}}}{f_{\theta_k}(x)} \Gamma(x) - S_k \right)$$

$$R_{k+1} = R_k + \beta_k \left(\frac{1}{N} \sum_{x \in \Lambda_k} \frac{e^{H(x)/T_{k+1}}}{f_{\theta_k}(x)} - R_k \right).$$

Step 3: Compute a new parameter θ_{k+1} as $m(\theta_{k+1}) = \alpha_k \frac{S_{k+1}}{R_{k+1}} + (1 - \alpha_k) m(\theta_k)$. Set $k = k + 1$ and reiterate from Step 1 until a stopping rule is satisfied.

3.1 Global Convergence of MARS-SA

We make the following assumptions, where A1 and A2 are regularity conditions on H, whereas A3–A5 are conditions on the algorithm input parameters.
Assumptions:

A1. *For a given constant* $\varepsilon < H(x^*)$, $\nu\big(\{x \in \mathbb{X} : H(x) \geq \varepsilon\}\big) > 0$.
A2. *For any* $\delta > 0$, $\sup_{x \in A_\delta} H(x) < H(x^*)$, *where* $A_\delta := \{x \in \mathbb{X} : \|x - x^*\| \geq \delta\}$.
A3. $\beta_k \in (0, 1]$, $\frac{1}{\beta_{k+1}} - \frac{1}{\beta_k} \leq 1 \; \forall\, k$, $\sum_{k=0}^{\infty} \beta_k = \infty$, *and* $\sum_{k=0}^{\infty} \beta_k^2 < \infty$; $\alpha_k \in (0, 1] \; \forall\, k$, $\alpha_k \to 0$ *as* $k \to \infty$, *and* $\sum_{k=0}^{\infty} \alpha_k = \infty$; $\alpha_k \beta_k = O(k^{-(1+\delta)})$, *where* $\delta \in (\frac{1}{2}, 1]$.
A4. $T_k \geq T_{k+1} > 0 \; \forall\, k$, $T_k \to T^* \geq 0$, *and* $\frac{1}{\beta_k}\big(\frac{1}{T_{k+1}} - \frac{1}{T_k}\big) \to 0$ *as* $k \to \infty$.
A5. $\sup_k E_{g_{k+1}}\big[\frac{g_{k+1}(X)}{f_{\theta_k}(X)}\big]$ *is bounded w.p.1.*

We have the following main convergence result for MARS-SA. Its proof can be found in [12].

Theorem 1. *If all Assumptions A1–A5 are satisfied, then*

$$m(\theta_k) \to E_{g^*}[\Gamma(X)] \;\; \text{w.p.1},$$

where the limit is taken component-wise and g^* *is the limiting Boltzmann distribution parameterized by* T^*.

The interpretation of this result depends on the specific form of the parameterized distribution family used in the algorithm. In many cases, the result implies that the sequence of parameterized sampling distribution $\{f_{\theta_k}\}$ generated by the algorithm will converge to a limiting distribution with all probability mass concentrated on the global optimizer x^* of (1); see, e.g., the discussion in [8] and the examples therein.

4 Numerical Experiments

We consider some computational experiments on a set of four benchmark test functions frequently used in the global optimization literature.

(1) Weighted Sphere function ($n = 40$, $-10 \leq x_i \leq 10$, $i = 1, \ldots, n$): $H_1(x) = -1 - \sum_{i=1}^{n} i x_i^2$, where $x^* = (0, \ldots, 0)^T$ and $H_1(x^*) = -1$.
(2) Powell function ($n = 40$, $-10 \leq x_i \leq 10$, $i = 1 \ldots, n$): $H_2(x) = -\sum_{i=1}^{(n-2)/2} \big[(x_{2i-1} + 10x_{2i})^2 + 5(x_{2i+1} - x_{2i+2})^2 + (x_{2i} - 2x_{2i+1})^4 + 10(x_{2i-1} - x_{2i+2})^4\big] - 1$, where $x^* = (0, \ldots, 0)^T$ and $H_2(x^*) = -1$.
(3) Levy function ($n = 40$, $-10 \leq x_i \leq 10$, $i = 1 \ldots, n$): $H_3(x) = -10\sin^2(\pi x_1) - \sum_{i=1}^{n-1} 100 x_i^2 (1 + 10\sin^2(\pi x_{i+1})) - 100(x_n - 1)^2 - 1$, where $x^* = (0, \ldots, 0, 1)^T$, $H_3(x^*) = -1$.
(4) Pinter's function ($n = 20$, $-10 \leq x_i \leq 10$, $i = 1, \ldots, n$): $H_4(x) = -\sum_{i=1}^{n} i x_i^2 - \sum_{i=1}^{n} 20 i \sin^2\big(x_{i-1}\sin x_i - x_i + \sin x_{i+1}\big) - \sum_{i=1}^{n} i \log_{10}\big(1 + i(x_{i-1}^2 - 2x_i + 3x_{i+1} - \cos x_i + 1)^2\big) - 1$, where $x_0 = x_n$, $x_{n+1} = x_1$, $x^* = (0, \ldots, 0)^T$, $H_4(x^*) = -1$.

Since all feasible regions are characterized by box constraints of the form $a_i \leq x_i \leq b_i$ for all $i = 1, \ldots, n$, the MARS-SA algorithm is implemented using truncated multivariate normal distributions with independent components. The following parameter setting is used in the experiment: a constant sample size $N = 4$, step-size sequences $\alpha_k = 1/(k + 100)^{0.601}$, $\beta_k = 1347/(k + 3000)^{0.9}$, and a logarithmic cooling schedule $T_k = |H(x_k^*)|/\ln(1 + k)$, where x_k^* denotes the current best solution found at the kth iteration of the algorithm. The performance of MARS-SA is compared with that of MARS. The same set of parameter values is used in implementing MARS, except that we have considered two different sample sizes: a constant sample size $N = 4$ and a polynomially increasing sample size used in [8] $N_k = \max\{4, \lfloor k^{0.502} \rfloor\}$.

For each test problem, we performed 50 independent replication runs of both algorithms. Figure 1 plots the averaged current best objective function values as a function of the number of function evaluations consumed thus far. Test results indicate the convergence of the proposed MARS-SA algorithm even when the per-iteration sample size is set to $N = 4$. Moreover, the performance of MARS-SA consistently dominates those of MARS with both constant and polynomially increasing sample sizes. In our experiments, we found that MARS with polynomial sample sizes converges to the global optimal solutions in all runs when the number of algorithm iterations is sufficiently large; however, it may fail to locate a near optimal solution within the prescribed number of function evaluations. On the other hand, MARS with $N = 4$ shows adequate average performance within the allowed number of function evaluations, but the algorithm may occasionally stagnate at non-optimal solutions.

5 Conclusions

We have proposed an approach for improving the efficiency of model-based optimization algorithms. The idea is to replace the sample average approximation used in implementing model-based algorithms with a recursive stochastic averaging

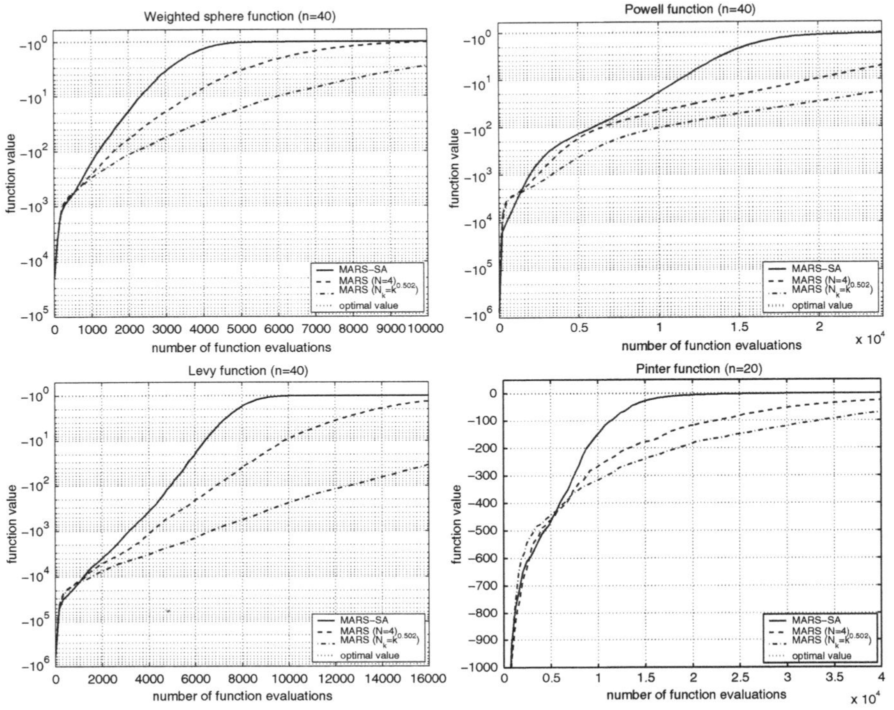

Fig. 1 Averaged performance of MARS-SA and MARS on test functions H_1 to H_4

procedure. To illustrate the idea, we have introduced a specific algorithm called MARS-SA and presented the global convergence property of the algorithm when the per iteration sample size is fixed at a constant value. Our preliminary numerical results also indicate that the algorithm may lead to improved performance over the original MARS algorithm.

Acknowledgements This work was supported by the National Science Foundation under Grants CMMI-1130273 and CMMI-1130761.

References

1. Fu, M.C., Hu, J., Marcus, S.I.: Model-based randomized methods for global optimization. In: Proceedings of the 17th International Symposium on Mathematical Theory of Networks and Systems, pp. 355–363 (2006)
2. Zlochin, M., Birattari, M., Meuleau, N., Dorigo, M.: Model-based search for combinatorial optimization: a critical survey. Ann. Oper. Res. **131**, 373–395 (2004)
3. Dorigo, M., Gambardella, L.M.: Ant colony system: a cooperative learning approach to the traveling salesman problem. IEEE Trans. Evol. Comput. **1**, 53–66 (1997)

4. Larrañaga, P., Lozano, J.A. (eds.): Estimation of Distribution Algorithms: A New Tool for Evolutionary Computation. Kluwer Academic, Boston (2002)
5. Rubinstein, R.Y., Kroese, D.P.: The Cross-Entropy Method: A Unified Approach to Combinatorial Optimization, Monte-Carlo Simulation, and Machine Learning. Springer, New York (2004)
6. Zabinsky, Z.B.: Stochastic Adaptive Search for Global Optimization. Kluwer Academic, Dordrecht (2003)
7. Hu, J., Fu, M.C., Marcus, S.I.: A model reference adaptive search algorithm for global optimization. Oper. Res. **55**, 549–568 (2007)
8. Hu, J., Hu, P.: Annealing adaptive search, cross-entropy, and stochastic approximation in global optimization. Naval Res. Logist. **58**, 457–477 (2011)
9. Hu, J., Hu, P., Chang, H.S.: A stochastic approximation framework for a class of randomized optimization algorithms. IEEE Trans. Autom. Control **57**, 165–178 (2012)
10. Wolpert, D.H.: Finding bounded rational equilibria part I: Iterative focusing. In: Vincent, T. (ed.) Proceedings of the Eleventh International Symposium on Dynamic Games and Applications, Tucson (2004)
11. Romeijn, H.E., Smith, R.L.: Simulated annealing and adaptive search in global optimization. Probab. Eng. Inf. Sci. **8**, 571–590 (1994)
12. Hu, J., Zhou, E.: Model-based Annealing Random Search with Stochastic Averaging. Working Paper (2013)

Uncertainty Relationship Analysis for Multi-Parametric Programming in Optimization

Tianxing Cai and Qiang Xu

Abstract Uncertainties exist at all levels of the industrial design and manufacturing. Hitherto, all the studies that handle multi-parametric programming (mp-LP, mp-QP, mp-NLP, mp-MILP, and mp-MINLP) treat uncertainties to be independent of each other; while under some circumstances, there might exist some kinds of quantitative relationship among them. There is still a lack of research studies on the relationship between these uncertainties, which can help simply the complexity of multi-parametric optimization problems in terms of reducing the dimension of uncertainty space or the region of uncertainty space.

This paper presents multiple types of relationships among uncertainty parameters, which can be generalized into two categories: strong relationship and weak relationship. The strong relationship can be used to reduce the dimension of uncertainty space while the weak relationship can be used to reduce the region of uncertainty space. With the combination of the above relationships, different kinds of multi-parametric programming problems can be solved more efficiently and effectively toward global optimality.

Keywords Multi-parametric programming • Optimization • Uncertainty relationship analysis

1 Introduction

Multi-parametric programming is a technique for solving one type of optimization problems, where some of the parameters may vary between specified lower and upper bounds. These parameters can be regarded as uncertainties. The main characteristic of multi-parametric programming is its ability to obtain [1–7].

For presentation in 2013 WCGO

T. Cai • Q. Xu (✉)
Dan F. Smith Department of Chemical Engineering, Lamar University, Beaumont, TX 77710, USA
e-mail: Qiang.xu@lamar.edu

D. Gao et al. (eds.), *Advances in Global Optimization*, Springer Proceedings in Mathematics & Statistics 95, DOI 10.1007/978-3-319-08377-3_43

1. The objective and optimization variable as functions of the varying parameters, and
2. The regions in the space of the parameters where these functions are valid.

Actually uncertainty and variability are inherent characteristics at all levels of industrial design and manufacturing. Many problems in the natural sciences and engineering are also rife with sources of uncertainties. Computer simulation modeling is the most commonly used approach to study problems in uncertainty quantification [8–10]. It tries to determine how likely certain outcomes will be if some aspects of the system are not exactly known. Small differences of what ?? in the manufacturing will lead to different results that can only be predicted in a statistical sense.

Uncertainty can appear in mathematical models and experimental measurements in various contexts. One way to categorize the sources of uncertainty is to consider [11]:

1. Parameter uncertainty, which comes from the model parameters that are inputs to the computer model (mathematical model) but whose exact values are unknown to experimentalists and cannot be controlled in physical experiments.
2. Structural uncertainty, also known as model inadequacy, model bias, or model discrepancy, which comes from the lack of knowledge of the underlying true physics. It depends on how accurately a mathematical model describes the true system for a real-life situation, considering the fact that models are almost always only approximations to reality.
3. Algorithmic uncertainty, also known as numerical uncertainty, which comes from numerical errors and numerical approximations per implementation of the computer model. Most models are too complicated to solve exactly.
4. Parametric variability, which comes from the variability of input variables of the model. For example, the dimensions of a work piece in a process of manufacture may not be exactly as designed and instructed, which would cause variability in its performance.
5. Experimental uncertainty, also known as observation error, which comes from the variability of experimental measurements. The experimental uncertainty is inevitable and can be noticed by repeating a measurement for many times using exactly the same settings for all inputs/variables.
6. Interpolation uncertainty, which comes as a lack of available data collected from computer model simulations and/or experimental measurements. For other input settings that don't have simulation data or experimental measurements, one must interpolate or extrapolate in order to predict the corresponding responses.

Another way of categorization is to classify uncertainty into two categories [12, 13]:

1. Aleatoric uncertainty, also known as statistical uncertainty, which is unknowns that differ each time we run the same experiment. Aleatoric uncertainties are therefore something an experimenter cannot do anything about: they exist, and they cannot be suppressed by more accurate measurements.

2. Epistemic uncertainty, also known as systematic uncertainty, which is due to things we could in principle know but don't in practice. This may be because we have not measured a quantity sufficiently accurately, or because our model neglects certain effects, or because particular data are deliberately hidden.

During traditional multi-parametric programming, such as mp-LP, mp-QP, mp-NLP, mp-MILP, and mp-MINLP, uncertainties are all considered as fully independent, which means that there is no any quantitative relationships. Thus, the programming will explore a super-rectangular uncertainty domain for an optimization problem. This paper presents possible multiple types of relationships among uncertainty parameters, which can be generalized into two categories: strong relationship and weak relationship. The strong relationship can be used to reduce the dimension of uncertainty space while the weak relationship can be used to reduce the region of uncertainty space. With the combination of the above relationships, different kinds of multi-parametric programming problems can be solved more efficiently and effectively toward global optimality. This can help simply the complexity of multi-parametric optimization problem in terms of reducing the dimension of uncertainty space or the region of uncertainty space.

2 Uncertainty Relationship

In this paper, four types of uncertainty relationship are defined: Cascade Uncertainties, Reducible Uncertainties, Internal Controlled Uncertainties, and Coefficient with Uncertainties.

2.1 Cascade Uncertainties

The mathematical model with cascade uncertainties can be described as below. In this category, the uncertainty parameter vector φ is the transformation of another uncertainty parameter vector θ, which can be expressed as $\varphi = F(\theta)$.

$$
\begin{aligned}
&\min_x J = f(x, \varphi) \\
&\text{s.t. } g_i(x, \varphi) \le 0, \quad \forall i = 1, \ldots, p \\
&\quad h_j(x, \varphi) = 0, \quad \forall j = 1, \ldots, q \\
&\quad \varphi = F(\theta) \\
&\quad x \in X \subseteq R^n \\
&\quad \theta \in \Theta \subseteq R^m
\end{aligned}
$$

2.2 Reducible Uncertainties

The mathematical model with reducible uncertainties can be described as below. In this category, the uncertainty parameter vector φ can satisfy a series of equations, which can be expressed as $F(\theta) = 0$.

$$
\begin{aligned}
&\min_x J = f(x, \theta) \\
&\text{s.t. } g_i(x, \theta) \le 0, \quad \forall i = 1, \ldots, p \\
&\quad h_j(x, \theta) = 0, \quad \forall j = 1, \ldots, q \\
&\quad F(\theta) = 0 \\
&\quad x \in X \subseteq R^n \\
&\quad \theta \in \Theta \subseteq R^m
\end{aligned}
$$

2.3 Internal Controlled Uncertainties

The mathematical model with internal controlled uncertainties can be described as below. In this category, the uncertainty parameter vector φ can be participated into two parts, θ_I and θ_D. Here the domain of subset θ_D is controlled by the subset θ_I, which is identified by the transformation $T(\Theta_I)$.

$$
\begin{aligned}
&\min_x J = f(x, \theta) \\
&\text{s.t. } g_i(x, \theta) \le 0, \quad \forall i = 1, \ldots, p \\
&\quad h_j(x, \theta) = 0, \quad \forall j = 1, \ldots, q \\
&\quad x \in X \subseteq R^n \\
&\quad \theta_I \in \Theta_I \subseteq R_I^m \\
&\quad \theta_D \in \Theta_D = T(\Theta_I)
\end{aligned}
$$

2.4 Coefficient with Uncertainties

The mathematical model with coefficients with uncertainties can be described as below. In this category, the uncertainty parameter will also play a role in the coefficients of decision variables.

$$
\begin{aligned}
&\min_x J = f(\theta x, \theta) \\
&\text{s.t. } g_i(\theta x, \theta) \le 0, \quad \forall i = 1, \ldots, p \\
&\quad h_j(\theta x, \theta) = 0, \quad \forall j = 1, \ldots, q \\
&\quad x \in X \subseteq R^n \\
&\quad \theta \in \Theta \subseteq R^m
\end{aligned}
$$

3 Methodology

Once the programming model has been set up, the model parameters can be verified whether parameters are deterministic parameters/variables. If it is a deterministic parameter/variable, there is no need to conduct further uncertainty relationship analysis on this parameter; if this parameter is a aleatory or epistemic parameter, the uncertainty relationship need to be analyzed through correlation analysis by neural network or regression. The above step will try to give the quantitative expression to link the uncertainty parameters. Next, the classification of possible uncertainty relationship can be determined. Then we can use the uncertainty relationship analysis for each type of programming problem with uncertainties (Fig. 1).

4 Case Studies

The first studied case is aiming to solve a control problem.

4.1 Mathematical Model for Control Algorithm

$$
\begin{aligned}
x(k+1) &= Ax(k) + Bu(k) + f \\
y(k) &= Cx(k) + Du(k) + g
\end{aligned}
$$

Here, the coefficient matrix for the uncertain parameters is listed:

$$
A = \begin{pmatrix} -1.7 & 1.1 & 0 \\ 0.2 & 0.5 & 2.1 \\ 0 & -0.1 & -1.1 \end{pmatrix}, \quad B = \begin{pmatrix} 0.1 & 0 \\ 0.1 & 1 \\ 0.1 & 0 \end{pmatrix}, \quad C = \begin{pmatrix} 1 & 0 & 0 \\ 0 & 2 & 0 \\ 0 & 0 & 3 \end{pmatrix}, \quad D = \begin{pmatrix} 0 & 0 \\ 0 & 1 \\ 0 & 0 \end{pmatrix}
$$

the output constraints are:

$$\begin{pmatrix}-5\\-5\\-5\end{pmatrix} \le y(k) \le \begin{pmatrix}5\\5\\5\end{pmatrix}, \quad \begin{pmatrix}-4\\-4\\-4\end{pmatrix} \le y(k+1) - y(k) \le \begin{pmatrix}4\\4\\4\end{pmatrix},$$

the input constraints are:

$$\begin{pmatrix}-5\\-5\end{pmatrix} \le u(k) \le \begin{pmatrix}5\\5\end{pmatrix}, \quad \begin{pmatrix}-3\\-3\end{pmatrix} \le u(k+1) - u(k) \le \begin{pmatrix}3\\3\end{pmatrix},$$

The original multi-parametric programming will provide the controller partition with 198 regions, which has been plotted in Fig. 2. The simulation result has been provided in Fig. 3.

Since the uncertain parameters $x(k)$ are subjective to the functions of step input $u(k)$, the uncertainty relationship analysis has been proceeded with 1,000 available

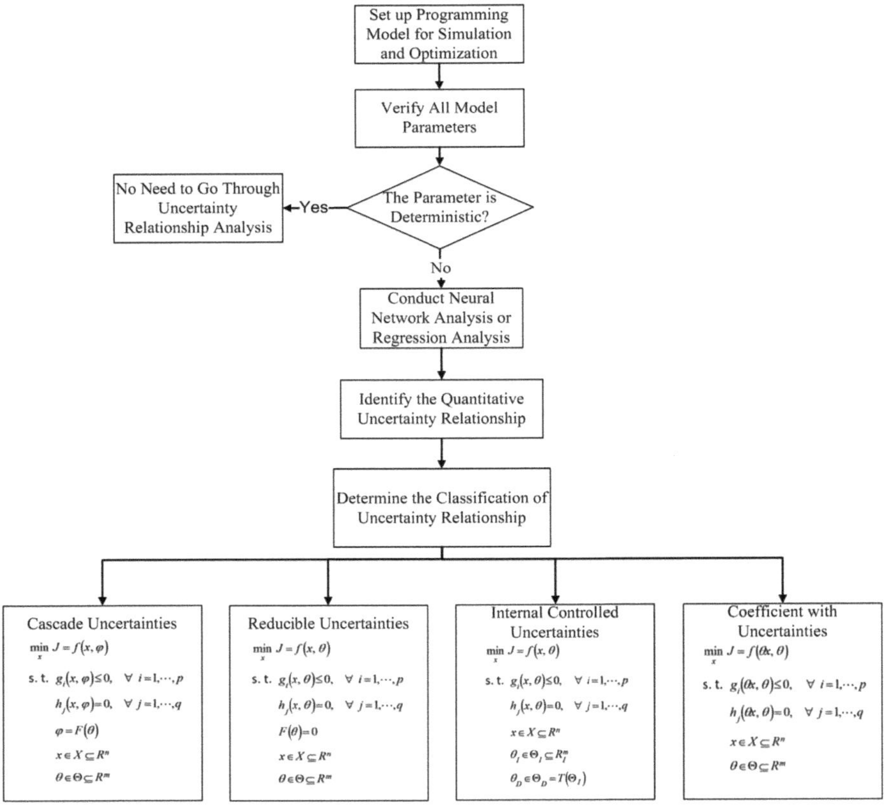

Fig. 1 Methodology framework

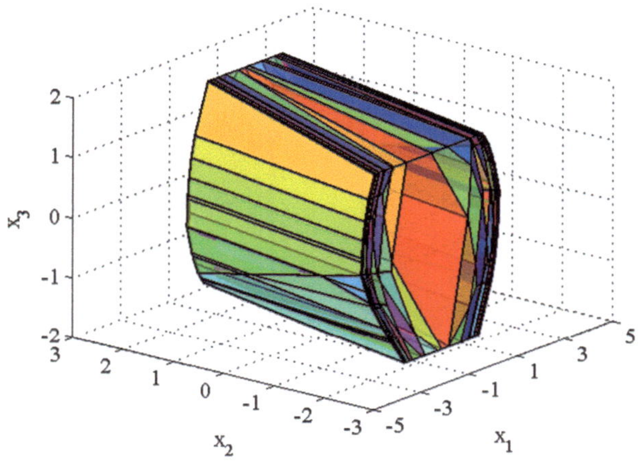

Fig. 2 Controller partition with 198 regions

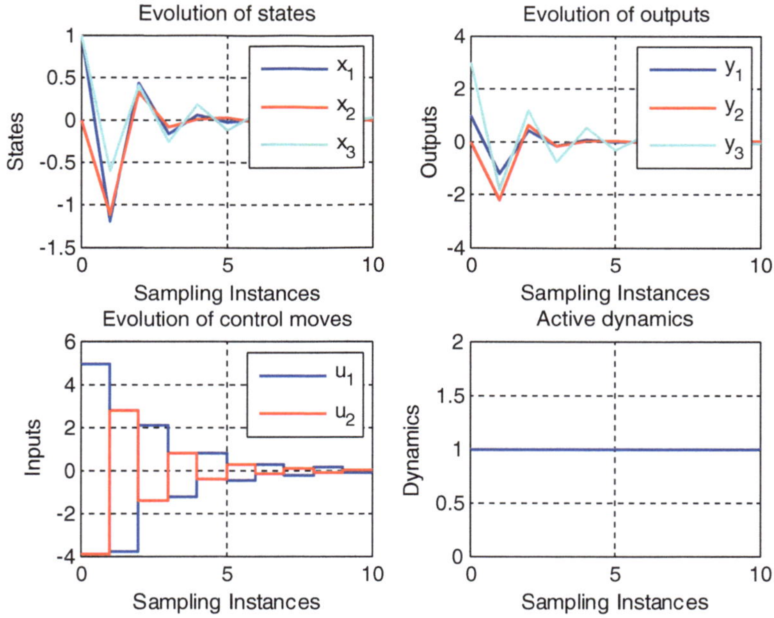

Fig. 3 Simulation results

data sets, whose distribution has been represented by below data distribution and data series of Fig. 4.

By uncertainty relationship analysis, the third uncertainty parameter can be approximated to $x_3 = 0.8x_1 + 0.4x_2$. This helps to reduce the uncertainty domain

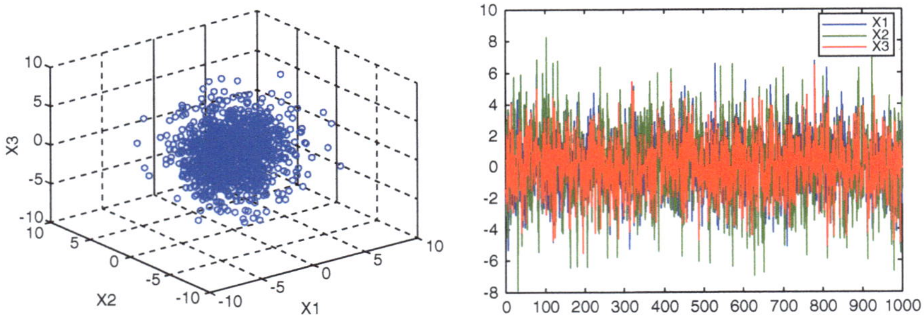

Fig. 4 Data series of uncertainty parameters

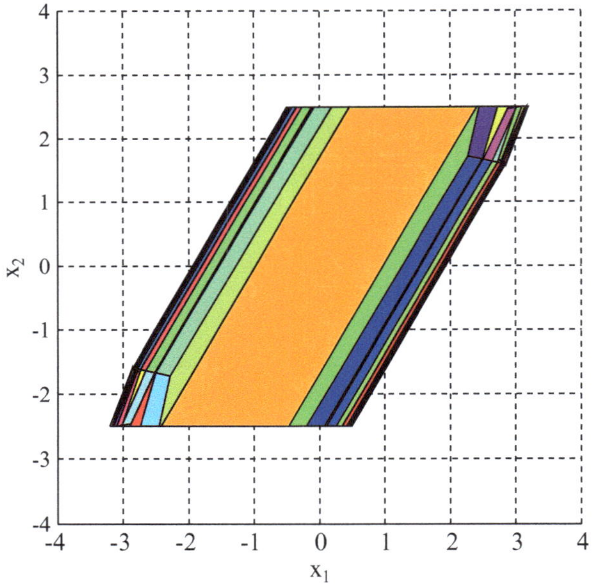

Fig. 5 Controller partition with 65 regions

space from 3 dimension to 2 dimension. The model has been refined to get the programming result as below. The controller partition, value of control action U_1, and U_2value function have been plotted through Figs. 5, 6, and 7 over 65 regions. Thus, the number of total partition regions has been reduced from 198 to 65 by almost two thirds. The computation time has been reduced accordingly.

For simplification, the solving result for second studied case has been provided in Fig. 8 while the programming procedure has been omitted. The application of uncertainty relationship analysis has helped to reduce the number of controller partition regions from 25 to 11.

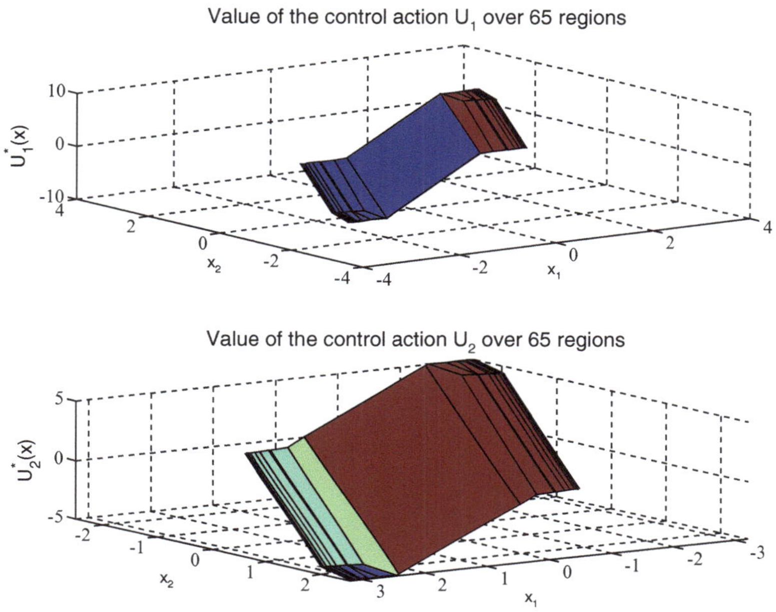

Fig. 6 Value of control action U_1 and U_2

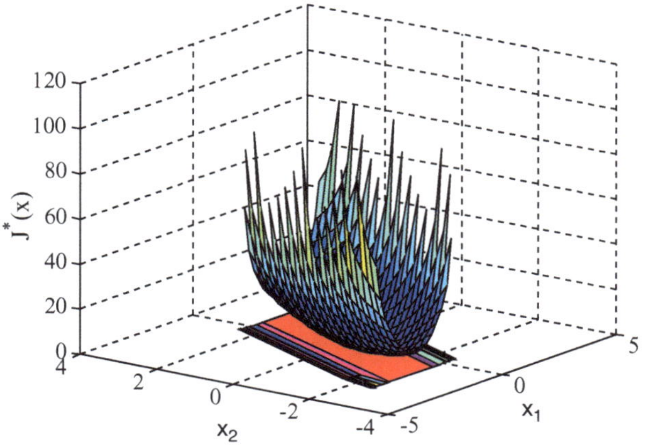

Fig. 7 Value function over 65 regions

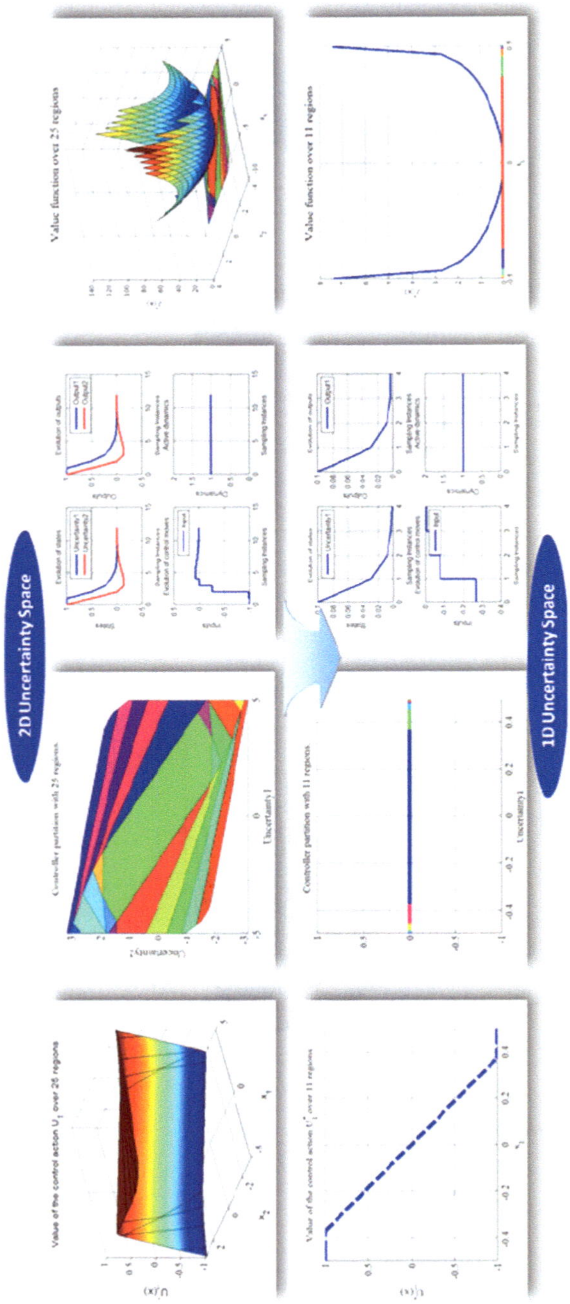

Fig. 8 Uncertainty relationship analysis result for case two

5 Conclusion

Through uncertainty relationship analysis, the dimension of uncertainty space or the region of uncertainty space might be reduced. This might help to reduce the solving effort in the parametric programming for large scale optimization and control system.

References

1. Acevedo, J., Pistikopoulos, E.N.: Parametric MINLP algorithm for process synthesis problems under uncertainty. Ind. Eng. Chem. Res. **35**, 147 (1996)
2. Acevedo, J., Pistikopoulos, E.N.: An algorithm for multiparametric mixed-integer linear programming problems. Oper. Res. Lett. **24**, 139 (1999)
3. Dua, V., Pistikopoulos, E.N.: Algorithms for the solution of multiparametric mixed-integer nonlinear optimization problems. Ind. Eng. Chem. Res. **38**, 3976 (1999)
4. Fiacco, A.V.: Introduction to Sensitivity and Stability Analysis in Nonlinear Programming. Academic, New York (1983)
5. Gal, T.: Postoptimal Analyses, Parametric Programming, and Related Topics. deGruyter, New York (1995)
6. Papalexandri, K., Dimkou, T.I.: Ind. Eng. Chem. Res. **37**, 1866 (1998)
7. Pertsinidis, A., Grossmann, I.E., McRae, G.J.: Comput. Chem. Eng. **22**, S205 (1998)
8. Sacks, J., Welch, W.J., Mitchell, T.J., Wynn, H.P.: Design and analysis of computer experiments. Stat. Sci. **4**(4), 409–423 (1989)
9. Iman, R.L., Helton, J.C.: An investigation of uncertainty and sensitivity analysis techniques for computer models. Risk Anal. **8**(1), 71–90 (1988)
10. Walker, W.E., Harremoës, P., Rotmans, J., van der Sluijs, J.P., van Asselt, M.B.A., Janssen, P., Krayer von Krauss, M.P.: Defining uncertainty: a conceptual basis for uncertainty management in model-based decision support. Integr. Assess. **4**(1), 5–17 (2003)
11. Kennedy, M.C., O'Hagan, A.: Bayesian calibration of computer models. J. R. Stat. Soc. Ser. B (Stat. Methodol.) **63**(3), 425–464 (2001)
12. Der Kiureghiana, A., Ditlevsen, O.: Aleatory or epistemic? Does it matter? Struct. Saf. **31**(2), 105–112 (2009)
13. Hermann, G.M.: Quantifying uncertainty: modern computational representation of probability and applications, Extreme Man-Made and Natural Hazards in Dynamics of Structures NATO Security through Science Series, pp. 105–135, Springer (2007)

DCBA-MPI: A Simulation Based Technique in Optimizing an Accurate Malmquist Productivity Index

Qiang Deng, Wai Peng Wong, and Chee Wooi Hooy

Abstract This paper describes a simulation based methodology utilizing simulation optimization technique in measuring an accurate Malmqusit productivity index (MPI) score. Given the high level of stochastic data in real environment, a novel methodology known as DCBA-MPI has been developed. An example of the mode application is shown in banking institutions. In addition to the novel approach presented, this article provides a new insight into the application domain of MPI measurement as well as the way one conducts productivity study in stochastic environment.

Keywords Malmquist productivity index • Stochastic simulation optimization • OCBA

1 Introduction

The measurement of productivity has widespread applications since [1] introduced the empirical notion of the frontier production function, which reveals the best-performance technology available among a sample of observed decision making units (DMUs). This frontier acts as a measure of the performance with the DMU converse inputs into outputs. In this conversion, growth accounting measures of total factor productivity (TFP) have been shifting to an index, i.e., the Malmquist productivity index (MPI) [2]. MPI is a popular tool for performance measuring in numerous areas (e.g., airports, banks, factories, universities, and hotels).

MPI is based on the observed data over two time periods and it implicitly requires the data should be known exactly. However, in real world, there are various sources that led to the uncertainness (e.g., human error, technical malfunction, behavioral bias); this may lead to the stochastic nature of data [3]. Moreover, if the collected data are not representative or missing, the results will be erroneous and misleading. Hence, the effects and characteristics of the data is vital source of impact that engenders confidence in the results of the MPI analysis. The issue is not new but has consistently banned the development of the TFP measuring and evaluating.

Q. Deng (✉) • W.P. Wong • C.W. Hooy
Universiti Sains Malaysia, 11800 Penang, Malaysia
e-mail: dq11_man112@student.usm.my; wongwp@usm.my; cwhooy@usm.my

D. Gao et al. (eds.), *Advances in Global Optimization*, Springer Proceedings in Mathematics & Statistics 95, DOI 10.1007/978-3-319-08377-3_44

In order to face the issue of stochastic data, some approaches have been launched. Lotf et al. [4] proposed a method for obtaining MPI on interval data. Due to the interval data are not exact and definite, but they lie in an interval, then MPI is changed as an interval, but this method is unable to give a credible conclusion based on the statistical precision. Deng and Wong [5] developed a method which named Stochastic-MPI based on the concept of general distributions and the Monte-Carlo simulation. The Stochastic-MPI can generate the interval MPI by using computational power to derive empirical distributions for MPI measures. Nevertheless, [5] can give the result with distributions from the statistical precision, the interval of the MPI is still too wide to make accurate evaluations for the proposed DMUs. The traditional Monte-Carlo simulation will waste much computation time on the non-critical alternatives. This is because in the conventional simulation process, each process of collecting data has been set equally with the same number of replications [6]. However, it's worth mentioning that some techniques of computing optimization have been developed to speed up the efficiency of the simulation. Chen [7] developed the method of Optimal Computing Budget Allocation (OCBA). This method can reduce the total computation time effectively for collecting the simulation data. Wong [8] have introduced OCBA into the area of efficiency evaluation, and they generated a mode of data collection budget allocation (DCBA) to predict more accurate efficiency scores. While for estimating the accurate MPI in the stochastic environment, the related literatures are neglected.

Therefore, in this paper, we aim to develop a mode to estimate the more accurate MPI. In what follows, we briefly expound the methodology of DCBA-MPI, which generates the interval MPI by using computational power to optimize the accuracy of the MPI scores. In Sect. 3, we show an application of our mode to estimate MPI in the banking industry of Malaysia. Our conclusions are in Sect. 4.

2 The Proposed Mode

Originated from the basic MPI model, the DCBA-MPI aims to provide an accurate measurement of the TFP index when the data are uncertain through the allocation of data collection effort. It has two steps: the first step is to obtain the MPI scores, and then, followed by step two, that is to improve the accuracy of the MPI scores. This approach can be used to handle uncertainties in data. First, we provide a short description of MPI [9].

2.1 Step 1: Basic MPI Measurement

In a multiple-input and multiple-output production system, where inputs x^t are used to produce outputs y^t in time period $t, t = 1, 2, \ldots, \mathrm{T}$ this can be defined as:

$$P^t(x^t) = \{y^t : y^t \text{ can be produced by } x^t \text{ at time } t\} \quad (1)$$

The TFP index of Malmquist can be defined by using distance function $D_o(x, y)$, where an output distance function is used to consider a maximum proportional expansion of the output y, given the inputs x at time t [10].

$$D_o^t(y^t, x^t) = \inf\{\lambda : y^t/\lambda \in P^t(x)\} \tag{2}$$

where $\lambda \in (0, 1]$ and $D_o^t(y, x) \leq 1$ if and only if $y \in P^t(x)$. The value of the distance function is the reciprocal of the Farrell technical efficiency [11]. The calculation of the distance can be computed by the following mode (3).

$$\begin{aligned} & D_o^t(x^t, y^t)^{-1} = \max\ \lambda_1^* \\ & \text{s.t.} \\ & \lambda_1^* y_i^t \leq \sum_{i=1}^{M} \lambda_i y_i^t \qquad i = 1, \ldots, M \\ & \sum_{i=M+1}^{S} \lambda_i x_i^t \leq x_i^t \qquad i = M+1, \ldots, M+N \\ & \lambda_i \geq 0 \qquad i = 1, \ldots, M+N \end{aligned} \tag{3}$$

A distance function which measures the maximal proportional change in output required to make (x^t, y^t) feasible in relation to the technology at t+1 which is called $D_o^t(x^{t+1}, y^{t+1})$. The Malmquist productivity change index is defined in Eq. (4).

$$M_0(x^{t+1}, y^{t+1}, x^t, y^t) = \left[\left(\frac{D_0^t(x^{t+1}, y^{t+1})}{D_0^t(x^t, y^t)}\right)\left(\frac{D_0^{t+1}(x^{t+1}, y^{t+1})}{D_0^{t+1}(x^t, y^t)}\right)\right]^{1/2} \tag{4}$$

In Eq. (4), it needs to calculate four distances, which is to solve four different linear-programming problems.

The issue in this step is that it does not cater for uncertainties in the data. In the next step 2, we aim to address the uncertainties of data in MPI measures by using the simulation optimization technique.

2.2 *Step 2: DCBA: Enhancement for Accuracy*

"How to allocate the budget effectively in order to obtain an accurate MPI score," is the main principle in this step. The distribution of the efficiency score, which ultimately identifies the tendency of where the true efficiency values normally lie, is determined by the amount of data collected for the inputs/outputs.

2.2.1 Conceptualization Based on Simulation Optimization

The issue of "How to allocate the budget effectively?" has been addressed in the field of simulation optimization, where Optimal Computing Budget Allocation (OCBA) has been developed to determine the simulation replications to be allocated to each simulation model in order to identify the best design using the least amount of computing budget [12]. Here we employed the concept of this technique to develop our DCBA model.

2.2.2 The DCBA Model

$$\begin{aligned} \min\ & F(\mathbf{n}) = E\left[(\tilde{\theta}(\mathbf{X}') - \theta(\mathbf{X}))^2\right] \\ \text{s.t.}\ & \sum_{i \in K} n_i = N \end{aligned} \tag{5}$$

The above is DCBA model derived from the OCBA concept. The accuracy of the MPI is measured through Mean Square Error (MSE). The objective function in the above model, $F(\mathbf{n})$ is defined as the MSE of the MPI score for allocation design $\mathbf{n}$ where $\mathbf{X}'$ is the belief of the inputs/outputs after additional data are collected following the allocation design $\mathbf{n}$; N is the total computing budget (data points) for allocating to k variables (the number of stochastic inputs and outputs). $\theta(\mathbf{X})$ is the MPI score computed by using (4) (i.e., $\theta(\mathbf{X}_D)$), for simplicity, we discard the notation D from $(\mathbf{X}_D)$ and $\tilde{\theta}(\mathbf{X}')$ represents the belief for the true MPI scores.

Note that the model cannot be solved directly, as the solution is not in closed-form formula. Inherently, the above problem is a non-convex discrete optimization problem on convex discrete optimization problem where there is no good structure exists for us to develop simple efficient algorithm. One way to estimate $F(\mathbf{n})$ is to derive $\mathbf{X}'$ through the Bayesian framework. Through the quantification of $\mathbf{X}'$ using Bayesian, F can be estimated for a given value of $\mathbf{n}$ as follows:

$$F(n) \approx \frac{1}{M} \sum_{i=1}^{M} (\theta(\hat{\mathbf{X}}_i) - \theta(\mathbf{X}))^2 \tag{6}$$

where $\hat{\mathbf{X}}_i$ is the realization of the inputs/outputs $\mathbf{X}'$ in the replication i of the Monte Carlo run for allocation design $\mathbf{n}$ and M is the cardinality of random data set. This method thus is a simulation based technique. Recall that an allocation design is given as $\mathbf{n} = [n_i]_{i \in K}$, K is the combined inputs and /outputs, D is the total number of inputs/outputs), N is the total number of data collections, and $K = \{1, \ldots, D\}$.

2.2.3 Allocation of Data Collection Plan

We need to evaluate the designs after the searching design process to get the best design (data collection plan), while in this design evaluation process, there still exist time wastes if we equally allocate the computing budget for each selected design. Hence, we apply OCBA [12], a simulation optimization technique in evaluating and choosing the best design with smallest MSE. OCBA helps to reduce the simulation time in the process of design evaluation, as it can intellectually choose to spend much more time and effort in evaluating a design which has potential to get lower MSE. Here, we select the non-sequential data collection plan as it is much simpler for explanation purpose.

To run the OCBA mode, additional notations required are as follows:

$T_{\mathbf{n}_i}$ = number of simulation replications allocated to design $\mathbf{n}_i$, $\bar{B}_{\mathbf{n}_i}$ = sample mean of the MSE for design $\mathbf{n}_i$, $E_{\mathbf{n}_i}$ = variance of the MSE for design $\mathbf{n}_i$, m = number of top designs to be selected, and H = total simulation runs (or computing budget, i.e. used to evaluate the designs). We limit the size of $\mathbf{n}$ to l in each simulation run. From literature of OCBA, the computing allocation budget (or simulation runs) for each design can be determined through the relationship (7) below.

$$\frac{T_{\mathbf{n}_i}^{t+1}}{T_{\mathbf{n}_j}^{t+1}} = \left(\frac{E_{\mathbf{n}_i}/G_{\mathbf{n}_i}}{E_{\mathbf{n}_j}/G_{\mathbf{n}_j}}\right), \quad i, j \in \{1, 2, \ldots, l\} \text{ and } i \neq j \tag{7}$$

where $G_{\mathbf{n}_i} = \bar{B}_{\mathbf{n}_i} - (\bar{B}_{\mathbf{n}_{i\,m}} + \bar{B}_{\mathbf{n}_{i\,m+1}})/2)^2$, $\bar{B}_{\mathbf{n}_{i\,m}}$ and $\bar{B}_{\mathbf{n}_{i\,m+1}}$ are the sample means of the MSE for design $\mathbf{n}_i$ in the m and m+1 top number of designs selected, respectively.

We first simulate all l designs with t_o replications. As simulation proceeds, the sample means and sample variances of each design are computed from the data already collected up to that stage. The simulation budget is then increased by Δ and Eq. (5) is applied to determine the new simulation runs allocation. Further simulation replications are then performed based on the allocation and the procedure is repeated until the total runs H is exhausted. The procedure of the OCBA-m allocation is as follows:

Step 1: Set t= 1 and perform t_o simulation replications for all designs. Set $T_{\mathbf{n}_1}^t = T_{\mathbf{n}_2}^t = \cdots = T_{\mathbf{n}_l}^t = t_o$.
Step 2: Calculate $\bar{B}_{\mathbf{n}_i}$, $E_{\mathbf{n}_i}$, and $G_{\mathbf{n}_i}$ for i= 1, ..., l.
Step 3: Allocate. Increase the computing budget by Δ and calculate the new budget allocation $T_{\mathbf{n}_1}^{t+1}$, $T_{\mathbf{n}_2}^{t+1}$, $\cdots$, $T_{\mathbf{n}_l}^{t+1}$ according to (7).
Step 4: Simulate. Perform additional $T_{\mathbf{n}_i}^{t+1} - T_{\mathbf{n}_i}^{t}$ simulations for design ni for i = 1, ..., l.
Step 5: Termination. If $\sum_{i=1}^{l} T_{\mathbf{n}_i}^t < H$, set t ← t + 1 and return to Step 2; otherwise, stop.

3 An Application of the DCBA-MPI

Consider an example in Malaysian banking industry. Ten banks have been chosen as DMUs which use two inputs (number of employees and capital) to produce one output (profit), both inputs and outputs are interval data. The data are shown in Table 1, and it displays the 2 years banking data with average, lower bound and upper bound for each bank, where the monetary amounts are billion Malaysia Ringgit (100RM≅33USD).

In order to run the simulations, we set the parameters in DCBA model as follows: D=3, which represents the total number of inputs and outputs; N=100, which refers to the total number of data collections; t_0=10, which represents the initial simulation replications; l=20, which represents the number of designs being evaluated for each simulation run. H=200, which means the total simulation replications or computing budget; $\Delta = 20$, which is the computing budget increment per each MSE evaluation.

After the simulation, a feasible solution $\beta=[n_1, n_2, n_3, n_4, n_5, n_6]$ represents the additional number of data to be collected for the three variables (input1, input2, output1) in two periods, respectively. Meanwhile, to avoid the computing chanciness to impact the simulation result, the entire process is executed 1,000 times to obtain the standard deviations and confidence intervals of the MPI scores. The left part (column 2) of Table 2 shows the MPI ranges calculated by the method in [4], and the right part (columns 3–10) shows the result by DCBA-MPI.

The method in [4] only can get the bounds of the productivity score for this example, and their MPI ranges are too wide to make one DMU classify from another. Our result shows narrower ranges of productivity score, the average improvement in the accuracy of MPI ranges is above 84.7 %. More importantly, our method not only can get the statistic information for the MPI, but it also can find the best simulation design with time reduction and direct to the real data collection. As a result, the productivity scores for banks 1, 2, 3, 5, 6, and 9 lie in the ranges of [1.235, 1.276], [0.094,0.097], [0.724,0.742], [0.882,0.918], [0.849,0.875], and [1.411,1.482], which have the accuracy improvement in narrow ranges with 93.59 %, 93.10 %, 93.70 %, 91.48 %, 92.02 %, and 90.08 %, respectively.

Notably, the productivity scores for bank 8 is close to unity, despite that the data are stochastic. This is because that almost the whole range of the interval data of the bank lies on the production frontier. Meanwhile, the best data collection design with MSE for simulation has been presented in Table 2, we find that the narrower is the MPI range, the smaller is the MSE. This indicates that the MSE, which has been chosen to be a numerical measure of accuracy level for MPI score in the DCBA-MPI mode, is verified as a critical factor of reference value for the level of accuracy, and as we increase the simulation times, the value of the MSE starts to decrease. We also find that the DMU with narrower MPI range has got lower standard deviation; in the statistics, the standard deviation represents the closeness of simulated scores to the mean values and hence it can express the stability of the simulation process. That is to say, the lower standard deviation represents the higher reliability of the result. In this application, even for the bank 9 that has the highest standard deviation with the value of 0.018, it only takes about 1 % of its mean MPI score.

Table 1 The data of ten banks in Malaysia from 2009 to 2010

	2009			2010		
Bank[a]	Employees	Capital	Profit	Employees	Capital	Profit
1	[9849, 10694]	[4814.7, 5188.6]	[746.9, 805.9]	[9807, 10613]	[3705.0, 3990.0]	[725.3, 775.9]
Ave.	10224	5000.63	784.29	10224	3857.31	759.19
2	[21721, 23867]	[16681.6, 17449.4]	[3008.9, 3186.1]	[21903, 23174]	[21881.4, 23093.4]	[371.2, 392.0]
Ave.	22856	17200.32	3118.58	22856	22510.44	383.08
3	[14273, 15212]	[11446.0, 12129.1]	[2144.0, 2229.4]	[14312, 14688]	[14370.4, 14748.4]	[1882.2, 2000.9]
Ave.	14518	11623	2185	14518	14599	1937
4	[13680, 14840]	[9192.5, 9789.0]	[2843.0, 3000.9]	[13611, 14559]	[10334.5, 10606.7]	[2695.4, 2889.6]
Ave.	14273	9392	2898	14273	10442	2789
5	[8574, 9279]	[6039.6, 6533.3]	[1287.5, 1379.3]	[8685, 9053]	[6988.7, 7329.0]	[1327.5, 1400.0]
Ave.	8855	6266.1	1317.15	8855	7180.04	1357.34
6	[5460, 5763]	[4689.0, 5013.8]	[926.5, 966.7]	[5485, 5873]	[5240.5, 5383.4]	[871.1, 903.0]
Ave.	5669	4923	949	5669	5319	886
7	[4550, 4819]	[3439.3, 3633.7]	[1150.1, 1211.4]	[4424, 4828]	[3729.7, 3986.2]	[781.5, 817.3]
Ave.	4648	3554	1185	4648	3803	805
8	[2698, 2820]	[3215.2, 3405.5]	[1019.8, 1055.5]	[2656, 2914]	[3121.7, 3325.9]	[886.2, 926.7]
Ave.	2785	3355.18	1038.22	2785	3275.3	895.67
9	[5742, 6110]	[3187.3, 3410.7]	[221.8, 240.6]	[5705, 6177]	[3592.4, 3832.4]	[364.8, 387.6]
Ave.	5986	3347.68	231.63	5986	3658.11	378.8
10	[2806, 3057]	[3038.3, 3260.7]	[819.8, 876.6]	[2843, 3078]	[3624.5, 3808.0]	[746.2, 804.7]
Ave.	2945	3112.18	848.42	2945	3661.57	784.92

[a] *[]* lower bound and upper bound, *Ave.* average

Table 2 The result of the simulations

Bank	MPI range[a]	Mean	Data collection plan						MSE	std[b]	MPI range[c]	Accuracy improvement[d] (%)
1	[0.975,1.620]	1.255	26	65	29	21	11	48	3.76E−02	0.009	[1.235,1.276]	93.59
2	[0.077,0.118]	0.096	33	48	19	34	34	32	8.14E−02	0.001	[0.094,0.097]	93.10
3	[0.600,0.889]	0.720	28	63	9	31	12	57	7.71E−02	0.004	[0.724,0.742]	93.70
4	[0.786,1.047]	0.920	28	60	12	47	46	7	6.26E−03	0.007	[0.906,0.934]	88.93
5	[0.717,1.137]	0.904	45	39	16	29	40	31	8.67E−03	0.006	[0.882,0.918]	91.48
6	[0.710,1.041]	0.870	65	12	23	26	36	38	1.64E−02	0.006	[0.849,0.875]	92.02
7	[0.628,1.421]	0.728	56	4	40	2	68	30	7.42E−02	0.014	[0.705,1.351]	18.65
8	[0.987,1.000]	1.000	59	15	26	20	76	4	1.09E−17	0.000	[0.998,1.000]	85.07
9	[1.133,1.848]	1.497	62	36	1	1	45	55	2.41E−02	0.018	[1.411,1.482]	90.08
10	[0.672,1.068]	0.829	58	28	14	27	71	2	1.23E−03	0.011	[0.801,0.858]	85.46

[a]The MPI range calculated by the method in [4]

[b]*std* standard deviation

[c]The MPI range calculated by DCBA-MPI

[d]The improvement of accuracy in MPI by narrowing the ranges

4 Conclusions

This work developed a method known as the DCBA-MPI for measuring the productivity index of DMUs in a more practically feasible way. The proposed method starts by measuring the MPI scores, after which it improves the accuracy of the score through data collection. The effort in data collection is allocated intelligently using the technique of computing simulation optimization. The proposed method was designed to tackle the limitations of the conventional MPI measurement modus operandi. To end, the salient point is that MPI scores are obtained in a more confident manner, as to which even the decision makers are uncertain about the data, the MPI scores can still be estimated accurately and performance analysis can be conducted smoothly.

Acknowledgements The authors acknowledge the grant support by [1001/PMGT/816224]. The first author acknowledges the fellowship support from the Institute of Postgraduate Studies, Universiti Sains Malaysia.

References

1. Farrel, M.J.: The measurement of productive efficiency. J. R. Stat. Soc. **120**(3), 253–290 (1957)
2. Fried, H.O., Lovell, C.A.K., Schmidt, S.S.: The Measurement of Productive Efficiency and Productivity Growth. Oxford University Press, New York (2008)
3. Dyson, R.G., Shale, E.A.: Data envelopment analysis, operational research and uncertainty. J. Oper. Res. Soc. **61**, 25–34 (2010)
4. Lotf, F.H., Navabakhsh, M., Balf, F.R., Jahantighey, M.A., Abolghasemzadeh, Sh.: Application of malmquist productivity index on interval data in education groups. Int. Math. Forum **16**, 10–17 (2006)
5. Deng, Q., Wong, W.P.: Stochastic mpi for measuring total factor productivity index: with an illustration of Malaysian commercial banks. In: International Conference on Arts, Economics and Literature (ICAEL'2012), pp. 122–125 (2012)
6. Chen, C.H., He, D., Fu, M., Lee, L.H.: Efficient simulation budget allocation for selecting an optimal subset. INFORMS J. Comput. **20**(4), 579–595 (2008)
7. Chen, C.H.: An effective approach to smartly allocate computing budget for discrete event simulation and control. In: Proceedings of the 34th IEEE Conference on Decision, pp. 2598–2605 (1995)
8. Wong, W.P., Jaruphongsa, W., Lee, L.H.: Budget allocation for effective data collection in predicting an accurate dea efficiency score. IEEE Trans. Autom. Control **56**(6), 1235–1246 (2011)
9. Lovell, C.A.K.: The decomposition of malmquist productivity indexes. J. Prod. Anal. **20**, 437–458 (2003)
10. Shephard, R.W.: Theory of Cost and Production Functions. Princeton University Press, Princeton (1970)
11. Färe, R., Grosskopf, S., Norris, M., Zhang, Z.: Productivity growth, technical progress, and efficiency change in industrialized countries. Am. Econ. Rev. **84**(1), 66–83 (1994)
12. Chen, C.H., Lee, L.H.: Stochastic Simulation Optimization: An Optimal Computing Budget Allocation. World Scientific, River Edge (2010)

The Robust Constant and Its Applications in Global Optimization

Zheng Peng, Donghua Wu, and Wenxing Zhu

Abstract Robust analysis is important for designing and analyzing algorithm for global optimization. In this paper, we introduced a new concept, robust constant, to quantitatively characterize robustness of measurable sets and measurable functions. The new concept is consistent with the robustness proposed in literature. This paper also showed that robust constant had significant value in the analysis of some random search algorithms for solving global optimization problem.

Keywords Global optimization • Robustness • Robust constant • Global random search methods

1 Introduction

Consider an unconstrained minimization problem of the form

$$c^* = \min_{x \in R^n} f(x) \tag{1.1}$$

where $f : R^n \to R$ is a summable function but not necessary convex. For the regularity, we assume that f is lower bounded and for arbitrarily given $x_0 \in R^n$, the level set $\{x \in R^n : f(x) < f(x_0)\}$ is bounded.

In general, we need the smoothness assumption on the objective function f such that some gradient-based optimization methods, e.g., the steepest descent

Z. Peng
College of Mathematics and Computer Science, Fuzhou University, Fuzhou 350108, China
e-mail: pzheng@fzu.edu.cn.

D. Wu
Department of Shanghai University, Shanghai 200444, China
e-mail: dhwu@shu.edu.cn

W. Zhu (✉)
Center of Discrete Mathematics and Theoretical Computer Science, Fuzhou University, Fuzhou 350108, China
e-mail: wxzhu@fzu.edu.cn

D. Gao et al. (eds.), *Advances in Global Optimization*, Springer Proceedings in Mathematics & Statistics 95, DOI 10.1007/978-3-319-08377-3_45

method and Newton-type methods, can be used to find a local minimizer. Unless the problem (1.1) has some special structures, for example, convex programming or fractional linear programming, finding a global minimizer of the problem (1.1) is an NP-hard problem, see Vavasis [1]. However, finding a global minimizer of a general nonconvex objective function is a common task in real-world applications. Thus, various global minimization techniques have been developed. The interested readers are referred to [2–5] for excellent survey papers.

Among the developed global minimization methods, Global Random Search (GRS) has better convergence in theory. Pure Random Search (PRS) and Pure Adaptive Search (PAS) are two classical methods of GRS.

The PRS method was originally proposed by Brooks [6]. After that, a number of hybrid or accelerated random search methods have been developed, see [7–9]. Let $H_0 = \{x \in R^n : f(x) < f(x_0)\}$, where $x_0 \in \operatorname{dom} f$ is an initial guesser of an optimal solution. If $f(x_0) > c^*$, then H_0 is a nonempty subset of R^n. At the k-th iteration, PRS produces a candidate x_k such that

$$f(x_k) = \min\left\{f(x_i), i = 1, 2, \ldots, N,\ x_i \text{ i.i.d.on } H_0\right\}. \tag{1.2}$$

The PRS method is extremely robust and easy to implement, but the convergence is very slow.

The PAS method studied by Zabinsky and Smith [10] is the "theoretical optimum" in GRS method. For a given x_k, let $H_k = \{x \in R^n : f(x) < f(x_k)\}$ be the level set of the k-th iteration, the PAS generates x_{k+1} uniformly distributed in H_k. By this mechanism we have $f(x_{k+1}) < f(x_k)$. It has been shown in [10] that if f is Lipschitz on R^n then the convergence rate of the PAS is exponential. However, the PAS is impossible to implement in general since identifying the current level set H_k is intrinsically harder than finding the optimal solution.

The integral level-set method (ILSM), originally proposed by Zheng et al. [11, 12], can also be considered as a GRS method. It constructed two sequences in the ILSM: a sequence of level value $\{c_k\}$ and a sequence of the corresponding level set $\{H_{c_k}\}$, they are

$$c_{k+1} = \frac{1}{\mu(H_{c_k})} \int_{H_{c_k}} f(x) d\mu, \tag{1.3}$$

$$H_{c_{k+1}} = \left\{x \in R^n : f(x) < c_{k+1}\right\}, \tag{1.4}$$

where μ is the Lebesgue measure on R^n. Under some suitable conditions, it can be proved that the generated level value sequence $\{c_k\}$ converges to optimal value, and, respectively, the corresponding level set sequence $\{H_{c_k}\}$ converges to optimal solution set. The ILSM has the same difficulty as that of PAS, which is, the level set H_{c_k} is hard to be determined. So in the implementable algorithm of the ILSM, the authors calculated the integration in (1.3) and determined the level set in (1.4) by the Monte-Carlo method. To do so, it leads to a drawback that convergence of the implementable algorithm cannot be guaranteed. To overcome these drawbacks, Yao et al. [13] presented an optimality condition and algorithm with deviation integral for integral global optimization method. Peng et al. [14] proposed a Level-Value Estimation method basing on the idea of the ILSM, and implemented the proposed method by the Monte-Carlo method with important sampling, and proved the convergence.

To overcome the drawback of identifying the current level set H_{c_k}, Peng et al. [15] proposed a modified integral level-set method (MILSM) basing on importance sampling. Let

$$F_k(x) = \begin{cases} c_k, & \text{if } f(x) > c_k, \\ f(x), & \text{otherwise,} \end{cases} \tag{1.5}$$

the MILSM updates the level value by

$$c_{k+1} = \frac{1}{N}\sum_{t=1}^{N} F_k(X_t) \tag{1.6}$$

where X_t is independently identically distributed (iid for short) from the distribution with the density $g_k(x)$ on R^n. The efficiency of the MILSM depends heavily on the distribution characterized by the density function $g_k(x)$. The Cross-Entropy (CE) method [16–18] provides a novel idea for the choice and updating rule of the density function. However, numerous computational experiments show that, by the importance sampling technique, the effectual samples depend not only on the sample density $g_k(x)$ but also on the current level set H_{c_k}.

To characterize the property of level set H_{c_k}, Zheng [19] introduced robust analysis to global optimization. In his paper, Zheng introduced some concepts on robust set and robust function for qualitative description. To quantitatively analyze the robustness of robust set and robust function, we introduce a new concept in this paper, i.e., robust constant.

Let us revisit the MILSM proposed in [15]. Assume that $g_k(x) = N(x, m_k, \sigma_k^2)$, the coordinate-normal density function, where m_k and σ_k are chosen by the Cross-Entropy method. Suppose the density function $g_k(x)$ is "good enough" at the k-th iteration in some sense, and the MILSM generates T samples $\{X_i, i = 1, 2, \ldots, T\}$ at the current iteration, then the decrement of level value, $\Delta c_k = c_k - c_{k+1}$, fully depends on the effectual samples defined by $S_k = \{X_i : F(X_i) = f(X_i), i = 1, 2, \ldots, T\}$. For example, suppose that the objective function $f(x)$ and the density function $p(x)$ are shown in Fig. 1, then the level set with the value c is

$$H_c = \{x : f(x) < c\} = AB_1 + A_2B_2 + A_3B_3 + A_4B_4 + A_5B_5 + A_6B_6 + A_7B.$$

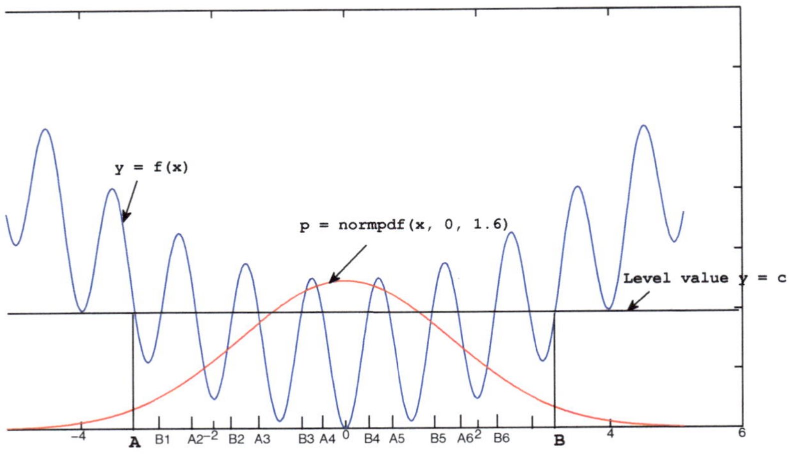

Fig. 1 The level set and density function

Let $\rho = \int_{AB} g_k(x)dx$, the number expected of effectual samples is

$$T_{\text{res}} = \rho T \times \frac{\mu(H_c)}{\mu(AB)}. \tag{1.7}$$

It is worth to emphasize that T_{res} depends on $\dfrac{\mu(H_c)}{\mu(AB)}$ in which AB is essentially the closed convex hull of the level set H_c.

Inspired by the above observation, we give the concept of robust constant in Sect. 2. As an example of applications of the robust constant, Sect. 3 analyzes a modified pure adaptive method by using this concept. The analysis shows that, with the utilization of global robust constant, we can give a simple and checkable stop criterion of the proposed GRS method, and prove its rationality. Finally, Sect. 4 gives some concluding remarks.

2 Robust Constant

Let X be a topological space and D be a subset of X. By Zheng [19], we have

Definition 2.1 ([19]). A set D is robust if and only if

$$\text{cl}D = \text{cl}(\text{int } D) \tag{2.1}$$

where cl denotes closure of a set.

Obviously, by Definition 2.1 the empty-set $\emptyset$ is a robust set.

Definition 2.2 ([19]). A function $f : X \to R$ is robust if and only if its level set $H_c = \{x \in X : f(x) < c\}$ is robust for all $c \in (-\infty, +\infty)$, where X is a robust set.

Let X be a normal topological space, Ω be a σ-field of subsets of X.

Definition 2.3 ([19]). A measure space (X, Ω, μ) is said to be a Q-measure space if:

M1. each open set is in Ω;
M2. the measure of each nonempty open set is positive;
M3. the measure of each compact set is bounded.

Let (X, Ω, μ) be a Q-measure space. For a given measurable set $S \subset X$, let clco$S$ denote the closed convex hull of S. We give the concept of robust constant as follows.

Definition 2.4. Let S be a robust set, the robust constant of S is given by

$$R(S) = \frac{\mu(S)}{\mu(\text{clco}S)}. \tag{2.2}$$

The robust constant of an empty-set is set to be $R(\emptyset) = 1$.

Definition 2.5. Let $f : X \to R$ be a robust function where X is a robust set. The robust constant of f with respect to (w.r.t.) level value c is given by

$$R(f, c) = \frac{\mu(H_c)}{\mu(\text{clco}H_c)}, \tag{2.3}$$

where $H_c = \{x \in X : f(x) < c\}$.

It is obvious by the definition that $R(f, c) = R(H_c)$. In what follows, we give some examples to better understand the robust constant.

Example 2.1. Let $X \in R^n$ be a closed convex set. Since clco$X = X$, the robust constant of X is $R(X) = 1$.

Example 2.2. Let $f : R^n \to R$ be a strictly convex and lower-bounded function. For all $c \in R$, since $H_c = \{x \in R^n : f(x) < c\}$ is a convex set and $\mu(H_c) = \mu(\text{clco}H_c)$, we have $R(f, c) = 1$.

Example 2.3. Let $X = [0, 1]$ and $f : X \to R$ be defined by

$$f(x) = \begin{cases} 1, \text{ if } x \text{ is irrational}, \\ 0, \text{ if } x \text{ is rantional}. \end{cases}$$

Then we have

$$R(f, c) = \begin{cases} 1, \text{ if } c \geq 1, \\ 0, \text{ otherwise}. \end{cases}$$

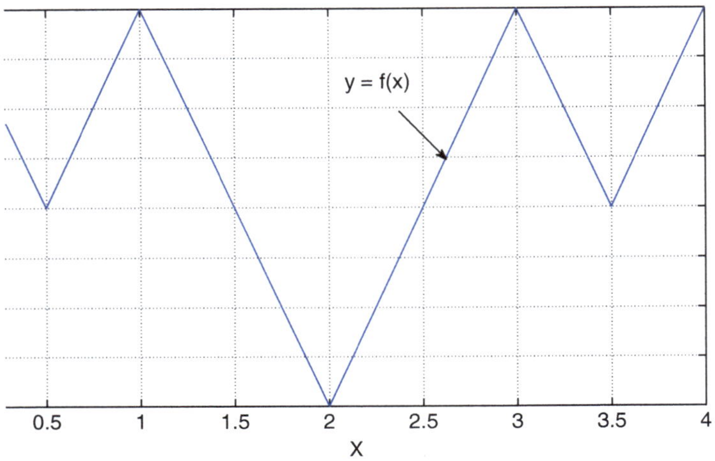

Fig. 2 Example 2.5 $f : [0, 4] \to R$

Example 2.4. Let $X = [0, 4\pi]$ and $f : X \to R$ be defined by

$$f(x) = \begin{cases} \sin x, & x \in [0, 2\pi), \\ \frac{1}{\pi}x - 2, & x \in [2\pi, 4\pi]. \end{cases}$$

Then we have: $R(f, c) = 1$ for all $c \geq 1$ or $c < 0$, and $R(f, \frac{1}{2}) = \frac{11}{15}$, and so on.

Undoubtedly, the robust constant of a function (or a set) is hard to compute in general. However, it has significance in theory for global optimization, as Lipschitz constant does for convex analysis. For a global optimization problem, we are also interesting in global robust constant. We have

Definition 2.6. A function $f : R^n \to R$ is globally robust with constant $r > 0$ if

$$R(f, c) \geq r, \quad \forall c \in (-\infty, +\infty). \tag{2.4}$$

Example 2.5. Suppose the function $f : [0, 4] \to R$ is displayed in Fig. 2. It is obvious that

$$R(f, c) \geq \min R(f, c) = R(f, 2) = \frac{1}{3} = r.$$

Thus the global robust constant of the function f is $r = \frac{1}{3}$.

Indeed, the new concept is consistent with the robustness in literature. The consistency is shown as follows.

Theorem 2.1. *If $f : X \to R$ is a robust function where X is a robust set, then $R(f,c) > 0$ for all $c \in (-\infty, +\infty)$.*

Proof. The function $f : X \to R$ is robust, which means that $H_c = \{x \in X : f(x) < c\}$ is robust for all $c \in (-\infty, +\infty)$ by Definition 2.2. Thus, by Definition 2.1, we have

$$\text{cl}(H_c) = \text{cl}(\text{int}H_c). \tag{2.5}$$

If $H_c = \emptyset$, then $R(f,c) = R(H_c) = R(\emptyset) = 1 > 0$. If $H_c \neq \emptyset$, then by (2.5) we have

$$\text{cl}(\text{int}H_c) = \text{cl}(H_c) \neq \emptyset$$

which means that $\text{int}H_c \neq \emptyset$, and consequentially, $\mu(\text{int}H_c) > 0$ by Definition 2.3. Thus $\mu(H_c) \geq \mu(\text{int}H_c) > 0$, which deduces that $R(f,c) > 0$.

3 Applications

Robust constant is a useful concept in some important areas, such as random search technique for numerical global optimization and simulation optimization. As an example of these applications, we use the new concept, i.e., robust constant, to analyze a modified pure adaptive search (MPAS) method for global optimization.

For the problem (1.1), it is well known that: if $H_c = \emptyset$, then $c < c^* = \min f(x)$. Suppose $f(x)$ is robust on R^n, and the scheme (1.5)–(1.6) is used to search for the minimum of f, where X_t $(t = 1, 2, \ldots, N)$ are iid random samples with density function $g(x)$ characterized by the mean u and variance σ^2. A sample X is said to be effectual to the level value c if and only if $f(X) < c$. Then, the rate of effectual sample is given by

$$\rho = R(f,c) \times \int_{\text{clco}H_c} g(x)d\mu. \tag{3.1}$$

Among the N samples, the number expected of effectual samples is $N^{\text{res}} = \rho N$.

Based on this observation, we propose a MPAS method for global optimization problem (1.1), and analyze the proposed method by robust constant.

Algorithm 3.2. A Modified Pure Adaptive Search Method, MPAS

s0. Let X_0 be a random point in R^n and $c_0 = f(X_0)$, and let $g_0(x)$ be a probability density function with mean vector u_0 and variance vector σ_0^2. Set $\varepsilon > 0$ and $k = 0$.

s1. Generate a sample set $S_k = \{X_t, t = 1, 2, \ldots, N_k\}$ iid from distribution with density $g_k(x)$. Let

$$c_{k+1} = \frac{1}{N_k} \sum_{t=1}^{N_k} F_k(X_t) \tag{3.2}$$

where $F_k(x)$ is defined by (1.5). Let $\hat{H}_k = \{X_t : f(X_t) < c_k, t = 1, 2, \ldots, N_k\}$ be the effectual sample set, and $n_k = |\hat{H}_k|$ be the number of elements in the set.

s2. If $|c_{k+1} - c_k| < \varepsilon$, stop.

s3. Let

$$u_{k+1} = \frac{1}{n_k} \sum_{X_t \in \hat{H}_k} X_t, \quad \text{and} \quad \sigma_{k+1,i}^2 = \frac{1}{n_k} \sum_{X_t \in \hat{H}_k} (X_{t,i} - u_{k+1,i})^2, \quad i = 1, 2, \ldots, n. \tag{3.3}$$

Smooth these parameters by

$$u_{k+1} = \alpha u_{k+1} + (1-\alpha) u_k, \quad \sigma_{k+1,i} = \beta_k \sigma_{k+1,i} + (1-\beta_k)\sigma_{k,i}, \quad i = 1, 2, \ldots, n, \tag{3.4}$$

where $0.5 < \alpha < 0.9, \quad 0.8 < \beta < 0.99$ and $\beta_k = \beta - \beta(1 - \frac{1}{k})^q$, q is an integer (typically between 5 and 10), see Rubinstein et al. [16] and [18].

Construct the new density function $g_{k+1}(x)$ with the mean vector u_{k+1} and variance σ_{k+1}^2, let $k := k + 1$, go to s1.

Remark 3.1. The step s1 accepts the new level value c_{k+1} if it satisfies

$$c_{k+1} - c_k \leq \lambda_k (c_k - c_{k-1}), \tag{3.5}$$

where $\lambda_k \in (0, 1)$ is step length at the k-th iteration. The sample size N_k can be updated adaptively according to some information at the current iteration.

It is easy to prove the convergence to global optimum in probability of the proposed algorithm, and we omit the proof here. In what follows, we analyze the rationality of stop criterion, i.e., $|c_{k+1} - c_k| < \varepsilon$, of the MPAS method.

For the objective function $f(x)$ of problem (1.1) and a given level value c, suppose that $H_c \neq \emptyset$, a sample density function $g(x)$ is said to be "good" for the level value c if

$$\int_{\mathrm{clco} H_c} g(x) d\mu \geq p_0 > 0.$$

Theorem 3.1. *Suppose $f : R^n \to R$ is globally robust with the constant r, and sample density functions $g_k(x)$ are "good" for all k. Then, there exists a positive constant λ_B such that step length $\lambda_k \geq \lambda_B$ for every iteration k of the MPAS method.*

Proof. At the k-th iteration, we have the sample set S_k. We partition S_k into two parts: one is $S_k^1 = \left\{X_t \in S_k : F_k(X_t) > c_k - \eta(c_{k-1} - c_k)\right\}$, and the other is $S_k^2 = \left\{X_t \in S_k : F_k(X_t) \leq c_k - \eta(c_{k-1} - c_k)\right\}$, where $\eta \in (0,1)$ is a constant. Let $N_k^1 = |S_k^1|$ and $N_k^2 = |S_k^2|$, then $N_k = N_k^1 + N_k^2$. Based on the fact that $F_k(x) \leq c_k$, we have

$$
\begin{aligned}
c_{k+1} &= \frac{1}{N_k} \sum_{X_t \in S_k} F_k(X_t) \\
&= \frac{N_k^1}{N_k}\left(\frac{1}{N_k^1} \sum_{X_t \in S_k^1} F_k(X_t)\right) + \frac{N_k^2}{N_k}\left(\frac{1}{N_k^2} \sum_{X_t \in S_k^2} F_k(X_t)\right) \\
&\leq \frac{N_k^1}{N_k} c_k + \frac{N_k^2}{N_k}\Big(c_k - \eta(c_{k-1} - c_k)\Big) \\
&= c_k - \eta \frac{N_k^2}{N_k}(c_{k-1} - c_k).
\end{aligned}
$$

Let $\lambda_k = \eta \dfrac{N_k^2}{N_k}$, we have

$$
c_{k+1} \leq c_k - \lambda_k (c_{k-1} - c_k). \tag{3.6}
$$

On the other hand, let $\hat{c}_k = c_k - \eta(c_{k-1} - c_k)$, then by (3.1) we have $N_k^2 = \rho_k N_k$, where

$$
\rho_k = R(f, \hat{c}_k) \times \int_{\mathrm{clco} H_{\hat{c}_k}} g_k(x) d\mu.
$$

Thus, we obtain

$$
\lambda_k = \eta \rho_k = \eta \times R(f, \hat{c}_k) \times \int_{\mathrm{clco} H_{\hat{c}_k}} g_k(x) d\mu. \tag{3.7}
$$

Since $f : R^n \to R$ is globally robust with the constant r, i.e., $R(f, \hat{c}_k) \geq r$ for all k, and the sample density function $g_k(x)$ is "good" such that

$$
\int_{\mathrm{clco} H_{\hat{c}_k}} g_k(x) d\mu \geq p_0 > 0, \tag{3.8}
$$

we have

$$
\lambda_k \geq \eta r p_0 := \lambda_B > 0. \tag{3.9}
$$

Theorem 3.2. *If the objective function $f(x)$ of the problem* (1.1) *and the sample density function $g_k(x)$ satisfy the same conditions of Theorem 3.1, and the level value sequence $\{c_k\}$ is generated by the MPAS method, then we have*

$$\lim_{k\to\infty}\left(c_{k-1}-c_k\right)=0. \tag{3.10}$$

Proof. By (1.5) and (3.2), it is easy to show that $c_{k-1}-c_k \geq 0$ for all k. Adding (3.6) from $k=1$ to ∞, and let $c_\infty := \lim_{k\to\infty} c_k$, we get

$$c_\infty \leq c_1 - \sum_{k=1}^{\infty}\lambda_k(c_{k-1}-c_k). \tag{3.11}$$

Since the objective function f is lower bounded, we suppose $f(x) \geq b$ for all $x \in R^n$. Then we have $c_\infty \geq b$. By (3.9) and (3.11), we get

$$\lambda_B\sum_{k=1}^{\infty}(c_{k-1}-c_k) \leq \sum_{k=1}^{\infty}\lambda_k(c_{k-1}-c_k) \leq c_1-c_\infty \leq c_1-b<+\infty, \tag{3.12}$$

which deduces (3.10) directly.

Theorem 3.2 gives a simple and utilizable stop criterion of the MPAS method. Indeed, under some suitable assumptions, $c_{k-1}-c_k=0$ gives a necessary condition of global optimality of the unconstrained global optimization problem.

Remark 3.2. In engineering applications, one can update the sample-size N_k by the following rule: at the k-th iteration, generate N_k samples according to the distribution with density $g_k(x)$, denote the sample set by $S_k = \{X_t, t = 1,2,\ldots,N_k.\}$, and compute c_{k+1} by (3.2). If the acceptance criterion (3.5) is satisfied, then accept c_{k+1} as the new level value; otherwise, generate $\lceil \frac{N_k}{10}\rceil$ samples, let $N_{k+1} := N_{k+1} + \lceil \frac{N_k}{10}\rceil$, and accept c_{k+1} until the condition (3.5) is satisfied. At the $(k+1)$-th iteration, we first reserve the effectual samples w.r.t. c_{k+1} of the previous iteration, and denote it by $S_k^{\text{res}} = \{X_t \in H_{c_{k+1}} : X_t \in S^k\}$. Then generate N_{k+1} samples according to the distribution with density $g_{k+1}(x)$, denote the sample set by $\tilde{S}_{k+1}$, let $S_{k+1} = \tilde{S}_{k+1} \cup S_k^{\text{res}}$ and $N_{k+1} := N_{k+1} + |S_k^{\text{res}}|$. The rest is to compute c_{k+2} and check whether it satisfies the criterion (3.5) or not, and so on.

4 Conclusions

In this paper, we have introduced a new concept, robust constant, for quantitative description of robustness of measurable sets and measurable functions. Robust constant is a useful concept for analyzing random search methods for global optimization and simulation optimization.

It is well known that, global optimization problem is NP-hard, but finding a global optimizer of a general function is a common task in real-world applications. In engineering optimization, GRS techniques are the most useful method for finding a global optimizer. However, many GRS methods do not have a suitable criterion to stop their iteration, since it is also an NP-hard problem to verify that a feasible (maybe local) solution is a global solution. As an example of applications of the robust constant, we have proposed a modified pure adaptive method for unconstrained global optimization, and given a suitable stop criterion of the proposed method and analyzed the rationality. This is essentially a utilization of a necessary condition of global optimality in some sense.

Acknowledgements This work is supported by Natural Science Foundation of China (61170308), Natural Science Foundation of FuJian Province (2011J01008) and the talent foundation of Fuzhou University (XRC-1043).

References

1. Vavasis, S.A.: Complexity issues in global optimization: a survey. In: Handbook of Global Optimization, pp. 27–41. Kluwer, Dordrecht (1995)
2. Schoen, F.: Stochastic techniques for global optimization: a survey of recent advances. J. Glob. Optim. **1**(3), 207–228 (1991)
3. Floudas, C.A., Gounaris, C.E.: A review of recent advances in global optimization. J. Glob. Optim. **45**(1), 3–38 (2009)
4. Pardalos, P.M., Romeijn, H.E., Tuy, H.: Recent developments and trends in global optimization. J. Comput. Appl. Math. **124**, 209–228 (2000)
5. Thomas, W.: Global optimization algorithms—theory and application (2008). Online available at http://www.it-weise.de/
6. Brooks, S.H.: A discussion of random methods for seeking maxima. Oper. Res. **6**, 244–251 (1958)
7. Appel, M.J., Labarre, R., Radulovic, D.: On accelerated random search. SIAM J. Optim. **14**, 708–731 (2003)
8. Kabirian, A.: Hybrid probabilistic search methods for simulation optimization. J. Ind. Syst. Eng. **2**(4), 259–270 (2009)
9. Radulovia, D.: Pure random search with exponential rate of convergency. Optimization **59**(2), 289–303 (2010)
10. Zabinsky, Z.B., Smith, R.L.: Pure adaptive search in global optimization. Math. Program. **53**, 323–338 (1992)
11. Chew, S., Zheng, Q.: Integral Global Optimization (Theory, Implementation and Applications). Springer, Berlin (1988)
12. Zheng, Q., Zhuang, D.M.: Integral global minimization: algorithms, implementations and numerical tests. J. Glob. Optim. **7**, 421–454 (1995)
13. Yao, Y.R., Chen, L., Zheng, Q.: Optimality condition and algorithm with deviation integral for global optimization. J. Math. Anal. Appl. **357**(2), 371–384 (2009)
14. Peng, Z., Wu, D., Zheng, Q.: A level-value estimation method and stochastic implementation for global optimization. J. Optim. Theory Appl. **156**(2), 493–523 (2013)
15. Peng, Z., Shen, Y., Wu, D.: A modified integral global optimization method and its asymptotic convergence. Acta Math. Appl. Sin. **25**(2), 283–290 (2009)

16. Rubinstein, R.Y.: The cross-entropy method for combinatorial and continuous optimization. Methodol. Comput. Appl. Probab. **2**, 127–190 (1999)
17. De Boer, P.T., Kroese, D.P., et al.: A tutorial on the cross-entropy method. Ann. Oper. Res. **134**, 19–67 (2005)
18. Kroese, D.P., Porotsky, S., Rubinstein, R.Y.: The cross-entropy method for continuous multi-extremal optimization. Methodol. Comput. Appl. Probab. **8**, 383–407 (2006)
19. Zheng, Q.: Robust analysis and global optimization. Ann. Oper. Res. **24**, 273–286 (1990)

Part VIII
Complex Simulation and Supply Chain Analysis

Closed-Loop Supply Chain Network Equilibrium with Environmental Indicators

Shi-Qin Xu, Guo-Shan Liu, and Ji-Ye Han

Abstract In this paper, we propose a closed-loop supply chain network equilibrium model with environmental indicators through variational inequality theory, which is composed by raw material suppliers, manufacturers, retailers, demand markets, and recovery centers. In view of sustainable development, the Ministry of Environmental Protection legislates by imposing emission penalties to prevent the manufacturers from violating laws and regulations but offering premiums to stimulate the recovery centers to recycle used products. The penalties and premiums should be greater than the corresponding shadow prices as environmental indicators determined by the model, which is constructive for decision making of the authorities. We describe their behavior, derive optimality conditions, and establish the variational inequality in accordance with the closed-loop supply chain network equilibrium conditions. Based on the existence of solution to the model under reasonable assumptions, a numerical example is provided for illustration.

Keywords Closed-loop supply chain • Network equilibrium • Environmental indicators • Variational inequality

1 Introduction

There is no doubt that a considerable number of studies are involved in closed-loop supply chains and we can refer [1, 2] and the references therein. The goal of subsequent closed-loop supply chain networks is to maximize the value originated from collecting for remanufacturing to make full use of the essential components from recycled products. Such frameworks are derived from contributions in supply chain networks composed by tiers of decision-makers including manufacturers, distributors, retailers, and demand markets with the governing equilibrium conditions, that is, *network equilibrium* see [3, 4] and [5]. In particular, the previous models

S.-Q. Xu (✉) • G.-S. Liu
School of Business, Renmin University of China, Beijing, China
e-mail: xushiqin85@126.com

J.-Y. Han
Institute of Applied Mathematics, Chinese Academy of Sciences, Beijing, China

D. Gao et al. (eds.), *Advances in Global Optimization*, Springer Proceedings in Mathematics & Statistics 95, DOI 10.1007/978-3-319-08377-3_46

involved the variational inequality theory which had widely used in various research fields such as supply chain networks [6, 7], transportation networks [8], financial networks [1], knowledge networks [9], and so forth.

According to the concept of equilibrium, [1] presented a variational inequality formulation for a closed-loop supply chain network equilibrium model. Based on the above model, [10] established the oligopolistic closed-loop supply chain in which recovery centers received subsidies from government organizations. In addition, [2] has discussed a closed-loop supply chain network combined with competition, distribution channel investment, and uncertainties.

In view of the severe environmental pollution, the Ministry of Environmental Protection (*MEP*) has placed more emphasis on how to implement relevant policies such as premium and penalty mechanisms to relieve the pressure. In this paper, we consider that the *MEP* sets upper limits of emission on manufacturers to control their discharges in the form of the legislation whereas it offers lower limits of recycled products to recovery centers in the form of incentives. Therefore, nonnegative emission penalties are imposed on the manufacturers to guarantee the environmental target, and nonnegative premiums are offered to recovery centers to stimulate them to recover used products as much as possible. These penalties and premiums should be greater than corresponding shadow prices which are Lagrangian multipliers with regard to the constraints as environmental indicators.

Based on [2, 11], consequently, after integrating variational inequality theory as well as policies of the *MEP*, we formulate a theoretical framework of the closed-loop supply chain network equilibrium with environmental indicators. The model is composed of five tiers of decision-makers, that is, raw material suppliers, manufacturers, retailers, demand markets, and recovery centers that are responsible for collecting used products. Then, we describe the optimizing behavior of the various decision-makers, derive the governing optimality conditions, and obtain the variational inequality relevant to the closed-loop supply chain network equilibrium model with environmental indicators. After providing the existence property of the solution, a numerical example is presented for illustrative purpose.

2 The Closed-Loop Supply Chain Network Equilibrium with Environmental Indicators

The closed-loop supply chain network with environmental indicators is depicted in Fig. 1.

We denote a typical raw material supplier by i $(i = 1, 2, \ldots, I)$, a typical manufacturer by j, $(j = 1, 2, \ldots, J)$, a typical retailer by k $(k = 1, 2, \ldots, K)$, a typical demand market by l $(l = 1, 2, \ldots, L)$, and a typical recovery center by r $(r = 1, 2, \ldots, R)$, respectively. It is assumed that members in the network compete in a noncooperative mode, which means that each decision-maker in the

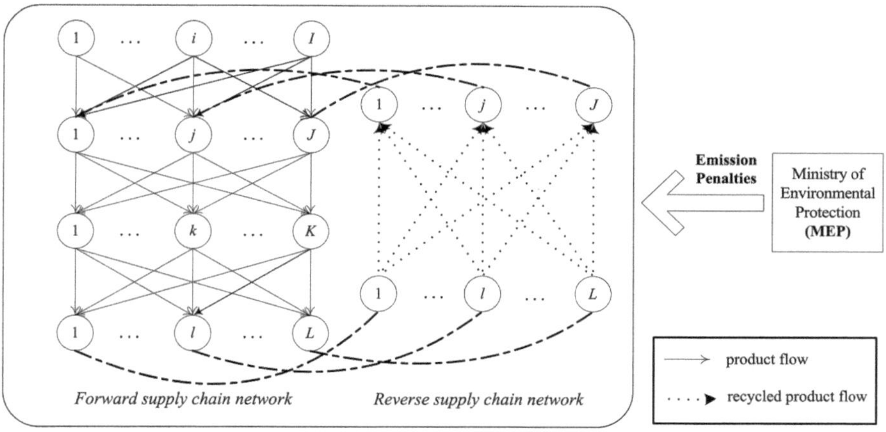

Fig. 1 The structure of the closed-loop supply chain network with environmental indicators

network determines its optimal quantities, given those of competitors. Definitions of variables, functions, and parameters in the closed-loop supply chain network are summarized below:

q_{ij}: nonnegative amount of the raw material from raw material supplier i to manufacturer j. Group shipments between material suppliers and manufacturers into the column vector $Q_1 \in R_+^{IJ}$.

q_{jk}: nonnegative amount of product shipment transacted with retailer k by manufacturer j. Group product flows between manufacturers and retailers into the column vector $Q_2 \in R_+^{JK}$.

q_{kl}: nonnegative product shipment from retailer k to demand market l. Group flows between retailers and demand markets into the column vector $Q_3 \in R_+^{KL}$.

q_{lr}: nonnegative product shipment from demand market l to recovery center r. Group flows between demand markets and recovery centers into the column vector $Q_4 \in R_+^{LR}$.

$\tilde{q}_{jr}$: nonnegative amount of reusable material flow from recovery center r to manufacturer j. Group flows between recovery centers and manufacturers into the column vector $Q_5 \in R_+^{JR}$.

ρ_l: demand price of the product at demand market l. Group prices into the column vector $\rho \in R_+^L$.

β_v: fraction of a unit of the raw material transformed into the new product, $0 \le \beta_v \le 1$.

β_u: fraction of a unit of reusable material transformed into the remanufactured product, $0 \le \beta_u \le 1$.

f_i: procurement cost of raw material supplier, $f_i = f_i(Q_1)$, $\forall\, i$.

f_j^v: production cost of manufacturer j associated with the raw material, $f_j^v = f_j^v(\beta_v, q_j^v)$ where q_j^v is the total amount of manufacturer j's product made from the raw material.

f_j^u: remanufacturing cost of manufacturer j with the useful materials extracted from recycled products, $f_j^u = f_j^u\left(\beta_u, \tilde{q}_j^u\right)$, where $\tilde{q}_j^u$ is the total amount of manufacturer j's remanufactured product.
c_{ij}: cost of transacting with manufacturer j by raw material supplier i, $c_{ij} = c_{ij}(q_{ij})$.
$\widehat{c}_{ij}$: cost of transacting with raw material supplier i by manufacturer j, $\widehat{c}_{ij} = \widehat{c}_{ij}\left(q_{ij}\right)$.
c_{jk}: cost of transacting with retailer k by manufacturer j, $c_{jk} = c_{jk}(q_{jk})$.
$\widehat{c}_{jk}$: cost of transacting with manufacturer j by retailer k, $\widehat{c}_{jk} = \widehat{c}_{jk}\left(q_{jk}\right)$.
c_{kl}: cost of transacting with demand market l by retailer k, $c_{kl} = c_{kl}(q_{kl})$.
c_k: handling cost of retailer k including the display cost and the cost of transacting, $c_k = c_k(Q_2)$.
$\widehat{c}_{kl}$: cost of transacting with retailer k from the perspective of demand market l, $\widehat{c}_{kl} = \widehat{c}_{kl}\left(q_{kl}\right)$.
c_{lr}: cost of transacting with recovery center r from the perspective of market l, $c_{lr} = c_{lr}(q_{lr})$.
$\widehat{c}_{lr}$: cost of transacting with demand market l from the perspective of recovery center r, $\widehat{c}_{lr} = \widehat{c}_{lr}\left(q_{lr}\right)$.
c_{jr}: cost of transacting with recovery center r by manufacturer j, $c_{jr} = c_{jr}\left(\tilde{q}_{jr}\right)$.
$\widehat{c}_{jr}$: cost of transacting with manufacturer j by recovery center r, $\widehat{c}_{jr} = \widehat{c}_{jr}\left(\tilde{q}_{jr}\right)$.
c_r: cost of inspection and separation from recovery center r, $c_r\left(q_r\right)$, $q_r = \sum_{l=1}^{L} q_{lr}$.
δ_j: disposal fee charged by landfills for per unit of emission from manufacturer j.
s_r: subsidy from the *MEP* to recovery center r for recycling per unit of product from demand markets.
θ_r: fraction of a unit of usable recycled product that can be transformed to reusable material for recovery center r.
E_j: upper limits of emission set by the *MEP* on different manufacturers.
Γ_r: lower limits of recycled products set by the *MEP* on different recovery centers.

2.1 The Behavior of the Raw Material Suppliers and Their Optimality Conditions

Let ρ_{ij} denote the price charged for the raw material by supplier i to manufacturer j and then i's profit maximization problem is expressed as

$$\text{Max} \sum_{j=1}^{J} \rho_{ij} q_{ij} - f_i\left(Q_1\right) - \sum_{j=1}^{J} c_{ij}\left(q_{ij}\right) \tag{1}$$

$$s.t. \quad q_{ij} \geq 0, \qquad \forall i, j. \tag{2}$$

It is necessary to assume that the functions, $f_i(Q_1)$ and $c_{ij}(q_{ij})$, are convex and continuously differentiable with regard to their respective decision variables. Then, the optimality conditions for all suppliers can be described simultaneously using the following variational inequality: determine $Q_1^* \geq 0$ satisfying

$$\sum_{i=1}^{I}\sum_{j=1}^{J}\left[\frac{\partial c_{ij}\left(q_{ij}^*\right)}{\partial q_{ij}}+\frac{\partial f_i\left(Q_1^*\right)}{\partial q_{ij}}-\rho_{ij}^*\right]\times\left(q_{ij}-q_{ij}^*\right)\geq 0 \tag{3}$$

2.2 The Behavior of the Manufacturers and Their Optimality Conditions

Let ρ_{jk} denote the selling price charged for the product by manufacturer j to retailer k and $\tilde{\rho}_{jr}$ denote the price for the recycled product by manufacturer j to recovery center r. Then the economic profit maximization problem faced by manufacturer j can be expressed as

$$\begin{aligned}\text{Max}\ &\sum_{k=1}^{K}\rho_{jk}q_{jk}-\sum_{i=1}^{I}\rho_{ij}q_{ij}-\sum_{l=1}^{L}\tilde{\rho}_{jr}\tilde{q}_{jr}-f_j^u\left(\beta_u,\tilde{q}_j^u\right)\\&-f_j^v\left(\beta_v,q_j^v\right)-\sum_{i=1}^{I}\widehat{c}_{ij}\left(q_{ij}\right)\\&-\sum_{k=1}^{K}c_{jk}\left(q_{jk}\right)\ -\sum_{i=1}^{I}c_{jr}\left(\tilde{q}_{jr}\right)\\&-\delta_j\cdot\left[(1-\beta_u)\sum_{l=1}^{L}\tilde{q}_{jr}+(1-\beta_v)\sum_{i=1}^{I}q_{ij}\right]\end{aligned} \tag{4}$$

$$s.t.\quad \sum_{k=1}^{K}q_{jk}\leq\beta_v\sum_{i=1}^{I}q_{ij}+\beta_u\sum_{r=1}^{R}\tilde{q}_{jr} \tag{5}$$

$$(1-\beta_u)\sum_{r=1}^{R}\tilde{q}_{jr}+(1-\beta_v)\sum_{i=1}^{I}q_{ij}\leq \mathrm{E}_j \tag{6}$$

$$q_{ij}\geq 0,\quad q_{jk}\geq 0,\quad \tilde{q}_{jr}\geq 0,\quad 1\geq\beta_u\geq 0,\quad 1\geq\beta_v\geq 0. \tag{7}$$

It is assumed that the functions, f_j^u, f_j^v, $\widehat{c}_{ij}$, c_{jk}, and c_{jr}, are convex and continuously differentiable and then the optimality conditions for all manufacturers are expressed as the inequality: determine $(Q_1^*, Q_2^*, Q_4^*) \geq 0$ satisfying

$$\sum_{i=1}^{I}\sum_{j=1}^{J}\left[\frac{\partial f_j^v\left(\beta_v, q_j^{v*}\right)}{\partial q_{ij}}+\frac{\partial \widehat{c}_{ij}\left(q_{ij}^*\right)}{\partial q_{ij}}+\delta_j\left(1-\beta_v\right)+\rho_{ij}^*-\beta_v\lambda_j^*+\left(1-\beta_v\right)\sigma_j^*\right]$$
$$\times\left(q_{ij}-q_{ij}^*\right)+\sum_{j=1}^{J}\sum_{k=1}^{K}\left[\frac{\partial c_{jk}\left(q_{jk}^*\right)}{\partial q_{jk}}+\lambda_j^*-\rho_{jk}^*\right]\times\left(q_{jk}-q_{jk}^*\right)$$
$$+\sum_{j=1}^{J}\left[\beta_v\sum_{i=1}^{I}q_{ij}^*+\beta_u\sum_{r=1}^{R}\tilde{q}_{jr}^*-\sum_{k=1}^{K}q_{jk}^*\right]\times\left(\lambda_j-\lambda_j^*\right)$$
$$+\sum_{j=1}^{J}\sum_{r=1}^{R}\left[\frac{\partial f_j^u\left(\beta_u, \tilde{q}_j u^*\right)}{\partial \tilde{q}_{jr}}+\frac{\partial c_{jl}\left(\tilde{q}_{jr}^*\right)}{\partial \tilde{q}_{jr}}+\tilde{\rho}_{jr}^*+\delta_j\left(1-\beta_u\right)\right.$$
$$\left.-\beta_u\lambda_j^*+\left(1-\beta_u\right)\sigma_j^*\right]\times\left(\tilde{q}_{jr}-\tilde{q}_{jr}^*\right)$$
$$+\sum_{j=1}^{J}\left[\mathrm{E}_j-\left(1-\beta_u\right)\sum_{r=1}^{R}\tilde{q}_{jr}^*-\left(1-\beta_v\right)\sum_{i=1}^{I}q_{ij}^*\right]\times\left(\sigma_j-\sigma_j^*\right)\geq 0 \tag{8}$$

It can be shown that λ_j^* and σ_j^* are Lagrange multipliers related to constraint (5) and (6) respectively. To be more specific, σ_j^* is also a shadow price which has an interpretation as the maximum price paid by manufacturer j that is willing to pay for an extra unit of the given upper limit of emission. To prevent manufacturers from exceeding such limits, the emission penalties set by the MEP should be greater than the respective shadow prices.

2.3 *The Behavior of the Retailers and Their Optimality Conditions*

Let ρ_{kl} denote the selling price of per unit product from retailer j to consumers at demand market l, and then j's optimization problem of profit maximization can be shown as

$$\text{Max}\ \sum_{l=1}^{L}\rho_{kl}q_{kl}-\sum_{j=1}^{J}\rho_{jk}q_{jk}-\sum_{j=1}^{J}\widehat{c}_{jk}\left(q_{jk}\right)-\sum_{l}^{L}c_{kl}\left(q_{kl}\right)\ -c_k\left(Q_2\right) \tag{9}$$

$$s.t. \quad \sum_{l=1}^{L} q_{kl} \leq \sum_{j=1}^{J} q_{jk} \tag{10}$$

$$q_{jk} \geq 0, \quad q_{kl} \geq 0. \tag{11}$$

It is necessary to assume that the functions, c_k, $\widehat{c}_{jk}$, and c_{kl}, are convex and continuously differentiable and hence the optimality conditions of all retailers can be described simultaneously using the variational inequality: determine $(Q_2^*, Q_3^*, \mu^*) \geq 0$ satisfying

$$\sum_{j=1}^{J}\sum_{k=1}^{K}\left[\frac{\partial \widehat{c}_{jk}\left(q_{jk}^*\right)}{\partial q_{jk}} + \rho_{jk}^* + \frac{\partial c_k\left(Q_2^*\right)}{\partial q_{jk}} - \mu_k^*\right] \times \left(q_{jk} - q_{jk}^*\right)$$
$$+ \sum_{k=1}^{K}\left[\frac{\partial c_{kl}\left(q_{kl}^*\right)}{\partial q_{kl}} + \mu_k^* - \rho_{kl}^*\right] \times \left(q_{kl} - q_{kl}^*\right) + \sum_{k=1}^{K}\left[\sum_{j=1}^{J} q_{jk}^* - \sum_{l=1}^{L} q_{kl}^*\right]$$
$$\times \left(\mu_k - \mu_k^*\right) \geq 0 \tag{12}$$

In this formulation, μ_k is the Lagrange multiplier associated with constraint (10) for retailer k and μ is the column vector of all the retailers' multipliers.

2.4 The Consumers at the Demand Markets and the Equilibrium Conditions

In the forward supply chain, denote the demand at demand market l by d_l which is a continuous function $d_l = d_l(\rho)$, $\forall\, l$. The equilibrium conditions for consumers at demand market l take the following form:

$$\rho_{kl}^* + \widehat{c}_{kl}\left(q_{kl}^*\right) \begin{cases} = \rho_l^*, & if \quad q_{kl}^* \geq 0 \\ \geq \rho_l^*, & if \quad q_{kl}^* = 0 \end{cases} \tag{13}$$

$$d_l\left(\rho^*\right) \begin{cases} = \sum_{k=1}^{K} q_{kl}^*, & if \ \ \rho_l^* \geq 0 \\ \leq \sum_{k=1}^{K} q_{kl}^*, & if \ \ \rho_l^* = 0 \end{cases} \tag{14}$$

Condition (13) states that consumers at demand market l will purchase the product from retailer k if the price charged by retailer k plus the cost of transacting from the consumers does not exceed the price that those consumers are willing to pay. Condition (14) states that if the equilibrium price the consumers are willing to pay for the product is positive, then the quantity of product consumed at demand market l is equal to the demand at that market.

In the reverse supply chain, consumer aversion at demand market l is described as a monotone increasing function $a_l(Q_4)$. Consumers at market l will choose to return recycled products in terms of the buy-back price and the amount of recycled products must not exceed the amount obtained from the retailers, which can be expressed by condition (15) and (16).

$$a_l\left(Q_4^*\right)\begin{cases}=\tilde{\rho}_{lr}^*, & if \quad \tilde{q}_{lr}^*\geq 0\\ \geq\tilde{\rho}_{lr}^*, & if \quad \tilde{q}_{lr}^*=0\end{cases} \tag{15}$$

$$s.t. \quad \sum_{r=1}^{R}\tilde{q}_{lr}\leq\sum_{k=1}^{K}q_{kl} \tag{16}$$

The above equilibrium conditions are equivalent to the following variational inequality: determine $(Q_3^*, Q_4^*, \rho^*, \gamma^*)\geq 0$ satisfying

$$\begin{aligned}&\sum_{k=1}^{K}\sum_{l=1}^{L}\left[\rho_{kl}^*+\widehat{c}_{kl}\left(q_{kl}^*\right)-\rho_l^*-\gamma_l^*\right]\times\left(q_{kl}-q_{kl}^*\right)\\&+\sum_{l=1}^{L}\sum_{r=1}^{R}\left[a_l\left(Q_4^*\right)-\rho_{lr}^*+\gamma_l^*\right]\times\left(q_{lr}-q_{lr}^*\right)\\&+\sum_{l=1}^{L}\left[\sum_{k=1}^{K}q_{kl}^*-d_l\left(\rho^*\right)\right]\times\left(\rho_l-\rho_l^*\right)+\sum_{l=1}^{L}\left[\sum_{k=1}^{K}q_{kl}^*-\sum_{r=1}^{R}q_{lr}^*\right]\times\left(\gamma_l-\gamma_l^*\right)\geq 0\end{aligned} \tag{17}$$

In this formulation, γ_l^* is the Lagrange multiplier associated with constraint (16) for demand market l and γ is the L-dimensional column vector of such multipliers for all markets.

2.5 The Recovery Centers and the Equilibrium Conditions

It is assumed that the recovery centers are required to collect recyclable products from demand markets, which are inspected and separated in order to obtain reusable materials sent to the manufacturers. Given the above description, each recovery center r wishes to maximize its profit:

$$\text{Max} \sum_{j=1}^{J} \tilde{\rho}_{jr}\tilde{q}_{jr} - \sum_{l=1}^{L} \rho_{lr}q_{lr} - \sum_{j=1}^{J} \widehat{c}_{jr}\left(\tilde{q}_{jr}\right) - \sum_{l=1}^{L} c_{lr}\left(q_{lr}\right) - c_r\left(q_r\right)$$
$$- \overline{\rho} \cdot (1-\theta_r) \sum_{j=1}^{J} \tilde{q}_{jr} + s_r \cdot \sum_{l=1}^{L} q_{lr} \tag{18}$$

$$s.t. \quad \sum_{j=1}^{J} \tilde{q}_{jr} \leq \theta_r \sum_{l=1}^{L} q_{lr} \tag{19}$$

$$\Gamma_r \leq \sum_{l=1}^{L} q_{lr} \tag{20}$$

Assume that the recycling cost and the transaction cost functions are continuous and convex. Consequently, the optimality conditions of all recovery centers can be described simultaneously using the variational inequality: determine $(Q_5^*, Q_4^*, \upsilon^*, \zeta^*) \geq 0$ satisfying

$$\sum_{j=1}^{J}\sum_{r=1}^{R}\left[\frac{\partial c_{jr}\left(\tilde{q}_{jr}^*\right)}{\partial \tilde{q}_{jr}^*} + \upsilon_r^* - \tilde{\rho}_{jr}^*\right] \times \left(\tilde{q}_{jr} - \tilde{q}_{jr}^*\right)$$
$$+\sum_{l=1}^{L}\sum_{r=1}^{R}\left[\frac{\partial c_{lr}\left(q_{lr}^*\right)}{\partial q_{lr}} + \frac{\partial \widehat{c}_r\left(q_r^*\right)}{\partial q_{lr}} + \rho_{lr}^* + \overline{\rho}(1-\theta_r) - \theta_r\upsilon_r^* - \varsigma_r^* - s_r\right]$$
$$\times \left(q_{lr} - q_{lr}^*\right) + \sum_{r=1}^{R}\left[\theta_r \cdot \sum_{l=1}^{L} q_{lr}^* - \sum_{j=1}^{J}\tilde{q}_{jr}^*\right] \times \left(\upsilon_r - \upsilon_r^*\right)$$
$$+\sum_{r=1}^{R}\left[\sum_{l=1}^{L} q_{lr}^* - \Gamma_r\right] \times \left(\varsigma_r - \varsigma_r^*\right) \geq 0 \tag{21}$$

In this formulation, υ_r and ς_r are Lagrange multipliers relevant to constraint (19) and constraint (20) for recovery center r, and υ and ς are the column vectors of all the recovery centers' corresponding multipliers.

3 The Closed-Loop Supply Chain Network Equilibrium with Environmental Indicators

Definition 1. The closed-loop supply chain network equilibrium with environmental indicators. The equilibrium state of the closed-loop supply chain network with environmental indicators is one where the product flows between five-tier decision-makers coincide and the product outputs, product shipments and prices satisfy the sum of the optimality conditions (3), (8), (12), (17) and the equilibrium conditions (21).

Theorem 1 *The equilibrium conditions governing the closed-loop supply chain network with environmental indicators are equivalent to the solution to the variational inequality given by: determine* $X^* = (Q_1^*, Q_2^*, Q_3^*, Q_4^*, Q_5^*, \rho^*, \lambda^*, \sigma^*, \mu^*, \gamma^*, \upsilon^*, \zeta^*) \in \Omega$ *satisfying*

$$\sum_{i=1}^{I}\sum_{j=1}^{J}\left[\frac{\partial c_{ij}\left(q_{ij}^*\right)}{\partial q_{ij}} + \frac{\partial f_i\left(Q_1^*\right)}{\partial q_{ij}} + \frac{\partial f_j^v\left(\beta_v, q_j^{v*}\right)}{\partial q_{ij}} + \frac{\partial \widehat{c}_{ij}\left(q_{ij}^*\right)}{\partial q_{ij}}\right.$$

$$\left. + \delta_j\left(1-\beta_v\right) - \beta_v\lambda_j^* + \left(1-\beta_v\right)\sigma_j^*\right] \times \left(q_{ij} - q_{ij}^*\right)$$

$$+ \sum_{j=1}^{J}\sum_{k=1}^{K}\left[\frac{\partial c_{jk}\left(q_{jk}^*\right)}{\partial q_{jk}} + \lambda_j^* + \frac{\partial \widehat{c}_{jk}\left(q_{jk}^*\right)}{\partial q_{jk}} + \frac{\partial c_k\left(Q_2^*\right)}{\partial q_{jk}} - \mu_k^*\right] \times \left(q_{jk} - q_{jk}^*\right)$$

$$+ \sum_{k=1}^{K}\sum_{l=1}^{L}\left[\frac{\partial c_{kl}\left(q_{kl}^*\right)}{\partial q_{kl}} + \mu_k^* + \widehat{c}_{kl}\left(q_{kl}^*\right) - \rho_l^* - \gamma_l^*\right] \times \left(q_{kl} - q_{kl}^*\right)$$

$$+ \sum_{l=1}^{L}\sum_{r=1}^{R}\left[a_l\left(Q_4^*\right) + \gamma_l^* + \frac{\partial c_{lr}\left(q_{lr}^*\right)}{\partial q_{lr}} + \frac{\partial c_r^u\left(q_r^{u*}\right)}{\partial q_{lr}}\right.$$

$$\left. + \overline{\rho}\left(1-\theta_r\right) - \theta_r\upsilon_r^* - \varsigma_r^* - s_r\right] \times \left(q_{lr} - q_{lr}^*\right)$$

$$+ \sum_{j=1}^{J}\sum_{r=1}^{R}\left[\frac{\partial f_j^u\left(\beta_u, \tilde{q}_j^u\right)}{\partial \tilde{q}_{jr}} + \frac{\partial c_{jr}\left(\tilde{q}_{jr}^*\right)}{\partial \tilde{q}_{jr}} + \delta_j\left(1-\beta_u\right) - \beta_u\lambda_j^*\right.$$

$$\left. + \left(1-\beta_u\right)\sigma_j^* + \frac{\partial \widehat{c}_{jr}\left(\tilde{q}_{jr}^*\right)}{\partial \tilde{q}_{jr}} + \upsilon_r^*\right] \times \left(\tilde{q}_{jr} - \tilde{q}_{jr}^*\right)$$

$$+\sum_{l=1}^{L}\left[\sum_{k=1}^{K}q_{kl}^{*}-d_{l}\left(\rho^{*}\right)\right]\times\left(\rho_{l}-\rho_{l}^{*}\right)+\sum_{j=1}^{J}\left[\beta_{v}\sum_{i=1}^{I}q_{ij}^{*}+\beta_{u}\sum_{r=1}^{R}\tilde{q}_{jr}^{*}-\sum_{k=1}^{K}q_{jk}^{*}\right]$$
$$\times\left(\lambda_{j}-\lambda_{j}^{*}\right)+\sum_{j=1}^{J}\left[\mathrm{E}_{j}-(1-\beta_{u})\sum_{r=1}^{R}\tilde{q}_{jr}^{*}-(1-\beta_{v})\sum_{i=1}^{I}q_{ij}^{*}\right]\times\left(\sigma_{j}-\sigma_{j}^{*}\right)$$
$$+\sum_{k=1}^{K}\left[\sum_{j=1}^{J}q_{jk}^{*}-\sum_{l=1}^{L}q_{kl}^{*}\right]\times\left(\mu_{k}-\mu_{k}^{*}\right)+\sum_{l=1}^{L}\left[\sum_{k=1}^{K}q_{kl}^{*}-\sum_{r=1}^{R}q_{lr}^{*}\right]\times\left(\gamma_{l}-\gamma_{l}^{*}\right)$$
$$+\sum_{r=1}^{R}\left[\theta_{r}\cdot\sum_{l=1}^{L}q_{lr}^{*}-\sum_{j=1}^{J}\tilde{q}_{jr}^{*}\right]\times\left(\upsilon_{r}-\upsilon_{r}^{*}\right)+\sum_{r=1}^{R}\left[\sum_{l=1}^{L}q_{lr}^{*}-\Gamma_{r}\right]\times\left(\varsigma_{r}-\varsigma_{r}^{*}\right)\geq 0 \tag{22}$$

For easy reference in the subsequent sections, variational inequality (22) can be rewritten in standard variational inequality form as follows: determine $X^{*}\in\Omega$ satisfying:

$$\left\langle F\left(X^{*}\right),\ X-X^{*}\right\rangle\geq 0,\quad\forall X\in\Omega. \tag{23}$$

where $F(X)=(F_{ij},F_{jk},F_{kl},F_{lr},F_{jr},F_{l},F_{j1},F_{j2},F_{k},F_{l},F_{r1},F_{r2})_{i=1,2\ldots,I;\,j=1,2\ldots,J;\,k=1,2\ldots,K;\,l=1,2;\,r=1,2\ldots,R}$ and the specific components of F are given by the function terms preceding the multiplication signs in (22). The term $<\cdot,\cdot>$ denotes the inner product in N-dimensional Euclidean space.

From Nagurney [8], a variational inequality admits at least one solution if the entering function F is continuous, and the feasible set is compact. While F in (23) may be not continuous, the feasible region Ω is not compact. However, it is possible to impose a weak condition on Ω to guarantee existence. Let

$$\begin{aligned}\Omega_{b}=\{&(Q_{1},Q_{2},Q_{3},Q_{4},Q_{5},\rho,\lambda,\sigma,\mu,\gamma,\upsilon,\zeta)\ |0\leq Q_{1}\leq b_{1},\ 0\leq Q_{2}\leq b_{2},\\&0\leq Q_{3}\leq b_{3},0\leq Q_{4}\leq b_{4},\ 0\leq Q_{5}\leq b_{5},0\leq\rho\leq b_{6},0\leq\lambda\leq b_{7},\\&0\leq\sigma\leq b_{8},0\leq\mu\leq b_{9},0\leq\gamma\leq b_{10},0\leq\upsilon\leq b_{11},0\leq\zeta\leq b_{12}\}\end{aligned}$$

where $b=(b_{1},b_{2},b_{3},b_{4},b_{5},b_{6},b_{7},b_{8},b_{9},b_{10},b_{11},b_{12})\geq 0$ and $Q_{1}\leq b_{1},Q_{2}\leq b_{2}$, $Q_{3}\leq b_{3},Q_{4}\leq b_{4},Q_{5}\leq b_{5}$, $\rho\leq b_{6}$, $0\leq\lambda\leq b_{7}$, $0\leq\sigma\leq b_{8}$, $0\leq\mu\leq b_{9}$, $0\leq\gamma\leq b_{10}$, $0\leq\upsilon\leq b_{11},0\leq\zeta\leq b_{12}$ mean that $q_{ij}\leq b_{1},q_{jk}\leq b_{2}$, $q_{kl}\leq b_{3}$, $q_{lr}\leq b_{4}$, $\tilde{q}_{jr}\leq b_{5},\rho_{l}\leq b_{6},\lambda_{j}\leq b_{7},\sigma_{j}\leq b_{8},\mu_{k}\leq b_{9},\gamma_{l}\leq b_{10},\upsilon_{r}\leq b_{11},\zeta_{r}\leq b_{12}$.. Then, Ω_{b} is a bounded, closed convex subset of $R^{IJ+JK+KL+LR+JR+L+2J+K+L+2R}$. Thus, the variational inequality:

$$\left\langle F\left(X^{b}\right),X-X^{b}\right\rangle\geq 0,\quad\forall X\in\Omega_{b} \tag{24}$$

admits at least one solution. Therefore, following Nagurney et al. [3] we have:

Lemma 1. *Variational inequality* (23) *admits a solution if and only if there exists a* $b > 0$ *such that variational inequality* (24) *admits a solution in* Ω_b *with*

$$\begin{aligned} &Q_1^b \leq b_1, Q_2^b \leq b_2, Q_3^b \leq b_3, Q_4^b \leq b_4, Q_5^b \leq b_5, \rho^b < b_6, \\ &\lambda^b \leq b_7, \sigma^b \leq b_8, \mu^b \leq b_9, \gamma^b \leq b_{10}, \upsilon^b \leq b_{11}, \zeta^b \leq b_{12}. \end{aligned} \tag{25}$$

Theorem 2 (Existence). *Suppose there exist positive constants* M, N, R *with* $R > M$ *such that*:

$$\begin{aligned} &\frac{\partial c_{ij}\left(q_{ij}\right)}{\partial q_{ij}} + \frac{\partial f_i\left(Q_1\right)}{\partial q_{ij}} + \frac{\partial f_j^v\left(\beta_v, q_j^v\right)}{\partial q_{ij}} + \frac{\partial \widehat{c}_{ij}\left(q_{ij}\right)}{\partial q_{ij}} \geq R, \\ &\forall Q_1 \text{ with } q_{ij} \geq N, \forall i, j. \end{aligned} \tag{26}$$

$$\frac{\partial c_{jk}\left(q_{jk}\right)}{\partial q_{jk}} + \frac{\partial \widehat{c}_{jk}\left(q_{jk}\right)}{\partial q_{jk}} + \frac{\partial c_k\left(Q_2\right)}{\partial q_{jk}} \geq R, \ \forall Q_2 \text{ with } q_{jk} \geq N, \ \forall j, k. \tag{27}$$

$$\frac{\partial c_{kl}\left(q_{kl}\right)}{\partial q_{kl}} + \widehat{c}_{kl}\left(q_{kl}\right) \geq R, \quad \forall Q_3 \text{ with } q_{kl} \geq N, \forall k, l. \tag{28}$$

$$a_l\left(Q_4\right) + \frac{\partial c_{lr}\left(q_{lr}\right)}{\partial q_{lr}} + \frac{\partial c_r\left(q_r\right)}{\partial q_{lr}} \geq R, \forall Q_4 \text{ with } q_{lr} \geq N, \forall l, r. \tag{29}$$

$$\frac{\partial f_j^u\left(\beta_u, \tilde{q}_j^u\right)}{\partial \tilde{q}_{jr}} + \frac{\partial c_{jr}\left(\tilde{q}_{jr}\right)}{\partial \tilde{q}_{jr}} + \frac{\partial c_{jr}\left(\tilde{q}_{jr}\right)}{\partial \tilde{q}_{jr}} \geq R, \forall Q_5 \text{ with } \tilde{q}_{jr} \geq N, \ \forall j, r. \tag{30}$$

$$d_l\left(\rho\right) \leq N, \quad \forall \rho \text{ with } \rho_l > M, \forall l. \tag{31}$$

Then variational inequality (22) *admits at least one solution.*

Proof Follows using analogous arguments as the proof of existence for Theorem 2 in [11]. □

4 A Numerical Example

In this section, we present a numerical example to illustrate effects of the legislation proposed by the *MEP* on its equilibrium solution, which is constructed with two raw material suppliers, two manufacturers, two retailers, two demand markets, and two recovery centers, that is, $I = J = K = L = R = 2$.

The procurement cost functions of the two raw material suppliers are:

$$f_1(Q_1) = (q_{11} + q_{12})^2 + 0.5(q_{11} + q_{12})(q_{21} + q_{22}) + (q_{11} + q_{12});$$
$$f_2(Q_1) = (q_{21} + q_{22})^2 + 0.5(q_{11} + q_{12})(q_{21} + q_{22}) + (q_{21} + q_{22}).$$

The production cost functions of the two manufacturers are given by:

$$f_1^v(\beta_v, q_1^v) = 1.5[\beta_v(q_{11} + q_{21})]^2 + 2[\beta_v(q_{11} + q_{21})]\ [\beta_v(q_{12} + q_{22})];$$
$$f_2^v(\beta_v, q_2^v) = 1.5[\beta_v(q_{12} + q_{22})]^2 + 2[\beta_v(q_{11} + q_{21})]\ [\beta_v(q_{12} + q_{22})].$$

The remanufacturing cost functions associated with recycled products for the two manufacturers are given by:

$$f_1^u(\beta_u, \tilde{q}_1^u) = 0.5[\beta_u(\tilde{q}_{11} + \tilde{q}_{12})]^2 + \beta_u(\tilde{q}_{11} + \tilde{q}_{12}) + 2;$$
$$f_2^u(\beta_u, \tilde{q}_2^u) = 0.5[\beta_u(\tilde{q}_{21} + \tilde{q}_{22})]^2 + \beta_u(\tilde{q}_{21} + \tilde{q}_{22}) + 2.$$

The handling cost functions of the two retailers are given by:

$$c_k(Q_2) = 0.5\left(\sum_{j=1}^{2}\sum_{k=1}^{2} q_{jk}\right)^2 + 3,\ k = 1,\ 2.$$

The transacting cost paid by recovery centers to manufacturers is:

$$\widehat{c}_{jr}(\tilde{q}_{jr}) = 0.5(\tilde{q}_{jr})^2 + 2, \quad j = 1,\ 2;\ k = 1,\ 2.$$

The demand functions at the two demand markets are given, respectively, by:

$$d_1 = -2\rho_1 - 1.5\rho_2 + 1{,}000; \quad d_2 = -2\rho_2 - 1.5\rho_1 + 1{,}000.$$

The aversion functions for consumers are given by: $a_l(Q_4) = 0.5\left(\sum_{k=1}^{2}\sum_{l=1}^{2} q_{kl}\right) + 5$, $l = 1, 2$.

The cost of inspection and separation from recovery centers are set as: $c_r(q_r) = 2\left(\sum_{l=1}^{2} q_{lr}\right)^2, \quad r = 1, 2.$

All the other cost functions are set to zero (e.g., [10]) and the parameters are set by $\beta_v = 0.6$, $\beta_u = 0.4$, $\delta_1 = \delta_2 = 1$ and $s_1 = s_2 = 2$. In addition, we assume that $\mathsf{E}_1 = 20$, $\mathsf{E}_2 = 25$, $\Gamma_1 = 30$, $\Gamma_2 = 30$. Then, the *Lingo* converges in 1365 iterations yielding the following equilibrium patterns presented as follows.

The equilibrium raw material transactions between raw material suppliers and manufacturers q_{ij}:

$$q_{11}^* = 24.027, \quad q_{12}^* = 20.973, \quad q_{21}^* = 18.167, \quad q_{22}^* = 26.833;$$

The product transactions between manufacturers and retailers q_{jk}:

$$q_{11}^* = 16.147, \quad q_{12}^* = 11.250, \quad q_{21}^* = 21.352, \quad q_{22}^* = 11.250;$$

The product transaction shipments between retailers and demand markets q_{kl}:

$$q_{11}^* = 30.000, \quad q_{12}^* = 7.500, \quad q_{21}^* = 0.000, \quad q_{22}^* = 22.500;$$

The recycled product shipments between demand markets and recovery centers q_{lr}:

$$q_{11}^* = 18.000, \quad q_{12}^* = 12.000, \quad q_{21}^* = 12.000, \quad q_{22}^* = 18.000;$$

The reusable material flows from recovery centers to manufacturers $\tilde{q}_{jr}$:

$$\tilde{q}_{11}^* = 2.602, \tilde{q}_{12}^* = 2.602, \tilde{q}_{21}^* = 4.898, \tilde{q}_{22}^* = 4.898;$$

The equilibrium prices at the two demand markets ρ_l: $\rho_1^* = \rho_2^* = 277.143$;
The Lagrange multipliers:

$$\lambda_1^* = \lambda_2^* = 572.800; \quad \sigma_1^* = 374.476, \quad \sigma_2^* = 369.425; \quad \mu_1^* = \mu_2^* = 632.800;$$
$$\gamma_1^* = \gamma_2^* = 355.657; \quad \upsilon_1^* = \upsilon_2^* = 0.000; \quad \zeta_1^* = \zeta_2^* = 509.157.$$

The emissions generated by both manufacturers are given by:

$$(1 - \beta_u) \sum_{r=1}^{R} \tilde{q}_{jr} + (1 - \beta_v) \sum_{i=1}^{I} q_{ij}.$$

According to the above expression, the two manufacturers' emissions volumes are equal to:

$$0.6 \times (2.602 + 2.602) + 0.4 \times (24.027 + 18.167) = 20.000;$$
$$0.6 \times (4.898 + 4.898) + 0.4 \times (20.973 + 26.833) = 25.000.$$

It can be also observed that these two figures are all equal to their respective upper limits of emission. In such a case, the emission penalty the *MEP* imposes on the first manufacturer should be greater than 374.476; at the same time, the emission penalty the *MEP* imposes on the second manufacturer should be greater than 369.425. On the other hand, the premium the *MEP* offers on the two recovery centers should be both greater than 509.157. It has an essential managerial insight that the *MEP* can, according to different shadow prices serving as environmental indicators under different situations, adjust penalties and/or premiums, and transform corresponding limits to achieve its expected environmental target. It is worth noting the shadow prices which have a crucial impact on the decision making of the *MEP* for environmental protection. If a particular shadow price is equal to zero, then emission penalty and premium are set as zero. In contrast, if the shadow price is positive, then the penalty or premium should also be set as a positive number which is greater than the corresponding shadow price.

5 Conclusion

In the paper, we propose a framework of the closed-loop network equilibrium with environmental indicators consisting of raw material suppliers, manufacturers, retailers, demand markets, and recovery centers. It is assumed that the *MEP* legislates that the manufacturers' emissions should be less than their respective upper limits of emission to restrain pollution and protect environment. To achieve this objective, the *MEP* imposes distinct nonnegative emission penalties on the manufacturers and offers different nonnegative premiums to the recovery centers by means of the shadow prices serving as environmental indicators for the *MEP*.

In particular, we describe their optimal behavior for pursuing maximum profit within the constraint of upper limits of emission as well as incentives of lower limits of used products. Then, we establish the optimality conditions of decision-makers which are equivalent to a variational inequality. Existence of the solution under suitable assumptions on the underlying functions is presented and finally an illustrative example is provided to verify the rationality of the model. For further research, the paper may include the consideration about how to apply algorithms to concrete numerical examples.

Acknowledgement This research is supported by Nature Science Foundation of China Grant (11171348).

References

1. Daniel, P.: Variational inequalities for evolutionary financial equilibrium. In: Nagurney, A. (ed.) Innovations in Financial and Economic Networks. Edward Elgar Publishing, Cheltenham, England (2003)
2. Qiang, Q., Ke, K., Anderson, T., Dong, J.: The closed-loop supply chain network with competition, distribution channel investment, and uncertainties. OMEGA-Int. J. Manage. S. **41**, 186–194 (2013)
3. Nagurney, A., Dong, J., Zhang, D.: A supply chain network equilibrium model. Transport. Res. E **38**, 282–303 (2002)
4. Nagurney, A., Loo, J., Dong, J., Zhang, D.: Supply chain networks and electronic commerce: a theoretical perspective. Netnomics **4**, 187–220 (2002)
5. Nagurney, A., Toyasaki, F.: Supply chain supernetworks and environmental criteria. Transport. Res. D **8**, 185–213 (2003)
6. Nagurney A., Cruz J., Toyasaki F.: Statics and dynamics of global supply chain networks with environmental decision-making. Pareto Optimality, Game Theory and Equilibria, pp. 803–836 (2008)

7. Cruz, J.M.: Dynamics of supply chain networks with corporate social responsibility through integrated environmental decision-making. Eur. J. Oper. Res. **184**, 1005–1031 (2008)
8. Nagurney, A.: On the relationship between supply chain and transportation network equilibria: a supernetwork equivalence with computations. Transport. Res. E **42**, 293–316 (2006)
9. Nagurney, A., Dong, J.: Management of knowledge intensive systems as supernetworks: modelling, analysis, computations, and applications. Math. Comput. Model **42**, 397–417 (2005)
10. Yang, G., Wang, Z., Li, X.: The optimization of the closed-loop supply chain network. Transport. Res. E **45**, 16–28 (2009)
11. Hammond, D., Beullens, P.: Closed-loop supply chain network equilibrium under legislation. Eur. J. Oper. Res. **183**, 895–908 (2007)

Dynamic Impacts of Social Expectation and Macroeconomic Factor on Shanghai Stock Market: An Application of Vector Error Correction Model

Zou Ao

Abstract Stock movement often shows inconformity with macroeconomic situation and waves more intensively than expected. In this paper, we mainly focus on the influence of social expectation and macroeconomic issue on Shanghai stock market. By establishing a vector error correction model (i.e., VECM), we are able to find out the relationship among them.

In the model, we assume that the macroeconomic situation can be fairly represented by the data of total retail sales of consumer goods and total fixed asset investment. As for social expectation, it can be indicated by the leading index in the economic climate index system. Also, the performance of Shanghai stock market can be regarded as the variation of the Shanghai composite index. Afterwards, we build a VECM connecting these four elements synthetically. Depending on the Granger causality tests and variance decomposition, we can figure out the dynamic impacts of another three factors on Shanghai composite index.

It turns out that expected effect (i.e., leading index) exerts more influence on stock fluctuation than macroeconomic factors. More importantly, all of those three elements fail to nicely explain the variation of Shanghai composite index, which shows that Shanghai stock market's efficiency is really weak.

Keywords VECM • Social Expectation • Granger Causality Test • Variance Decomposition

1 Introduction

In recent years, China's stock markets stay in downturn and stock prices shake intensively. In the meantime, however, China's macroeconomic fundamentals, including economic growth rate, individual income growth rate and inflation rate, lie in a healthy and positive state. This is because macroeconomic issues are not able to explain the stock market's variation to a fair degree.With assistance of

Z. Ao (✉)
School of Economics and Management, Wuhan University, Wuhan, China
e-mail: 1217168046@qq.com

D. Gao et al. (eds.), *Advances in Global Optimization*, Springer Proceedings in Mathematics & Statistics 95, DOI 10.1007/978-3-319-08377-3_47

historical data, we conclude that stock market's condition can poorly reflect the change of substantial economy. And more than China, the stock markets of the rest of the world are also not consistent with their own economic condition. Thus, the change of macroeconomic condition is absolutely not the unique reason for the fluctuation of stock prices.

So, what are other elements that concerned with the stock market? Generally speaking, people's anticipation of future economic fundamentals can significantly affect their judgments when purchasing or selling stocks, therefore influencing stock prices to a substantial extent. Considering the effect of feedback loop of market mechanism, the variation range of stock prices is more intensive than entity economy. Therefore, social expectation on future economic state can greatly account for the alteration of stock market index. As a result, we can divide the factors influencing stock market's operation into two parts: entity part and expectation part. By means of empirical analysis of our vector error correction model (VECM), we are able to figure out their separate influencing extent and make it clear which part is more reasonable to lead to the fluctuation of stock market. In the end, based on the explanatory power of these elements, we have the ability to make a preliminary judgment on Shanghai stock market's efficiency.

2 Methodology

2.1 *Empirical Data*

2.1.1 Data of the Variation of Shanghai Stock Market

The Shanghai stock composite index can be used to comprehensively represent China's stock markets' operating condition.

We select the monthly data of Shanghai composite index from January, 2005 to August, 2012, which is available in the website of the Hithink Flush Information Network.

2.1.2 Data of the Social Expectation

In the climate index system, the leading index is typically aimed to forecast the future economic state, reflecting the whole society's expectation on future economic healthy condition. It is an integrated index making up of 12 financial indexes and its components would change based on actual economic affairs.

We choose the monthly data of the leading index from January, 2005 to August, 2012. Similarly, that data can also be accessible in the website of the Hithink Flush Information Network.

2.1.3 Data of the Macroeconomic Factors

Actually, there are many indicators that can represent the macroeconomic status. In a general way, macro-economy is regarded to be composed of three parts: total social consumptions, total social investment, and government expenditures. Since we have no access to monthly data of government purchasing, we rely on total consumption and investment to roughly reflect macroeconomic conditions.

The indicators of total consumption and investment are, respectively, the total retail sales of consumer goods and the total fixed asset investment. We select the monthly data, originating from the Hithink Flush Information Network as well, of these two indicators from January 2005 to August 2012.

2.2 Methods Introduction

2.2.1 Granger Causality Test and Variance Decomposition

In this paper, we establish a co-integration connecting Shanghai composite index, leading index, total retail sales, and total fixed asset sales, and then build a VECM including the co-integration series and other four series [1].

After the establishment of the VECM, we are aimed to apply this model to analyze the influence of other three elements on composite index, finding out which element helps to explain the variation of stock prices and quantifies their influential power.

With the help of Granger causality test, we are capable of figuring out which variable can affect the Shanghai composite index in statistical sense. If one series is the Granger reason of composite index, this series would exert substantial effects on Shanghai stock market [2, 3].

Furthermore, by applying the technique of variance decomposition, we have the ability to precisely measure the influential power of every single element on Shanghai composite index [2, 3].

3 Empirical Analysis

3.1 Data Processing

In this model, SH indicates Shanghai composite index; XX represents leading index; XF stands for total retail sales; TZ means total fixed asset investment.

Since composite index and leading index are relative numbers, we should take the logarithm for them to reduce the heteroscedasticity. Hence, SH and XX become LSH and LXX. Also, the amount of total retail sales and total fixed asset investment are under the effect of the end of year so that these series might possess seasonality.

The results of seasonality test are as in Table 1.

Table 1 Seasonality test of LSH, LXX, XF and TZ

Variable	Seasonality test (at the 0.1 % level)	Adjusted variable
LSH	Unseasonal	LSH
LXX	Seasonal	LXX_SA
XF	Seasonal	LXF_SA
TZ	Seasonal	LTZ_SA

Table 2 Test results of unit root

	Test result of level-values			Test result of first differential value		
Variables	Test forms (C,T,L)	ADF values	P values	Test forms (C,T,L)	ADF values	P values
LSH	(C,00)	−1.695	0.4302	(C,01)	−4.919	0.0001
LXX_SA	(C,0,1)	−2.795	0.0629	(C,0,0)	−4.529	0.0004
XF_SA	(C,T,2)	−1.565	0.7988	(C,01)	−9.753	0
TZ_SA	(C,T,5)	0.0652	0.9965	(C,0,0)	−11.515	0

In the table, C, T and L respectively represent intercept term, tendency term and lag intervals

3.2 *Unit Root Test*

Now, let these four series undergo unit root tests in specific forms, the results are listed in Table 2.

As the table shows, at the 5 % level, all the variables are not stationary, but the differential values of variables are all stationary. Hence, variables are integrated of order one, and we can further test the co-integration relationship among variables.

3.3 *Co-integration Test*

Considering that XF_SA and TZ_SA both possess tendency term and intercept term, indicating that some series have certain tendency, namely tendency term in linear space. But LSH and LXX_SA do not contain tendency term so that there is not tendency term in the co-integration space made up of those four variables. Therefore, we should choose the third form when conducting Johansen co-integration test [4].

When it comes to the lag order of co-integration, we can refer to the lag intervals of VAR model composed of these four variables. Since VECM can be seen as the differential form of the VAR model, the lag order of VECM would be one less than the VAR model.

Then, we build a VAR model made up of these four variables and find out the optimal lag order. Since the optimal lag interval selected by each criteria is different, we should give priority to the result of AIC and SC. When lag is 1, AIC = 19.998SC = 20.577; when lag is 2, AIC = 19.743SC = 20.784; when

Table 3 VAR lag order selection criteria

VAR Lag Order Selection Criteria						
Endogenous variables: LSH LXX_SA TZ_SA XF_SA						
Exogenous variables: C						
Date: 01/17/13 Time: 03:02						
Sample: 2005M01 2012M08						
Included observations: 84						
Lag	LogL	LR	FPE	AIC	SC	HQ
0	−1,310.240	NA	4.57e+08	31.29143	31.40718	31.33796
1	−819.9466	922.2187	5,697.503	19.99873	20.57749*	20.23139
2	−793.2125	47.73936	4,424.665	19.74316	20.78493	20.16194*
3	−773.6481	33.07323	4,092.080	19.65829	21.16308	20.26320
4	−764.0686	15.28155	4,828.606	19.81116	21.77896	20.60220
5	−741.4650	33.90545	4,212.124	19.65393	22.08474	20.63110
6	−717.4285	33.76552*	3,588.052*	19.46258	22.35641	20.62588
7	−703.0699	18.80289	3,898.982	19.50167	22.85851	20.85109
8	−681.9371	25.66128	3,664.632	19.37946*	23.19931	20.91500

Asterisks can indicate the optimal lag order determined by these five different selection criteria. Under normal circumstances, we should give priority to the results of AIC and SC.

lag is 3, AIC = 19.658SC = 21.163. Taken together, when we choose lag interval as 2, these two indicators are optimal. Therefore, the lagged difference of the VAR is 2 and then VECM is 1.

Establish a group of four variables and then conduct co-integration test, choose the third form of Johansen test, select the lag interval as (1,1). The test result is listed in Table 4.

Trace test and Max-eigenvalue test all indicate 1 cointegrating equation.

3.4 *Application of VECM*

3.4.1 Granger Causality Test

In order to find out whether leading index, total retail sales, and total fixed investment significantly affect Shanghai composite index, we should turn to Granger causality test for the group made up of these four elements. The best lag number for the Granger test generally equal to the lagged difference of the VAR model mentioned above, namely 2.

The causalities of LSH with LXX_SA, XF_SA, and TZ_SA are separately listed. The test results represent that only LXX_SA is the Granger reason of LSH, indicating that LXX_SA are reasonable to explain the variation of LSH in Granger sense. From the economic perspective, social expectation, expressed by leading index, constitutes the powerful reason for the fluctuation of stock prices but macroeconomic factors fail to impact stock market significantly. Trace test and Maxeigenvalue test all indicate one cointegrating equation (Table 4).

Table 4 Test of co-integration

Sample (adjusted): 2005M03 2012M08				
Included observations: 90 after adjustments				
Trend assumption: Linear deterministic trend Series: LXX_SA LSH TZ_SA XF_SA				
Lags interval (in first differences): 1 to 1				
Unrestricted Cointegration Rank Test (Trace)				
Hypothesized No. of CE(s)	Eigenvalue	Trace statistic	0.05 critical value	Prob.**
None*	0.333514	58.85511	47.85613	0.0033
At most 1	0.159112	22.33891	29.79707	0.2800
At most 2	0.066534	6.742220	15.49471	0.6078
At most 3	0.006045	0.545681	3.841466	0.4601

Trace test indicates one cointegrating eqn(s) at the 0.05 level
* Rejection of the hypothesis at the 0.05 level; ** MacKinnon–Haug–Michelis (1999) p-values

Unrestricted Cointegration Rank Test (Maximum Eigenvalue)				
Hypothesized No. of CE(s)	Eigenvalue	Max-Eigen statistic	0.05 critical value	Prob.**
None*	0.333514	36.51620	27.58434	0.0028
At most 1	0.159112	15.59669	21.13162	0.2493
At most 2	0.066534	6.196540	14.26460	0.5881
At most 3	0.006045	0.545681	3.841466	0.4601

Max-eigenvalue test indicates one cointegrating eqn(s) at the 0.05 level
* Rejection of the hypothesis at the 0.05 level; ** MacKinnon–Haug–Michelis (1999) p-values

Table 5 Granger causality test

Pairwise Granger Causality Tests			
Date: 01/17/13 Time: 04:40			
Sample: 2005M01 2012M08			
Lags: 2			
Null hypothesis	Obs	F-statistic	Prob.
LXX_SA does not Granger Cause LSH	90	3.63478	0.0306
LSH does not Granger Cause LXX_SA		0.23461	0.7914
TZ_SA does not Granger Cause LSH	90	0.68546	0.5066
LSH does not Granger Cause TZ_SA		0.02935	0.9711
XF_SA does not Granger Cause LSH	90	0.56672	0.5695
LSH does not Granger Cause XF_SA		0.63173	0.5341

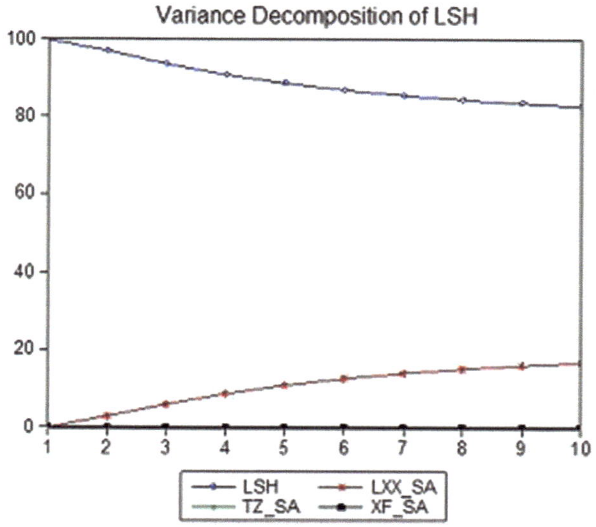

Fig. 1 The variance decomposition of LSH

3.4.2 Variance Decomposition

Other than Granger causality test, we can count on the technique of variance decomposition to measure particular influential force of other three series on LSH. The specific figure is as below (Fig. 1).

The result of variance decomposition of LSH indicates that most of LSH's volatility, about 80 %, originates from itself. The impact of LXX_SA on LSH might constitute 20 % in all influencing factors. As to XF_SA and TZ_SA, their effects are so weak that can be neglected.

The conclusion drawn from the variance decomposition tallies with the consequence of the Granger causality test. Social expectation has a fair share in the variance of Shanghai composite index, affecting stock market much more greatly than macroeconomic issues. It should be pointed out that the influential force of LXX_SA, XF_SA, and TZ_SA is really limited. In other words, the fluctuation of composite index is mainly arisen from its own factor or other elements that we fail to take into consideration.

4 Conclusion

- **The impact of social expectation on stock market is much more significant than macroeconomic elements.**

  The results of Granger causality test and variance decomposition clearly show that leading index is powerful enough to influence the stock movement. However, the delegates of macroeconomic factors, namely total retail sales and total fixed asset investment, possess very weak explanatory power for the fluctuation of stock price.

- **The factors influencing stock market are diverse.**
  The total influential force of these variables above is just around 20 %, relatively small portion. This is because that there might be some other hidden important factors for the variance of stock market. At present we cannot clearly figure out those hidden elements, but it is evident that the influencing forces are diverse, coming from various fields.

References

1. Liqin, H.: The Experimentation of Financial Time Series. Wuhan University Press, Wuhan (2012)
2. Menezes, R., Dionísio, A., Hassani, H.: On the globalization of stock markets: an application of vector error correction model, mutual information and singular spectrum analysis to the G7 countries. Q. Rev. Econ. Finance **52**(4), 369–384 (2012)
3. Maysami, R.C., Koh, T.S.: A vector error correction model of the Singapore stock market. Int. Rev. Econ. Finance **9**(1), 79–96 (2000)
4. Hess, M.K.: Dynamic and asymmetric impacts of macroeconomic fundamentals on an integrated stock market. J. Int. Financial Mark. Inst. Money **14**(5), 455–471 (2004)

Comparative Research of Financial Model in Supply Chain

Jin Jin, Ziqiu Wei, and Guoshan Liu

Abstract Supply chain financial services have been studied for years. However, the literatures have been silent on comparative research of every financial model, especially the model of third party logistics (3PL) firms as credit providers in Cash-strapped supply chains. This paper investigates an extended supply chain model with a Cash-strapped retailer, a supplier, a bank, and a 3PL firm, in which the retailer has insufficient initial budget and may borrow or obtain trade credit from bank, supplier, or a 3PL firm. Our analysis indicates that the 3PL firm model yields higher profits not only for the 3PL firm but also for the supplier, the retailer, and the entire supply chain.

Keywords Supply chain • Financial services • Third party logistics (3PL)

1 Introduction

There is always shortage of funds in the supply chain. Many suppliers and retailers are facing funding gap, and need to find solutions urgently, otherwise it will hinder the growth of these companies seriously. Lack of funds is the first unfavorable factor in the development of small and medium-sized firms in China [1]. More than 80 % of the small and medium-sized firms rely on self-financing, while more than 90 % of the small and medium-sized firms are depended on the internal financing channel in the initial growth period, and the financing difficulty of small and medium-sized firms widespread [2].

J. Jin
Chinese People's Public Security University, Beijing 100038, China

Z. Wei (✉)
School of Economics and Management, Hebei University of Science and Technology, Shijiazhuang 050018, China
e-mail: lovelywzq@163.com

G. Liu
School of Business, Renmin University of China, Beijing 100872, PR China
e-mail: liuguoshan@gmail.com

D. Gao et al. (eds.), *Advances in Global Optimization*, Springer Proceedings in Mathematics & Statistics 95, DOI 10.1007/978-3-319-08377-3_48

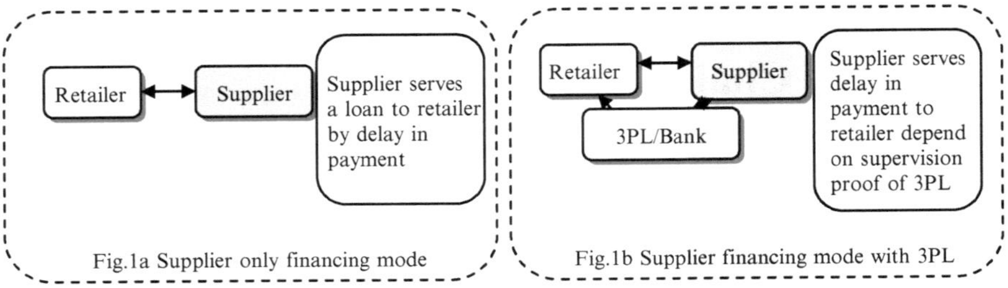

Fig. 1 Supplier-based financing mode [3–5]

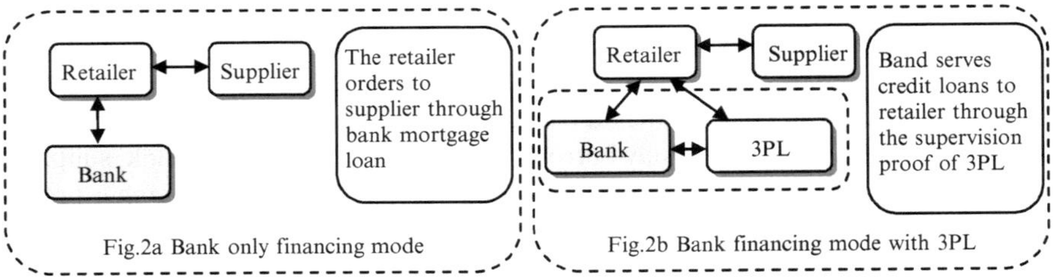

Fig. 2 Bank-based financing mode [6–8]

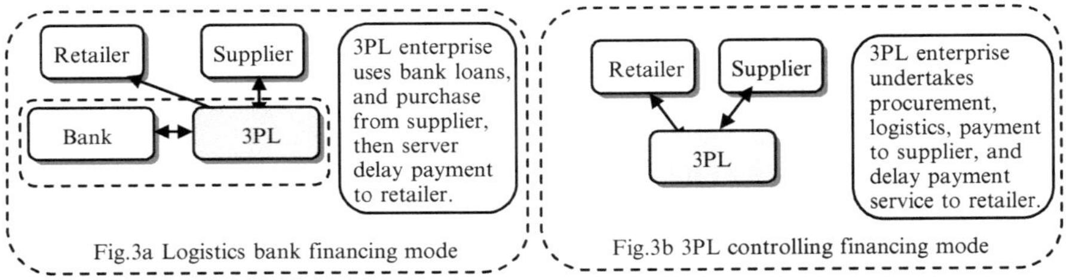

Fig. 3 3PL-based financing mode [9, 10]

There are many kinds of financing methods in the case of shortage of funds, such as internal financing from upstream and downstream firms in supply chain by accounts payable, mortgage of goods, and so on, and financial support from various external financial organizations of supply chain. Summary of the various forms of financing in supply chain can be seen in Figs. 1, 2, and 3. In the figure, for the convenience of comparison, we assume that the retailer faces the problem of financing difficulty. In such case, the retailer can choose to obtain trade credit from supplier [11], which is called the supplier-based financing mode (Fig. 1); also can choose bank to provide a loan, which is called the bank-based financing mode (Fig. 2); or choose the third party logistics (3PL) firm to provide logistics and financial Integrated, which is called 3PL-based financing mode (Fig. 3). The key difference between these three kinds of financing modes is that retailer accesses to credit financial trade directly from which side. In the first two kinds of financing modes, supplier and bank will invite 3PL firm into the cooperative system to assist regulatory retailer reputation in order to reduce the risk of cooperation (Figs. 1b and 2b). 3PL firm has two kinds of roles in the supply chain: logistics service role, 3PL firm only provides logistics services; comprehensive role, 3PL firm not only provides logistics services, while also provides financing service. For example, Figs. 1b and 2b models have 3PL firm to participate in, but 3PL firm only serves to provide logistics service, it's only a supporting role; while in Fig. 3, the 3PL firm not only provides logistics services, but also directly or indirectly provides financial services, to be the master in the whole supply chain finance operation. These financing modes can solve the problem of shortage of funds for retailer, but the benefits of each party and the whole supply chain efficiency gains are not the same due to the subjects, the financing object, and applications are different for each model.

Although the researches on the areas of bank financing and trade credit are abundant in recent years [6, 12], but there is little literature analysis on 3PL firm-based supply chain financing mode systematically, and on effects of this kind of supply chain financing mode. Therefore, research problems are in front of us: what impacts on the capital shortage in the supply chain by the mode of supply chain finance based on 3PL firm exactly? What are the differences between it and other financing modes? What do especially the supply chain firms and supply chain overall benefit? Which model is better? Based on these questions, this study will build financing models of supplier-based (supplier model), bank-based (bank model), and 3PL firm-based (3PL model), and have a comparative study. We expect to get a conclusion that 3PL model is the most superior model. That's because 3PL firms can provide logistics and financial services at the same time, it can supervise the transaction, and coordinate the supply chain more effectively, so as to obtain more symmetrical information and provide lower loan interest rates. As the result, the retailer's order quantity will be more, and whole supply chain benefit is more effective. This study will be the complement of supply chain finance theory with 3PL firm involved, and explore the status of 3PL mode, and provide reference for supply chain firms, to help small and medium-sized firms to solve the bottleneck problem of the shortage of funds. All of these reflect the significance and necessity of the study.

2 Model Building

The basic situation is that there are four Parties in a supply chain network, they are the supplier, retailer, 3PL firm, and bank; the retailer has shortage of funds, and he can get help from any other three parties.

Retailer makes order to supplier according to stochastic market demand D, and the random function obeys to the cumulative distribution function F and density f. In this problem, the transportation cost is clearly defined. We assume that the supplier, retailer, bank, and 3PL firm are using the letters s, r, b, l as superscript.

The demand distribution function $F(D)$ satisfies conditions as follows:

1. It is absolutely continuous with density $f(D) > 0$, $in(a, b)$, $where$ $0 \leq a \leq b \leq \infty$;
2. $\overline{D}$ has a finite mean.
3. $\overline{F}(D) = 1 - F(D)$.

We assume the order quantity of retailer is $Q(B)$, B stands for the initial funding. The product unit price is P, which is standardized as 1 in order to simplify the calculation. The wholesale price is w_s, and unit logistics freight rate is w_l. Therefore, the unit ordering cost of retailer is $w = w_s + w_l$. The unit cost of the product is c_s, and the unit cost of logistics is c_l. The operation cost of 3PL firm is $C_{3pl} = w_p + c_l$. Similarly, the cost of supplier is $C_{su} = w_l + c_s$. At the same time, we assume that $c_l < w_l$, $c_s < w_s$, $(1 + r)w < 1$, $C_{3pl} < w$, $C_{su} < w$, r is credit interest rate. The total amount of the loan is $(wQ(B) - B)^+$, where the "+" means nonnegative number. When demand realized, retailer needs to return $(wQ(B) - B)^+(1 + r(B))$. However, if the repayment ability of retailer cannot satisfy, just as $\min\{D; Q(B)\} < (wQ(B) - B)^+(1 + r(B))$, the retailer declared bankruptcy, while the credit side gets $\min\{D; Q(B)\}$ instead of $(wQ(B) - B)^+(1 + r(B))$.

We use the letter subscripts su, ba, and 3pl to represent the supplier model, bank model, and 3PL model respectively. Reference to previous literature [12, 13]), we use Stackelberg two stage game model. The credit sides (suppliers, banks, or 3PL firm) give credit interest rate in the first stage, and then the retailer decides the optimal order quantity in the second stage. In these three models, the retailer's decisions are basically consistent, so we will discuss the optimal order quantity of retailer firstly.

Cash-strapped retailer makes loans to financial institutions, the loan amount as $(wQ(B) - B)^+$. The retailer's initial capital B is common information. If the credit interest rate is $r(B)$, then the profit function of retailer is,

$$\begin{aligned}\Pi^r(B) &= \max_{Q(B)\geq 0} E\left\{[\min[D, Q(B)] - (wQ(B) - B)(1 + r(B))]^+ - B\right\} \\ &= \max_{Q(B)\geq 0}\left\{\int_{(wQ(B)-B)(1+r(B))}^{Q(B)} \overline{F}(D)dD - B\right\} \end{aligned} \tag{1}$$

If $\min[D, Q(B)] > (wQ(B) - B)(1 + r(B))$, retailer can pay back the loan and interest; otherwise, the retailer can only pay $\min[D, Q(B)]$. The optimal order quantity to solve the problem will draw the following conclusions:

Cash-strapped retailer will determine the optimal order quantity according to a given credit interest rate as follows:

$$Q^*(B) = \begin{cases} \overline{F}^{-1}(w), & \text{if} \quad B \geq w\overline{F}^{-1}(w), \\ B/w, & \text{if} \quad w\overline{F}^{-1}(w(1 + \tilde{r}(B))) < B < w\overline{F}^{-1}(w), \\ \widehat{Q}(B), & \text{if} \quad B \leq w\overline{F}^{-1}(w(1 + \tilde{r}(B))) \end{cases} \tag{2}$$

Where $\widehat{Q}(B)$ is determined by $\overline{F}(Q(B)) = w(1 + r(B))\overline{F}[(wQ(B) - B)(1 + r(B))]$, and $\widehat{Q}(B)$ is unique and decreases in $r(B) \in [0, \tilde{r}(B)]$.

The conclusion is from the literature of Dada and Hu [12]. As can be seen, optimal order quantity of cash-strapped retailer depends on his initial capital and credit interest rate. In the initial capital adequacy ($B \geq w\overline{F}^{-1}(w)$) case, retailers would not borrow from financial institutions, which is equivalent to the classical newsvendor model without capital constraint condition; in the second case, the initial funding for retailers is not enough to order the ideal quantity of goods, however, the credit interest rate is very high, the retailer is not worth borrowing and just uses the existing initial money; in the third case, the initial funds are less, the interest rate provided by financing institutions is within acceptable range, so retailer is willing to have the loan $\left(w\widehat{Q}(B) - B\right)^{+}$, which will decrease with the interest rate. Our study is for the third case.

2.1 Supplier-Based Financing Models

Many suppliers will provide trade credit directly to the cash-strapped retailers in practice cooperation. For example, the HP had begun to provide trade credit services to its retailers from 1998 [14]. The retailer dares not to hide initial funding information to supplier in supplier model whether or not there is 3PL firm logistics supervision, which is because that the honest reputation will help retailer continue to get trade credit from supplier.

In the first stage of the Stackelberg game, supplier offers a trade credit contract $(w, r_{su}(B))$; in the second stage, retailer's order is $Q^{*}_{su}(B)$. We have already got the retailer's decision, so we only discuss the supplier's profit model:

$$\begin{aligned}
\Pi^{s}_{su}(B) &= \max_{0 \leq r_{su}(B) \leq \tilde{r}(B)} E\left\{\min\left[\min\left[D, Q^{*}_{su}(B)\right], \left(wQ^{*}_{su}(B) - B\right)(1 + r_{su}(B))\right]\right. \\
&\quad \left. + B - C_{su}Q^{*}_{su}(B)\right\} \\
&= \max_{0 \leq r_{su}(B) \leq \tilde{r}(B)} \left[(w_s - c_s)\, Q^{*}_{su}(B) + \left(wQ^{*}_{su}(B) - B\right)^{+} r_{su}(B) \right. \\
&\quad \left. + \int_{0}^{\left(wQ^{*}_{su}(B) - B\right)(1 + r_{su}(B))} \left[D - \left(wQ^{*}_{su}(B) - B\right)(1 + r_{su}(B))\right] dF(D) \right]
\end{aligned}$$

$$\overline{F}\left(Q^{*}_{su}(B)\right) = w(1 + r_{su}(B))\overline{F}\left[\left(wQ^{*}_{su}(B) - B\right)(1 + r_{su}(B))\right] \quad (3)$$

Where $\left(wQ_{su}^{*}(B)-B\right)^{+}r_{su}(B)+\int_{0}^{\left(wQ_{su}^{*}(B)-B\right)(1+r_{su}(B))}\left[D-\left(wQ_{su}^{*}(B)-B\right)\right.$ $\left.(1+r_{su}(B))\right]dF(D)$ denotes financial income, $(w_s-c_s)Q_{su}^{*}(B)$ denotes sales hboxrevenue.

2.2 Bank-Based Financing Model

In fact, the cash-strapped retailer conceals his real situation of initial funds to bank, and will tell the bank he has higher initial funds. This is because that the bank will decrease interest rates to attract retailer to increase order quantity when bank knows retailer has higher initial capital. However, the bank will be a great risk in this case and the expected return is hard to reach. If the bank cannot grasp the real information of retailer and be afraid of risk, the bank will refuse to provide the loan to retailer. In this contradictory condition, the pledge of goods appears that retailer will have goods mortgaged to the bank until the repayment. So there creates another service content of regulation to the mortgage, which the bank is not good at, and the alliance firm of 3PL will assist bank to finish it.

The bank's profit can be expressed as:

$$\begin{aligned}\Pi_{ba}^{b}(B)&=\max_{0\le r_{ba}(B)\le\tilde{r}(B)}\mathrm{E}\left\{\min\left[\min\left[D,\widehat{Q}_{t}(B)\right],\left(w\widehat{Q}_{ba}(B)-B\right)(1+r_{ba}(B))\right]\right.\\&\quad\left.-\left(w\widehat{Q}_{ba}(B)-B\right)\right\}\\&=\max_{0\le r_{ba}(B)\le\tilde{r}(B)}\left\{\int_{0}^{\left(w\widehat{Q}_{ba}(B)-B\right)(1+r_{ba}(B))}\overline{F}(D)dD-\left(w\widehat{Q}_{ba}(B)-B\right)\right\}\end{aligned}\tag{4}$$

If $\min\left[D,\widehat{Q}_{ba}(B)\right]\ge\left(w\widehat{Q}_{ba}(B)-B\right)(1+r_{ba}(B))$, then the bank gets $\left(w\widehat{Q}_{ba}(B)-B\right)r_{ba}(B)$; otherwise, executes liquidation of retailers and gains $\min\left[D,\widehat{Q}_{ba}(B)\right]-\left(w\widehat{Q}_{ba}(B)-B\right)$.

2.3 Financing Model Based on 3PL Firm

In this mode, 3PL firm can not only regulatory transport of goods effectively, but also offers the trade credit service for cash-strapped retailers.

Profit model for 3PL firm:

$$\begin{aligned}\Pi^{l}_{3pl}(B) &= \max_{0\le r_{3pl}(B)\le \tilde{r}(B)} \mathrm{E}\Big\{\min\Big[\min\Big[D, Q^{*}_{3pl}(B)\Big], \Big(wQ^{*}_{3pl}(B)-B\Big) \\ &\quad \big(1+r_{3pl}(B)\big)\Big] + B - C_{3pl}Q^{*}_{3pl}(B)\Big\} \\ &= \max_{0\le r_{cl}(B)\le \tilde{r}(B)} \mathrm{E}\Big\{\big(w-C_{3pl}\big)\,Q^{*}_{3pl}(B) + \Big(wQ^{*}_{3pl}(B)-B\Big)^{+} r_{3pl}(B) \\ &\quad -\Big(\min\Big[D, Q^{*}_{3pl}(B)\Big] - \Big(wQ^{*}_{3pl}(B)-B\Big)\big(1+r_{3pl}(B)\big)\Big)^{-}\Big\}\end{aligned} \tag{5}$$

Superscript"–"represents nonpositive. The above formula contains two parts: financial income $(wQ^{*}_{3pl}(B)-B)^{+}r_{3pl}(B)-(\min[D,Q^{*}_{3pl}(B)]-(wQ^{*}_{3pl}(B)-B)(1+r_{3pl}(B)))^{-}$ and operating income $(w-C_{3pl})Q^{*}_{3pl}(B)$. When $w>C_{3pl}$, operating profit is positive. However, if the demand uncertainty is too big, the financial revenue will not be optimistic, and then the retailer will not finish repayment according to the interest. Therefore, when a trade occurs, a 3PL firm tends to choose smaller $r_{3pl}(B)$ to promote the operating performance on one hand; on the other hand, 3PL firm is willing to give a higher $r_{3pl}(B)$ to raise financial revenue. Therefore, 3PL firm will consider these two factors at the same time when signs the contract. In the actual operation of the process, the 3PL model has obvious advantages, because 3PL firm can reduce logistics cost c_l using scale economic effect.

3 Comparison of Three Kinds of Financing Models

3.1 *Supplier Model Is Better Than the Bank Model*

The supplier and the retailer share the risk in the supplier model, so the supplier can make lower interest rates than bank. Then the retailer increases the order quantity because of lower interest rates, the supplier also gets benefit because of higher quantity, and the supply chain will be more effect and better for the win–win.

3.2 *The 3PL Model Is Better Than the Supplier Model If $w_l - c_l > w_p - c_p$*

$w_l - c_l > w_p - c_p$ means that the marginal profit of 3PL firm in 3PL model is higher than the marginal profit of supplier in supplier mode. Research shows that retailers will get higher benefit from the higher marginal profit Party. The reason is that higher marginal profit can share more risk to trade credit provider, then the trade credit provider can set lower interest rates, which can increase the order quantity, all of which can promote the supply chain profit. If the marginal profit of supplier is lower than 3PL firm, the retailer will choose 3PL firm to carry out trade credit cooperation, and vice versa.

3.2.1 3PL Model Is Better Than the Bank Model Not Only for Each Party But Also for the Whole Profits of Supply Chain

In the 3PL model, 3PL firm shares the demand uncertainty risk when he provides the trade credit and logistics services at the same time. He can make lower interest rates than bank because of the lower product wholesale price and lower logistics cost, and the retailer gets benefit from lower interest rates; then the supplier benefits from the higher order quantity, while 3PL firm benefits from integration services of financial and traditional logistics. All of these show that integration services can coordinate fund shortage of supply chain effectively, and create win–win results during all parties of the supply chain. The results of this study provide theoretical support for 3PL firms to expand their business to the financial area.

4 Conclusions

In this study, we conducted an in-depth analysis of different financing mode for capital constrained supply chain. Cash-strapped retailer can obtain financial support from supplier, bank, or 3PL firm. Our analysis shows that the 3PL model has the advantage not only for 3PL firm but also for supplier and retailer. In comparison with the supplier model, we find that the marginal profit is the key factor on determining the model advantage. However, both the 3PL model and the supplier model have obvious advantages than bank model to promote cash-strapped retailer to more produce order quantity.

References

1. Lin, H.C., Guan, H.X.: A comparative study of competitiveness of small and medium-sized firms in different industries of China. Chin. Soc. Sci. **3**, 48–58 (2005)
2. Vandenberg, P.: Adapting to the financial landscape: evidence from small forms in Nairobi. World Dev. **31**(11), 1829–1843 (2003)
3. Sadlovska, V., Enslow, B.: Supply Chain Finance Benchmark Report. Aberdeen Group, Boston (2006)
4. Chen, X.F.: Trade credit contract with limited liability. In: International Conference of System Science, Management Science & System Dynamics (2007)
5. Tang, S.Y.: Logistics finance practice research. China Federation Logistics Purchasing **5**, 13–18 (2005)
6. Buzacott, J., Zhang, R.: Inventory management with asset-based financing. Manage. Sci. **50**(9), 1274–1292 (2004)

7. Caldentey, R., Chen, X.F.: The role of financial services in procurement contract. In: Working Paper, INFORMS, and Stern School of Business in New York University, Submitted to MSOM (2010)
8. Ren, W.C.: From the "material banks" to "logistics banks". China Federation Logistics Purchasing **9**, 18–19 (2006)
9. David, B.: Logistics financiers. J. Commerce **4**, 40–42 (2004)
10. Chen, X.F., Cai, G.S.: Joint logistics and financial services by a 3PL firm. Eur. J. Oper. Res. **214**, 579–587 (2011)
11. Rajan, R., Zingales, L.: D we know about capital structure? Some evidence from international data. J. Finance **50**(5), 1421–1460 (1995)
12. Dada, M., Hu, Q.: Financing newssupplier inventory. Oper. Res. Lett. **36**(5), 569–573 (2008)
13. Kouvelis, P., Zhao, W.: Financing the newssupplier: supplier vs. bank, optimal rates, and alternative schemes. In: Working Paper, Olin Business School, Washington University, St. Louis (2008)
14. Zhou, J., Groenevelt H.: Impacts of financial collaboration in a three-party supply chain. In: Working Paper, The Simon School, University of Rochester (2008)

Intuitive Haptics Interface with Accurate Force Estimation and Reflection at Nanoscale

Asim Bhatti, Burhan Khan, Saeid Nahavandi, Samer Hanoun, and David Gao

Abstract Technologies, such as Atomic Force Microscopy (AFM), have proven to be one of the most versatile research equipments in the field of nanotechnology by providing physical access to the materials at nanoscale. Working principles of AFM involve physical interaction with the sample at nanometre scale to estimate the topography of the sample surface. Size of the cantilever tip, within the range of few nanometres diameter, and inherent elasticity of the cantilever allow it to bend in response to the changes in the sample surface leading to accurate estimation of the sample topography. Despite the capabilities of the AFM, there is a lack of intuitive user interfaces that could allow interaction with the materials at nanoscale, analogous to the way we are accustomed to at macro level. To bridge this gap of intuitive interface design and development, a haptics interface is designed in conjunction with Bruker Nanos AFM. Interaction with the materials at nanoscale is characterised by estimating the forces experienced by the cantilever tip employing geometric deformation principles. Estimated forces are reflected to the user, in a controlled manner, through haptics interface. Established mathematical framework for force estimation can be adopted for AFM operations in air as well as in liquid mediums.

1 Introduction

Nanotechnology is playing key role in the advancements of fields such as materials [1], microbiology [2,3], nano-medicine [4–6], nano-robotics [7,8] and environment [9,10]. A number of technologies have emerged over the last few decades that have revolutionised the way we interact with the nanoworld. Atomic Force Microscope

A. Bhatti (✉) • B. Khan • S. Nahavandi • S. Hanoun
Deakin University, 75 Pigdons Road, Waurn Ponds, VIC 3217, Australia
e-mail: asim.bhatti@deakin.edu.au; burhan.khan@deakin.edu.au; samer.hanoun@deakin.edu.au; saeid.nahavandi@deakin.edu.au

D. Gao
School of Science, Information Technology and Engineering, University of Ballarat, Ballarat, VIC 3350, Australia
e-mail: d.gao@ballarat.edu.au

D. Gao et al. (eds.), *Advances in Global Optimization*, Springer Proceedings in Mathematics & Statistics 95, DOI 10.1007/978-3-319-08377-3_49

(AFM) is one of those revolutionising technologies that allows physical interaction with the materials at nanometre scale providing opportunity to understand the physical dynamics of the material at the atomic level. Improved knowledge of physical dynamics of nanoscale structures in natural systems can have great impact on our everyday lives. For instance, improved understanding of bioavailability and biodegradation will lead to the development of nanotechnologies useful in preventing or mitigating environmental harms and development of customized cure of diseases. Research in biotechnology could open up doors for the development of advanced drug delivery systems, new ways to treat diseases and repair damaged tissues and cells [11, 12]. Study and intuitive interaction with micro-organisms can help in enhanced understanding of the evolution [13].

Despite the great potential of nanotechnology field in aforementioned domains, the tools that facilitate the research are not free of shortcomings. We believe improved user interfaces that could provide direct interaction with the materials at nanoscale similar to what we are accustomed to as macro scale would enhance our learning capabilities. Direct interaction involves experiencing physical dynamics in response to interaction such as surface topography, friction, stiffness, elasticity, deformation and plasticity. In this work we are constrained to AFM technology that allows physical interaction with materials at nanoscale. To address the issue of improved user interface, we propose an intuitive haptics interface for AFM to allow direct interaction with the materials at nanoscale. This requires the estimation of forces that the AFM cantilever tip undergoes during scanning and reflecting it to the user in an intuitive and interacting way.

In this work, we develop an intuitive haptics interface allowing user to interact with the materials at nanoscale experiencing scaled up forces that AFM cantilever undergoes while scanning the material surface. In this development we have employed SensAble Omni haptics device and Bruker Nanos AFM. Physical forces that AFM cantilever experiences are estimated using the cantilever deformation information provided by the AFM controller. Two different force estimation representations are provided in this work to accommodate the operation of the AFM in dry and liquid mediums, thanks to the work of Sader [14]. Topographic information generated by the AFM is acquired by the developed haptics interface to allow the user to interact with the material at nanometre scale experiencing the forces that tip has experienced during scanning. In the next Sect. 2 we provide force estimation framework that haptics framework employs to provide force sensations to the user. Section 3 provides haptics framework and user interface allowing direct interaction with the sample at nanoscale.

2 Force Characterisation Using Cantilever Deformation

To estimate the dynamic forces experienced by the cantilever during interaction with the sample material requires characterisation of the cantilever deformation in response to sample surface topography, as is shown in Fig. 1. The deformation of

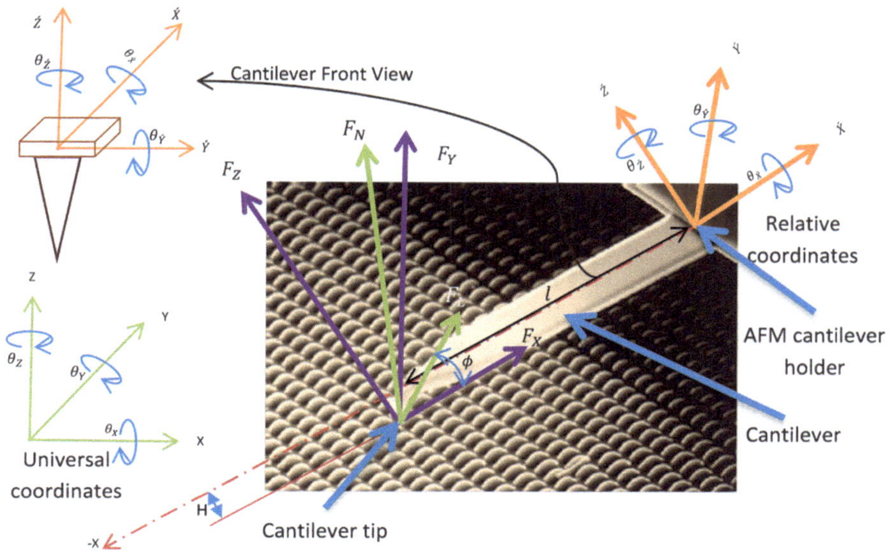

Fig. 1 Force estimation framework employing geometric deformation of AFM cantilever

the cantilever happens in six degrees of freedom, i.e. $[X, Y, Z, \theta_{\acute{X}}, \theta_{\acute{Y}}, \theta_{\acute{Z}}]$. Forces exerted on the cantilever can be expressed by two force vector components that are normal F_N and lateral F_L force vectors. Lateral force component is in fact angular force component that can be represented in terms of Torque around $[\acute{X}, \acute{Y}, \acute{Z}]$ axis. Each of the force vectors consists of three sub force components as

$$F_N = [F_{\acute{X}}, F_{\acute{Y}}, F_{\acute{Z}}] \tag{1}$$

$$F_L = \tau_L = [F_{\theta_{\acute{X}}}, F_{\theta_{\acute{Y}}}, F_{\theta_{\acute{Z}}}] = [\tau_{\acute{X}}, \tau_{\acute{Y}}, \tau_{\acute{Z}}] \tag{2}$$

Graphical representation of the force vectors F_N and F_L is shown in Fig. 1, with respect to universal and relative frame of reference. The deflection and elastic deformation of the cantilever can be represented by an elastic spring and the forces can be defined using Hooke's law as

$$F = k\delta \tag{3}$$

where F represents normal force component, i.e. F_N, δ represents deflection in Z direction, i.e. δ_Z, and k represents spring constant of the cantilever. Expression (3) can be rewritten as

$$F_N = k\delta_Z \tag{4}$$

Spring constant k is based on the material of the cantilever and can be expressed in static (k_s) as well as dynamic (k_d) forms. Dynamic representation of spring constant k_d becomes important when the AFM operates in fluidic medium to incorporate the influence of fluidic on the deformation of the cantilever. Static spring constant, i.e. k_s can be expressed using Young's Modulus as

$$k_s = \frac{Ewt^3}{4l^3} \tag{5}$$

where E represents the Young's Modulus, whereas w, t and l represent dimensional parameters of the cantilever that are width, thickness and length, respectively, assuming cantilever as a thin beam cantilever with rectangular cross section. As the materials that we are interested to interact with are the biological and the medium that we are interested to operate in is liquid therefore we have adopted the expression of dynamic spring constant from [sader reference] as

$$k_d = \rho b^2 l \Lambda(Re) w_R^2 Q \tag{6}$$

where ρ is the density of the fluid that cantilever is operating in, b and l represent cantilever width and length, w_R radial resonance frequency of the cantilever and Q quality factor. $\Lambda(Re)$ represents the hydrodynamic function as a function of Reynolds number Re. Reynolds number Re defines a dimensionless number providing the measure of the ratio between inertial and viscous forces. Reynolds number Re quantifies the relative importance of these two types of forces for given flow conditions. $\Lambda(Re)$ can be represented explicitly as

$$\Lambda(Re) = \frac{L_f^3}{b^2 l} \Omega(\beta) \tag{7}$$

where b and l are cantilever dimension parameters as in expression (6), whereas L_f represents traveled length of the fluid. $\Omega(\beta)$ is a dimensionless function and can be expressed in terms of energy dissipation E_d during one oscillation cycle of the cantilever with respect to oscillation amplitude A_o as

$$\Omega(\beta) = \frac{1}{2\pi\rho L_f^3 w_R} \frac{\partial^2 E_d}{\partial A_o^2} \tag{8}$$

Referring back to expression (2), F_N can easily be estimated using static k_s or dynamic k_d spring constants using (5) or (6), respectively, and known cantilever deflection δ_Z. Cartesian force components generate torques at the AFM cantilever holder position, shown in Fig. 1 and can be expressed as

$$\begin{aligned}
\tau_{\acute{X}} &= F_{\acute{Y}} H \\
\tau_{\acute{Y}} &= F_{\acute{Z}} l + F_{\acute{X}} H \\
\tau_{\acute{Z}} &= F_{\acute{Y}} l
\end{aligned}$$

As the torque $\tau_{\acute{Y}}$ is caused by F_N or combined contribution of $F_{\acute{Z}}$ and $F_{\acute{X}}$, we can redefine $\tau_{\acute{Y}}$ in terms of F_N as

$$\tau_{\acute{Y}} = F_N \times l = F_{\acute{Z}} l + F_{\acute{X}} H \tag{9}$$

by rearranging (9) we have

$$F_{\acute{Z}} = \frac{\tau_{\acute{Y}} - F_{\acute{X}} H}{l} = F_N - \frac{H}{l} F_{\acute{X}} \tag{10}$$

Generally the term $\frac{H}{l}$ is very small therefore we can safely assume $F_{\acute{Z}} \approx F_N$. Suppose the lateral motion of cantilever tip is at an angle ϕ with respect to the $\acute{X}$ axis, lateral force F_L opposite to the direction of motion can be expressed as

$$F_L = \frac{F_{\acute{Y}}}{sin(\phi)} \tag{11}$$

Similarly $F_{\acute{X}}$ can be expressed in terms of $F_{\acute{Y}}$ as

$$F_{\acute{X}} = F_{\acute{Y}} tan\, \phi \tag{12}$$

where $F_{\acute{Y}}$ can be expressed employing torsion constant of the cantilever [15] as

$$F_{\acute{Y}} = \frac{k_{tor}}{H} \theta_X \tag{13}$$

3 Haptics User Interface

Haptics interface developed for AFM allows the users to experience the force dynamics of the cantilever tip. A simplified block diagram, as shown in Fig. 2, highlights information flow loop between AFM and the haptics interface. Force and torques value that AFM cantilever undergoes during sample interaction are estimated employing the expressions (4)–(6). Image processing techniques are employed to smoothen the data before importing into the haptics framework [16–18]. Smoothing is necessary to remove high frequency components of the data, giving rise to improved haptics experience. Smoothing also helps in creating three dimensional objects with relatively smaller number of vertices, making the data faster to load and unload into the haptics environment.

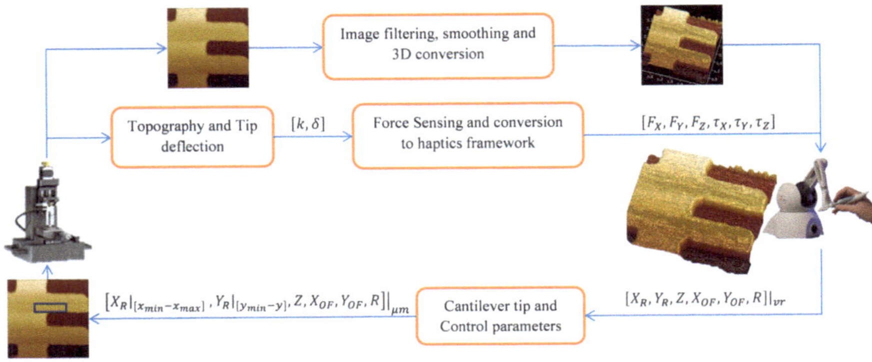

Fig. 2 Information flow diagram of haptics interface with AFM

Haptics interface developed, as shown in Fig. 3, provides an intuitive interaction with the topographic information estimated by the AFM. Force that the AFM cantilever tip has undergone during the scanning processes is estimated using the expressions (4)–(6) from Sect. 2. User can further select the region on the image frame shown at the right-hand side of the right display to require new zoomed topographic data directly from the AFM. This provides a far more realistic and natural interface to interact with the materials at nanoscale. Future version of the interface will allow direct and discrete interaction with the cantilever tip to interact with the sample in contrast to acquiring the topographic data of the complete scan. This direct interaction will facilitate interactions, such as nano-indentation and cutting.

4 Conclusion

An improved haptics interface is presented allowing interaction with the materials at nanometer scale. AFM technology is employed to estimate the material dynamics at nanoscale by characterising AFM cantilever tip deformations. Force and torques estimated from the AFM cantilever deformations are reflected to the user in a controlled manner.

Future focus of this work is to establish a framework that could allow not just accurate estimation and reflection of the material dynamics to the user but to provide an intuitive interface to control the tip to perform manipulation tasks such as indentation and cutting. Intuitive interface with accurate force reelection will open up new horizons for investigation and discoveries.

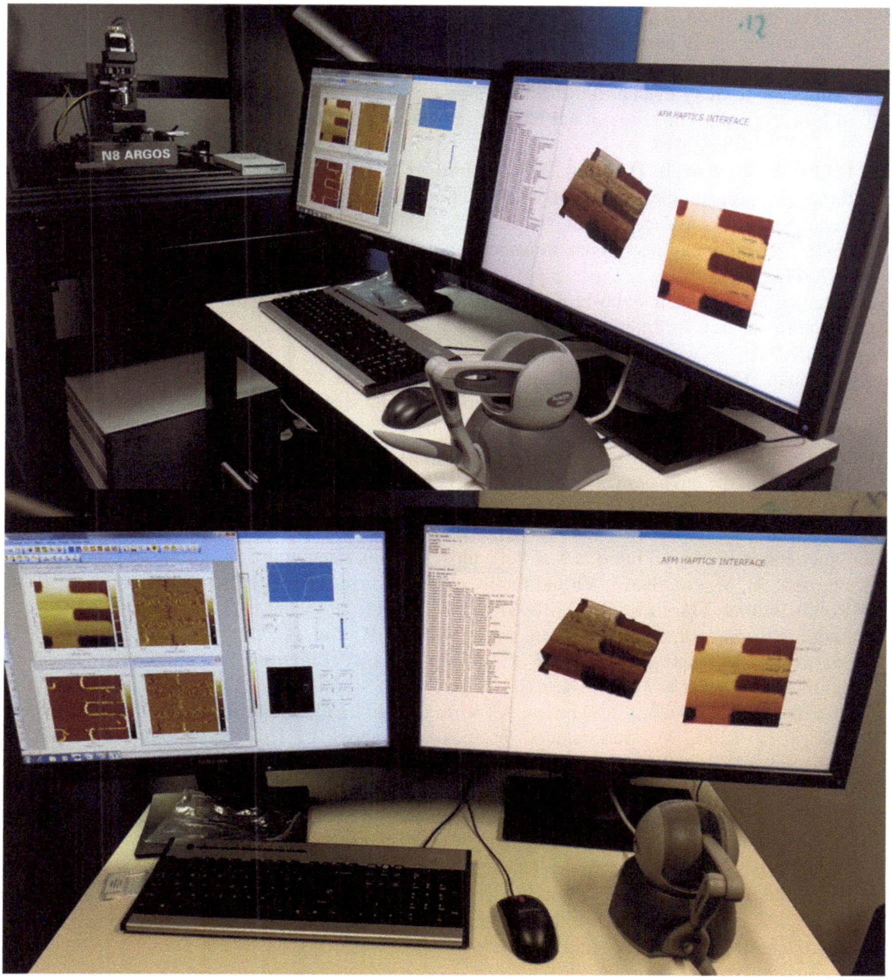

Fig. 3 Haptics interface for AFM with accurate force estimations and reflections

References

1. Fedoseyev, A.I., Turowski, M., Raman, A., Taylor, E.W., Hubbard, S., Polly, S., Balandin, A.A.: Investigation and modeling of space radiation effects in quantum dot solar cells. In: 35th IEEE Photovoltaic Specialists Conference (PVSC), pp. 2533–2536 (2010)
2. Ortoleva, P., Adhangale, P., Cheluvaraja, S., Fontus, M., Shreif, Z.: Deriving principles of microbiology by multiscaling laws of molecular physics. IEEE Eng. Med. Biol. Mag. **28**(2), 70–79 (2009)
3. Breguet, J.-M., Driesen, W., Kaegi, F., Cimprich, T.: Applications of piezo-actuated micro-robots in micro-biology and material science. In: International Conference on Mechatronics and Automation, ICMA 2007, pp. 57–62 (2007)

4. Atakan, B., Akan, O.B., Balasubramaniam, S.: Body area nanonetworks with molecular communications in nanomedicine. IEEE Commun. Mag. **50**(1), 28–34 (2012)
5. Prina-Mello, A., Crosbie-Staunton, K., Salas, G., del Puerto Morales, M., Volkov, Y.: Multi-parametric toxicity evaluation of SPIONs by high content screening technique: identification of biocompatible multifunctional nanoparticles for nanomedicine. IEEE Trans. Magn. **49**(1), Part 2, 377–382 (2013)
6. Hinkal, G.W., Farrell, D., Hook, S.S., Panaro, N.J., Ptak, K., Grodzinski, P.: Cancer therapy through nanomedicine. IEEE Nanotechnol. Mag. **5**(2), 6–12 (2011)
7. Lenaghan, S.C., Wang, Y., Xi, N., Fukuda, T., Tarn, T., Hamel, W.R., Zhang, M.: Grand challenges in bioengineered nanorobotics for cancer therapy. IEEE Trans. Biomed. Eng. **60**(3), 667–673 (2013)
8. Peteu, S.F.: Micro-to nano-biosensors and actuators integrated for responsive delivery of countermeasures. In: International Conference Semiconductor (CAS2010), vol. 1, pp. 179–190 (2010)
9. Ramanan, G., Pandian, A., Raghunandan, A.: Effects of nanotechnology on health and environment. In: International Conference on Nanoscience, Engineering and Technology (ICONSET2011), pp. 280–284 (2011)
10. Mayor, P., Bradley, P., Micheli, G.D.: Nano-Tera.ch: engineering complex systems for health, security, and the environment [IEEE News]. IEEE Solid State Circuits Mag. **2**(3), 87–92 (2010)
11. Muthu, M.S., Singh, S.: Targeted nanomedicines effective treatment modalities for cancer, AIDS and brain disorders. Nanomedicine **4**(1), 105–118 (2008)
12. Pancrazio, J.J.: Neural interfaces at the nanoscale. Nanomedicine **3**(6), 823–830 (2008)
13. Ferreira, A., Mavroidis, C.: Virtual reality and haptics for nano robotics: a review study. IEEE Robot. Autom. Mag. **13**, 78–92 (2006)
14. Sader, J.E., Sanelli, J.A., Adamson, B.D., Monty, J.P., Wei, X., Crawford, S.A., Friend, J.R., Marusic, I., Mulvaney, P., Bieske, E.J.: Spring constant calibration of atomic force microscope cantilevers of arbitrary shape. Rev. Sci. Instrum. **83**, 1–16 (2012)
15. Sharma, G., Mavroidis, C., Ferreira, A.: Virtual reality and haptics in nano- and bionanotechnology. In: Rieth, M., Schommers, W. (eds.) Handbook of Theoretical and Computational Nanotechnology, vol. X, pp. 1–33. American Scientific, Stevenson Ranch (2005)
16. Bhatti, A., Nahavandi, S.: Wavelets/multiwavelets analysis of AFM images for haptics enabled force feedback framework. In: Proceedings of the International Conference on Nanotechnology: Fundamentals and Applications, Ottawa, ON, 4–6 August 2010
17. Bhatti, A., Nahavandi, S., Hossny, M.: Haptics enabled offline AFM image analysis. In: ICIVC 2009, International Conference on Image and Vision Computing, pp. 446–451 (2009)
18. Bhatti, A., Nahavandi, S.: Depth estimation using multiwavelet analysis based stereo vision approach. Int. J. Wavelets Multiresolut. Inf. Process. **6**(3), 481–497 (2008)

Research on Eliminating Harmonic in Power System Based on Wavelet Theory

Huiyan Zhang, Qingwei Zhu, and Jihong Zhang

Abstract With the development of applications of high power electronic device as a harmonic wave source in power system increase, and harmonic wave which is the most common interference signals in power system often leads to error, malfunction of electronic equipment. In this paper, a method to determine the wavelet threshold based on wavelet theory with particle swarm optimization is provided to remove the noises which have a good ability to adapt the voltage waveform distortion conditions and to improve the real-time performance of harmonic signal processing. Simulation results show that this method is feasible and effective.

1 Introduction

Nonlinear electrical equipments are the main harmonic source of the power system. With the development on energy saving demand, a lot of power electronic devices and nonlinear elements are applied widely in the power system which make the harmonic pollution problem more serious and complex [1–3]. The traditional methods of harmonic detection methods are mainly Fast Fourier Transform (FFT) and Short Time Fourier Transform (STFT) [4–6]. For the difficulty of the FFT algorithm on synchronous sampling and integer period truncation during the power system harmonic analysis, which would cause spectral leakage and affect the results of harmonic analysis. By adding windows and the use of interpolation correction algorithm, the STFT has higher accuracy of the calculation while resolution accuracy is limited by the time-frequency fixed-width window.

Wavelet analysis is an important tool for time domain analysis it overcomes the shortcoming of Fourier analysis in the frequency domain completely localized while without localized processing in the time domain which particularly suitable for mutation signal analysis and processing in harmonic detection and analysis. Wavelet

H. Zhang (✉) • Q. Zhu
College of Computer and Information Engineering, Beijing Technology and Business University, Beijing, China
e-mail: zhanghuiyan369@126.com

J. Zhang
International Business School, Beijing Foreign Studies University, Beijing, China

D. Gao et al. (eds.), *Advances in Global Optimization*, Springer Proceedings in Mathematics & Statistics 95, DOI 10.1007/978-3-319-08377-3_50

analysis of the good time-frequency characteristics could be used for separating on the envelope of voltage fluctuation and flicker signal to get accurate amplitude and frequency of flicker signal, and could accurately detect the occurrence and termination moment of flicker signal. In addition according to the needs of different signals, the variation of wavelet algorithm or a combination of other various algorithms could improve the real-time and precision of harmonic measurement.

Today, the main literatures of wavelet power harmonics and power system transient characteristic analysis are: references [7, 8] which paid attention to the aliasing phenomenon on the family of wavelet functions in the signal analysis, a method of using frequency domain interpolation or new discrete wavelet transform to eliminate aliasing wavelet effectively and can detect the harmonic signal accurately; references [9, 10] which combining with statistical concept of entropy use wavelet energy distribution along the scale to effectively extract the characteristics of different types of transient signal references [11, 12] which provide the adaptive algorithms to adaptively select the optimal wavelet decomposition level or threshold methods to improve the adaptability of the algorithm and the performances of filtering.

This paper concerns to determine the layer number of signal decomposition in wavelet filter algorithm, and a optimized algorithm of adaptive determine the layer number is provided to remove the noises which make a good ability to adapt the power system waveform distortion conditions and to improve the real-time performance of harmonic signal processing. Simulation results show that this method shows a better significance in practical application.

2 Principle of Wavelet Algorithm

2.1 *Wavelet Theory*

The practical power frequency signal is about 50 Hz with positive and negative 5 % change range, and we could detect simultaneously the harmonic introduced by nonlinear power electronic equipment, random noise signal, and the decaying dc component accompanied by the high power load switching. Therefore the detected signal can be expressed as:

$$s(t) = x(t) + dd(t) + noise(t) \tag{1}$$

where *dd(t)* indicates the decaying dc component, *noise(t)* indicates the random noise signal, and the *x(t)* indicates practical power frequency signal with random white noise.

The white noise wavelet coefficients are still white noise after wavelet decomposition, which average power inverse with scales, while the level amplitude decreases as wavelet decomposition level increase. Therefore, by setting a threshold value to make zero of the frequency components with the small coefficient, we could remove the noise greatly and obtain signal *x(t)* which contains the harmonic component.

Signal *x(t)* can be expressed as:

$$x(t) = \sum_k (cA_0(k)\phi_{j,k}) = \sum_k (cA_1(k)\phi_{j-1,k}) + \sum_k (cD_1(k)\omega_{j-1,k}) \quad (2)$$

where $A_1(k)$ and $D_1(k)$ are the coefficients of the scale metric space *j-1* which are obtained after the decomposing of the coefficient $A_0(k)$ of the scale metric space *j*; $\phi_{j,k}$ and $\omega_{j,k}$ are orthogonal basis. Analogously we could reconstruct $A_0(k)$ by $A_1(k)$ and $D_1(k)$.

And these coefficients could be calculated by using the formula coefficient inner product. Clearly the algorithm of discrete wavelet decomposition was relatively simple and reconstruction is just the reverse process of decomposition.

2.2 Principle of Removing Harmonic Wavelet

Generally the noise signal contained in a higher frequency and after the wavelet decomposition by setting a reasonable threshold to adjust the weights of the wavelet coefficients and then reconstruct the signal without noises.

To remove the harmonic of the power system, should keep the right frequency band signal at a smaller frequency range which would remove abundantly times harmonic.

Specific steps:

1. One-dimensional signal is wavelet decomposed select a wavelet decomposition level number and then calculate the decomposition.
2. Determine the threshold of high-frequency wavelet coefficients to remove high-frequency noise by making a soft thresholds processing.
3. One-dimensional wavelets are reconstructed by the lowest level of high frequency wavelet-based reconstruction of one-dimensional wavelet.

3 Wavelet Filter Adaptive Algorithm

3.1 Adaptive Slicing Algorithm

It is seen from the wavelet decomposition process that the number of wavelet coefficients of each layer N_j is mutative. If the sample size does not meet the minimum requirements for precision, it is impossible to correctly infer the overall. Adopting the Kolmogorov–Smirnov test methods [12,13], the test statistic is defined as follows:

Let $F^*(D)$ is the cumulative normal distribution function, where the mean $\mu = \overline{D}$ and variance σ^2; noise sample statistics cumulative distribution function $F_{N_j}(D)$.

Definition: $\delta = MAX_D|F^*(D) - F_{N_j}(D)|$ represents the maximum vertical distance and indicates the bilateral test statistic.

Null hypothesis H_0: sample from an overall normally distributed population;

Alternative hypothesis H_1: sample does not belong to the overall normal distribution.

Comparing δ with the critical value which could be obtained by statistic quantile table to determine to accept or reject the null hypothesis.

After the wavelet decomposition the wavelet coefficients of white noises are still the white noise and their average power is inversely proportional to the scale with the amplitude decreased as the increasing of the layer number of wavelet decomposition. Conversely the energy of useful signal is compressed to the wavelet coefficient which is relatively fewer with larger values; the magnitude increases as the wavelet decomposition levels increase. By setting the threshold to zero on a small coefficient to remove noise, obtain the pure and harmonic estimate of wavelet coefficients.

3.2 *Algorithm of Dynamic Adaptive Removal Harmonics*

Since the highest power system harmonic frequency harmonic signals is unknown and time-varying while the range of desired signal frequency is identified. For the sampling frequency is not well accurately estimated in this case, the algorithm in Sect. 3.1 could not be applied directly to determine the stratified number.

An algorithm of dynamic adaptive removal harmonics is provided in the paper:

1. Signal spectrum is obtained by Fourier decomposition to determine the corresponding maximum values of harmonic frequencies based on a reasonable amplitude level calibration line;
2. Wavelet decomposition and white noise detection to make the threshold denoising;
3. Wavelet reconstruction to get the pure signals with those harmonic signals;
4. Combined with Shannon sampling theorem and wavelet hierarchical algorithm to determine the adaptive wavelet layers to remove harmonic hierarchically and the fundamental signal wavelet coefficients be preserved within a reasonable band;
5. Wavelet reconstruction and output the desired signal.

If first filter the harmonic and then filter the noise which would destroy the characteristics of the random noise and make little effect threshold results.

4 A Numerical Example

In order to verify the effectiveness of the proposed method, Matlab software simulation platform is applied. The simulation input signal is:

$$s(t) = sin(2\pi f_0 t) + 50 \sum_{n_M} sin(2k\pi f_0 t) + 100e^{-t} + noise(t) \quad (1)$$

where n_M is the highest harmonic number, frequency f_0 is range from 45 to 55 Hz, and *noise(t)* is white noise function.

By inputting different frequency harmonic signals for simulation analysis, when the $n_M = 5$ the *SNR=22.5574* and *MSE=0.2741*. Figure 1 shows the simulation graphics.

It is important to select the appropriate decomposition level to achieve better denoising effect, and reasonable bandwidth could be determined by dynamic adaptive algorithm which is suitable for the time-varying power harmonic signals. To improve the SNR and lower the MSE which would adapt power system harmonic signal removal, improve the accuracy of the fundamental signal.

Table 1 indicates that even in the extreme conditions as expressed in Eq. (1) where weights of harmonic components are great, stratified number of wavelet decomposition is no more than 4 and sampling points is no more than 250 which

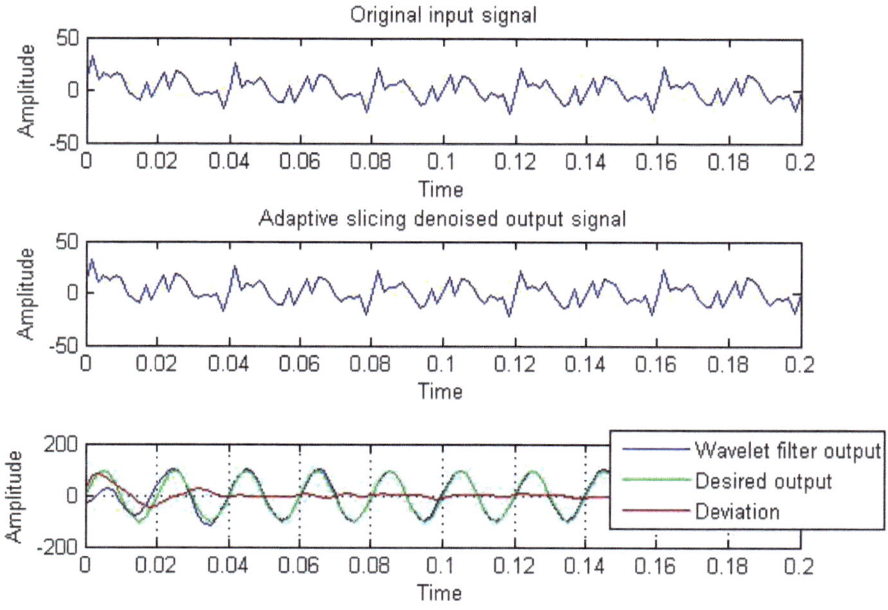

Fig. 1 The signal de-noising effect diagram with the fifth harmonic of the highest harmonic

Table 1 Signal denoising analysis table with different highest harmonic

Classes	SNR	MSE	Stratified number	Sampling points
$n_M = 5$	22.5574	0.2741	3	120
$n_M = 6$	22.0354	0.2791	3	120
$n_M = 7$	22.8701	0.2219	4	231
$n_M = 8$	22.7578	0.2391	4	246
$n_M = 9$	22.3911	0.2847	4	225

is five times the time-varying practical system frequency. When the weights of harmonic components are reducing with small ranges of the practical power system frequency, the performance SNR and MSE would improve.

5 Conclusion

For non-stationary time-varying waveform distortion in power system, the filtering algorithm based on wavelet transform is an important method. For the additive white Gaussian noise, a novel method to determine the optimal decomposition level in wavelet transform based on the Kolmogorov–Smirnov test improved by adding Fourier Transform to determine the maximum frequency harmonics. And later adaptively set the optimal decomposition level and could improve the performance of the traditional method for the harmonic signals processing in the power system.

Acknowledgements This work was supported by Scientific Research Common Program of Beijing Municipal Commission of Education *KM* 201210011005 and College Students Scientific Research and Undertaking Starting Action Project (2013 014213 000067).

References

1. Zeng, L., Li, X., et al.: Research on harmonic suppression in power system based on improved adaptive filter. Energy Procedia **16**, 1479–1486 (2012)
2. Ketabi, A., Sheibani, M.R., et al.: Power quality meters placement using seeker optimization algorithm for harmonic state estimation. Int. J. Electr. Power Energy Syst. **40**(1), 54–61 (2012)
3. Wenchun, Z., Weiming, M., An, H.: FFT algorithm with high accuracy for harmonic analysis in the electric machine. Proc. CSEE **21**(12), 83–87 (2001)
4. Hao, P., Dong-xia, L., Yun-xiao, Z., et al.: An improved algorithm for harmonic analysis of power system using fft technique. Proc. CSEE **23**(6), 50–54 (2003)
5. En-rang, Z., Run-xian, Y., Sen, Z.: Study on problems about detecting harmonic based on FFT in power system. RELAY **1**(34), 52–57 (2006)
6. Hui, G., Cheng-hua, F. Chun-fang, H.: Analysis of voltage inter-harmonics based on STFT. Telecommun. Electric Power Syst. **29**(4), 66–70 (2008)

7. Tian-jun, D., Guang-ju, C., Yong, L.: A novel method for power system harmonic detection based on wavelet transform with aliasing compensation. Proc. CSEE **25**(3), 54–59, 134 (2005)
8. Tian-jun, D., Guang-ju, C., Yong-le, X., et al.: Harmonic detection based on fr uency domain interpolation anti-aliasing shannon wavelet packet transform. Power Syst. Technol. **29**(11), 14–19 (2005)
9. Haoru, L., Jikai, C., Haoyu, L., et al.: Research for approach of power system harmonic detection based on Shannon wavelet energy entropy and FFT. Ind. Instrum. Autom. **1**, 6–10, 33 (2010)
10. Zheng-you, H., Xiao-qin, C.: A study of electric power system transient signals identification method based on multi-scales energy statistic and wavelet energy entropy. Proc. CSEE **26**(10), 33–39 (2006)
11. Wen-liao, D., Ru-min, Z., Yan-ming, L.: Adaptive selection of optimal ecomposition level in filtering algorithm based on wavelet transform. J. Optoelectron. Laser **21**(9), 1408–1411 (2010)
12. Jixian, Z., Qiuhai, Z., Ya Ping, D.: The determination of the threshold and the decomposition order in threshold de-noising method based on wavelet transform. Proc. CSEE **24**(2), 118–122 (2004)
13. Henderson, A.R.: Testing experimental data for univariate normality. Clin. Chim. Acta **366**, 112–129 (2006)

Synchronization of Hyperchaotic Memristor-Based Chua's Circuits

Junjian Huang, Pengcheng Wei, Yingxian Zhu, Bei Yan, Wei Xiong, and Yunbing Hu

Abstract This paper further investigates the problem of synchronization of hyperchaotic memristor-based Chua's circuits. An active control method is employed to design a controller to achieve the global synchronization of two identical memristor-based systems. Based on Lyapunov stability theory, a sufficient condition is given to guarantee the stability of the synchronization error system.

Keywords Memristor-based systems • Synchronization • Chua's circuit

1 Introduction

The existence of the memristor as the fourth fundamental circuit element included along with the resistor, capacitor, and inductor was predicated by Chua in 1971 [1]. Until 2008, the Hewlett-Packard (HP) research team announced that they had realized a prototype of memristor-based on nanotechnology [2]. Many researchers focus on the memristor because of its potential applications in programmable logic, signal processing, neural networks, control systems, reconfigurable computing, Brain-computer interfaces, and RFID [3–9].

Recently, the research on circuits based on memristor is becoming a focal topic [10–19]. Itoh and Chua presented a fourth-order memristor-based Chua's oscillator by replacing Chua's diode with an active two-terminal circuit consisting of a conductance and a flux-controlled memristor [10]. Pershin and Di Ventra introduced an approach to use memristors in programmable analog circuits [11]. Rak and

J. Huang (✉)
College of Computer Science, Chongqing University, Chongqing 400030, China

Department of Computer Science, Chongqing University of Education, Chongqing 400067, China
e-mail: hmomu@sina.com

P. Wei • Y. Zhu • B. Yan
Department of Computer Science, Chongqing University of Education, Chongqing 400067, China

W. Xiong • Y. Hu
College of Computer, Chongqing College of Electronic Engineering, Chongqing 401331, China

D. Gao et al. (eds.), *Advances in Global Optimization*, Springer Proceedings in Mathematics & Statistics 95, DOI 10.1007/978-3-319-08377-3_51

Cserey presented a new simulation program with integrated circuit emphasis macromodel of the recently physically implemented memristor [12]. Petras presented fractional-order memristor-based Chua's circuit [13]. The hyperchaotic behavior in memristor-based Chua's circuit is performed with the help of nonlinear tools [19]. In this letter, we will study the problem of synchronization of memristor-based Chua's systems. Based on the feedback control method, we design a controller to guarantee the exponentially stability of the synchronization error system.

The rest of the paper is organized as follows. In Sect. 2, a memristor-based system is introduced. And using feedback control method, a general convergence criterion for stabilization of synchronization error system is established. Conclusions are finally drawn in Sect. 3.

2 Problem Formulation and Preliminaries

Referring to [19], Andrew L. Fitch proposed a circuit by adding an inductor in parallel with conductance—G and fourth-order memristor-based canonical oscillato. The equations for the circuit are described by

$$\begin{cases} \frac{dq_{l2}(t)}{dt} = \frac{1}{L_2}\phi_{c2}(t), \\ \frac{d\phi_{c2}(t)}{dt} = \frac{1}{C_2}\left(-q_{l2}(t) - G\phi_{c2}(t) - q_{l1}(t)\right), \\ \frac{dq_{l1}(t)}{dt} = \frac{1}{L_1}\left(\phi_{c2}(t) - q_{l2}(t)R - \phi_{c1}(t)\right), \\ \frac{d\phi_{c1}(t)}{dt} = \frac{1}{C_1}\left(-q_{l1}(t)R - a\phi_{c1}(t) - b\phi_{c1}{}^3(t)\right). \end{cases} \tag{1}$$

The system Eq. (1) can be reorganized by

$$\begin{cases} \frac{d\phi_{c1}(t)}{dt} = -\tau v_1(t), \\ \frac{dv_1(t)}{dt} = \frac{1}{C_1}\left(i_{L1}(t) - W\left(\phi_{c1}(t)\right)v_1(t)\right), \\ \frac{dv_2(t)}{dt} = \frac{1}{C_2}\left(Gv_2(t) - i_{L1}(t) - i_{L2}(t)\right), \\ \frac{di_{L1}(t)}{dt} = \frac{1}{L_1}\left(v_2(t) - v_1(t) - Ri_{L1}(t)\right) \\ \frac{di_{L2}(t)}{dt} = \frac{1}{L_2}v_2(t). \end{cases} \tag{2}$$

where τ is an integration constant which introduced to rescale the values of voltage into practical range, and $W(\phi_{c1}(t)) = a + 3b\phi_{c1}{}^2(t)$.

Letting $x_1(t) = \phi_{c1}(t)$, $x_2(t) = v_1(t)$, $x_3(t) = v_2(t)$, $x_4(t) = i_{L1}(t)$, $x_5(t) = i_{L2}(t)$, system (2) can be further rewritten as

$$\begin{cases} \dot{x}_1(t) = -\tau x_2(t), \\ \dot{x}_2(t) = \frac{1}{C_1}\left(x_4(t) - ax_2(t) - 3bx_1{}^2(t)x_2(t)\right), \\ \dot{x}_3(t) = \frac{1}{C_2}\left(Gx_3(t) - x_4(t) - x_5(t)\right), \\ \dot{x}_4(t) = \frac{1}{L_1}\left(x_2(t) - x_3(t) - Rx_4(t)\right), \\ \dot{x}_5(t) = \frac{1}{L_2}x_3(t). \end{cases} \tag{3}$$

When $L_1 = 10mH$, $L_2 = 60mH$, $C_1 = 6.8nF$, $C_2 = 15nF$, $G = 0.0005S$, $a = 0.00067$, $b = 0.000029$, $R = 65\Omega$, $\tau = 26,000$, existence of hyperchaos.

The system (3) can be rewritten with linear part and nonlinear part as follows:

$$\dot{x}(t) = Ax(t) + f(x(t)), \tag{4}$$

where $x(t) = (x_1(t), x_2(t), x_3(t), x_4(t), x_5(t))^T$,

$$A = \begin{bmatrix} 0 & -\tau & 0 & 0 & 0 \\ 0 & \frac{-a}{C_1} & 0 & \frac{1}{C_1} & 0 \\ 0 & 0 & \frac{G}{C_1} & \frac{-1}{C_2} & \frac{-1}{C_2} \\ 0 & \frac{1}{L_1} & \frac{-1}{L_1} & \frac{-R}{L_1} & 0 \\ 0 & 0 & \frac{-1}{L_2} & 0 & 0 \end{bmatrix}, \quad A = \begin{pmatrix} 0 & -\tau & 0 & 0 & 0 \\ 0 & -a/C_1 & 0 & 1/C_1 & 0 \\ 0 & 0 & G/C_2 & -1/C_2 & -1/C_2 \\ 0 & 1/L_1 & -1/L_1 & -R/L_1 & 0 \\ 0 & 0 & 1/L_2 & 0 & 0 \end{pmatrix}$$

and

$$f(x(t)) = \begin{pmatrix} 0 \\ -3bx_1{}^2x_2/C_1 \\ 0 \\ 0 \\ 0 \end{pmatrix}.$$

As for vector function $f(x)$, assume that for any $x, y \in \Omega$ we have

$$|f_i(x) - f_i(y)| \le L_{\max}|x - y|, \; i = 1, 2, 3, 4 \tag{5}$$

The above condition is considered as the uniform Lipszchitz condition, and $L_{\max} > 0$ refers to the uniform Lipschitz constant.

We construct the response system as below:

$$\dot{y}(t) = Ay(t) + f(y(t)) + u(t) \tag{6}$$

where $y(t) = (y_1(t), y_2(t), y_3(t), y_4(t))^T$ is the response state, $u(t)$ is the control gain defined by:

$$u(t) = k(y(t) - x(t)).$$

where $\tau > 0$ is the propagation delay, k denotes control strength. Let $e(t) = y(t) - x(t)$ be the synchronization error between the systems (2) and (3), then yields the error system

$$\begin{aligned} \dot{e}(t) &= \dot{y}(t) - \dot{x}(t) \\ &= Ae(t) + f(y) - f(x(t)) + u(t) \\ &= Ae(t) + f(y) - f(x(t)) + ke(t) \end{aligned} \tag{7}$$

We now state our main results.

Theorem 1. *Suppose that there exist positive constants* s_1, g_1 *such that* $A + A^T + 2kI + s_1 I + {s_1}^{-1} {L_{\max}}^2 + g_1 I \leq 0$,

Then, the synchronization error system (7) *is globally exponentially stable, and the systems* (4) *and* (6) *are globally exponentially synchronized.*

Proof. Choose the Lyapunov function as follows

$$V(t) = e(t)^T e(t). \tag{8}$$

Then the differentiation of V along trajectories of (7) is

$$\begin{aligned}
\dot{V}(t) &= e(t)^T \dot{e}(t) + \dot{e}(t)^T e(t) \\
&= e(t)^T \left[Ae(t) + f(y(t)) - f(x(t)) + ke(t)\right] + \\
&\quad \left[Ae(t) + f(y(t)) - f(x(t)) + ke(t)\right]^T e(t) \\
&\leq e(t)^T \left[A + A^T + 2kI\right] e(t) + s_1 e(t)^T e(t) \\
&\quad + {s_1}^{-1}\left[f(y(t)) - f(x(t))\right]^T \left[f(y(t)) - f(x(t))\right] \\
&\leq e(t)^T \left[A + A^T + 2kI + s_1 I\right] e(t) + {s_1}^{-1} {L_{\max}}^2 e(t)^T e(t) \\
&= e(t)^T \left[A + A^T + 2kI + s_1 I + {s_1}^{-1} {L_{\max}}^2 + g_1 I\right] e(t) - g_1 e(t)^T e(t) \\
&\leq -g_1 e(t)^T e(t) \\
&= -g_1 V(t)
\end{aligned}$$

According to Lyapunov theory, the inequality $\dot{V}(t) \leq -g_1 V(t)$ indicate $V(t)$ converges to zero exponentially. Furthermore, we can conclude that the synchronization error systems $e(t)$ converges to zero globally and exponentially with a rate g_1, and the synchronization between with system (4) and system (6) can be obtained. This completes the proof. □

3 Conclusions

In this paper, the synchronization problem of memristor-based chaotic system has been discussed. A feedback controller was designed to stabilize the synchronization error system globally exponentially.

Acknowledgements The work described in this paper was partially supported by NSFC (Grant No. 60974020) and Natural Science Foundation Project of CQ CSTC (Grant No. cstc2011jjA40005), and the Foundation of Chongqing Education Committee (Grant No. KJ121505).

References

1. Chua, L.O.: Memristor-the missing circuit element. IEEE Trans. Circuit Theory **18**, 507 (1971)
2. Strukov, D.B., Snider, G.S., Stewart, D.R., Williams, R.S.: The missing memristor found. Nature **453**, 80 (2008)
3. Ho, Y., Huang, G.M., Li, P.: Nonvolatile memristor memory: device characteristics and design implications. In: Proceedings of IEEE/ACM International Conference Computer-Aided Design Digest of Technical Papers, 485 (2009)
4. Borghetti, J., Snider, G.S., Kuekes, P.J., Jang, J.J., Stewart, D.R., Williams, R.S.: 'Memristive' switches enable 'stateful' logic opera via material implication. Nature **468**, 873 (2010)
5. Raja, T., Mourad, S.: Digital logic implementation in memristor-based crossbar: a tutorial. In: Proceedings of IEEE International Symposium on Electron Design, Test and Application, 303 (2010)
6. Jo, S.H., Chang, T., Ebong, I., Bhadviya, B.B., Mazumder, P., Lu, W.: Nanoscale memristor device as synapse in neuromorphic systems. Nano Lett. **10**, 1297 (2010)
7. Pershin, Y.V., Fontaine, S.L., Ventra, M.D.: Memresistive model of amoeba's learning. Phys. Rev. E **80**, 021926-1 (2009)
8. Pershin, Y.V., Ventra, M.D.: Experimental demonstration of associative memory with memristive neural networks. Neural Netw. **23**, 881 (2010)
9. Afifi, A., Ayatollahi, A., Raissi, F.: Implementation of biologically plausible spiking neural network models on the memristor crossbar-based CMOS/nano circuits. In: Proceedings of IEEE European Conference on Circuits Theory Design, 563 (2009)
10. Itoh, M., Chua, L.O.: Memristor oscillators. Int. J. Bifurcat. Chaos **18**, 3183 (2008)
11. Pershin, Y.V., Di Ventra, M.: Practical approach to programmable analog circuits with memristors. IEEE Trans. Circuits Syst. **57**, 1857 (2010)
12. Rak, A., Cserey, G.: Macromodeling of the memristor in SPICE. IEEE Trans. Computer-Aided Des. Integr. Circuits Syst. **29**, 632 (2010)
13. Petras, I.: Fractional-order memristor-based Chua' circuit. IEEE Trans.Circuits Syst. **57**, 975 (2010)
14. Muthuswamy, B., Chua, L.O.: Simplest chaotic circuit. Int. J. Bifurcat. Chaos **20**, 1567 (2010)
15. Muthuswamy, B.: Implementing memristor based chaotic circuits. Int. J. Bifurcat. Chaos **20**, 1335 (2010)
16. Witrisal, K.: Memristor-based stored-reference receiver C the UWB solution. Electron. Lett. **45**, 713 (2009)
17. Muthuswamy, B., Kokate, P.P.: Memristor based chaotic circuits. IETE Tech. Rev. **26**, 415 (2009)
18. Bao, B.C., Xu, J.P., Liu, Z.: Initial state dependent dynamical behaviors in memristor based chaotic circuit. Chin. Phys. Lett. **27**, 070504 (2010)
19. Fitch, A.L., Yu, D.S., Iu, H.H.C., Sreeram, V.: Hyperchaos in a memristor-based modified canonical Chua's circuit. Int. J. Bifurcat. Chaos **22**, 1250133 (2012)

Complex Simulation of Stockyard Mining Operations

Vu Thanh Le, Michael Johnstone, James Zhang, Burhan Khan, Doug Creighton, Samer Hanoun, and Saeid Nahavandi

Abstract Conflicts between resources in stockyards cause mining companies millions of dollars a year. An effective planning strategy needs to be established in order to reduce these operational conflicts. In this research a stockyard simulation model of a mining operation is proposed. The simulation uses discrete event and continuous strategies to create a high detail level of visualization and animation that closely resemble actual stockyard operation. The proposed simulation model is tightly integrated with a stockpile planner and it is used to evaluate the feasibility of a given production plan. The high detail visualization of the simulation model allows planner to determine the source of conflict, which can be used to guide the elimination of these conflicts.

Keywords Stockyard simulation • Discrete-continuous • Mining operations • High fidelity • Visualization • Stacker • Reclaimer

1 Introduction

Increased worldwide competition and stochastic in demand pattern have put great pressure on the coal supply chain. This challenges the ability of stockyard planners to create robust schedule in order to meet the increase in throughput level.

Previous researches have shown that mathematical methods can be used to create schedule and manage stockyards and mineral supply chains operation [1, 2]. These methods allow solution to be generated quickly and are an essential tool to help planners with their daily planning operation [1]. It has been shown that visually appealing tools are more readily accepted by end users. In domains such as manufacturing [3] or airport operations [4], visual simulations have shown to be an effective tool for assisting the operational planning and scheduling process.

V.T. Le (✉) • M. Johnstone • J. Zhang • B. Khan • D. Creighton • S. Hanoun • S. Nahavandi
Centre for Intelligent Systems Research, Deakin University, Geelong, VIC, Australia
e-mail: vu.le@deakin.edu.au; michael.johnstone@deakin.edu.au; james.zhang@deakin.edu.au; burhan.khan@deakin.edu.au; douglas.creighton@deakin.edu.au; samer.hanoun@deakin.edu.au; saeid.nahavandi@deakin.edu.au

D. Gao et al. (eds.), *Advances in Global Optimization*, Springer Proceedings in Mathematics & Statistics 95, DOI 10.1007/978-3-319-08377-3_52

Discrete event simulation (DES) has shown to be an effective tool to model system variation in a high level of detail. DES is able to model complex real-world systems that are usually difficult to describe mathematically. However, a model of a complex system is costly to build [5, 6]. Due to multiple sub-components interacting with one another to form one complex system within a stockyard operation, modeling can be a challenging process.

Existing modeling methodology and optimization for a stockyard operation is detailed in [7]. More recently, a model combined with an optimizer was developed [8], with a focus on the blending of inputs into a supply chain to produce the desired output. The authors address the challenging problem of multi objective optimization by integrating a solution derived offline with a simulation model.

The fundamental complexity in simulation and modeling stockyards is due to the magnitude of the resources involved. In a conventional DES model, coal would be assumed to be a part travelling along a conveyor and transferred to other sections of the system in a batch. However, this is a biased view of the coal handling process, in which coal is continuously flowing. Combining of continuous and discrete simulation is necessary to provide a better view of the system as a whole [9].

The detail required to model stockyard operation demands a hybrid DES model. Therefore, a method to convert the continuous aspects of the system into a series of events that accurately represent the continuous process is required. This problem was studied in Paluszczyszyn [10] in the context of a water network system. By discretizing certain processes, the authors were able to generate almost identical results to their previous method, but with fewer data points and in less time. Alternative methods such as discrete element and smoothed-particle hydrodynamic methods [11–14] can accurately represent coal stacking, reclaiming and flow behavior. However, these methods rely on high end graphic processing unit (GPU) for rendering. Their visualization accuracy and rendering performance is dependent on the granular size of particles.

In this paper, a high fidelity modeling methodology for stockyard mining operations is presented. The paper explores core components inside the model to accurately represent coal stacking and reclaiming operations. As conflicts between stackers and reclaimers result in production feed shutdown, causing major production losses, these situations are undesirable. This paper details our simulation framework and its operation in relation with a stockpile planner, to detect production conflicts. The simulation modeling methodology can be used by planners to evaluate their production plan, which supports decision making and production planning processes.

This paper is structured as follows. Section 2 provides an overview of the stockyard mining operation. Section 3 provides a description of the stockyard mining simulation system and the components that are used in the model construction to resemble real-world process. Section 4 provides snapshot of output results produced by the simulation system and the display presentation to aid coal production planning process. Section 5 provides our concluding remarks.

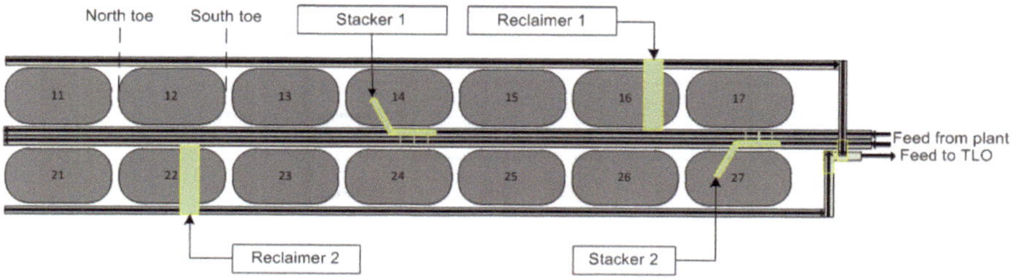

Fig. 1 Stockyard stockpile configuration

2 Stockyard Mining Operation

The stockyard mining operation under consideration consists of two parallel stockpiles, with two stackers running along a central rail and two reclaimers running on opposing outside rails. Each stockpile is separated into zones to accommodate the stacking of different coal types as illustrated in Fig. 1. Dependent on supply from the mine plant, stackers are required to stack different coal types to designated stockpile. Similarly, demands from the port are met by reclaiming coal from assigned stockpiles. In this stockyard configuration, only a single resource can operate within a stockpile at any given time. Priority is given to reclaimers to ensure that reclaiming time targets are met. Conflicts will require the plant to be shut down in order to relocate the stacker. The aim of the planner is to minimize the number reclaimer/stacker conflicts. Lost production will occur if an operating stacker is required to be relocated or have its boom slewed to allow a reclaimer to go pass. The need for an efficient method that allows planner to identify static and dynamic resource conflicts is crucial for business operation.

3 Stockyard Simulation System

A stockyard simulation system is a dynamic simulation environment that provides a detailed visualization of scheduled coal stacking, reclaiming and train load out (TLO) or railing operations. The simulation model is designed as a user feedback system. It aims to provide guidance and aid planners to effectively carry out the production planning and scheduling process. The accuracy of this process provides a number of cost improvement/saving opportunities and benefits. This methodology allows operational planners to refine their plan over the planning horizon to reduce the chance of stacker and reclaimer crossover conflict. This reduces production downtime, thus productivity will be improved. If conflicts caused as little as 15 min of lost production per week, over a year the loss will equate to approximately \$2,221,596AUD per annum for a low grade coal as of Feb 2013.

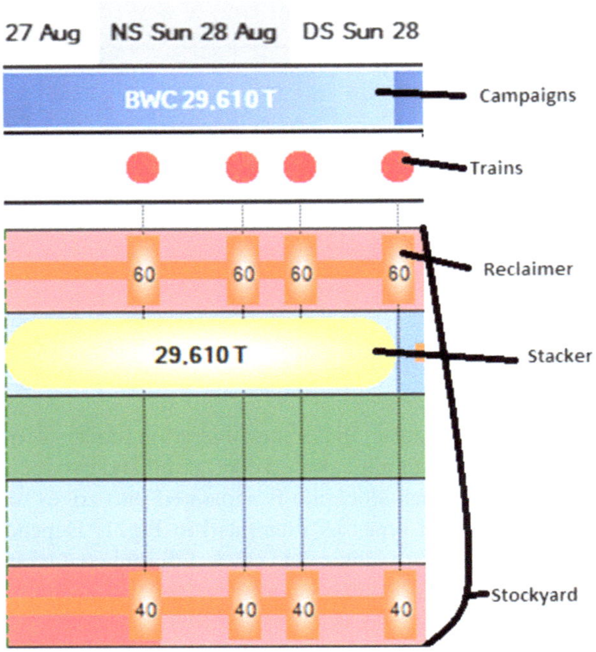

Fig. 2 The stockpile planner interface with a campaign of 29,610T allocated to a stockpile using a stacker. Four trains are shown, loaded with the percentage of train capacity specified at each reclaimer. Campaign, train, and stockpile *colors* indicate different coal types

In order for the simulation to be an effective tool to aid the production planning process, the model must closely resemble the actual stockyard operations and provide realistic detailed visualization of the operation. It also needs to integrate with the scheduling software to provide production feasibility feedback.

3.1 Stockpile Planner

The main task of the stockpile planner involves allocating stacking and reclaiming activities. Initially, campaigns, trains details, and stockyard volumes are either manually configured or imported from a database. Next, the stockyard planner is given the ability to allocate resources in chronological order. Finally, the stockyard planner can save the created plan in the database for future editions and for later committing it to execution. The stockpile planner does not resolve or detect any stockyard machinery collisions, crossover conflicts, and travel times. These are managed by the simulation algorithms. Figure 2 shows a snapshot of the planning operations.

3.2 *Simulation of Stockyard Mining Operation*

The stockyard mining simulation model consists of multiple components defined as controllers. During the model building and loading process, resource parameters such as speed is defined to all conveyor sections and mobile resources including stackers and reclaimers. When the simulation begins, stackers and reclaimers are initialized into their predefined location and state. All stockpiles in the stockyard are adjusted to the simulation state and level defined by the stockpile planner. The production controller loads all stacker feeds and stacking information, while the TLO controller prepared train and coal reclaiming detail.

The simulation begins at the coal controller as it requests production and TLO details from the production and reclaiming data. The production and TLO controllers generate and transform stackers production plan and reclaimers' load out information as communication messages. These messages are referred to as campaign messages and are direct to the coal controller, see Fig. 3.

The coal controller acts as a message manager and dispatcher. First, it will send the message to the collision controller. Once the collision controller resolves and confirms that there is no further conflict for the current campaign, the message is returned to the coal controller with a different message state. Once retrieving the message, the coal controller coordinates message distribution to the stacker, reclaimer, and coal pile controller. A stacking message is sent to the stacker controller, a TLO message is sent to the reclaimer controller, while coal animation behavior will dispatch to the coal pile controller.

After the stacker and reclaimer have completed their given tasks, they send a request message to the coal controller. Finally, the coal controller sends these requests to the production and TLO controllers and the process continues.

The stockyard mining simulation model in this study is developed using the DES environment Delmia QuestTM. We have introduced continuous behavior into the components in the model to allow a close resemblence to the actual system. Coal flow animation on conveyor, coal stacking, and reclaiming are some of the continuous properties that enhance the realism of the model. Vehicle kinematics and basic collision detection are some other features that the simulation model in this research inherited as demonstrated in Fig. 4.

4 Results

In average, it takes the simulation model less than 30 s to simulate 7 days production. The stockyard simulation system generates an output stamp file that includes the detailed information of stacker and reclaimer events, stockpile tonnage, and detail on the source of conflicts. The system provides a weekly summary of

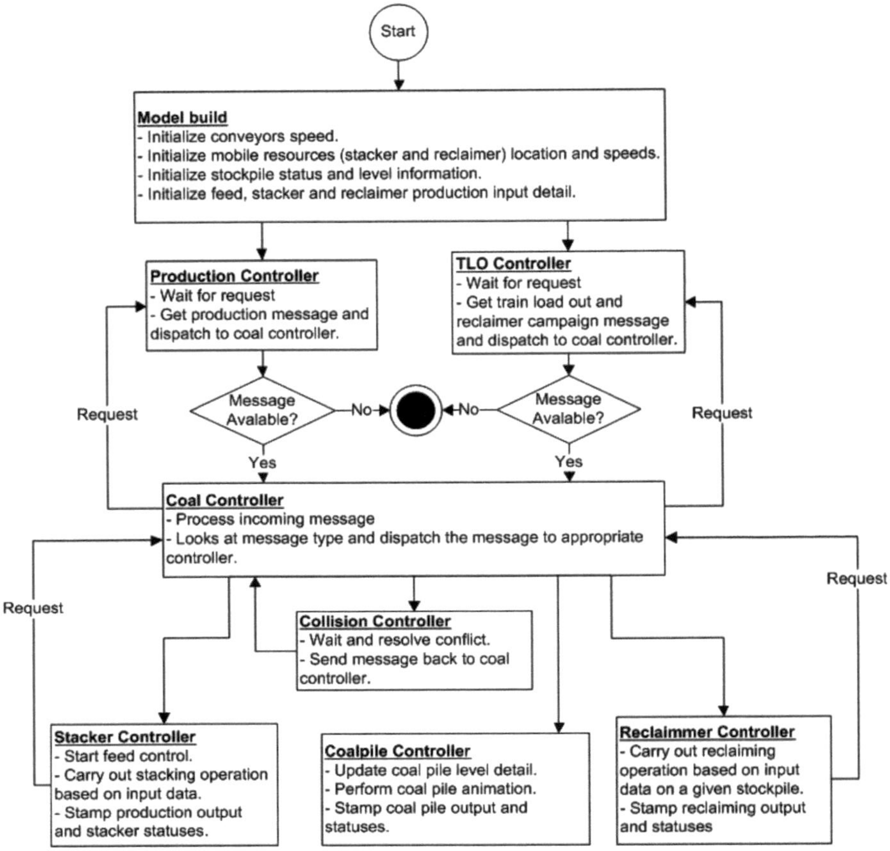

Fig. 3 Interaction between different controllers in the stockyard mining operation simulation model

coal production and amounts of coal railed, a summary of daily tonnage railed, a summary of daily stacking production of different coal types, a summary of stockpile coal inventory level plotted against the normal and alarmed operating level, a summary of cumulative production against forecast data, and finally a summary of cumulative tonnage railed/reclaimed compares to forecast trend as demonstrated in Fig. 5.

The simulation system also provides a summary of daily stockpile events such as conflicts and resource utilization, Fig. 6. The detailed information of the conflict type is also recorded and present in the event logger through the stockpile planner interface of our stockyard simulation system.

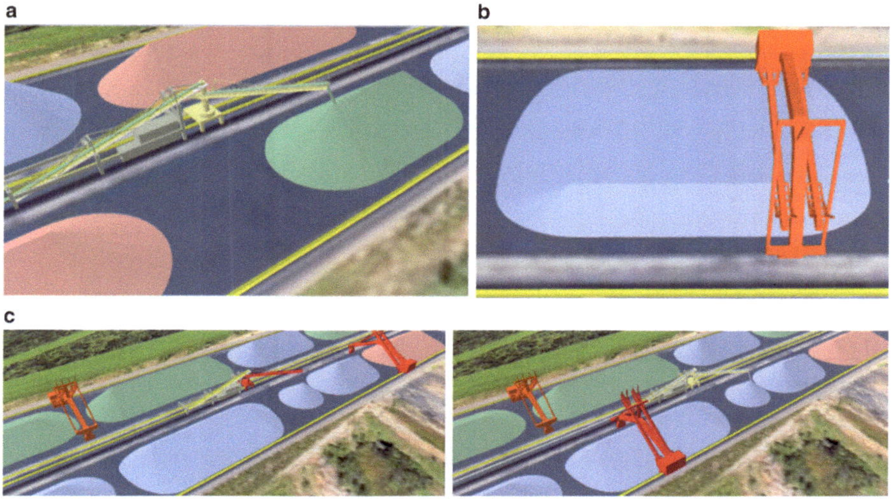

Fig. 4 Snapshot of the variation in visualization detail of stockyard mining operations simulation model. (**a**) Coal stacking. (**b**) Coal reclaiming. (**c**) Reclaimer passing stacker conflict

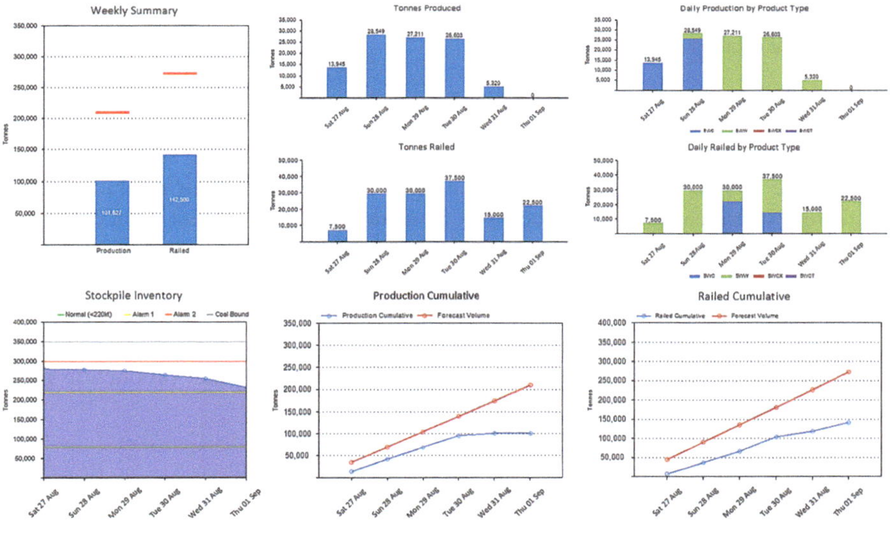

Fig. 5 Simulation system output showing weekly production summary

5 Conclusion

In this paper, a high fidelity model of stockyard simulation system that links with a stockpile planner is presented. The system provides several benefits compared to traditional mathematical models. It can be used by production planners as a tool to a

Stockpile Events	Week	Sat	Sun	Mon	Tue	Wed	Thu	Fri
Lost Feed Count	3	1	0	1	0	1	0	-
Divert Feed Tonnes	0	0	0	0	0	0	0	-
Lost Feed Tonnes	26,464	550	172	0	133	12,858	12,751	-
Feed Switch Count	3	0	0	2	0	1	0	-
New Stack Count	5	1	2	1	1	0	0	-
Completed Stack Count	5	0	2	1	1	1	0	-
RCL to Base Count	0	0	0	0	0	0	0	-
RCL Long Travel during TLO	0	0	0	0	0	0	0	-
STK Long Travel (Metres)	898	191	6	91	0	277	0	-
RCL Long Travel (Metres)	2,444	301	57	95	482	100	97	-
Train Count	19	1	4	4	5	2	3	-

b

Mobile Resource	Week	Sat	Sun	Mon	Tue	Wed	Thu	Fri
Reclaimer 1 Utilisation	27.67 %	10.08 %	34.65 %	34.05 %	44.45 %	17.24 %	25.56 %	-
Reclaimer 2 Utilisation	27.85 %	11.36 %	34.49 %	35.3 %	42.82 %	16.98 %	26.14 %	-
Stacker 1 Utilisation	45.63 %	49.4 %	63.39 %	60.64 %	80.35 %	19.98 %	0 %	-
Stacker 2 Utilisation	51.46 %	1.38 %	36.71 %	100 %	100 %	70.67 %	0 %	-
Reclaimer 1 Availability	72.33 %	89.92 %	65.35 %	65.95 %	55.55 %	82.76 %	74.44 %	-
Reclaimer 2 Availability	72.15 %	88.64 %	65.51 %	64.7 %	57.18 %	83.02 %	73.86 %	-
Stacker 1 Availability	54.37 %	50.6 %	36.61 %	39.36 %	19.65 %	80.02 %	100 %	-
Stacker 2 Availability	48.54 %	98.62 %	63.29 %	0 %	0 %	29.33 %	100 %	-

Fig. 6 Snapshot of production information. (**a**) Stockpile conflict event summary. (**b**) Stackers and reclaimers utilization

evaluate the feasibility of their production plan. The method reduces the complexity for a planner to determine the source of problems within their schedule. It also allows the planner to effectively communicate the results of their production plan with fellow colleagues. It can be used as a training environment to accustomise planners to different scenarios such that the planning process becomes automatic. It can be used as a tool to train new planners and guide them to create an effective production plan that minimizes risks and production conflicts.

Future work will investigate in development of real time control that allows us to determine the differences and correlation between the planner production plans with the simulation results information.

Acknowledgements This research is supported by the Centre for Intelligent Systems Research in Deakin University.

References

1. Garcia-Flores, R., Singh, G., Ernst, A., Welgama, P.: In: 19th International Congress on Modelling and Simulation (MODSIM2011) (Modelling and Simulation Society of Australia and New Zealand, 2001), pp. 311–317 (2011)
2. Eivazy, H., Askari Nasab, H.: Int. J. Min. Miner. Eng. **4**(2), 89 (2012)
3. Creighton, D., Nahavandi, S.: Robot. Comput. Integr. Manuf. **19**, 469 (2003)

4. Nahavandi, S., Creighton, D., Johnstone, M., Le, V.T., Zhang, J.: Simulation-based knowledge management in airport operations. In: Fathi, M. (Ed.), Integration of Practice-Oriented Knowledge Technology: Trends and Prospectives, pp. 83–95. Springer, Heidelberg (2013)
5. Banks, J.: In: Winter Simulation Conference Proceedings, vol. 1, pp. 7–13 (1999)
6. Law, A.M.: Simulation Modeling and Analysis, 4th edn. McGraw-Hill, New York (2007)
7. Marasini, R., Dawood, N.: In: 2002 Winter Simulation Conference, pp. 1731–1736 (2002)
8. Sandeman, T., Fricke, C., Bodon, P., Stanford, C.: In: Proceedings of the 2010 Winter Simulation Conference, pp. 1898–1910 (2010)
9. Özgün, O., Barlas, Y.: In: Proceedings of the 27th International Conference of the System Dynamics Society (2009)
10. Paluszczyszyn, D., Skworcow, P., Ulanicki, B.: In: 14th Water Distribution Systems Analysis Conference (2012)
11. Randles, P.W., Libersky, L.D.: Comput. Methods Appl. Mech. Eng. **139**(14), 375 (1996)
12. Cleary, P.W., Prakash, M.: Philos. Trans. A Math. Phys. Eng. Sci. **362**(1822), 2003 (2004)
13. Kessler, F., Prenner, M.: FME Trans. **37**(4), 185 (2009)
14. Fernandez, J.W., Cleary, P.W., Sinnott, M.D., Morrison, R.D.: Miner. Eng. **24**(8), 741 (2011)

MIX
Papier aus verantwortungsvollen Quellen
Paper from responsible sources
FSC® C105338

If you have any concerns about our products,
you can contact us on
ProductSafety@springernature.com

In case Publisher is established outside the EU,
the EU authorized representative is:
Springer Nature Customer Service Center GmbH
Europaplatz 3, 69115 Heidelberg, Germany

Printed by Libri Plureos GmbH
in Hamburg, Germany